Interactions of Pacific tuna fisheries

FAO
FISHERIES
TECHNICAL
PAPER

336/2

Volume 2
Papers on biology and fisheries

Proceedings of the First FAO Expert Consultation
on interactions of Pacific Tuna Fisheries
3-11 December 1991
Nouméa, New Caledonia

edited by
Richard S. Shomura
Hawaii Institute of Marine Biology
University of Hawaii, Manoa
Honolulu, Hawaii, USA
Jacek Majkowski
FAO Fisheries Department
Sarah Langi
Neiafu, Vava'u
Tonga

Food
and
Agriculture
Organization
of
the
United
Nations

Rome, 1994

M-43
ISBN 92-5-103454-0

CORRIGENDUM

FAO Fisheries Technical Paper No. 336/2 - 1994
Interactions of Pacific tuna fisheries
Volume 2: Papers on biology and fisheries

In "A review of the biology and fisheries for northern bluefin tuna, *Thunnus thynnus,* in the Pacific Ocean", by William H. Bayliff, the headings in the bottom half of Tables 13a (p. 285) and 13b (p. 286) should be **Age 2**, **Age 3** and **Age 4** rather than Age 3, Age 4 and Age 5.

M-43/T1817E/1/1.94/4500
ISSN 0429-9345
ISBN 92-5-103454-0

PREPARATION OF THIS DOCUMENT

This publication results from the First FAO Expert Consultation on Interactions of Pacific Tuna Fisheries hosted in Noumea, New Caledonia by the South Pacific Commission in cooperation with the Institut Français de Recherche Scientifique pour le Développement en Coopération from 3 to 11 December 1991. The Consultation was organized by the FAO Trust Fund project: "Cooperative Research on Interactions of Pacific Tuna Fisheries" in close collaboration with regional and national institutions involved in tuna fisheries research in the Pacific (see Acknowledgements).

The information presented at the Consultation was compiled by TUNET, a network of ten Working Groups organized by the FAO project. That information was contributed by scientists of the regional and national institutions studying tuna stocks and fisheries mainly in the Pacific, but also outside of the region.

<u>Distribution</u>

FAO Fisheries Department
FAO Regional Fisheries Officers
FAO fisheries projects and programmes
International fisheries organizations
National fisheries departments
National fisheries research laboratories
Members of TUNET (FAO's network of working groups
        studying tuna fisheries interactions in the Pacific)

Shomura, R.S.; Majkowski, J.;Langi, S. (eds.)
Interactions of Pacific tuna fisheries. Proceedings of the first FAO Expert
Consultation on Interactions of Pacific Tuna Fisheries. 3-11 December 1991.
Noumea, New Caledonia. Volume 2: papers on biology and fisheries.
FAO Fisheries Technical Paper. No. 336, Vol.2. Rome, FAO. 1993. 439p.

ABSTRACT

This publication presents papers and discussions of the First FAO Expert
Consultation on Interactions of Pacific Tuna Fisheries held in Noumea, New
Caledonia from 3 to 11 December 1991. The objectives of the Consultation
included:
- the identification and documentation of concern related to interactions
  among fisheries directed at tuna and tuna-like species in the Pacific,
- the classification of these interactions,
- the review of information on them and methods for their study, and
- the formulation of recommendations for future research.

Volume 1 contains:
- the Summary Report of the Consultation,
- a review paper on methods for studying interactions in tuna fisheries,
- thirteen papers presenting new methods and case studies on such
  interactions, and
- seven reviews on fisheries interactions related to individual stocks of
  Pacific tuna and tuna-like species.

Volume 2 includes:
- eleven review papers on the biology, population dynamics and fisheries
  associated with the stocks of Pacific tuna, which are supplemented by
- four additional papers on specific fisheries.

The information contained herein demonstrates the potential for interactions
occurring between and among the fisheries directed at tunas and tuna-like
species. Empirical evidence for such interactions, however, has been
available for only few fisheries, and these interactions have been quantified
for even fewer fisheries. It is unclear whether interactions are insignificant
among fisheries directed at tuna and tuna-like species or whether scientists
are unable to detect these interactions possibly due to various changes to
fisheries and resources, resulting in a too-variable background which conceals
the effects of interactions.

v

ACKNOWLEDGEMENTS

The editors of this document would like to thank Convenors of TUNET's Working Groups, Session Chairmen, Rapporteurs and participants of the Consultation for their effort and collaboration in preparing the Summary Report of the Consultation (see Volume 1 of this document). Thanks are also due to authors and referees of the papers for their valuable contributions. Valuable assistance and advice were received from Dr William Bayliff, Ms Mary Lynne Godfrey, and Mr Robert Harman in the editing and from Mr Harman in the final formatting and lay-out of the document.

The Consultation and its Proceedings were made possible through the close cooperation of tuna scientists in the Pacific region. Funds for the organization of the Consultation were provided by the Government of Japan and by FAO. Technical expertise, data, and computer facilities for the preparatory work and the Consultation were contributed by many institutions, especially:
- Commonwealth Scientific and Industrial Research Organization (Hobart, Australia),
- FAO/UNDP Regional Fisheries Support Programme (Suva, Fiji),
- Indo-Pacific Tuna Programme (Colombo, Sri Lanka),
- Inter-American Tropical Tuna Commission (La Jolla, USA),
- International Commission for the Conservation of Atlantic Tunas (Madrid, Spain),
- Institut Français de Recherche Scientifique pour le Développement en Cooperation (Noumea, New Caledonia),
- Ministry of Agriculture and Fisheries (Wellington, New Zealand),
- National Marine Fisheries Service (La Jolla and Honolulu, USA),
- National Research Institute of Far Seas Fisheries (Shimizu, Japan),
- South Pacific Commission (Noumea, New Caledonia), and
- South Pacific Forum Fisheries Agency (Honiara, Solomon Islands).

National research laboratories of many countries of Latin America, Southeast Asia, and the South Pacific also contributed significantly to the work before and during the Consultation.

Particular thanks are extended to local organizers from the institutions hosting the Consultation, Dr Anthony Lewis of the South Pacific Commission in Noumea, New Caledonia and Mr Renaud Pianet of the Institut Francais de Recherche Scientifique pour le Développement en Coopération also in Noumea.

The editors would like to acknowledge the assistance and encouragement of the staff of the Fishery Resources and Environment Division and the Operations Service of the FAO Fisheries Department (Rome, Italy), the FAO/UNDP Regional Fisheries Support Programme (Suva, Fiji), and the Indo-Pacific Tuna Programme (Colombo, Sri Lanka) and particularly to Dr John Caddy, Dr Serge Garcia, Mr Robert Gillett, Mr Andhi Isarankura, Dr Yasuhisa Kato, Mrs Christiane Lagrange-Hall, Mr Toshifumi Sakurai, Mr Mitsuo Yesaki, and Mr Hugh Walton.

INTERACTIONS OF PACIFIC TUNA FISHERIES

Proceedings of the First FAO Expert Consultation on
Interactions of Pacific Tuna Fisheries

PREFACE

Tunas and tuna-like species are extremely valuable commercially, especially albacore, bigeye, northern and southern bluefin, skipjack, and yellowfin tuna. Collectively, these species are referred to as principal market tuna species, and are prized for canning, sashimi (raw fish dishes), and other products. The lesser known tuna species, however, should not be discounted because they provide considerable in-country commerce and are important sources of protein in some parts of the world. In recent years, the Pacific Ocean has become the dominant ocean for tuna landings. Between 1980 and 1991, the annual catch of tuna and tuna-like species in the Pacific increased by 68% to about 3 million metric tons (mt). The 1991 Pacific total catch represented about 68% of the world's catch of these species. While these very high catches in the Pacific are impressive, some recent studies suggest that there is potential for still higher sustainable catches of some species.

The increases in the catches of Pacific tunas and tuna-like species have resulted from both intensification and expansion of existing fisheries, and the development of new fisheries. These changes have led to overlap of areas of operations of large and small-scale fisheries, as well as competition for the same tuna resources by large-scale fisheries using different gear. Detecting or predicting even this most direct type of interaction, however, is difficult and presents a serious research challenge. Presently, two or more tuna fisheries may be operating simultaneously on the same stock in overlapping geographical areas, targeting fish of similar sizes. In such a situation, changes in the fishing intensity or pattern of one fishery may affect the catches of the other fisheries. A further factor in fisheries interaction among tunas is the ability of many tuna species to undertake rapid, long distance movements or migrations across or even between oceans. Under these circumstances, fisheries operating in different exclusive economic zones and on the high seas may significantly affect each other.

The knowledge of fisheries interactions is essential for rational management of fisheries. The principal market tuna species and many tuna-like species are recognized by the United Nations Convention for the Law of the Sea (UNCLOS) as highly migratory. Recently, considerable attention has been directed to the need for rational management of fisheries for highly migratory species and resources that straddle adjacent exclusive economic zones (EEZs). Such management would enhance economic and social benefits to the countries involved in fishing, processing, and trade of these resources. Presently, small-scale tuna fisheries exist in many developing countries in the Pacific, and many of these fisheries operate in the same areas as the large industrial tuna fisheries (purse seine, pole and line, and longline).

From 6 to 8 May 1992, the International Conference on Responsible Fishing was held in Cancun, Mexico, leading to the Cancun Declaration. From 7 to 15 September 1992, FAO organized the Technical Consultation on High Seas Fishing held in Rome, Italy, to consider technical issues related to such fishing. As a consequence of these meetings, FAO is involved in addressing the issue of flag of convenience. This issue is of major relevance to tuna fisheries and their management because many tuna vessels use such flags of convenience to avoid restrictive measures imposed by certain countries. Also, FAO actively participates in the development of a Code of Conduct of Responsible Fishing, which will apply to both the high seas and economic exclusive zones.

In a broader context, fisheries issues were considered at the United Nations Conference on Environment and Development (UNCED) held in Rio de Janeiro, Brazil, from 3 to 14 June 1993. The outcome of this Conference is relevant to fisheries directed at tuna and tuna-like species. The existing programme of action on environment and development (referred to as Agenda 21 or the Rio Declaration) and the two Conventions on Biodiversity and Climate Change are now open for ratification. These initiatives represent an important commitment at the highest national political level to resolve a wide range of problems associated with rational use of marine resources.

According to UNCLOS, fisheries management needs to be based on the best available scientific information. The recent attention directed to fisheries management of highly migratory and high seas resources has pointed the need for scientific information on interactions of fisheries directed at tuna and tuna-like species. This need has also become evident at recent regional and international meetings of fisheries scientists and administrators.

In the Pacific, where most catch of tuna and tuna-like species is taken, there is an additional urgent need to integrate available information and to coordinate fisheries research. The Pacific is the only ocean where there is neither a single fisheries body nor a technical programme directed to tuna and tuna-like species that encompass the entire ocean. Some of the Pacific stocks of tuna and tuna-like species are only partly covered by existing fisheries bodies and programmes in terms of their areas of distribution. This situation promoted FAO to initiate a project: "Cooperative Research on Interactions of Pacific Tuna Fisheries" and to create a network of ten working groups of scientists (TUNET) to provide direction and to facilitate the implementation of the project.

To provide an information base for the execution of the project, FAO organized the First FAO Expert Consultation on Interactions of Pacific Tuna Fisheries hosted in Noumea, New Caledonia, by the South Pacific Commission with collaboration of the Institut Francais de Recherche Scientifique pour le Developpement en Cooperation (ORSTOM) from 3 to 11 December 1991. The Consultation was preceded by a preparatory meeting held in Noumea in late 1989. The success of the Consultation was due to the close collaboration and contribution of many other institutions; these institutions and the host organizations are duly acknowledged in the Summary Report presented in Volume I.

The objectives of the First FAO Expert Consultation on Interactions of Pacific Tuna Fisheries included describing the concerns related to interactions of Pacific fisheries directed at tuna and tuna-like species, classifying these interactions, reviewing all available information on them and the methods applied to their study, and making recommendations for future research.

The information presented in the proceedings demonstrates that there is a potential for interactions occurring between and among the fisheries directed at tunas and tuna-like species. Empirical evidence for such interactions, however, has been available for only few fisheries, and these interactions have been quantified for even fewer fisheries. It is unclear whether interactions are insignificant among fisheries directed at tuna and tuna-like species or whether scientists are unable to detect these interactions possibly due to various changes to fisheries and resources, resulting in a too-variable background which conceals the effects of interactions.

These "Proceedings of the First FAO Expert Consultation on Interactions of Pacific Tuna Fisheries" are provided in two volumes:

- Volume 1 contains the Summary Report of the Consultation. Volume 1 also includes a review paper on methods for studying interactions in tuna fisheries, thirteen papers presenting new methods and case studies on such interactions, and seven reviews on fisheries interactions related to individual stocks of Pacific tunas and tuna-like species.

- Volume 2 includes eleven review papers on the biology, population dynamics, and fisheries associated with the Pacific tuna resources. These reviews are supplemented by four additional papers on specific fisheries.

INTERACTIONS OF PACIFIC TUNA FISHERIES

Proceedings of the First FAO Expert Consultation
on Interactions of Pacific Tuna Fisheries

VOLUME 2

CONTENTS

page

A REVIEW OF THE BIOLOGY AND FISHERIES FOR SKIPJACK TUNA, *KATSUWONUS PELAMIS*, IN THE PACIFIC OCEAN

Alex Wild
Inter-American Tropical Tuna Commission
La Jolla, California

and

John Hampton
South Pacific Commission
Noumea, New Caledonia

1. INTRODUCTION

This paper is intended as a review of the biology, resource, and fisheries associated with skipjack tuna, *Katsuwonus pelamis*, in the Pacific Ocean. Because of the distribution of the fisheries, it has been convenient in several sections to discuss skipjack in the eastern Pacific Ocean (EPO) and western and central Pacific Ocean (WPO) separately. This division may, for the purpose of these discussions, be taken to occur at approximately 150°W. Certain sections contain information from the Indian and Atlantic Oceans in order to offer a wider perspective. The task of compiling the material was simplified considerably through the extensive use of three major sources of information: the synopses of biological data on skipjack by Forsbergh (1980) and Matsumoto *et al.* (1984), and a study on the environmental factors affecting their apparent abundance (Forsbergh, 1989). In Section 13, the material that refers to the EPO is intended to be both a general and an interaction-specific review for the purse-seine and baitboat fisheries.

2. CLASSIFICATION

The family Scombridae is composed of 15 genera comprising 49 species of epipelagic marine fishes, which include the mackerels, Spanish mackerels, bonitos and tunas (Collette and Nauen, 1983). The family is divisible into two subfamily categories, the Gasterochismatinae, which contains the single species *Gasterochisma melampus*, and the Scombrinae. On the basis of internal osteological characters the Scombrinae in turn have been divided into two groups of tribes: the primitive mackerels and Spanish mackerels, and the more advanced group consisting of the bonitos (Sardini) and tunas (Thunnini). Among other characteristics, the Sardini differ from the Thunnini in that the latter possess a subcutaneous, counter-current vascular system that permits members of the tribe to be warmer than the surrounding water (Collette *et al.*, 1984). In this respect the four genera of the Thunnini tribe are unique relative to other bony fishes. The three more primitive members of the tribe (*Auxis*, *Euthynnus*, and *Katsuwonus*), together with the yellowfin group (*Thunnus*), have central as well as lateral heat exchangers. The diagram below indicates the position of skipjack tuna within the family Scombridae after following Lindberg's (1971) scheme to the family level:

2

Phylum Chordata
   Subphylum Vertebrata
      Superclass Gnathostomata
         Class Teleostomi
            Subclass Actinopterygii
               Order Perciformes
                  Suborder Scombroidei
                     Family Scombridae
                        Sub-family Scombrinae
                           Tribe Thunnini
                             Genus *Katsuwonus*
                               Species *pelamis*

The classification of skipjack tuna in particular, and in relation to other tunas, is discussed by Matsumoto *et al.* (1984) and Collette *et al.* (1984), respectively.

3. EARLY LIFE HISTORY

Ripe skipjack ovarian eggs are spherical, smooth, transparent, and usually contain a single oil droplet (Brock, 1954; Yabe, 1954; Yoshida, 1966). The eggs are 0.80 to 1.17 mm in diameter with variable-sized oil droplets ranging from 0.22 to 0.45 mm. Because the size and appearance of the eggs of artificially-reared skipjack are similar to those of other artificially-reared scombrids (Harada *et al.* 1971; Ueyanagi *et al.*, 1973), it is not possible to identify those of skipjack in field collections of tuna eggs.

Net tows by Japanese research vessels recorded the distribution (Figure 1) of skipjack larvae in the world oceans (Ueyanagi, 1969; Nishikawa *et al.*, 1985). In particular, in the Pacific they are found near 35°N off Japan and as far south as 37°S off the southeastern part of Australia. This wide distribution is maintained eastwardly until slightly past the Hawaiian Islands in the northern hemisphere and the Society Islands to the south. Matsumoto *et al.* (1984) concluded that, "... the distribution then narrows abruptly toward the equator at about 145°W and remains within 10°-15° of latitude on either side of the equator to the Central and South American coasts." Klawe (1963) found few larvae in the eastern Pacific, while Ueyanagi (1969) and Ueyanagi *et al.* (1969) showed that they appeared to be concentrated in the equatorial region with an increase in abundance from east to west. After adjusting for differences between net tows, Matsumoto (1975) found that the greatest concentration occurred between 160°E and 140°W. Recent work by Leis *et al.* (1991) identified particularly high concentrations of skipjack and *Thunnus* spp. larvae in the vicinity of coral reefs of the islands of Moorea, Rangiora, and Takapoto in French Polynesia.

Skipjack larval distribution is strongly influenced by temperature. Forsbergh (1989) showed that the concentration of larvae in the Pacific approximately doubles with each 1°C increase in the sea-surface temperature from 24°-29°C, and decreases by a variable amount after 30°C. However, it is not known whether the relationship is brought about by the larvae, the spawners, or both. The larvae are therefore concentrated in the tropics, although the warm, poleward-flowing Kuroshio and East Australian Currents facilitate a seasonal (summer) expansion of the larval distribution into subtropical waters in the far WPO.

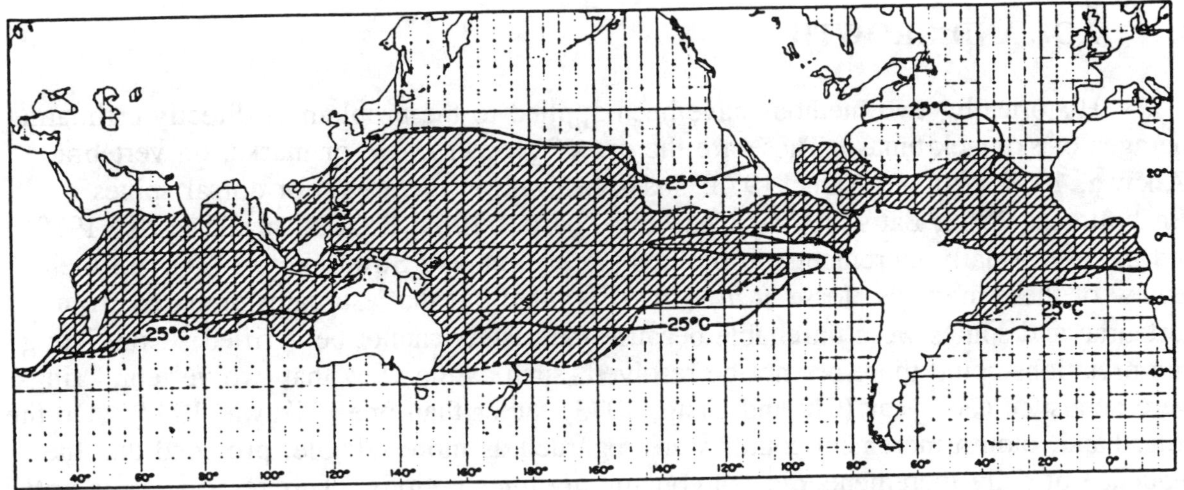

Figure 1. Distribution of skipjack tuna larvae, 1952-1975. (After Matsumoto, *et al.*, 1984).

In terms of depth, Klawe (1963) did not find any larvae below the mixed layer in the eastern Pacific, but there is a suggestion in Strasburg's (1960) closing-net sampling data from 11°N-5°S, 140°W that this may not always be the case. Upon examining Japanese data, Forsbergh (1980) concluded that 22 percent of the larvae were captured in surface tows and 55 percent in subsurface tows. The larvae are scarce at the surface during the day, and although they are more common there at night, there are still fewer than at deeper levels (Ueyanagi, 1969). Richards and Simmons (1971) found that, in the northwestern Gulf of Guinea and off Sierra Leone, the larvae of yellowfin (*T. albacares*) and bigeye (*T. obesus*) tunas migrate to the surface during the day while skipjack larvae migrate to the surface at night. In contrast to yellowfin, the swimbladder in skipjack larvae degenerates before the fish reaches a size of two cm (Richards and Dove, 1971).

Most juvenile skipjack have been collected from the stomachs of larger tunas and billfishes, from experimental mid-water trawls and by dipnetting at night under a light (Forsbergh, 1980). Higgins (1967) examined the records of 518 juveniles (1.2-30 cm) caught in the Pacific in the period from 1916 to 1966. Of the 372 caught in the WPO west of 180°, the majority were found in tuna stomachs; 135 were found in the central Pacific (180°-140°W) in either midwater trawls or stomachs, and 11 were caught by dipnetting in the EPO east of 140°W. Mori (1972) defined juvenile skipjack as those less than 15 cm long, and young skipjack as those between 15 and 35 cm. He examined the data from 3,778 juveniles and young recovered from the stomachs of billfishes caught by longlines in the Pacific and concluded, together with Yoshida (1971), that the distribution of juveniles was similar to that of the larvae. The apparent abundance of the juveniles was greatest west of 180° between 10°N and 10°S, and least east of 120°W. They were also more prevalent in areas with surface waters warmer than 24°C. On the other hand, skipjack young were more widely distributed and more common in areas with surface temperatures less than 24°C. Based on the distribution of baitboat-caught skipjack with juvenile skipjack in their stomachs, Argue *et al.* (1983) concluded that, in the vast area of the central Pacific, the abundance of juveniles was greatest during October-March between the equator and 25°S in two broad geographical areas, one including eastern Polynesia (130°-150°W) and the other adjacent to Papua New Guinea, the Solomon Islands and Vanuatu (140°-170°E).

4.   AGE AND GROWTH

Historically, two methods have been applied to the problem of directly estimating the ages of skipjack tunas. These are the use of growth rings, or marks, on vertebrae (Aikawa, 1937; Chi and Yang, 1973; Sosa-Nishizaki *et al.*, 1989) or dorsal spines (Shabotiniets, 1968; Batts, 1972a; Chur and Zharov, 1983; Sosa-Nishizaki *et al.*, 1989) and the use of daily increments on otoliths (Uchiyama and Struhsaker, 1981). In their review of these methods, Josse *et al.* (1979) pointed out that estimates prepared from vertebrae and spines were unreliable because the results cannot be verified. Conflicting interpretations could therefore not be resolved, such as the view that two vertebral rings were deposited each year (Chi and Yang, 1973) rather than one (Aikawa, 1937). On the other hand, otolith increment counts were regarded as more reliable, provided that the sequence of daily increments did not contain any interruptions. This reservation is critical with respect to skipjack tuna. In the eastern Pacific, Wild and Foreman (1980) found that 21 out of 26 skipjack failed to deposit an increment (I) each day (d) during their period at liberty in a mark-recapture experiment conducted at sea. The fish were initially captured, tagged, injected with oxytetracycline in order to label their otoliths, and released. For fish in the length range of 42-64 cm, the average deposition rate was only 0.76 I/d (range: 0.44-1.09 I/d) during periods of liberty from 17-249 days. These results contrast with the expected rate of one I/d and the rates for other tunas, *i.e.*,

Tuna	Length Range	Deposition Rate	Source
Albacore (*T. alalunga*)	51-97 cm	0.95 I/d	Laurs *et al.*, 1985
N. Bluefin (*T. thynnus*)	19-68 cm	daily	IATTC, 1987a
Yellowfin (E.Pacific)	30-148 cm	daily	Wild and Foreman, 1980; IATTC, 1987b
Yellowfin (C.Pacific)	25-40 cm (n=12)	daily	Yamanaka, 1990
Yellowfin (E.Atlantic)	47-64 cm (n=5)	daily	IATTC, 1987b

The low deposition rate for skipjack could have resulted from counting increments in the wrong location on the otolith, the sex of the fish, or because some of the increments were too small to be detected by the light microscope. Therefore, on fish recovered from a second tetracycline-injection experiment carried out in 1980-81, additional counting sites and the enhanced magnification (3,000-3,500x) of the scanning electron microscope were applied to the otoliths of each sex. Despite these refinements, the deposition rates for the sexes did not differ, and their pooled average rate of 0.83 I/d (n = 41) was not significantly different from the previous result (Wild and Wexler, IATTC; unpubl. data). Consequently, otolith increments are presently unreliable for estimating the age of skipjack tuna in the eastern Pacific. A count of the total number of increments on an otolith will underestimate the actual age and overestimate the growth rate.

5

For skipjack in the central Pacific it is known that increments are deposited daily
for at least the first five days after hatching (Radtke, 1983). To develop their growth
model for the central region, Uchiyama and Struhsaker (1981) concluded that captive
skipjack (n = 4) in the length range of 45-48 cm deposited daily increments over a period
of 5-30 days following the onset of satiation feeding. A check mark on the otolith,
induced by the stress of capture and adjustment to the holding tanks, was used to locate
the beginning of the period of deposition of wider increments caused by full feeding.
Because such stress marks are difficult to distinguish from other, similar marks on the
otolith, the authors noted that the assumption of daily deposition would eventually have to
be validated. With this proviso, the growth-equation parameters and the sizes-at-age of
skipjack in the central Pacific are recorded in Table 1. Selected examples of growth
curves developed for the Pacific Ocean also appear in Figure 2.

The measurement of skipjack growth is seemingly a more tractable problem than
direct age determination, but the two most popular methods of measuring growth, tag-
recapture experiments and modal progression of length frequencies, are not without
limitations. Joseph and Calkins (1969), for example, drew attention to the subjective
aspect of modal analysis after applying the method to skipjack in the eastern Pacific.
Based on data collected from the South Pacific, Josse *et al.* (1979) also characterized the
method as least reliable because of the uncertainty of demonstrating a "...clear
progression of modes for more than a few months. Apparent mode progressions may
give in a single region, in different years, growths which are rapid, slow, nil and even
negative." With respect to tagging, one of the most persistent and troublesome questions
is whether the procedure affects subsequent growth. The rapid buildup of blood lactate
due to the stress of capture and tagging probably increases mortality (Barrett and Connor,
1962), and tagging has been shown to reduce condition in southern bluefin tuna (*T.
maccoyii*) for a short period after tagging (Hampton, 1986). However, the longterm
effect on tuna growth is unknown. Schaefer (1961) concluded that, compared to untagged
fish in modal progressions, the growth rate in the EPO of yellowfin tagged with loop tags
was somewhat lower and their attrition rate higher. More recently, the average growth
rate of untagged yellowfin, caught north of the equator and in the length range from
40-135 cm, was estimated from otolith increments to be 1.06 mm/d with a 95-percent
confidence interval of 1.03-1.08 mm/d (Wild, 1986). For yellowfin tagged with dart
tags, Bayliff (1988) found that their overall growth rate in the EPO was 0.85 mm/d
(variance: 0.07). Considering that skipjack are more sensitive to handling than yellowfin,
it is unlikely that their response to tagging would be dissimilar (Forsbergh, 1980).
Through simulation studies, Sibert *et al.* (1983) also noted the effect of measurement
errors on the parameter estimates of the equation (von Bertalanffy, 1938) used most
frequently to analyze tagging data, *i.e.*,

$$l_2 = L_\infty(1\text{-exp}(-K \cdot \Delta t)) + l_1\text{exp}(-K \cdot \Delta t) \qquad (1)$$

where l_2 is the length at recapture, l_1 is the length at tagging, L_∞ is the asymptotic size, K
is a growth parameter and Δt is the elapsed time. An increase in the magnitude of the
error in l_2 increased the confidence limits of the parameter estimates of L_∞ and K, but
they were unbiased. However, small changes in Δt or the least reliable length
measurement, l_1, caused large changes in estimates of L_∞ and K, and frequently their joint
confidence region excluded their true values.

TABLE 1. Growth parameters of the von Bertalanffy (1938) equation for skipjack from major oceanic regions, and estimates of lengths at different ages prepared by various methods. Bracketed lengths represent extrapolations of the data. The range of the L and K values for the South Pacific include the island nations affiliated with the South Pacific Commission as well as New Zealand. Legend: f = fixed parameter; r = relative sizes; a = average.

| Source | Area | Growth parameters | | | Lengths (cm) at estimated ages (yr). | | | | | Range of data (cm) | Method |
		L (cm)	K (yr)	t_0 (yr)	1	2	3	4	5		
Pacific Ocean											
Aikawa (1937)	East of Japan	-	-	-	26	34	43	54	-	-	Vertebrae
Yao (1981), (Kawasaki's [1955a,b] data)	North & south Japan	76.6	0.60	-0.31	42	58	66	(71)	-	32 - 68	Length frequencies
Chi and Yang (1973)	Taiwan	103.6	0.302	-0.016	27	47	62	(73)	(81)	27 - 65	Vertebrae
Josse et al., (1979)	Papua New Guinea	65.5	0.945	-	-	-	-	-	-	40 - 60	Tagging
Sibert et al., (1983)	South Pacific (range)	62.5a 46.6- 171.0	2.00a 5.88- 0.11	-	-	-	-	-	-	-	Tagging
Brouard et al., (1984)	South Pacific	60.0	0.75	-	-	-	-	-	-	30 - 50	Length frequencies
Uchiyama and Struh-saker (1981)	Central Pacific	102.0	0.55	-0.02	44	68	(83)	(91)	-	3.7 - 80.3	Otolith increments
Joseph and Calkins (1969)	East. Pacific	88.1	0.431	-0.005	(31)	51	64	(72)	-	40 - 68	Tagging; grouped data
Bayliff (1988)	East. Pacific										
	North region	89.3	0.682	-	-	-	-	-	-	28 - 72	Tagging; grouped data
	South region	68.6	1.649	-	-	-	-	-	-	28 - 72	Tagging; grouped data
Atlantic Ocean											
Chur and Zharov (1983)	Gulf of Guinea	86.7	0.307	-0.317	(29)	44	55	64	70	35 - 74	Dorsal spines
Bard and Antoine (1986)	East.; equator	80(f)	0.322	-	40r	51r	59r	65r	-	40 - 65	Tagging
	East.; northern temperate	80(f)	0.601	-	40r	58r	68r	73r	-	40 - 65	Tagging
Cayré et al., (1986)	Senegal	62.0	2.08	-	-	-	-	-	-	-	Tagging
	Republic Cap Vert	60.0	1.537	-	-	-	-	-	-	-	Tagging
	West. Atlantic	79.6a	0.195a-4.329a		49	57	64	-	-	26 - 76	Dorsal spines
	males	87.2	0.154	-							
	females	78.4	0.195								
Indian Ocean											
Shabotiniets (1968)		-	-	-	-	-	40-45	40-60	-	-	Dorsal spines

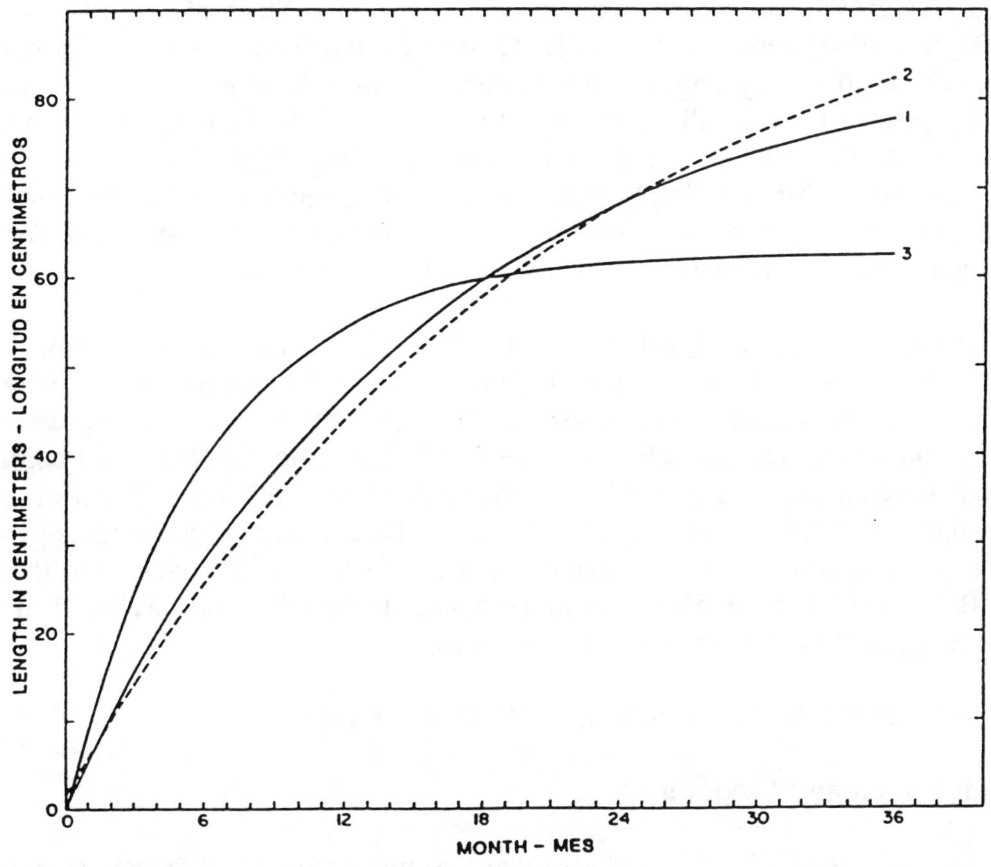

Figure 2. Examples of three, von Bertalanffy-type growth curves for skipjack tuna. Curve: 1) from tagging data in the EPO from Bayliff (1988); 2) from otolith increments in the central Pacific from Uchiyama and Struhsaker (1981); and 3) based on average parameter estimates derived from tagging data from the South Pacific Commission area. (After Forsbergh, 1989).

From the foregoing remarks it is clear that, in the absence of a validated method of determining the age of skipjack, the predicted sizes-at-age (Table 1 and Figure 2) are approximations. This conclusion, drawn earlier by Matsumoto *et al.* (1984) remains unaltered at present. Nevertheless, several comments can be made on skipjack growth, and the first one concerns its variability. Sibert *et al.* (1983) found that there were significant differences in the amount of growth and the parameters of the von Bertalanffy (1938) equation estimated for the same regions at different times (see Table 1). The differences between regions were also significant. Therefore, these variations in growth and unstable parameters were likely due to the temporal and spatial variation of environmental conditions. The broad range and values of K, *i.e.*, 1.54-3.13, in the Senegal and Cap Vert regions (Cayré *et al.*, 1986) also emphasize the variability in growth over short-term recovery periods ($\leq$ 60 d). Based on the previous comment that a short (1 cm) error in the release length of tagged fish can seriously bias parameter estimates, the results of many of the earlier experiments, *e.g.*, Schaefer *et al.*, 1961; Rothschild, 1967; Joseph and Calkins, 1969, may be compromised because the fish were measured only to the closest 5 cm. Bayliff (1988) eliminated such data from his study in the EPO and found that, overall, skipjack east of 100°W grew less rapidly than those west of 100°W. Provided that there are no interruptions in the record of otolith increments, the growth curves in the central (Uchiyama and Struhsaker, 1981) and eastern Pacific (Bayliff, 1988) appear to be similar (Figure 2), but the average growth rate is more rapid during the early stages in the western Pacific (Sibert *et al.*, 1983). The data in Table 1 represent but a small selection of the available information; a more extensive list appears in Bayliff (1988). The difficulties in estimating skipjack age, and in relating age to length, have substantial implications for the assessment of stocks. In view of these difficulties, the application of traditional age-structured models may be limited for skipjack, and alternative methods of assessment may be required.

The length-weight relationships in Table 2 are reproduced, in part, from Matsumoto *et al.* (1984), following the original outline by Nakamura and Uchiyama (1966). To facilitate parameter comparisons, all lengths and weights are expressed in centimeters and kilograms, respectively. The detailed statistics needed to set confidence limits about selected sizes are generally not available in the literature. It appears however that, with the exceptions of Evans *et al.* (1978) and Batts (1972a), the weights of 40-, 50- and 60-cm skipjack from different oceanic regions are similar. Skipjack from the western Atlantic (Batts, 1972a) in the 50-60 cm range appear to be relatively heavier than elsewhere and progressively so as length increases.

5.   MATURATION, SPAWNING, AND SEX RATIO

5.1   Maturation and Fecundity

Matsumoto *et al.* (1984) concluded that the minimum size of female skipjack at maturity is 40 cm, and that first spawning may occur in fish between 40-45 cm, or larger. They based their interpretation on several sources of information in which, for the Pacific, "... Marr (1948) recorded skipjack tuna as small as 40 cm fork length with spent ovaries from the Marshall Islands; Wade (1950) recorded fish in the 40.0 to 40.9 cm size class with ripe and spent ovaries from Philippine waters, and reported ... a female in the 34.0 to 34.9 cm size class having ripe ovaries; and Brock (1954) noted that the smallest fish that possessed maturing ova during the spawning season in Hawaiian waters were around 40 to 45 cm." In the eastern Pacific, Schaefer and Orange (1956) estimated that the minimum size at maturity in the vicinity of the Revillagigedo Islands, offshore from Baja California, was 55 cm and nearer 50 cm off Central America. In the Indian Ocean, Raju (1964a) estimated the size at sexual maturity around Minicoy Island at about 40 to 45 cm. His conclusion was based on the observation that remnants of mature ova were present in fish above 40 cm, but none in fish below that size. Similarly, on the basis of fish with spawned or spent ovaries, Simmons (1969) reported that the minimum size at maturity in the Atlantic Ocean was 41 cm. Stequert (1976) also noted the first spawning of skipjack tuna at 41 to 43 cm off the northwest coast of Madagascar.

In contrast to this rather narrow range at first maturity, the fecundity of female skipjack is quite variable in terms of both oceanic region and fish size. For example, although 50-cm and 60-cm fish in the eastern Atlantic (Batts, 1972b) and Pacific (Joseph, 1963), respectively, produce nearly equal numbers of ova (300 x 10^3), at a length of 70 cm the fecundity in the Pacific (970 x 10^3) exceeds that in the Atlantic (600 x 10^3) by a considerable margin (Figure 3). Within the Atlantic, increasing differences in fecundity with the size of fish are also demonstrated by regressions 4 and 5 in Figure 3. Caution is needed in interpreting this information, however, since Simmons' (1969) data represents only 13 fish from both sides of the Atlantic, whereas Batts' (1972b) data includes 30 fish from only the western Atlantic. The parameter estimates for most of the regressions in Figure 3 are reproduced in Table 3.

TABLE 2. Length-weight relationships for skipjack tuna. The parameters refer to the allometric relationship, weight = $a(FL)^b$, where FL is the fork length, or the distance from the tip of the upper jaw to the cartilaginous median of the caudal fork. This is the definition given by Marr and Schaefer (1949) to total length (TL), and in the table TL probably refers to FL, the most reasonable length measurement on tunas. (After Matsumoto, *et al.*, 1984).

Source	Location	No. of fish	Size range (cm)	Constant a	Coefficient of allometry b	Length	Calculated weight (kg) at length (cm) 40	50	60	Comment
Marcille and Stequert (1976)	Madagascar	848	41-62	1.13×10^{-5}	3.16	FL	1.30	2.62	4.67	Baitboat catches
Vooren (1976)	New Zealand	100	35-54	9.612×10^{-6}	3.19	FL	1.24	2.53	4.52	Purse-seine catches
Habib (1978)	" "	120	38-71	6.776×10^{-5}	3.29	FL	1.26	2.63	4.80	Purse-seine catches
Ronquillo (1963)	Philippines	189	34-66	1.482×10^{-5}	3.096	TL	1.35	2.70	4.74	Males; from Ronquillo (1963), Figure 25
" "	"	151	34-63	2.257×10^{-5}	2.965	TL	1.27	2.46	4.22	Females; from Ronquillo (1963), Figure 25
Evans[1] *et al.* (1978)	"	189	34-66	1.482×10^{-4}	2.20	TL	1.84	3.01	4.49	Handline catches
" " " "	"	44	17-62	3.727×10^{-3}	1.70	TL	1.97	2.88	3.93	Artesinal fishing gears
Kawasaki (1952)	N. E. of Japan	20	30-60	1.13×10^{-5}	3.16	FL	1.30	2.64	4.70	Baitboat catches
Tester and Nakamura (1957)	Hawaii	268	39-83	8.518×10^{-6}	3.216	FL	1.21	2.48	4.46	Calculated size range
Nakamura and Uchiyama (1966)	"	1298	33-88	4.812×10^{-6}	3.368	FL	1.20	2.54	4.69	Baitboat catches
Chatwin (1959)	E. Pacific	924	39-71	4.033×10^{-6}	3.413	FL	1.18	2.54	4.73	Baitboat and purse seine
Hennemuth (1959)	" "	1282	39-71	5.530×10^{-6}	3.336	FL	1.26	2.61	4.74	Includes data from Chatwin (1959)
Batts (1972a)	W. Atlantic	644	26-76	5.771×10^{-6}	3.353	FL	1.36	2.87	5.29	Troll catches
Lenarz (1974)	E. Atlantic	2554	36-64	5.611×10^{-6}	3.315	FL	1.15	2.41	4.40	-
Planet (1974)	" "	520	40-73	3.419×10^{-6}	3 456	FL	1.18	2.54	4.78	-

[1] Evans, L.C., A.D. uy and D.D. Tandog. 1978. The abundance, biology and distribution of tuna (Family Thunnidae) in Camiguin and nearby waters. Philippine Dept. Nat. Resour., Bureau Fish. Aquatic Resour., Prelim. Rep., 55 p.

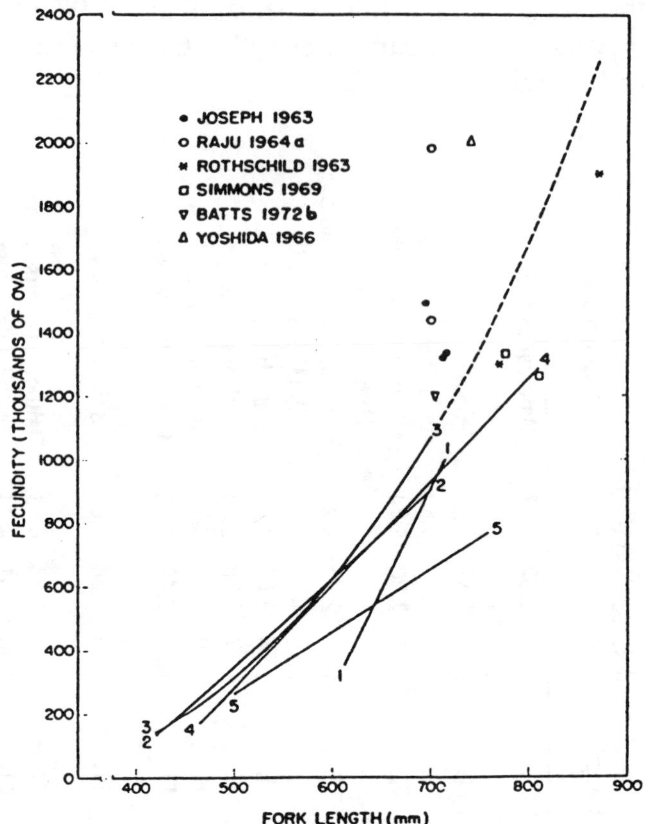

Figure 3. Regressions of fecundity on length and observed fecundity estimates of skipjack tuna above 69 cm. 1) Joseph (1963), eastern Pacific; 2) Raju (1964b), Indian Ocean; 3) fecundity-length^3, Raju (*op. cit.*); 4) Simmons (1969), eastern and western Atlantic combined; and 5) Batts (1972b), western ATlantic. (After Matsumoto *et al.*, 1984).

TABLE 3. A comparison of fecundity-length and fecundity-weight relationships of skipjack tuna from various locations. The relationships are all of the form $Y = a + b(FL)^x$ or $Y = a + bW$, where Y is the number of ova in thousands, FL is the fork length (mm) raised to the power x, and W (lbs) is the weight. The number in brackets after the relationship refers to the corresponding numbered curve in Figure 2. (After Matsumoto *et al.*, 1984).

Source	Location	Relationship		Parameters $a \times 10^{-3}$	b	Standard error of estimate ($\times 10^{-3}$)	Correlation coefficient
Joseph (1963)	E. Pacific	FL	(1)	-3,503	6.326	276.	-
		W		- 608.81	88.613	276.	-
Raju (1964b)	Indian Ocean	FL	(2)	-1,004.94	2.713	28.6	0.645
	(Minicoy Is.)	$(FL)^2$		- 292.26	2.54×10^{-3}	27.4	0.668
		$(FL)^3$	(3)	- 109.0	3.42×10^{-6}	24.7	0.741
		W		- 67.69	67.01	22.5	0.789
Simmons (1969)	E. and W. Atlantic	FL	(4)	-1,333.54	3.238	-	0.873
Batts (1972b)	W. Atlantic	FL	(5)	- 632.09	1.854	-	0.384
		W		164.2	29.92	-	0.384

5.2   Spawning

"Gonadal studies in the Pacific, Atlantic and Indian Oceans indicated that skipjack tuna spawn throughout the year in tropical waters near the equator and from spring to fall in subtropical waters" (Matsumoto *et al.*, 1984). The collection of spawning adults for such studies have taken place in association with commercial and experimental fisheries in diverse locations such as waters south of Japan, the Philippine, Marshall and Hawaiian Islands, New Caledonia, Marquesas-Tuamotu Archipelago, and off central America and Baja California. Since the larvae are generally found in waters with sea-surface temperatures > 24°C (Ueyanagi, 1969), the spawning period necessarily becomes shorter with increasing distance from the latitude of greatest temperature. As mentioned earlier (Section 3), the distribution of warm waters, and hence that of spawning and larvae, is affected by current transport. Warm waters extend to higher latitudes in the western Pacific because the water is transported poleward along the western shores. In the eastern Pacific, however, the warm water is restricted to lower latitudes because currents transport colder water toward the equator along the eastern shores (Forsbergh, 1989). On the basis of gonad indices, Naganuma (1979) decided that in the western Pacific between 15°N and 20°S smaller adults spawn during the warm season from October to March, and larger fish (> 60 cm) spawn north of 15°N in the northern summer. In the EPO, skipjack spawn off Central America at least during the late winter and spring months (Schaefer and Orange, 1956). Off Baja California near the Revillagigedo Islands, females with maturing ovaries were collected from April through November, although the main spawning season is indicated to be from July through November (Orange, 1961).

The use of gonad indices (GI), or the ratio of gonad weight to body weight, is a relatively rapid procedure to gauge the degree of maturity and it avoids the difficulty of measuring ova diameters. The accuracy of the index, however, has been questioned. Yoshida (1966), for example, found that the diameter was not related to GI at higher indices. Matsumoto *et al.* (1984) also concluded that the GI "...does not seem to be a sound measure of maturity of skipjack...", because of the wide range of values for the same developmental stage. The index is also biased because larger females develop larger ovaries in proportion to their body weight than do smaller females (de Vlaming, 1982). Schaefer (1987) suggested that the GI be validated with histology or öocyte diameter and adjusted for the size of individuals if it is to be used to interpret reproductive activity correctly.

In the vicinity of the Philippines (Buñag, 1956) and in the Indian Ocean (Raju, 1964a), multiple spawning of skipjack has been indicated by following the development of ova-diameter modes over time. Earlier, Brock (1954) concluded that skipjack near Hawaii may spawn several times during a season. Multimodal distributions of ova diameters during the spawning season, and the absence of spawned-out fish until after the season were used to support this idea. Based on data from Brock (1949) and Buñag (1956), Matsumoto *et al.* (1984) estimated that skipjack in Hawaiian waters spawned at least twice during a year: in late April or early May and again at the earliest in late July or early August. The first spawning occurred after a 2-month development of ova from the immature to the ripe stage (April). Following a reversion of the ovaries to an intermediate mature stage, at least 6 to 7 weeks were then required for a second ripening in September. Thereafter, additional spawnings were not ruled out provided that the fish sought warmer waters. Histological studies of the ovaries of 87 skipjack captured in

November and December in the South Pacific indicated that spawning occurred about every 1.18 days (Hunter *et al.*, 1986). Therefore, it is possible that individuals may spawn repeatedly over the duration of a seasonal spawning period.

5.3 Sex Ratio

The sex ratios of skipjack tuna obtained from various locations display a broad range of values (Table 4). Matsumoto *et al.* (1984) concluded that, in general, fisheries that rely on young, immature fish have ratios dominated by females, whereas those that capture mostly older and spawning-age fish have sex ratios that are mostly male. Due to the uncertainty of spawning frequency and sex ratios, the variation in numbers of ova and fish size, the possible reduction in fecundity after successive spawnings (Joseph, 1963), and the lack of information on the size composition and abundance of spawning stocks, the total egg production of skipjack tuna would be difficult to estimate (Matsumoto *et al.*, 1984).

6. STOCK STRUCTURE, DISTRIBUTION, AND MIGRATION

6.1 Subpopulations

Fujino *et al.* (1981) found significant differences in the serum naphthyl esterase (Es[1]) allele frequencies between skipjack from the Indian and Atlantic Oceans, and between those from the Indian Ocean and the western Pacific (Figure 4). They proposed that skipjack originated in the Indian Ocean, and by dispersal to other oceans together with reproductive isolation, they formed a total of four subpopulations located in the Indian, Atlantic, western Pacific, and central-eastern Pacific. They further demonstrated that in the Pacific the Es[1] frequency remained relatively low and constant from 80°W to 175°E, and thereafter in the southern hemisphere it exhibited a "step structure" by increasing from about 0.50 at 175°E to about 0.67 at 160°E (Figure 5). Based on these results they concluded that the population structure, as evidenced by the trend in Es[1] frequencies, was discontinuous rather than clinal, thereby supporting Fujino's (1970a,b; 1972) earlier statements regarding the existence of at least two subpopulations in the Pacific. In the northern hemisphere between 140°E and 175°E the Es[1] frequencies were both high and low. This variability was also interpreted to mean that fish from the central-eastern subpopulation also occurred off the east coast of Japan (Forsbergh, 1988). Richardson's (1978) electrophoretic studies supported the idea of two major subpopulations, but he also proposed that a third exists off New Zealand. Sharp (1978) hypothesized that there were at least five genetically-different subpopulations in the Pacific.

Schaefer (1963) proposed that skipjack in the eastern Pacific originate in the central Pacific. Morphometrics, blood types, relatively few tag returns, and the fact that little spawning occurs in the fishing areas of the eastern Pacific were used to support this idea (Forsbergh, 1988). Rothschild (1965) and Joseph and Calkins (1969) also noted that skipjack were infrequently captured in an area of warm water off southern Mexico at about 15°N. Historically, the fish to the northwest and southwest of this area were thought to be from two different subpopulations. This idea was revised in favour of a single subpopulation following the capture of a large amount of skipjack in the area in late 1955 and in 1956 (Broadhead and Barrett, 1964), and to the west of the area between

TABLE 4. Sex ratios for skipjack tuna from various locations. (After Matsumota *et al.*, 1984).

	Area	Period	Ratio male to female		Comment
			Range	Average	
Atlantic Ocean					
Batts (1972b)	North Carolina	1964 - 66	1.06:1 - 0.58:1	0.86:1	Recreational catch
Cayré (1979)	Gulf of Guinea	-	-	0.95:1	Purse-seine catch
Chur *et al.*, (1980)	E. tropical	1969 - 77	-	1.0:1	Purse-seine catch
Indian Ocean					
Marcille & Stequert (1976); Stequert (1976)	Madegascar	1974 - 75	1.01:1 - 0.72:1	0.83:1	Baitboats
Pacific Ocean					
Brock (1954)	Hawaii	1949 - 50	-	1.16:1	Baitboats
Habib (1978)	New Zealand	Dec. 1976-Mar. 1978	-	0.76:1	Purse-seine catch
Hu & Yang (1972)	Taiwan	-	3.01:1	1.00:1	40 - 49 cm
			to	1.31:1	50 - 59 "
		-	0.64:1	1.00:1	60 - 69 "
Marr (1948)	N. Marshall Is.	-	-	1.60:1	
Raju (1964b)	Philippines	-	-	>1.00:1	Males dominant in older groups
Schaefer & Orange (1957)	E. Pacific	-	-	0.73:1	Small fish; sex indeterminate for large proportion
Tester & Nakamura (1957)	Hawaii	-	-	1.57:1	Inshore trolling
Yoshida (1960)	Cent. S. Pacific	-	-	1.05:1	
Wade (1950)	Philippines	-	-	0.86:1	
Waldron (1963)	Japan	-	-	1.09:1	

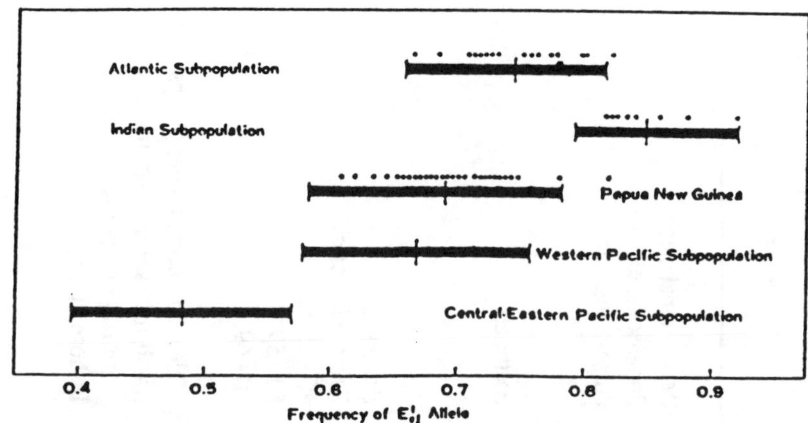

Figure 4. Rejection limits, at 5% significance level, of serum esterase allele frequency distributions for the populations from the Atlantic, Indian and Pacific Oceans. The two rejection limits at the right represent those for the two subpopulations identified by Fujino (1972). (After Fujino *et al.*, 1981).

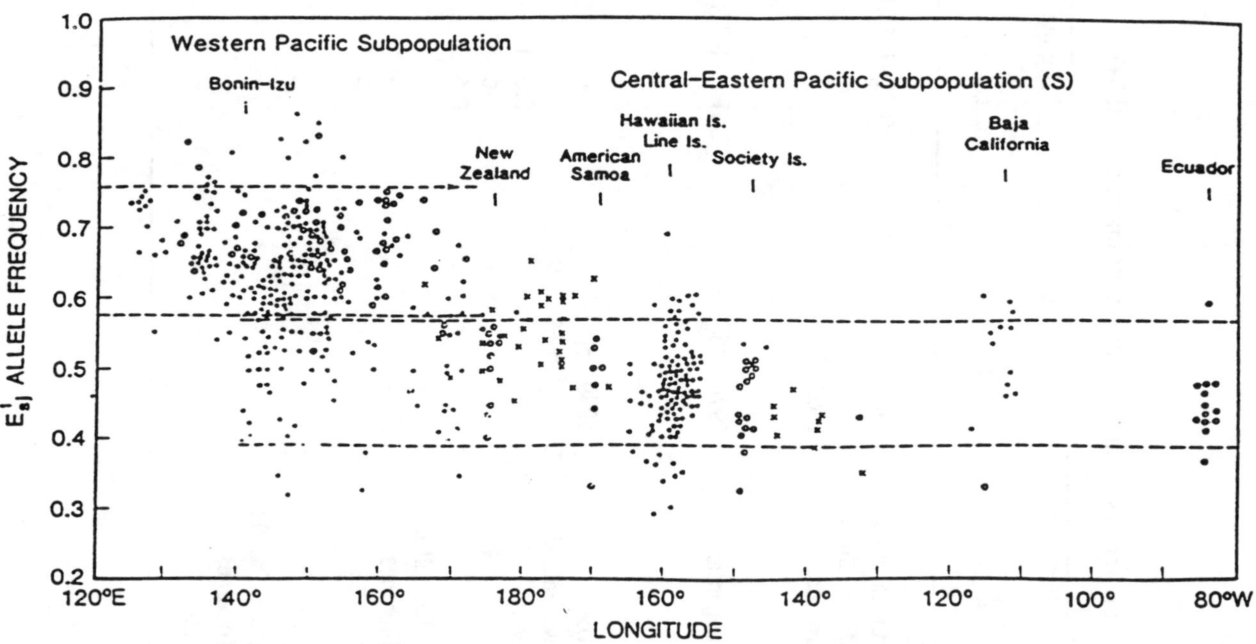

Figure 5. Gene frequenciies for serum esterase alleles (Es[1]) *versus* longitude. Small dots and open circles represent individual lots of samples from the northern and southern hemispheres, respectively (from Fujino *et al.*, 1981). Crosses represent samples from the South Pacific Commission (1981a). Bars on the right are the 95% confidence limits of Fujino's (1972) western Pacific (upper) and central-eastern Pacific (lower) subpopulations. (After Hunter *et al.*, 1986).

10°-20°N in 1980 and 1981 (IATTC 1981, 1982). Events such as these demonstrated that the distribution of skipjack could be continuous throughout the area. It was further suggested that there is a single group of skipjack distributed in an arc-shaped area surrounding the region of warm water off southern Mexico (IATTC, 1983; Bayliff, 1984). By means of tag returns, the details of which appear in Bayliff (1988, Appendix), it was also evident that the fish at the ends of the distribution mix to at least some extent on the spawning grounds of the central and western Pacific (IATTC, 1984).

The problem of whether there is more than one subpopulation of skipjack in the Pacific was also considered during workshops sponsored by the SPC (SPC, 1980, 1981a). The material examined for this purpose included the results of tagging experiments and data on the gene frequencies of Es^1, serum transferrin and erythrocyte guanine deaminase (Gda^1) in skipjack blood. Because of the large geographic distances involved, the conclusion was drawn that skipjack in the Pacific are not panmictic because all adults do not have an equal opportunity of breeding with each other throughout their oceanic distribution within a generation. In contrast to panmixia, the distribution of the Es^1 data in Figure 5 was interpreted as either a continuous or a stepped cline, demonstrating that skipjack follow, "...some form of population structuring across the Pacific...", (SPC, 1981a). To describe this structure, two major hypotheses were recognized as the most feasible, and the first one was Fujino's (1972) discrete subpopulation model mentioned above. Although the workshop could not identify any genetically-isolated subgroups separated by stable geographic boundaries, neither could the extent of the neighbouring regions be defined on the basis of existing genetic data. Consequently, the possibility of discrete subpopulations that reflect a stepped cline could not be ruled out. The second, or the continuous cline hypothesis, accepts that Pacific skipjack belong to a single population, but the chance of any two fish breeding is inversely proportional to the distance separating them. An isolation-by-distance model is therefore an integral part of this hypothesis.

Richardson (1983) examined 42 loci for protein variation in skipjack tuna by utilizing material from previously unsampled sites in the South Pacific (Figure 6). Having found only one new polymorphism, he concluded that "...the discriminatory power of the genetic approach to the identification of skipjack subpopulations...is still very limited." By combining these new samples with the previous results of Fujino (1972, 1976) and Sharp (1978), the important observation was made that "...the gene frequency for any well-studied stock in the western Pacific is heterogeneous." Heterogeneity was detected within and between stocks in areas A and B of Figure 6, and the discrete-subpopulation hypothesis neither accounts for these results nor the dispersion of tagged skipjack between areas (SPC, 1981a). The mobility of skipjack and the genetic heterogeneity within a stock also signify a degree of overlapping that is compatible with Sharp's (1978) view of multiple subpopulations. No evidence of homing for breeding purposes has been detected, and the absence of this requirement of the discrete-subpopulation hypothesis further detracts from its acceptance. Samples collected in the New Zealand fishery from single schools over a four-year period also demonstrate heterogeneity for the Es^1 allele among and between years, and the allele frequency also declined significantly with an increase in the average size of fish in the school (Richardson and Habib, 1987).

The isolation-by-distance model proposes that skipjack exist in a series of semi-isolated genetic neighbourhoods, each of which encloses a group of randomly-breeding adults (Richardson, 1983). Although such a model is consistent with the observed heterogeneity within an area as well as the clinal changes in gene frequency, it is difficult to estimate reliably the size of a neighbourhood or to know whether the size is the same in different locations. Amongst several other untested assumptions noted by Richardson (1983), the model also assumes that there are no barriers to movement other than distance. Esterase and, to a lesser extent, Gda^1 clines of different magnitude exist to the north and south of the North Equatorial Counter Current, however, and this suggests a restriction in genetic flow between the regions (Richardson, 1983). It is also known that sea-surface temperature and salinity show gradients across the Pacific, and the longitudinal clines exhibited by Es^1 and Gda^1 may be related to such an environmental variable (Endler, 1977; Lewis, 1981; SPC, 1981a). Consequently, the difficulties that are encountered in applying either the isolation-by-distance or discrete-subpopulation hypotheses prevent the choice of a single, descriptive model of the skipjack population at this time (Richardson, 1983; IATTC, 1984).

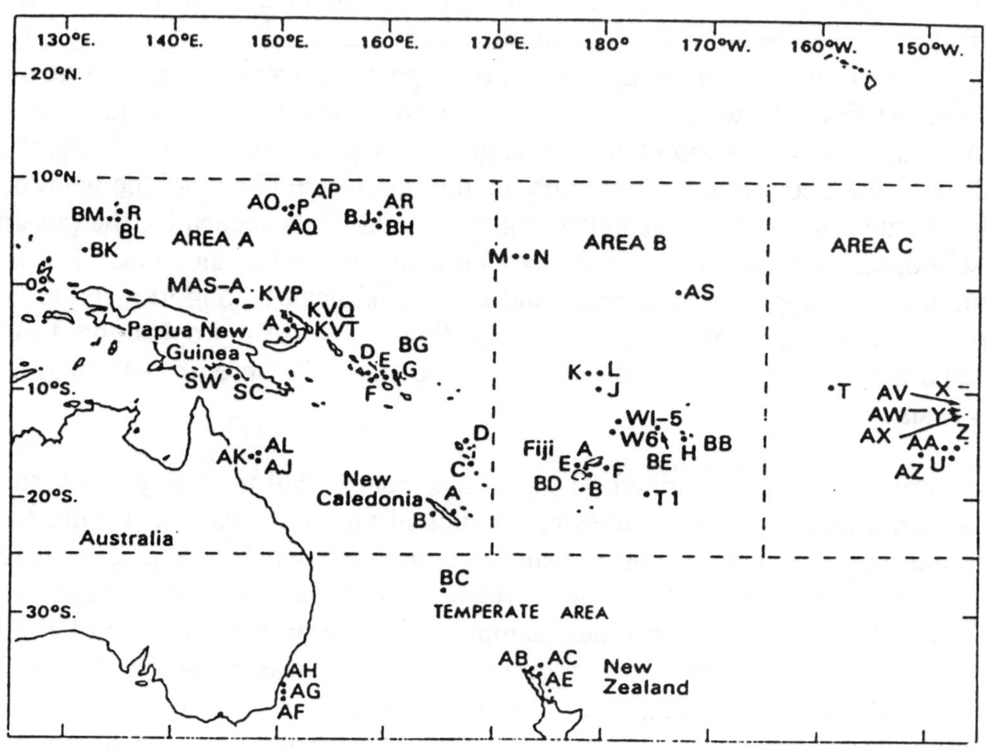

Figure 6. Sampling stations and area boundaries used to study protein variation in Pacific skipjack tuna. (After Richardson, 1983).

6.2 Distribution

In general, skipjack occur within the 15°C or warmer isotherm of the world oceans (Matsumoto *et al.*, 1984). In the western Pacific the widespread distribution of longline fishing demonstrates that skipjack have been captured as far north as 44°N off Japan and as far south as 37°S off Australia. The catches were highest from May to August off Japan, and from October to March off New Guinea (Forsbergh, 1980). Skipjack occur from October to April off northern New Zealand (Clement, 1976; 1978), while large concentrations have been seen as far south as southeastern Tasmania (Robins, 1952). In the eastern Pacific, skipjack have been fished along the west coast of the Americas from 34°N off southern California to 27°S off northern Chile. In the north, their range fluctuates with sea temperature from 25°N in February through April to 34°N in August through October (Williams, 1970). As the distribution of skipjack approaches the western Pacific it diverges from the equator to a greater extent than in the eastern

Pacific. As mentioned earlier, this is due to the westward transport of warm surface waters and their poleward displacement along the western shores. The average distribution of skipjack in the EPO during 1979-1987 for all purse-seine sets is depicted in Figure 7a. The average distribution of skipjack catch by purse seiners and baitboats in 1988-90 in the WPO, as indicated by data submitted to the SPC[1] is shown in Figure 7b.

Skipjack physiology and morphology play a major role in the determination of suitable habitat, and by extension, their distribution. In common with most small scombrids, skipjack lack a swim bladder. While this permits rapid vertical movements within the near-surface habitat, it also increases the minimum swimming speed required to maintain hydrostatic equilibrium. Sharp (1978) calculated that a 50-cm skipjack must swim 60.5 km/d just for hydrodynamic stability and respiration. Skipjack have a large red muscle mass that functions aerobically to sustain this basal swimming requirement. In addition, white muscle tissue, which can function both aerobically and anaerobically, enables short bursts of very high swimming speeds for pursuit of prey and escape from predators. High oxygen demands due to high metabolic rates and the resultant production of large amounts of metabolic heat by the red muscle mass are of primary importance in determining skipjack distribution. A minimum dissolved oxygen level of about 2.45 ml/l is required by skipjack to maintain the basal swimming speed (Sharp, 1978), and a somewhat higher level would be required (about 3.0-3.5 ml/l) to sustain their survival over extended periods. Dissolved oxygen usually approaches saturation (4.5 ml/l) in surface waters, but is often less than the minimum requirement in waters in and below the thermocline (Toole *et al.* 1988). This generally restricts skipjack to the mixed layer above the thermocline. Despite some ability to thermoregulate, large skipjack encounter the problem of retention of excessive metabolic heat, and thus need to remain in the vicinity of cooler water. In the tropics, suitable temperatures for these large fish would be found only at the thermocline and below. However, as noted above, such water may be an unsuitable habitat because of low dissolved oxygen. Using physiological data from tank experiments, Barkley *et al.* (1978) defined the hypothetical limits of skipjack habitat. They suggested that only small (<4 kg) skipjack can inhabit most surface tropical waters and that the habitat of large (>6.5 kg) skipjack in the tropics is in the vicinity of the thermocline, adjacent to cooler water. Where this is poorly oxygenated, large skipjack would be excluded. The gross features of skipjack distribution appear to broadly fit this hypothesis.

A long list of oceanographic and biological features is known to influence, either directly or indirectly, the distribution of skipjack within their overall limits. These include temperature, salinity, dissolved oxygen, thermocline structure, bottom topography, water transparency, current systems, water masses and biological productivity. Within the areas of greatest abundance, *i.e.* the tropics, temperature may play only a minor role in determining small-scale distribution, as thermal gradients in these areas are generally weak. In subtropical areas, skipjack fisheries exhibit seasonality which correlates well with surface temperature, *e.g.* in New Zealand and Australian waters. In the tropics, the distribution of suitable forage would seem to play a more

---

[1] Data provided to the SPC result mainly from bilateral access agreements between Pacific island countries and distant-water fishing nations, and therefore catches in international waters may be under-represented. The exception to this is that high-seas catches by USA purse seiners are provided as a condition of the Mulitlateral Treaty on fisheries.

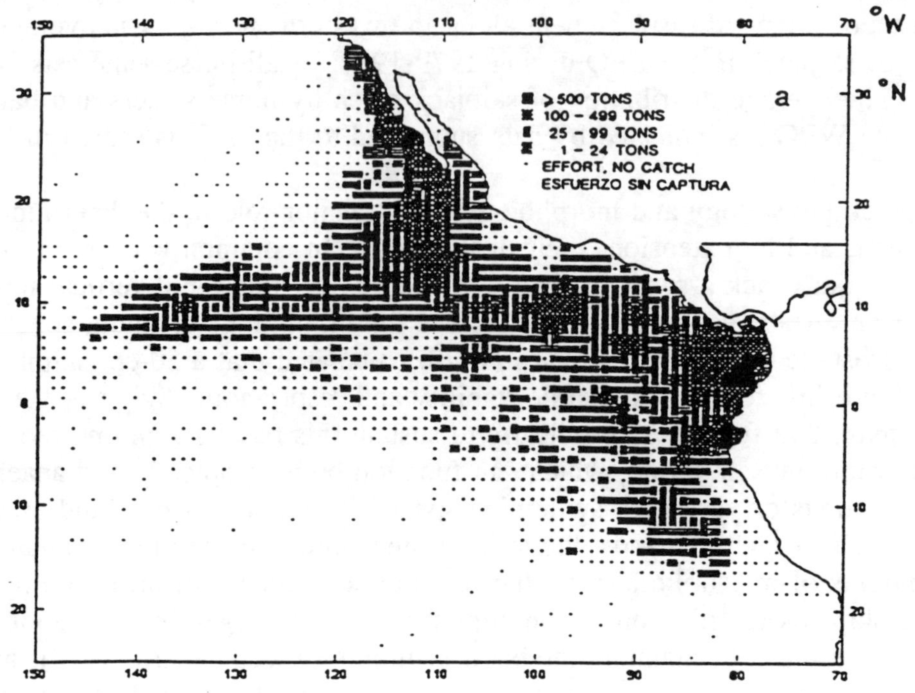

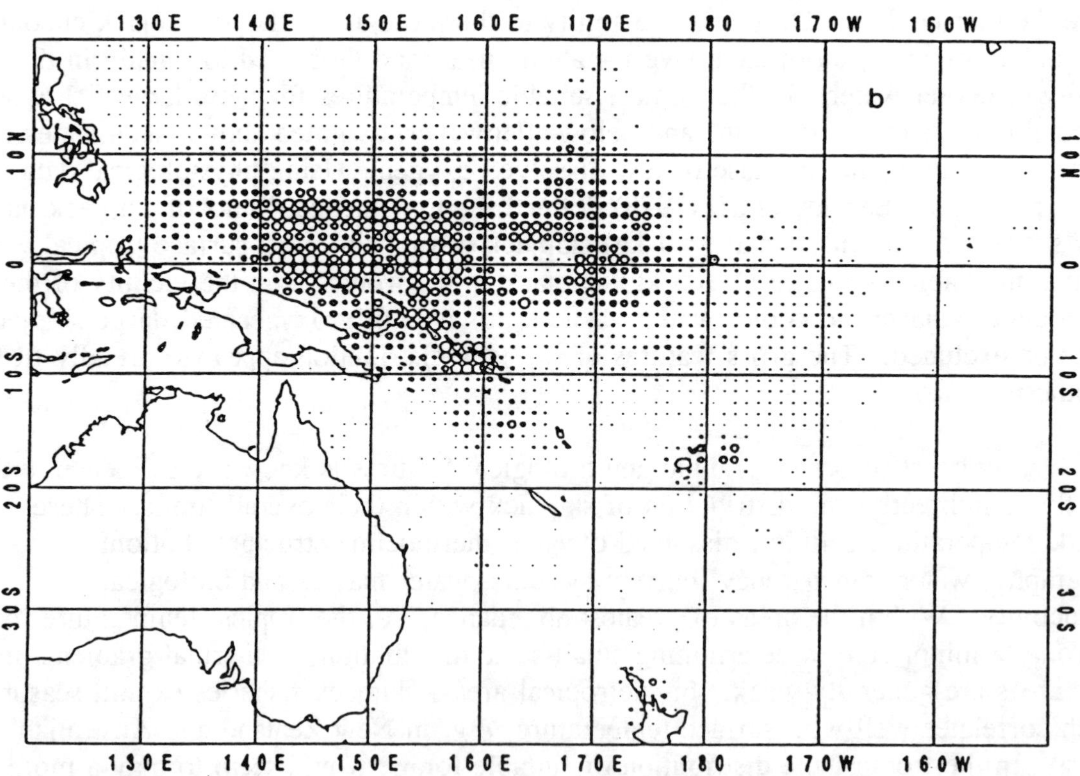

Figure 7. a) Average annual catches of skipjack in the EPO during 1979-1987 for all purse-seine trips supported by logbook data, (after IATTC, 1989); and b) Average distribution of skipjack in the WPO purse-seine and baitboat fisheries, 1988-1990. The maximum circle size represents a catch of 500 mt or more. (From the SPC Regional Fisheries Database).

important role in determining skipjack distribution. In fact, the patchy distribution of skipjack may be best correlated with the similar patchy distribution of zooplankton and micronecton (Lewis, 1981). Increased skipjack abundance around islands, seamounts, banks *etc.*, may well result from increased food supply near these areas. Similarly, the frequent occurrence of skipjack in the vicinity of upwellings and convergences may result from the increased productivity associated with deep-origin, nutrient-rich water and the concentration of drifting or weakly-swimming biota, respectively.

The small-scale distribution of juvenile and adult skipjack in the western Pacific is also influenced by their attraction to floating logs and other naturally-occurring flotsam. To an increasing extent, this is also true of artificial fish aggregating devices (FADs). Presently, FADs are moored in coastal waters of many Pacific Island countries to increase fishing success and minimize fuel costs for artisanal fishermen. In other locations, FADs play an important role in industrial baitboat and purse-seine fisheries, *e.g.* the Solomon Islands, Philippines, and Indonesia. Increasingly, large, industrial purse seiners are deploying moored or drifting FADs throughout the major fishing areas of the western tropical Pacific. Little is known regarding the dynamics of tuna attraction to and retention at FADs, both artificial and natural. As FAD deployment increases, changes in the small-scale distribution and vulnerability of skipjack and other small pelagics might be expected in some areas.

6.3   Migration: Eastern Pacific

Based on a variety of data sources, including genetic studies and tag returns, Rothschild (1965) hypothesized that a large fraction of the pre-recruit (35-45 cm) skipjack that disperse from the central Pacific migrate towards the coast of the Americas. One portion, the northern group, enters the Baja California-Revillagigedo Islands fishery, while the southern group proceeds to the fishery off Central America to northern Chile. The fish arrive in the EPO when they are 1 to 1.5 years old, and return to the central Pacific after several months, or when they are 2 to 2.5 years old (55 - 65 cm) (IATTC, 1979). It is also of interest to note that none of the 35,534 skipjack tagged in French Polynesia by the SPC and IATTC have been recovered in the eastern Pacific (Gillett and Kearney, 1983); it was considered that the fish were generally too large when tagged for them to have had sufficient time to undertake the migration and still be of a size vulnerable to the EPO skipjack fishery.

From tagging experiments, Fink and Bayliff (1970) noted that the fish in the northern group migrate from the Revillagigedo Islands to Cabo San Lucas, Baja California, during May and June, and then up the west coast of the peninsula from May to September (Figure 8a). The fish may penetrate as far north as 30°N depending, amongst other factors, on the temperature. During September to November or December they then migrate southward along the Baja California peninsula, and although most of the fish turn westward towards the central Pacific, the remainder continue towards the Revillagigedo Islands and arrive at year's end. Some fish that overwinter in the area are thought to either repeat the northern migration in the following year, return to the central Pacific or travel to waters near the southern tip of Mexico.

"Recruits to the southern region appear mostly in or near the Panama Bight (Figure 8b). Those which enter the fishery in the northern Panama Bight migrate rapidly northwest to Central America and south to Ecuador. The proportions which migrate in these two directions vary considerably from year to year, and may depend on differences in sea-surface temperatures. The fish appear to migrate from the Gulf of Guayaquil to the Ecuador coast during the first half of the year, and return during the second half. Some fish migrate further south to central Peru and to southern Peru and northern Chile" (Fink and Bayliff, 1970). There is evidence from more recent tagging experiments that skipjack from the southern group also return to the central Pacific (Bayliff, 1984). As mentioned above (Section 6.1), the area between the northern and southern groups is usually devoid of skipjack (Fink and Bayliff, 1970).

Williams (1972) supported Rothschild's (1965) hypothesis by proposing three migration models for skipjack from the central to the eastern Pacific. In the first one the fish actively swim eastward against the westward-flowing North and South Equatorial Currents (NEC and SEC) (Figure 9). The second model involves passive migration in which the fish are carried eastward by the North and South Equatorial Counter Currents (NECC and SECC). The third includes a pattern in which fish in the northern group are carried in the counterclockwise gyre of the northern equatorial water mass in the EPO, and most of the fish in the southern group are transported by a clockwise gyre formed by the NECC and SEC.

To evaluate these possibilities, Forsbergh (1988) reviewed data on geostrophic flow, current velocities, larval distribution, oxygen requirements, and swimming speeds in relation to fish length, as well as temperature profiles of the Cromwell, or Equatorial Undercurrent (EUC). From this information he drew several conclusions regarding the relationship between current patterns and their role in migration. For example, drogues placed in the NECC moved eastward for 4,500 km (2,428 mi) in 4 months (Wyrtki *et al.*, 1981), indicating that the NECC could be a major, passive migration route. This current also extends eastward to about 85°W from May to December, but reaches to only about 120°W in February and March (Wyrtki, 1967). However, the average monthly width of the NECC at 130°W is only 3°-6° of latitude (Wyrtki, 1965), whereas the latitudinal distribution of skipjack larvae in the central Pacific is about 40° (Nishikawa *et al.*, 1985). Consequently, only about 7.5-15 percent of the larvae could be carried eastward by the NECC, and the remainder, or major portion, could be transported westward by the NEC and SEC. More recently, Forsbergh (1989) found that estimates of relative apparent abundance of skipjack in the EPO were not correlated with the velocity of the NECC at earlier times. Although the latitudinal concentration of juveniles east of 180° appears to be similar in the NECC, NEC and SEC, there are insufficient data to determine the concentration of skipjack in relation to zonal currents (Nishikawa *et al.*, 1985; Mori, 1972). In contrast to the NECC, the eastward transport of juveniles by the SECC is negligible because the current does not always exist and it is weaker than the NECC. As another means of active transport, the EUC also flows eastward beneath the surface at the equator at depths between 25-300 m, but it probably contains few juveniles because the temperature is too low and the oxygen concentrations appear to be too low for skipjack > 35 cm (Forsbergh, 1988).

With regard to the active migration model, Forsbergh (1988) estimated that juveniles in the length ranges of 13-17, 26-34, and 38-51 cm could swim at 1, 2, and 3 kt, respectively, if they could maintain cruising speeds of 3-4 body lengths/sec. These velocities would be adequate to overcome the 0.3 kt average speed of the NEC, but

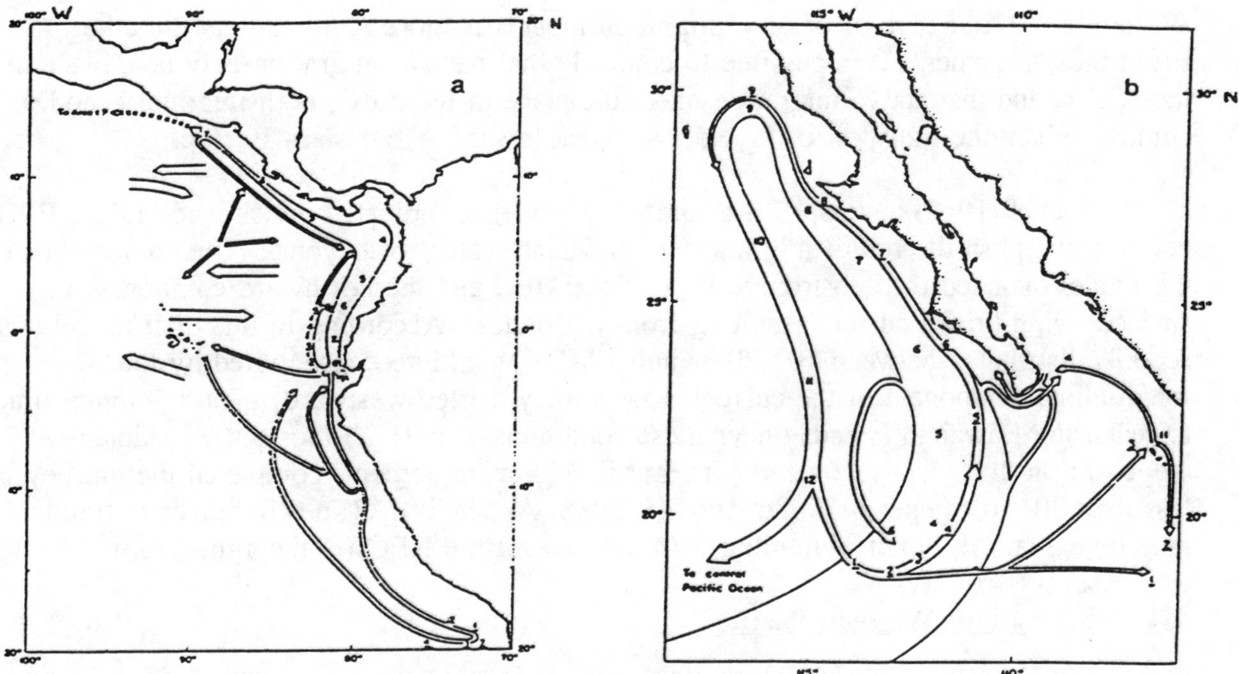

Figure 8. Migratory paths of skipjack entering (a) to the south and (b) to the north in the eastern Pacific. (After Fink and Bayliff, 1970).

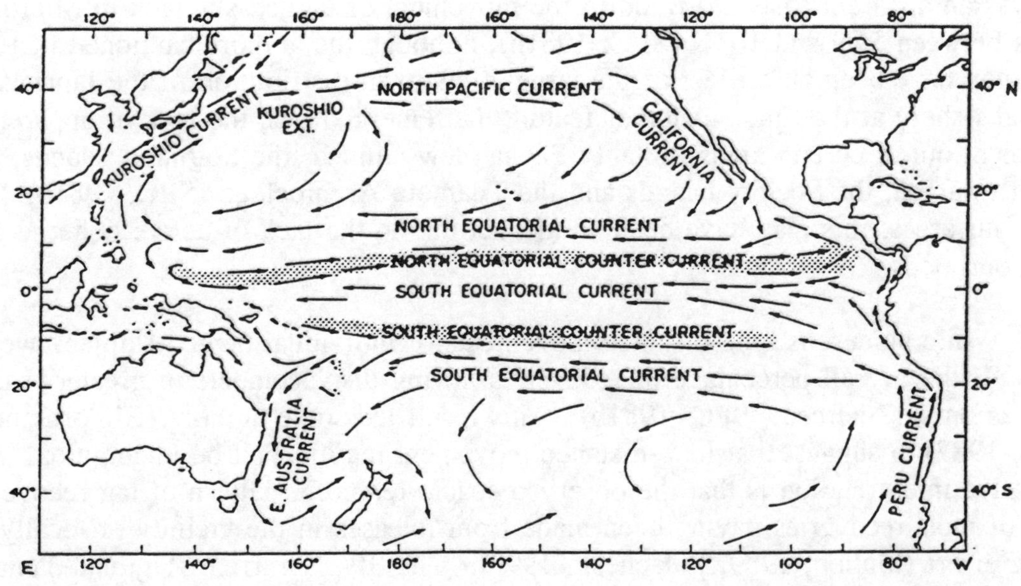

Figure 9. The major surface currents of the Pacific Ocean between 50°N and 50°S. Stippled areas indicate subsurface currents that are also found at the surface in some months. (After Suzuki *et al.*, 1978).

probably not that of the SEC (0.6-1.8 kt) on a consistent basis. From these and the previous observations, Forsbergh (1988) concluded that, "...it is unclear whether either of William's (1972) active or passive migration models is more realistic than the other." From these arguments it is possible to conclude that passive migration may take place in the NECC and that active migration may take place in the NEC; both migratory modes could result in the transport of juvenile skipjack towards the eastern Pacific.

Seckel (1972) developed a quantitative model to indicate how skipjack in the EPO could accomplish the return migration to Hawaiian waters or beyond. The components of the model included the geostrophic flow of the NEC and the northward component of surface, wind-driven current resulting from the trades. According to this drift model, fish schools distributed between $10°-20°N$ and $128°W$ would be concentrated by the longitudinal component of the current flow as they drifted westward. The minimum time to reach the Hawaiian Islands under these conditions was 21-23 mo. Of 23 skipjack tagged in the EPO and recovered off Hawaii, 17, or 78 percent, completed the journey in less than 20 mo (range: 8-15 mo; Bayliff, 1988, Appendix). The differences in timing may be explained by net swimming with the flow of the NEC (Forsbergh, 1988).

6.4 Migration: Western Pacific

A substantial body of information on skipjack migration, and movement in general, in the WPO was collected by the SPC during its Skipjack Survey and Assessment Programme (SSAP), carried out during 1977-80. Most of this information was provided by the return of 7,000 skipjack tags from approximately 140,000 releases. Analysis of these data has been hampered by the incompleteness of catch and effort data from the fisheries that recovered the tags; therefore the conclusions drawn to date with respect to movements have been largely qualitative in nature. One prominent feature of the SSAP data was an apparent eastward trend in the movement of tagged skipjack in Micronesian waters between $5°N$ and $10°N$ (SPC, 1981b), although the authors cautioned that this trend may have been an artifact of the largely unknown distribution of the Japanese baitboat fishery at the time. Another feature that emerged was the lack of apparent movement out of certain areas, notably Papua New Guinea, the Solomon Islands, Fiji, Western Samoa, the Society Islands and the Tuamotu Archipelago (SPC, 1981b). For the latter four areas, this may have been an artifact due to the lack of active fisheries in the neighbourhoods.

While numerous long-distance movements of individual tagged skipjack were observed, the overall percentage of recoveries having displacements of greater than 200 nmi was only 17 percent (SPC, 1981b). This result led some authors (Hilborn and Sibert, 1987) to suggest that long-distance movement may in fact be uncommon. The alternative interpretation is that the observed spacio-temporal pattern of tag recoveries is a result of most recoveries having been made from releases in the vicinity of locally-intense fishing effort (Mullen, 1989; Polacheck, 1990). The SPC's current Regional Tuna Tagging Project (RTTP) is providing new information on skipjack movement by tagging throughout the area $10°N-10°S$, $120°E-180°$. In these experiments, there is currently a higher percentage (27 percent at 1 June, 1991) of returns that have displacements of > 200 nm, and this percentage is likely to increase as longer times at liberty are sampled (SPC, unpubl. data). It is likely that this higher proportion of long-distance returns has resulted because of a more homogeneous distribution of fishing effort (mostly purse seine) than existed in 1977-80 (mostly baitboat), which would tend to support the views of Mullen (1989) and Polacheck (1990). This issue, which has important international management implications, will hopefully be resolved by the application of new, spatially-disaggregated movement models to the SSAP and RTTP tag data, along with appropriate data on the distribution of fishing effort and catch.

7. NATURAL MORTALITY

It is a difficult problem to estimate the instantaneous mortality rate (M) of most commercial species of fish, but it is particularly troublesome in the case of skipjack tuna. Part of the difficulty stems from the fact that, regardless of the method of evaluation or the fishery, an emigration term (E) must be factored into the overall rate of attrition. Particularly if tagging experiments are used to estimate mortality rates, several additional mortality components must be recognized besides the instantaneous fishing mortality rate, F. Bayliff and Mobrand (1972) noted that in a tagging experiment mortality can occur in two modes, actual and effective, as well as in two different time periods. Fish can either immediately die or shed their tags as a result of the tagging operation (Type 1 mortality), and the proportion surviving in each case is π and ρ, respectively. The proportion of fish recaptured with useful information (β) is frequently less than anticipated because recapture dates may be unknown or known imprecisely. Over the long term, fish may also die at continuous, instantaneous rates (Type 2 mortality) from carrying the tags (G), and through tag shedding (L). While the total mortality rate (Z) of untagged fish in a closed fishery is $Z = F + M$, in a fishery open to emigration the total attrition is expressed as $Z' = Z + E$. The total attrition rate for tagged skipjack (Z^*) in an open fishery is therefore $Z^* = F + X$, where $X = M + E + G + L$ (Joseph and Calkins, 1969; Bayliff, 1977). Estimates of F can usually be obtained from information on catch and effort and the recovery rate of tagged fish, but the best that can be achieved usually with the mortality components of M, E, G, and L is a partial separation.

From tagging experiments initiated off Baja California, Fink (1965) estimated that the proportion of yellowfin (y) surviving immediately after tagging was $(\pi\rho)_y = 0.85$, and Type-2 mortality[2] was $(G + L)_y = 1.06$. In the absence of any other information for skipjack (s) at the time, subtracting this last sum and $F = 2.10$ from $Z^* = 6.98$ yielded an estimate of $(M + E)_s = 3.82$. For the skipjack fishery off northern Perù, a similar estimate of $(M + E)_s = 3.65$ was obtained from values of $Z^* = 5.02$ and $F = 0.31$. Subsequently, Joseph and Calkins (1969) estimated from a skipjack tagging experiment that $(\pi\rho)_s = 0.68$. Bayliff and Mobrand (1972) utilized single and double-tagged experiments on yellowfin in the EPO to obtain estimates of $\rho_y = 0.9$ and $L_y = 0.28$. Based on Fink's (1965) result for $(G + L)_y$ above, one estimate of the long-term mortality rate from carrying tags is therefore $G_y = 1.06 - 0.28 = 0.78$. Since the last-mentioned value of ρ is greater than the previous product of $\pi\rho$ for skipjack, there is some indication that the immediate loss of tags and mortality of skipjack is greater than that of yellowfin. Joseph and Calkins (1969) were also able to estimate the combined rate of mortality, $X_s = 2.76$, from a single tagging experiment. Bayliff (1977) reassessed the data of Schaefer *et al.* (1961), Fink (1965) and Joseph and Calkins (1969), as well as more recent information, and concluded that the value of X_s was probably less than 3.0, thereby

---

[2] All instantaneous rates are expressed in annual rates.

supporting the earlier result. However, Bayliff (1977) was unable to establish a reliable difference between Z^*, which includes E, and attrition rates that presumably excluded E because the data base was limited (truncated) to tag returns of ≤ 6 mo. While it is reasonable to expect that Z^* would exceed the attrition rate from the truncated data, this result was achieved in only half of 26 experiments, and the differences between the two rates were often negligible. The test was not considered valid because of too few returns of fish at liberty in excess of 6 mo. It was also expected that the attrition rate of single-tagged skipjack would exceed that of double-tagged fish, but this result occurred in only one out of two such experiments. Bayliff (1977) concluded that irregularities in the semilogarithmic plots of tag returns per unit effort against time precluded the reliable estimation of attrition rates that ranged from 3.48 to 14.93 with an average value of $Z_s^* = 6.86$. These irregularities, in turn, were apparently caused by the failure of tagged and untagged fish to mix completely, and the uneven distribution of effort with respect to the fish.

Forsbergh (1987) drew attention to the small numbers of returns in most of the tagging experiments examined by Bayliff (1977) and the large variation in Z_s^* values. He therefore tried to estimate Z_s' directly from the catch rates of untagged fish of different age groups, thereby avoiding the additional mortalities attributed to tagging. By applying Robson and Chapman's (1961) method of catch-curve analysis to eight, consecutive quarterly intervals of catches from 1961-83, an average value was obtained of $Z_s' = 3.35$ (range: 1.56-5.56). This result was based on an assumed linear growth rate of 24 cm/yr (Forsbergh, 1989), which in turn was used to estimate the ages of fish in the length-frequency distribution. If growth was assumed to follow the von Bertalanffy (1938) model with $L_\infty = 86.0$ cm and $K = 0.79$ (IATTC, 1984), then $Z_s' = 3.25$. These rates for Z_s' are considerably less than Bayliff's (1977) value of 6.86, but Forsbergh (1987) attributed greater accuracy to the latter result because it was based on tagging experiments. Currently, the IATTC assumes that 75-90 percent of the fishable stock is lost through attrition each year (IATTC, 1989), and this is equivalent to $Z_s' = 1.39$-2.30. At this point, by assuming that either $G_y = G_s = 0.1$ (Bayliff, 1971), $L_y = L_s = 0.28$, or that $(G + L)_y = (G + L)_s = 1.06$, the following estimates of $(M + E)_s$ can be obtained by subtraction from the preceding information:

Source	Z^*	G	L	$(G+L)$	Z'	F	$(M+E)$
Bayliff (1977)	6.86	0.10	0.28	-	6.48	4.10	2.38
	6.86	-	-	1.06	5.80	4.10	1.70
Forsbergh (1987)	-	-	-	-	3.25	0.89	2.36
IATTC (1989)	-	-	-	-	1.39-2.30	-	-

Since so little is certain about skipjack in the EPO, Forsbergh (1987) also attempted to apply cohort analysis to quarterly catch data even though the assumption of a closed system is thought to be violated. Estimates of the recruits entering the fishery frequently failed to converge and the results were considered unreliable. The lack of convergence may have been due to variable immigration and recruitment during the first three quarters of the year and variable emigration during the last quarter.

As a result of the SSAP tagging carried out during 1977-1980 in the southwestern Pacific, Kleiber *et al.* (1983, 1987) estimated an aggregate value of $L_s = 0.09$ for the individual countries and territories participating in the South Pacific Commission's experiments. The results of double tagging also indicated that Type-2 tagging-mortality loss (G_s) was negligible, and therefore assumed to be zero. The product of Type-1 mortality and the midpoint between the worst and best estimates of θ was $(\pi\rho\theta)_s = 0.60$ in the aggregate. By subtracting $F_s = 0.08$ ($\pm$ 0.014) and L_s from $Z_s^* = 2.04$ ($\pm$ 0.24), the approximate, aggregate value of $(M + E)_s$ is 1.87. This value is not substantially different than those shown above for the EPO. Kleiber *et al.* (1987) cautioned that, with the exception of Z_s^* the confidence limits associated with such estimates would be larger if the uncertainties in $\pi\rho$ and θ were included. To augment these comparisons, it should be noted that for the Gulf of Guinea in the eastern Atlantic, Bard (1986) obtained estimates of $(\pi\rho)_s = 0.60$ and $\theta_s = 0.80$ from fish tagged in 1980-1981. By further assuming that $G_s = 0.10$-0.20, and $L_s = 0.10$, a value similar to that obtained in the South Pacific experiments by Kleiber *et al.* (1983), the sum of $(M + E + G)_s$ fell in the range from 1.78-1.88. Fonteneau (1986) estimated that $M_s = 0.60$-1.00 for eastern Atlantic skipjack, and subtracting these values from the previous sum yields $(E + G)_s = 0.78$-1.28.

8. OCEANIC FEATURES ASSOCIATED WITH THE SPECIES

The easterly trade winds drive the principal circulatory pattern in the upper waters of the Pacific (Figure 9), but the pattern is not symmetric with respect to the equator (Pickard, 1968). Of the westward flowing currents, the principal component of the SEC lies between 8°-10°S and 3°N, while the NEC is located between 8°-20°N. The eastward-flowing NECC is situated between these currents, and the generally weaker SECC, normally beneath the surface and under the SEC (IATTC, 1984; Philander, 1990), is also eastwardly directed. This major system of latitudinal flow extends approximately 12,000 km from the Gulf of Panama almost to the Philippines. Of the three major subsurface currents, the core of the Cromwell Current is located in the thermocline during normal, or non-El Niño episodes, and it extends westward from the Galapagos Islands for about 10,000 km. Under normal conditions the sea-surface temperature decreases from west to east, and the depth of the thermocline diminishes from about 150 m in the western Pacific to 40 m or less in the eastern Pacific (Philander, 1990). The distribution of larvae (Section 3), spawning activity (Section 5.2) and migration (Section 6.3) in relationship to these currents have been described earlier.

During El Niño episodes the southeasterly trade winds weaken in the eastern part of the South Pacific, and the tropical currents just described, including the California and Peru (Humboldt) Currents, change their usual positions and strengths (IATTC, 1984). In the eastern Pacific the sea surface rises in response to the decrease in the hydrospheric circulation rate, and coastal and equatorial temperatures also increase. As a result, the thermocline deepens and a flow of warmer surface waters extend over the normally upwelling regions off Ecuador and Peru (Miller and Laurs, 1975; Joseph and Miller, 1988). In addition to a weakening of the SEC, the SECC frequently emerges at the surface between 5°-10°S, and under particularly strong El Niño conditions, such as in the last quarter of 1982 and into 1983, a weakened Cromwell Current may be difficult to locate in transect data at the equator. Schaefer (1961) noted that during the El Niño of 1957-59 skipjack catches declined in the normally productive areas near the Gulf of

Guayaquil, Ecuador, and the Costa Rica Dome (centred at 9°N-90°W), but increased substantially off Peru. Warming trends in sea-surface temperatures (SST) were thought to be responsible, but Forsbergh (1989) showed that the SSTs and catches per unit of effort (CPUEs) in the areas were uncorrelated. The intense fishery in the Gulf of Guayaquil lies in the thermal front between the cold, upwelling waters of the Peru Current and the warmer waters north of the equator. The SSTs in the front change rapidly with latitude and position, and the intensity of the front varies with the season and year. Strong upwelling of the Peru Current produces a strong front, and it was thought to induce the formation of concentrations of skipjack forage. However, from 1961 to 1984 the logarithm (log) of CPUE *versus* SST was significantly correlated only in the first and fourth quarters (Forsbergh, 1989). In the northern part of the Panama Bight the abundance of skipjack appears to be related to upwelling which occurs from January to April; a period of forage production is then followed by a peak in skipjack catch rates in April and July (Forsbergh, 1969).

In terms of vertical distribution, schools of skipjack have been detected at depths of 140 m (Kimura *et al.*, 1952) and 120 m (Yamanaka *et al.*, 1966) by means of echo sounding devices, and at 98-152 m in Hawaiian waters by submarine (Strasburg *et al.*, 1968). Sonic tracking revealed that skipjack move between the surface and 263 m during the day, but remain within 75 m of the surface at night (Dizon *et al.*, 1978). In general, the vertical distribution is limited by the depth profile of the temperature and oxygen concentration with minimum values of 18°C and 3.0-3.5 ml/l, respectively, needed for long-term survival (Forsbergh, 1980). Sharp (1978) considered the lower limits of 15°C and 2.5 ml/l oxygen more appropriate. The upper temperature limit decreases with increasing size from 30°C or more for small fish to as low as 20°C for large fish (Barkley *et al.*, 1978). For these reasons young skipjack can tolerate the conditions in tropical surface waters, but the adult habitat is in the thermocline. As young fish grow they are forced to seek well-oxygenated waters associated with cooler, less stressful temperatures.

It is currently believed that the skipjack fishery in the EPO is supported principally by the immigration of recruits spawned earlier in the central Pacific and, possibly, in the western Pacific. The SST in this spawning area apparently has an effect on the catches in the EPO. Forsbergh (1989) found a positive correlation between the SST 18 months earlier and logCPUE for baitboats during the 1934-1960 period, and the logarithm of the catch rate of recruits (numbers per day) for purse seiners from 1961-1984. A wind-mixing index calculated as the cube of the wind speed 18 months earlier in the spawning area was also inversely correlated with logCPUE for purse seiners for the same period. The hypothesis supporting this negative relationship is that increased wind conditions can disrupt the stability of the mixed layer, thereby dispersing the food aggregations that support survival (Lasker, 1975; Owen, 1981a, 1981b). On the whole, though, attempts to relate the apparent abundance of skipjack with environmental conditions suggest that the relationship is complex and not obvious (Forsbergh, 1989). Some of the problems may be resolved if studies could be carried out on all population units that contribute to the fishery in the EPO rather than the portions which contribute only for part of their life span (IATTC, 1991).

9. INTERACTION WITH OTHER SPECIES

For purposes of potential management information, Orange *et al.* (1957) and Broadhead and Orange (1960) studied the frequency with which skipjack were caught together with yellowfin in mixed schools in the EPO. From baitboat data it appeared that about 51 percent of the yellowfin and 29 percent of the skipjack tonnages were caught in mixed schools, and for purse seiners the percentages were 6 and 16, respectively. These last figures are probably more representative of the fishery in the EPO today since it is now dominated by purse seiners. On average, Broadhead and Orange (1960) found that the sizes of skipjack from pure and mixed schools were similar. On the other hand, although yellowfin and skipjack in mixed schools were also alike, yellowfin in pure schools tended to be larger and more variable in size.

In the WPO, both skipjack and yellowfin are caught by purse seiners more frequently in mixed schools than is the case in the EPO. From all purse-seine data held by SPC, 37 percent of yellowfin and 47 percent of skipjack tonnages have been caught in mixed schools (SPC, unpubl. data). This is probably because of the greater emphasis placed on fishing schools associated with logs, or floating objects, and the virtual absence of dolphin-associated schools in the WPO.

The implication of "mixed" schools is that yellowfin and skipjack exist in composite aggregations. Yuen (1963) examined this concept by means of underwater observations and photography in a baitboat fishery. He found that the species maintained a significant degree of separation into schools or species groupings, and the random response of individuals from either group to the "chum", or bait, created the illusion of complete mixing in the catch. While skipjack and yellowfin may therefore prefer to school by species, their co-occurrence is probably the result of a mutual attraction to an external stimulus, such as food.

10. GENERAL DESCRIPTION OF THE FISHERIES

10.1 Eastern Pacific

During the early part of the 1950s, the surface fishery for tunas in the eastern Pacific was conducted primarily by baitboats in coastal waters and near offshore islands. Following the conversion of most of the larger vessels to purse seiners in the later part of the decade and early 1960s, the fishery began to expand offshore. By 1968 the fleet had penetrated the eastern Pacific as far west as 115°W, and by 1974 tuna were being caught near 150°W (Calkins, 1975; Peterson and Bayliff, 1985). Together with this expansion, the number of baitboats declined from 204 to 36 during the 1950-1988 period, and their capacity decreased from about 36,300 to 2,800 mt. At the same time the number of purse seiners increased from about 68 to 182 together with a capacity increase from 7,200 to 121,800 mt. With the exception of the region surrounding French Polynesia, the fishery in the EPO occupies the area to the east of 150°W between the latitudes of 39°N and 35°S. Besides baitboats and purse seiners, smaller amounts of skipjack are also caught near the equator and south of it by coastal dayboat operations involving jigboats and bolicheras. By 1990 approximately 10 nations had been involved in the surface fishery for this species.

The skipjack fishery in the EPO is concentrated primarily in two areas. The most important fishery in the south is east of 100°W and near Central America, northern South America, Cocos Island-Brito Bank and the Galapagos Islands. The secondary, northern fishery is west of 100°W and is near Baja California, the Revillagigedo Islands and Clipperton Island. With rare exceptions, the cell of warm water off southern Mexico that divides the two fishing regions is devoid of commercial quantities of skipjack. In the northern area the largest seasonal catches are taken during June to October, with a peak in August. The fishery in the southern region lasts all year and peaks can occur late in the year, usually in the fourth quarter. Overall, the fishery for skipjack tends to take place closer to shore than for yellowfin, but sporadically in recent years, such as in 1981, 1983-84 and 1988-89, increasing amounts of catches have been made further offshore. Catches and relative abundance in the south are generally twice as large as in the north, and the quantity caught can vary considerably from year to year (Figure 10). During the late 1950s large catches of skipjack were made south of 5°S, and from the late 1950s to the early 1970s large amounts were taken close to shore off Ecuador and northern Peru. During the early 1970s, however, the centre of abundance of the southern group appeared to shift towards the waters farther off Colombia and Central America. Beginning in the early 1980s the centre seemed to move back to Colombia and Ecuador. Compared to the average annual catches during 1979-87 (Figure 7a), the most substantial catches in 1989 were made off Central America and northern South America. The size composition of skipjack samples collected in the EPO during 1984 through 1988 (Figure 11) demonstrate a relatively narrow range of intermediate-sized fish, and support the hypothesis that the fishery is sustained by immigration and temporary residence of fish from outside the EPO.

10.2 <u>Western Pacific</u>

The tuna fisheries of the WPO are extremely diverse and might be conveniently categorized into fisheries based in island countries, fisheries operated by foreign-based, distant-water fishing nations (DWFNs) and fisheries based in Southeast Asia, primarily the Philippines and eastern Indonesia. In many Pacific island countries, tuna are captured by artisanal fishermen for consumption and, where markets are accessible, for sale. Skipjack, and to a lesser extent juvenile yellowfin, are the main species taken, mostly from small boats by trolling and poling, using lures made from mother-of-pearl shell. These artisanal tuna fisheries are of most significance to the smaller island countries, where shortage of land limits both agricultural production and exploitation of inshore marine resources. Despite their social importance, artisanal tuna fisheries are relatively minor in terms of total catch. Although reliable statistics are not available, total tuna catches by artisanal fishermen from Pacific islands are estimated to be less than 10,000 mt annually (Doulman and Kearney, 1987).

Most of the domestic tuna fishery development in the Pacific Island countries has taken place with the assistance of foreign countries, most notably Japan, either in the form of joint venture arrangements or as aid packages. A domestic baitboat fishery in Papua New Guinea began in 1970 in joint venture with Japan, averaging nearly 30,000 mt annually between 1971 and 1981 (SPC, 1983) before unfavourable economic conditions resulted in its termination. The other major domestic tuna fishery in the region is based in the Solomon Islands. This fishery began in the early 1970s, again as a joint venture with Japanese interests, and currently consists of a fleet of small baitboats and several

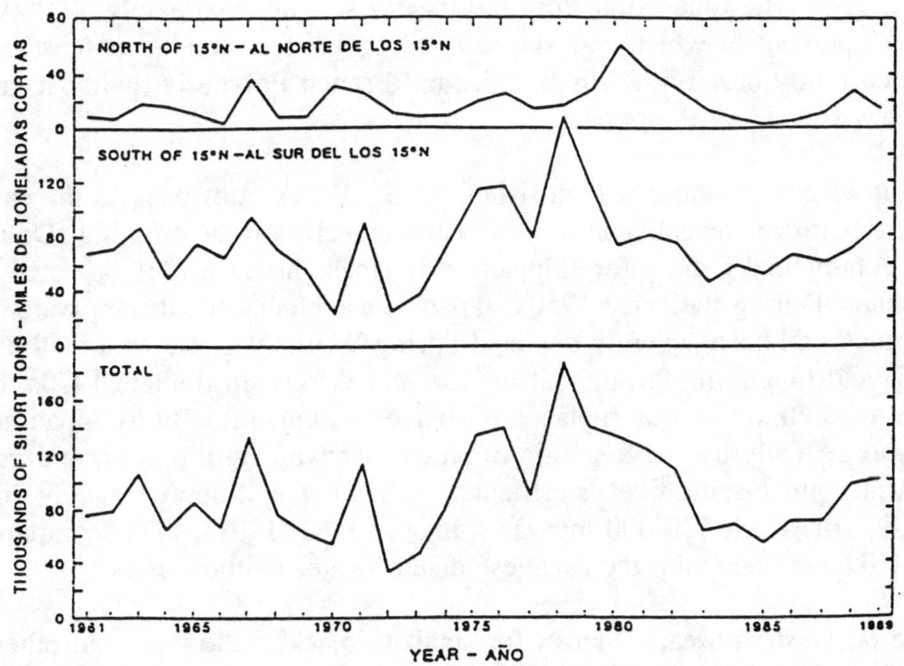

Figure 10. Estimated catches of skipjack in the EPO north and south of 15°N. (After IATTC, 1991).

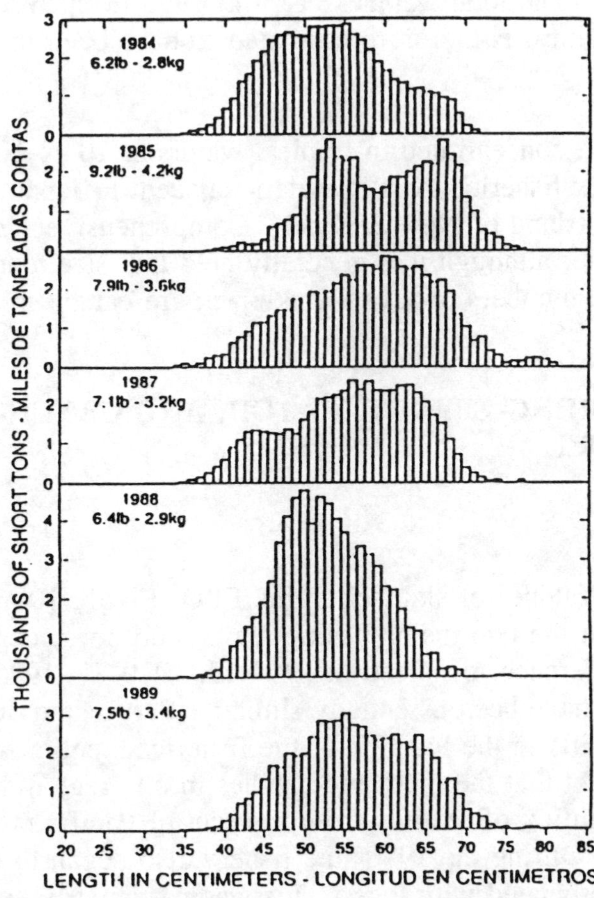

Figure 11. Estimated length composition of the skipjack catch by 1-cm intervals in the EPO. The values in the upper left corners of the panels are average weights. (After IATTC, 1991).

purse seiners. The total tuna catch from the fishery was approximately 30,000 mt in 1990, about 80 percent of which was skipjack (SPC, 1991). Smaller domestic tuna fisheries are currently based in Kiribati, Fiji, and French Polynesia (baitboat) and in Tonga and New Caledonia (longline).

Distant-water, commercial tuna fishing in the WPO dates back to the early 1920s, when Japanese baitboat fleets began to establish themselves in Micronesia (Doulman, 1987). The commercial fishery for skipjack in Japanese home waters operated well before this time. During the early 1950s, Japan re-established its distant-water fishing capability, which had been severely depleted during World War II, with baitboat fishing in Micronesia and longlining throughout the tropical WPO. In the late 1970s, baitboat fishing began to decline and was replaced by purse seining, initially by Japan and the USA, and more recently by the Republic of Korea, Taiwan, Philippines, and Indonesia. The distant-water purse-seine fleet is estimated to have caught approximately 700,000 mt of tuna in 1990, of which 520,000 mt was skipjack (SPC, 1991). This compares with 76,000 mt of skipjack taken by the Japanese distant-water baitboat fleet.

In the far western area, fisheries for small skipjack, yellowfin, and other scombrids have been developing in the Philippines and eastern Indonesia since the early 1970s. Tuna are captured by poling, trolling, handlining, ringnetting, purse seining, and other methods, mostly in association with FADs. In 1990, domestic fisheries in the Philippines and eastern Indonesia are estimated to have caught approximately 165,000 mt of skipjack (SPC, 1991). The total skipjack catch in 1990 in the WPO, including eastern Indonesia and the Philippines, is therefore estimated to have been in excess of 780,000 mt (SPC, 1991).

These fisheries are concentrated in tropical waters at 10°N-10°S (Figure 7b), although seasonal skipjack fisheries exist in waters adjacent to Japan, southeastern Australia, and the North Island of New Zealand. Comprehensive size composition data are generally not available, although it is generally held that size tends to increase from the far west, where large numbers of juvenile skipjack are caught in association with FADs, to the east.

11. TRENDS IN FISHING EFFORT, CATCH, AND CATCH-PER-UNIT OF FISHING EFFORT

11.1 Eastern Pacific

The trends in the catches of skipjack in the EPO (Figure 10 and Table 5) demonstrate that events in the two major fishing regions do not necessarily parallel one another. Substantial differences occurred in 1964-65, 1970-71, 1975, and in 1978-80, but since then the trends have been essentially similar. Part of the differences can be attributed to the major shifts in the location of the fisheries mentioned in Section 10.1. However, it is also thought that the increased catches in one region and the decline in the other may be due to the influx of recruits from the central Pacific favouring different regions (IATTC, 1978). During the 1970s the fishery also began to concentrate on small skipjack and yellowfin associated with logs. This was particularly evident in the fishery south of 15°N where the catches in 1975, 1976, and 1978 reached new levels of 107.6, 107.2, and 153.0 x 10³ mt, respectively. The decline in catches prior to and during the severe El Niño of 1982-83 was also more pronounced in the southern fishery as the vulnerability of fish diminished. Following the recovery of the fishery that began in 1984, the catches in both regions have increased again to the average levels attained during the 1960s and early 1970s.

Historically, the IATTC has progressively modified the standard against which the effort expended in the fishery is gauged. Originally, the standardization was made to Class-4 (182-272 mt capacity) baitboats when this type of gear dominated the fishery during the 1950s. Following the rapid conversion of the fleet to purse seiners by the early 1960s, Class-3 (92-181 mt) seiners became the new standard that persisted until 1967. By 1974, as a result of increased vessel construction that began in the mid-1960s, the fleet capacity of Class-3 seiners fell to 3 percent, while that of the most productive group, Class-6 (> 363 mt), rose to 81 percent. By 1983, Class-6 seiners represented at least 90 percent of the fleet and they were adopted as the new standard. Currently, raw Class-6 effort is used as the measure of effort in determining the catch per day's fishing (CPDF) without taking into account the minor contribution of smaller vessels. In addition, adjustments introduced in 1971, to account for improvements in fleet efficiency of vessels below Class-6, were discontinued when this class became the new standard.

The IATTC staff utilizes the sums of the logged catches (C) and effort (E) to derive the CPDF, and formerly the standardized CPSDF, as an index of the apparent abundance of skipjack. It is recognized that there are at least two problems associated with this approach. First, an index calculated as $\Sigma C/\Sigma E$, rather than as the average C/E weighted by area, $i.e.$, $\overline{C/E}$, is biased if the effort is not distributed equally over the fishing grounds (Griffith, 1960; Calkins, 1961). However, the early studies of Calkins (1961) and Joseph and Calkins (1969) indicated that, although the values of $\Sigma C/\Sigma E$ over time tended to be both larger and more variable than $\overline{C/E}$, the trends were quite similar. For this reason the CPSDF and CPDF are used directly as simple indices of abundance without weighting (Figure 12). Another application of the unweighted and weighted indices involved the ratio of $(\Sigma C/\Sigma E)/(\overline{C/E})$ as a concentration index (CI). Values greater than one supposedly indicate that the fishermen were able to concentrate their effort in areas with above-average densities of fish. During most of the 1960s, the CI values for skipjack were close to two and approximately twice as large as those of yellowfin (IATTC, 1968), thus indicating that skipjack were vulnerable more consistently in recognized areas than yellowfin. Despite the seemingly useful properties of CIs, however, the IATTC staff discontinued their use in 1969 because of the statistical ambiguity of ratios of ratios, and the difficulty in interpreting the meaningful effects on the fishery of indices greater or less than one.

The second problem concerning the index of abundance is that while the catches refer to skipjack alone, the effort represents that directed towards both yellowfin and skipjack. This situation arises because of the difficulty in obtaining a rational separation of effort in a fishery that essentially involves two species, yellowfin and skipjack, but is primarily directed towards yellowfin. To determine the relative apparent abundance in different years in such a situation, Joseph and Calkins (1969) first identified core regions of skipjack effort in the northern and southern fisheries in which the catches in 1° areas were at least 50 percent skipjack in successive years. These baseline regions were augmented with additional 1° areas of high catch specific to different years (i) before determining relative values of $(C/E)_i$ Similarly, and more recently, Forsbergh (1989) identified 22, 5° areas in which most of the skipjack were caught in the 1961-76 period before calculating the relative CPDFs. Some of the important conclusions that emerge from these studies and that of Calkins (1961) are that skipjack catches in successive years are highly variable (Table 5), sometimes differing by a factor of three, but the changes in catch do not appear to be related to effort in preceding years. In fact, and in contrast to yellowfin, a plot of CPDF *versus* days fishing (Figure 13) indicates that the CPDF is independent of effort. In this situation, considering that skipjack recruitment is derived from outside the EPO, the CPDF is probably indicative of apparent abundance in only the EPO, and is not useful for predicting future catch rates. For this reason the CPDF for skipjack may be more usefully interpreted as an economic indicator of fishing success. Caution is also urged in interpreting the CPDF values in Table 5 and Figure 12 because of two biases: the restricted, coastal operating range of baitboats during the 1960s, and the tendency of the fleet to concentrate on yellowfin in recent years.

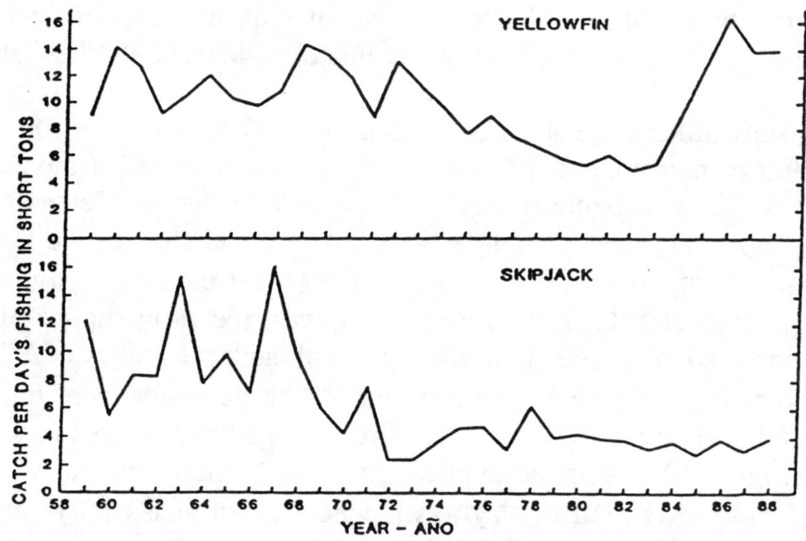

Figure 12. Catch per days fishing for yellowfin and skipjack in the EPO in Class-6 pure-seine units. The 1988 data are preliminary. (After IATTC, 1989).

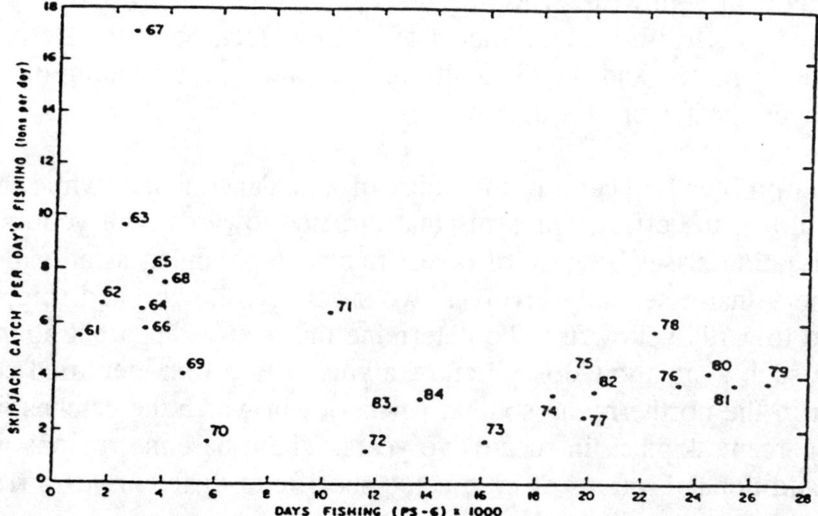

Figure 13. Plot of catch per days fishing (CPDF) and effort for skipjack in the EPO using data for all 5° areas. (After IATTC, 1985).

TABLE 5. Historic catches of skipjack in the EPO by the surface fleet, total effort, catch per days fishing (CPDF), and carrying capacity and number (No.) of vessels participating in the fishery. (Adapted from IATTC, 1991).

	Catch, mt x 10^{-3}					Number and carrying capacity				
	North	South	Total	Total		Purse seiner		Baitboat		Total
	of	of	east of	effort*	CPDFG@		mt		mt	mt
Year	15°N	15°N	150°W	10^{-3} days	(mt)	No.	x 10^{-3}	No.	x 10^{-3}	x 10^{-3}
1961	8.3	60.1	68.5	9.1	7.5	124	27.3	93	9.5	36.8
2	9.3	61.8	71.0	9.5	7.5	130	30.8	89	6.1	36.8
3	15.2	80.3	95.5	6.9	13.8	141	36.1	108	5.4	41.5
4	14.3	44.9	59.2	8.4	7.0	134	36.6	88	4.3	40.8
5	10.9	67.3	78.2	8.8	8.9	146	46.6	109	5.3	43.6
6	6.8	53.7	60.5	9.4	6.4	126	36.2	113	5.6	41.8
7	37.1	83.6	120.7	8.3	14.5	122	36.5	108	5.4	41.8
8	9.5	61.6	71.1	8.1	8.8	139	45.9	89	5.2	51.1
9	11.2	47.8	59.1	10.8	5.5	149	51.7	69	4.5	56.2
1970	28.0	28.0	56.1	14.6	3.8	162	61.2	49	3.9	65.1
1	22.5	82.2	104.7	15.4	6.8	185	80.6	102	5.1	85.6
2	9.1	24.3	33.3	15.2	2.2	206	102.0	108	6.1	108.0
3	7.5	36.4	44.0	20.3	2.2	216	145.4	106	6.3	125.9
4	7.9	70.9	78.8	24.3	3.2	230	133.0	111	7.1	140.4
5	16.3	107.5	123.8	29.5	4.2	249	148.6	102	6.7	155.3
6	19.0	107.2	126.2	29.2	4.3	250	160.1	99	6.4	166.6
7	11.0	75.3	86.4	29.9	2.9	250	162.2	79	4.9	167.1
8	16.7	153.0	169.8	30.2	5.6	262	164.0	68	4.5	168.6
9	23.7	108.3	132.0	36.2	3.6	268	166.3	45	3.6	170.3
1980	50.3	80.0	130.4	33.7	3.9	258	167.4	46	3.4	170.9
1	41.8	77.8	119.6	33.6	3.6	247	166.6	39	2.8	169.5
2	30.2	68.3	98.7	28.3	3.5	221	152.2	36	3.4	154.7
3	15.5	42.3	58.1	19.7	2.9	199	125.0	52	3.2	128.1
4	7.5	53.1	60.5	18.0	3.4	165	102.7	40	2.8	105.5
5	1.8	47.6	49.4	19.2	2.6	175	115.5	25	2.2	117.7
6	4.8	58.7	63.5	18.3	3.5	165	111.2	17	1.7	112.9
7	12.1	50.4	62.5	18.1	3.5	177	130.4	29	1.9	132.4
8	23.9	61.3	85.2	21.5	3.9	185	134.4	36	2.8	137.3
9	# 12.4	78.2	90.7	-	-	172	121.1	30	2.7	123.8

* Effort for all surface vessels in class -6 (>363 mt) purse-seine units
@ For Class-6 purse seiners
# All data for 1989 are preliminary

11.2 Western Pacific

The overall catch trend for skipjack in the WPO is shown in Figure 14. The catch has been increasing at a rapid and fairly constant rate since the early 1970s (SPC, 1991). A number of events have fueled this expansion. In the 1970s, the development of baitboat fisheries in Papua New Guinea and the Solomon Islands and the expansion of the Japanese distant-water, baitboat fishery led to the first large increases. In the 1980s, the development of large-scale purse seining in the WPO and the subsequent influx of vessels from several DWFNs resulted in further increases. In the 1990s, the policy of many canneries not to buy tuna caught in association with dolphins has resulted in the relocation of many purse seiners to the WPO, where problems of dolphin-associated tuna schools do not exist. This, along with tuna fishery development in Southeast Asia, has brought about further increases in the skipjack catch in the WPO. This trend is likely to continue for the next few years at least.

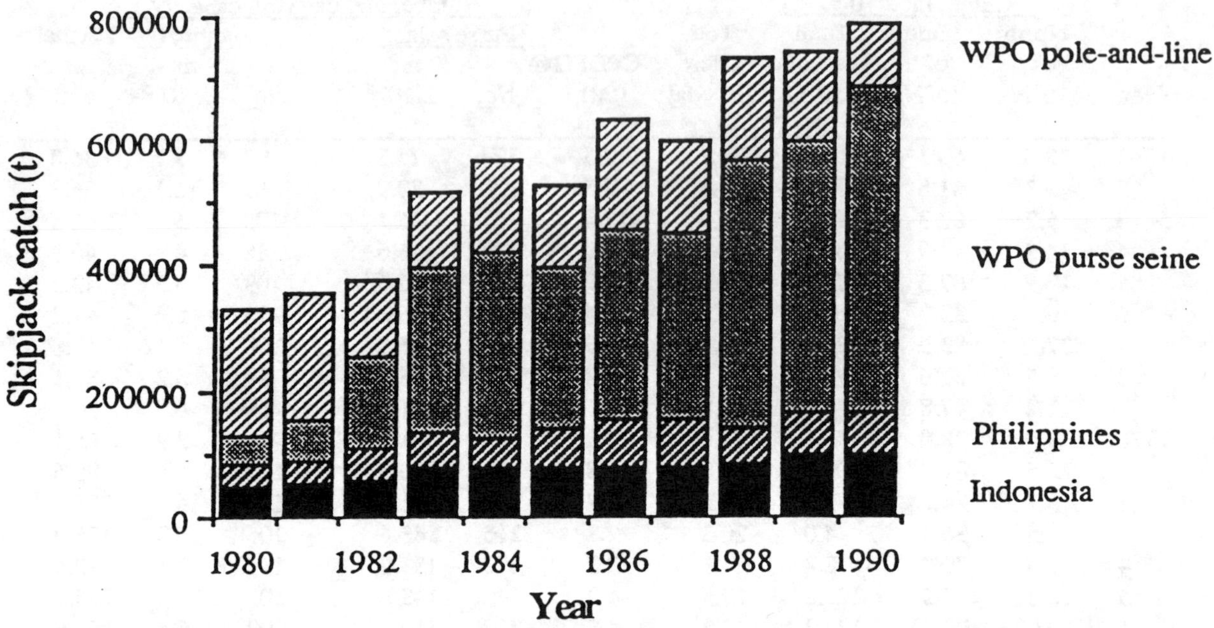

Figure 14. Catch trend for skipjack in the WPO. (After SPC, 1991).

The SPC routinely monitors total CPDF for both the baitboat and purse-seine fisheries for which it receives data. To date, abundance indices such as those described in the previous section for the EPO have not been routinely derived, although a preliminary study was reported in the SPC (1990). Here, indices were constructed using $(C/E)_j$ for Japanese purse seines, where the strata, j, were defined as areas of 2° of latitude by 5° of longitude within the overall area of 10°N-6°S, 130°-165°E, and time periods of either one or three months. These indices, along with raw CPDF, ($\Sigma C/\Sigma E$), suggest similar, steadily-increasing trends in apparent abundance (Figure 15). This is also true of CPDF in the Japanese distant-water, baitboat fishery. Whether these trends are real, *i.e.* they reflect skipjack abundance, or result from increasing fleet efficacy or other factors, has not yet been determined. In either case, it appears that the abundance of skipjack in the WPO is substantially higher than in the EPO, based on the observation of CPDF levels in the WPO 3-4 times those in the EPO (Figures 12 and 15) (although the effects of yellowfin targeting may bias such comparisons to some extent).

As in the EPO, similar difficulties exist in determining effective effort on skipjack in a purse-seine fishery that also targets yellowfin. In contrast to the EPO, however, calculated concentration indices suggest that the distribution of effort of at least Japanese purse seiners is influenced more by the apparent abundance of skipjack than that of yellowfin (SPC, 1990).

12.   POPULATION DYNAMICS

If the skipjack fishery in the EPO is open to both immigration and emigration, as is currently believed, then it is doubtful whether stock production modelling or cohort

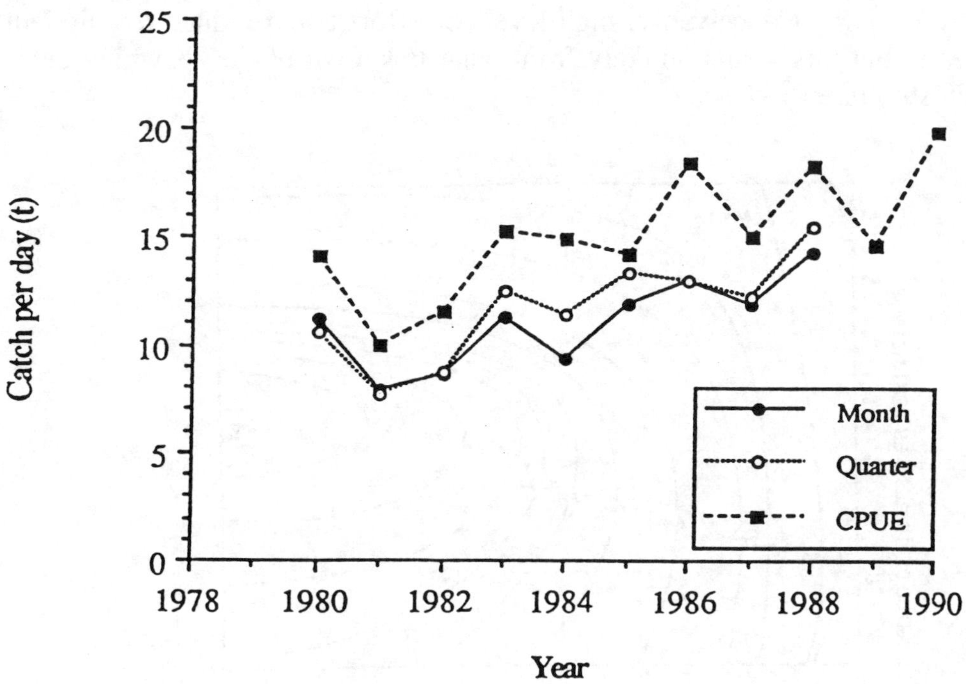

Figure 15. Various indicies of skipjack apparent abundance based on CPDF in the Japanese purse-seine fishery. "CPUE" refers to unadjusted CPDF, "Month" refers to average CPDF stratified by month and 2° latitude by 5° longitude rectranglar areas, and "Quarter" refers to average CPDF stratified by quarter and 2° latitude by 5° longitude recatangular areas. (After SPC, 1990).

analysis can produce useful results. Forsbergh (1987) drew attention to this point when values of F, and consequently recruitment estimates, failed to converge after applying Pope's (1972) cohort analysis to quarterly catches and various starting values and attrition rates. On the other hand, by incorporating information on growth rates, a length-weight relationship and approximate attrition rates into yield-per-recruit (Y/R) analysis, some insight can be gained into the dynamics of the fishery. For example, the yield-per-recruit isopleths in Figure 16 were constructed using a linear growth rate of 24 cm/yr (Forsbergh, 1989) and Hennemuth's (1959) length-weight formula from Table 2. The instantaneous attrition rates $M' = 1.5$ (upper panel) and $M' = 2.0$ (lower panel), which include both natural mortality and emigration, correspond to current estimates of the annual attrition rates of 78 and 86 percent, respectively. The effort levels represent multiples of the 1986-87 average which is set equal to one. The importance of the attrition rate with respect to the Y/R is apparent in that an increase of M' from 1.5 to 2.0 reduces the Y/R by nearly half. At the lower value of M' and recent effort levels there is also a marginal improvement in the Y/R by delaying the size at first capture until 45 cm, but at the higher attrition rate the Y/R is nearly constant in the size range from 30-40 cm. In general, fishing effort considerably greater than or sizes of entry considerably less than has been the case so far would be required to overfish the stock in the EPO in the Y/R sense (IATTC, 1991). The reason is that losses to the total weight of a cohort due to M' exceed the gains to it by growth, even when the fish are less than 50 cm and growing rapidly. In summary, the model does not indicate the need for management of the skipjack fishery in the EPO at present levels of effort. However, it is apparent that at the higher attrition rate in Figure 16 increased effort eventually leads to a decline in the marginal return of *Y/R*. Conceivably, high levels of effort could reduce recruitment in subsequent years, but this seems unlikely from what is known of the spawning behaviour and population structure.

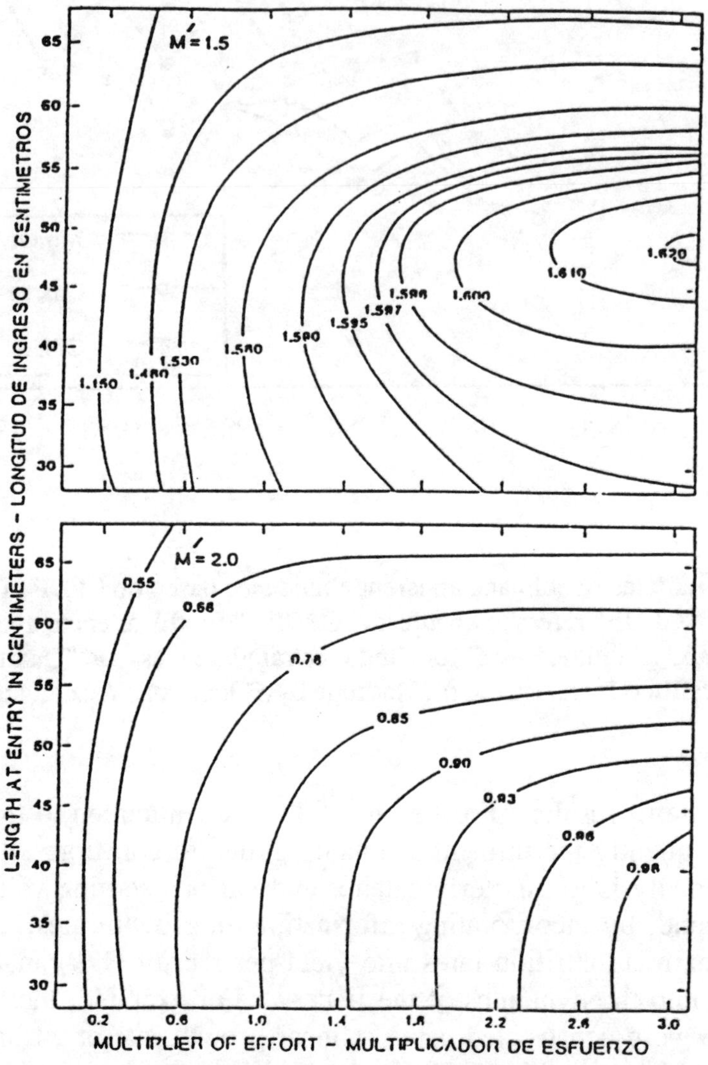

Figure 16. Yields (lbs) per recruit for skipjack with multiple levels of fishing effort in 1986-1987 (=1), length at entry into the fishery, and the sum of natural mortality and emigration (*M'*). (Adapted from IATTC, 1991).

In the WPO, comparable modelling of skipjack dynamics has not been carried out, due primarily to the lack of representative size composition and total catch and effort data, although this situation is gradually being rectified. Most studies relating to population dynamics and stock assessment have been based on the SSAP tagging data. The SPC produced a series of reports during the early 1980s that described the dynamics of skipjack stocks in the Exclusive Economic Zones (EEZs) of all its South Pacific member countries. In all, twenty such reports were produced along with a more general report dealing with the aggregate data set (Kleiber *et al.*, 1983; 1987). In the latter, total standing stock of skipjack in the WPO was estimated at 3×10^6 mt, with a total throughput of 6.2×10^6 mt per year. The overall harvest ratio (proportion of total attrition due to fishing mortality) was estimated at 0.04, while harvest ratios for seven subareas for which catch and effort data were available ranged from 0.02 to 0.46; only one exceeded 0.17. Kleiber *et al.* (1987) concluded that "...low harvest ratios over most of the study area during the period tags were at large imply a potential for increased skipjack catches in many subareas and in the whole study area." At that time the total skipjack catch in the WPO was approximately 230,000 mt per yr.

13. INTERACTIONS

During the 1950s and early 1960s, baitboats and purse seiners frequently competed for tuna resources in nearshore waters in the EPO. However, purse seiners make up over 97 percent of the current fleet capacity (Table 5), and gear competition for skipjack in particular is virtually nonexistent in the EPO. Similarly, purse seiners now account for most of the catching capacity in the WPO; in 1990, purse seiners caught 66 percent of the estimated total skipjack catch of 785,000 mt (SPC, 1991). However, the possibility of interaction between purse-seine and baitboat fisheries is of concern in countries like the Solomon Islands and Fiji, where domestic fisheries have existed for some time. In the case of the Solomon Islands, purse-seine fishing has increased in recent years, and the SPC has been collaborating with the Solomon Islands Government in a tagging experiment designed to estimate the magnitude of the purse-seine/baitboat interaction. The results of the study suggest that the effect of purse-seine catches on baitboat CPUE at recent levels of catch (baitboat $\approx$ 25,000 mt per year, purse seine $\approx$ 6,000 mt per year) is slight (SPC, 1992).

Although gear interactions are largely minor, interactions among geographical areas, particularly EEZs, are of concern in some areas. In the WPO, the main area of operation of the purse-seine and baitboat fisheries is composed primarily of the largely contiguous EEZs of Philippines, Indonesia, Palau, Federated States of Micronesia, Papua New Guinea, the Solomon Islands, Nauru, Kiribati, and the Marshall Islands. The degree of interaction among areas such as these will be determined by the size of the areas, the distances between them, skipjack movement rates, the natural mortality rate, and the intensity of the fisheries (SPC, 1988). As noted in Section 6.4, there has been some controversy regarding movement rates of skipjack and their possible effects on spatially-separated fisheries. However, some specific analyses have been carried out. Kleiber *et al.* (1984), using SSAP tagging data, calculated a series of interaction coefficients based on the proportions of total throughput in receiver EEZs derived from immigration from donor EEZs. These results are summarized in Table 6. Most of the coefficients are low, indicating that under conditions prevailing when the SSAP data were gathered, there was generally little potential for fishery interaction. Not surprisingly, most cases of significant exchange occurred between adjacent EEZs. In particular, the results suggested that 37 percent of throughput in the Marshall Islands EEZ at the time of tagging resulted from immigration from Federated States of Micronesia. Relatively high interaction coefficients were also observed for Northern Mariana Islands $\rightleftarrows$ Federated States of Micronesia and to a lesser extent Palau $\rightarrow$ Federated States of Micronesia and Papua New Guinea $\rightleftarrows$ Solomon Islands, indicating some potential for fishery interaction between those countries. The only case of a relatively high interaction coefficient for widely separated areas was New Zealand $\rightarrow$ Fiji; however, this may have been an artifact of the timing of tag releases into the highly seasonal New Zealand Fishery (Argue and Kearney, 1983).

TABLE 6. Coefficients of interaction between fisheries operating in various countries and territories in the WPO. Receiver countries are listed at the top of the table and doner countries down the left margin. (After Kleiber *et al.*, 1984).

To -->>	Papua New Guinea (PNG)	Solomon Islands (SOL)	Palau (PAL)	Federated States of Micronesia (FSM)	Marshall Islands (MAS)	Northern Mariana Islands (MAR)	Fiji (FIJ)	New Zealand (ZEA)	Western Samoa (WES)	French Polynesia (SOC)
PNG	-	2.6	0.8	1.4	0.5	-	-	-	-	-
SOL 1977	1.1	-	-	-	-	-	-	-	-	-
SOL 1980	3.7	-	-	-	-	-	-	-	-	-
PAL 1978	-	-	-	8.6	2.2	-	-	-	-	-
PAL 1980	1.6	0.4	-	3.5	1.3	0.7	-	-	-	-
FSM	0.7	0.9	-	-	37.0	10.8	-	-	-	-
MAS	-	-	-	-	-	-	-	-	-	-
MAR	-	-	-	17.4	-	-	-	-	-	-
FIJ 1978	-	-	-	-	-	-	-	0.6	-	-
ZEA	-	-	-	-	-	-	6.5	-	2.1	3.6
KIR(ibati)	-	-	-	>0.1	0.1	-	-	-	-	-

This relatively simple representation of interaction does not explicitly specify the controlling factors noted above. Sibert (1984) derived a more rigorous method to estimate interaction between two countries and applied the method to Papua New Guinea and the Solomon Islands, both of which had substantial baitboat fisheries for skipjack at the time of the tagging project. Exchange rates between the two EEZs, losses from natural mortality and movement to other areas, the proportions that remained resident and lived and the proportions that were caught locally on a monthly basis were estimated. The Solomon Islands' stock was found to be relatively stable with a low rate of natural mortality and emigration (resulting in high survival) and low rate of movement to Papau New Guinea. The Papua New Guinea stock was found to be more dynamic with a higher rate of natural mortality and emigration (lower survival), but with a low rate of movement to the Solomon Islands. Sibert (1984) estimated from these results that an increase in the catch in either EEZ of 1,000 mt would result in a decrease in the steady-state catch of the other of only 1-3 mt.

Incomplete catch and effort data availability has, until now, hindered a more thorough analysis of skipjack movement dynamics, and its interpretation with respect to interaction, in the WPO. This situation is now being rectified through a new, collaborative project under the auspices of the FAO Expert Consultation on Interactions of Pacific Ocean Tuna Fisheries.

14.   ACKNOWLEDGEMENTS

We would particularly like to thank Bill Bayliff, Kurt Schaefer, and Pat Tomlinson for generously sharing their insights on many interesting and controversial points in the manuscript. Tony Lewis and Albert Caton also provided valuable comments.

15. REFERENCES CITED

Aikawa, H. 1937. Notes on the shoal of bonito along the Pacific coast of Japan. *Bull.Jap.Soc.Sci.Fish.*, 6(1):13-21. [Engl. transl. in W.G. Van Campen, *Spec.Sci. Rep.U.S.Fish Wild.Serv. (Fish.)*, 83:32-50]

Argue, A.W., and R.E. Kearney. 1983. An assessment of the skipjack and baitfish resources of New Zealand. *Final Ctry.Rep.Skipjack Surv.Assess.Programme, S.Pac.Comm.*, (8):47 p.

Argue, A.W., F. Conand, and D. Wyman. 1983. Spatial and temporal distributions of juvenile tunas from stomachs of tunas caught by pole-and-line gear in the central and western Pacific Ocean. *Tech.Rep.Tuna Billfish Assess.Programme, S.Pac.Comm.*, (9): 47 p.

Bard, F.X. 1986. Analyse des taux de décroissance numérique des listaos marqués en Atlantique est. *In* Proceedings of the ICCAT Conference on the International Skipjack Year Program, edited by P.E.K. Symons, P.M. Miyake, and G.T. Sakagawa. Madrid, Spain, ICCAT:348-62.

Bard, F.X., and L. Antoine. 1986. Croissance du listao dans l'Atlantique. *In* Proceedings of the ICCAT Conference on the International Skipjack year Program, edited by P.E.K. Symons, P.M. Miyake, and G.T. Sakagawa. Madrid, Spain, ICCAT:301-08.

Barkley, R.A., W.H. Neill, and R.M. Gooding. 1978. Skipjack tuna, *Katsuwonus pelamis*, habitat based on temperature and oxygen requirements. *Fish.Bull.U.S.Fish. Wildl.Serv.* 76(3):653-62.

Barrett, I., and A.R. Connor. 1962. Blood lactate in yellowfin tuna, *Neothunnus macropterus*, and skipjack, *Katsuwonus pelamis*, following capture and tagging. *Bull.I-ATTC*, 6(6):233-80.

Batts, B.S. 1972a. Age and growth of skipjack tuna, *Katsuwonus pelamis* (Linnaeus), in North Carolina waters. *Chesapeake Sci.*, 13(4):237-44.

Batts, B.S. 1972b. Sexual maturity, fecundity, and sex ratios of the skipjack tuna, *Katsuwonus pelamis* (Linnaeus), in North Carolina waters. *Trans.Am.Fish.Soc.*, 101(4):626-37.

Bayliff, W.H. 1971. Estimates of the rates of mortality of yellowfin tuna in the eastern Pacific Ocean derived from tagging experiments. *Bull.I-ATTC*, 15(4):379-436.

Bayliff, W.H. 1977. Estimates of the rates of mortality of skipjack tuna in the eastern Pacific Ocean derived from tagging experiments. *Internal Rep.I-ATTC*, (10):59 p.

Bayliff, W.H. 1984. Migrations of yellowfin and skipjack tuna released in the central portion of the eastern Pacific Ocean, as determined by tagging investigations. *Internal Rep.I-ATTC*, (18):107 p.

Bayliff, W.H. 1988. Growth of skipjack, *Katsuwonus pelamis*, and yellowfin, *Thunnus albacares*, tunas in the eastern Pacific Ocean, as estimated from tagging data. *Bull.I-ATTC*, 19(4):311-85.

Bayliff, W.H., and L.M. Mobrand. 1972. Estimates of the rates of shedding of dart tags from yellowfin tuna. *Bull.I-ATTC*, 15(5):441-62.

von Bertalanffy, L. 1938. A quantitative theory of organic growth. *Hum.Biol.*, 10: 181-213.

Broadhead, G.C., and C.J. Orange. 1960. Species and size relationships within schools of yellowfin and skipjack tuna, as indicated by catches in the eastern tropical Pacific Ocean. *Bull.I-ATTC*, 4(7):449-92.

Broadhead, G.C., and I. Barrett. 1964. Some factors affecting the distribution and apparent abundance of yellowfin and skipjack tuna in the eastern Pacific Ocean. *Bull.I-ATTC*, 8(8):417-73.

Brock, V.E. 1949. A preliminary report on *Parathunnus sibi* in Hawaiian waters and a key to the tunas and tuna-like fishes of Hawaii. *Pac.Sci.*, 3(3):271-77.

Brock, V.E. 1954. Some aspects of the biology of the aku, *Katsuwonus pelamis*, in the Hawaiian Islands. *Pac.Sci.*, 8(1):94-104.

Brouard, F., R. Grandperrin and E. Cillaurren. 1984. Croissance des juenes thons jaunes (*Thunnus albacares*) et de bonites (Katsuwonus pelamis) dans de Pacifique tropical occidental. *Notes Doc.d'Oceangr.ORSTOM Port-Vila*, (10):23 p.

Buñag, D.M. 1956. Spawning habits of some Philippine tuna based on diameter measurements of the ovarian ova. *J.Philipp.Fish.*, 4(2):145-77.

Cayré, P. 1979. Determination de l'age de listao, Katsuwonus pelamis, debarqués à Dakar; Note préliminaire. *Collect.Vol.Sci.Pap.ICCAT*, 8:196-200.

Cayré, P., A. Fonteneau, and M.H. Santa-Rita Veira. 1986. Analyse des données de marquages et recaptures de listao (*Katsuwonus pelamis*) réalisés par le Sénégal et la République du Cap-Vert. *In* Proceedings of the ICCAT Conference on the International Skipjack Year Program, edited by P.E.K. Symons, P.M. Miyake, and G.T. Sakagawa. Madrid, Spain, ICCAT:309-16.

Calkins, T.P. 1961. Measures of population and concentration of fishing effort for yellowfin and skipjack tuna in the eastern Pacific Ocean, 1951-1959. *Bull.I-ATTC*, 6(3):71-152.

Calkins, T.P. 1975. Geographical distribution of yellowfin and skipjack tuna catches in the eastern Pacific Ocean and total catch statistics, 1971-1974. *Bull.I-ATTC*, 17(1):1-116.

Chatwin, B.M. 1959. The relationships between length and weight of yellowfin tuna (*Neothunnus macropterus*) and skipjack tuna (*Katsuwonus pelamis*) from the eastern tropical Pacific Ocean. *Bull.I-ATTC*, 3(7):307-52.

Chi, K.-S., and R.-T. Yang. 1973. Age and growth of skipjack tuna in the waters around the southern part of Taiwan. [Summ. in Chin.]. *Nat.Taiwan Univ.Sci.Rep.Acta Oceanogr.Taiwanica*, 3:199-221.

Chur, V.N., V.B. Grudinin, and V.L. Zharov. 1980. Data on length-age composition and gonad maturity stages of skipjack (*Katsuwonus pelamis*) of the eastern tropical Atlantic. *Collect.Vol.Sci.Pap.ICCAT*, 9:245-54.

Chur, V.N., and V.L. Zharov. 1983. Determination of age and growth of the skipjack tuna, *Katsuwonus pelamis* (Scombridae), from the southern part of the Gulf of Guinea. *J.Ichthyol.*, 23(3):53-67.

Clement, I.T. 1976. Distribution and abundance of skipjack, 1975-76. *Occas.Publ. Fish.Res.Div.New Zealand Minist.Agri.Fish.*, 11:36-9.

Clement, I.T. 1978. School fish sightings around New Zealand, 1976-77. *Occas.Publ. Fish.Res.Div.New Zealand Minist.Agri.Fish.*, 15:35-42.

Collette, B.B., and C.E. Nauen. 1983. FAO species catalogue. Vol. 2. Scombrids of the world. An annotated and illustrated catalogue of tunas, mackerels, bonitos and related species known to date. *FAO Fish.Synop.*, (125)Vol.2:137 p.

Collette, B.B., T. Potthoff, W.J. Richards, S. Ueyanagi, J.L. Russo, and Y. Nishikawa. 1984. Scombroidei: development and relationships. *In* Ontogony and Systematics of Fishes, edited by H.G. Moser, *et al.*, *Spec.Publ.Am.Soc.Ichthyol.Herpetol.*, 1:591-620.

Dizon, A.E., R.E. Brill, and H.S.H. Yuen. 1978. Correlations between environment, physiology, and activity and the effects of thermoregulation in skipjack tuna. *In* The physiological ecology of tunas, edited by G.D. Sharp and A.E. Dizon. New York, Academic Press, pp. 233-59.

Doulman, D.J. 1987. Distant-water fleet operations and regional fisheries cooperation. *In* The development of the tuna industry in the Pacific islands region: an analysis of options, edited by D.J. Doulman. Honolulu, Hawaii, East-West Center Press, pp. 33-52.

Doulman, D.J., and R.E. Kearney. 1987. Domestic tuna fisheries. *In* The development of the tuna industry in the Pacific islands region: an analysis of options, edited by D.J. Doulman. Honolulu, Hawaii, East-West Center Press, pp. 3-32.

Endler, J.A. 1977. Geographic variation, speciation and clines. New Jersey, Princeton University Press, 246 p.

Evans, L.C., A.D. Uy, and D.D. Tandog. 1978. The abundance, biology and distribution of tuna (Family Thunnidae) in Camiguin and nearby waters. *Bur.Fish.Aquat.Resour.Phillip.*, 55 p.

Fink, B.D. 1965. Estimations, from tagging experiments, of mortality rates and other parameters respecting yellowfin and skipjack tuna. *Bull.I-ATTC*, 10(1):3-82.

Fink, B.D., and W.H. Bayliff. 1970. Migrations of yellowfin and skipjack tuna in the eastern Pacific Ocean as determined by tagging experiments, 1952-1964. *Bull.I-ATTC*, 15(1):1-227.

Fonteneau, A. 1986. Etat des stocks de listao de l'Atlantique par analyse des cohorts, analyse de production par recrue et par le modèle global. *In* Proceedings of the ICCAT Conference on the International Skipjack Year Program, edited by P.E.K. Symons, P.M. Miyake, and G.T. Sakagawa. Madrid, Spain, ICCAT:208-33.

Forsbergh, E.D. 1969. On the climatology, oceanography and fisheries of the Panama Bight. *Bull.I-ATTC*, 14(2):45-385.

Forsbergh, E.D. 1980. Synopsis of biological data on the skipjack tuna, *Katsuwonus pelamis*, (Linnaeus, 1758), in the Pacific Ocean. *Spec.Rep.I-ATTC*, (2):295-360.

Forsbergh, E.D. 1987. Rates of attrition, cohort analysis, and stock production models for skipjack tuna, *Katsuwonus pelamis*, in the eastern Pacific Ocean. *Internal Rep.I-ATTC*, (20):37 p.

Forsbergh, E.D. 1988. A review of the question of subpopulations of skipjack tuna, *Katsuwonus pelamis*, in the Pacific Ocean, and of possible migration routes. *Internal Rep.I-ATTC*, (21):15 p.

Forsbergh, E.D. 1989. The influence of some environmental variables on the apparent abundance of skipjack tuna, *Katsuwonus pelamis*, in the eastern Pacific Ocean. *Bull.I-ATTC*, 19(6):433-569.

Fujino, K. 1970a. Immunological and biochemical genetics of tunas. *Trans.Am.Fish. Soc.*, 99(1):152-78.

Fujino, K. 1970b. Skipjack tuna subpopulation identified by genetic characteristics in the western Pacific. *In* The Kuroshio: a symposium on the Japan Current, edited by J.C. Marr. Honolulu, Hawaii, East-West Center Press, pp.385-93.

Fujino, K. 1972. Range of the skipjack tuna subpopulation in the western Pacific Ocean. *In* The Kuroshio II: proceedings of the second CSK symposium, edited by K. Suguwara. Tokyo, Saikon Pub. Co., Ltd., pp. 373-84.

Fujino, K. 1976. Subpopulation identification of skipjack tuna specimens from the southwestern Pacific Ocean. *Bull.Jap.Soc.Sci.Fish.*, 42(11):1229-35.

Fujino, K., K. Sasaki, and S. Okumura. 1981. Genetic diversity of skipjack tuna in the Atlantic, Indian and Pacific Oceans. *Bull.Jap.Soc.Sci.Fish.*, 47(2):215-22.

Gillett, R.D., and R.E. Kearney. 1983. An assessment of the skipjack and baitfish resources of French Polynesia. *Final Ctry.Rep.Skipjack Surv.Assess.Programme, S.Pac.Comm.*, (7):81 p.

Griffiths, R.C. 1960. A study of measures of population density and of concentration of fishing effort in the fishery for yellowfin tuna, *Neothunnus macropterus*, in the eastern tropical Pacific Ocean, from 1951 to 1956. *Bull.I-ATTC*, 4(3):39-98.

Habib, G. 1978. Skipjack biology and the 1976-77 purse-seine fishery. *In* Proceedings of the pelagic fisheries conference, July 1977, edited by G.V. Habib and P.E. Roberts. *Occas.Publ.Fish.Res.Div.New Zealand Minist.Agri.Fish.*, 15:17-26.

Hampton, J. 1986. Effect of tagging on the condition of southern bluefin tuna, *Thunnus maccoyii* (Castlenau). *Aust.J.Mar.Freshwat.Res.*, 37(6):699-705.

Harada, T., K. Mizuno, O. Murata, S. Miyashita, and H. Furutani. 1971. On the artificial fertilization and rearing of larvae in yellowfin tuna. *Bull.Fac.Agri.Kinki Univ.*, (4):145-51.

Hennemuth, R.C. 1959. Additional information on the length-weight relationship of skipjack tuna from the eastern tropical Pacific Ocean. *Bull.I-ATTC*, 4(2):25-37.

Higgins, B.E. 1967. The distribution of juveniles of four species of tunas in the Pacific Ocean. *Proc.Indo-Pac.Fish.Coun.*, 12(2):79-99.

Hilborn, R., and J.R. Sibert. 1988. Is international management of tuna necessary? *Mar. Policy*, 12(1):31-9.

Hu, F., and R.-T. Yang. 1972. A preliminary study on sexual maturity and fecundity of skipjack tuna. *J.Fish.Soc.Taiwan*, 1:88-98.

Hunter, J.R., B.J. Macewicz, and J.R. Sibert. 1986. The spawning frequency of skipjack tuna, *Katsuwonus pelamis*, from the south Pacific. *Fish.Bull.NOAA-NMFS*, 84(4):895-903.

Inter-American Tropical Tuna Commission. 1968. Annual report of the Inter-American Tropical Tuna Commission, 1967. *Annu.Rep.I-ATTC*, (1967):143 p.

Inter-American Tropical Tuna Commission. 1978. Annual report of the Inter-American Tropical Tuna Commission, 1977. *Annu.Rep.I-ATTC*, (1977):155 p.

Inter-American Tropical Tuna Commission. 1979. Annual report of the Inter-American Tropical Tuna Commission, 1978. *Annu.Rep.I-ATTC*, (1978):163 p.

Inter-American Tropical Tuna Commission. 1981. Annual report of the Inter-American Tropical Tuna Commission, 1980. *Annu.Rep.I-ATTC*, (1980):234 p.

Inter-American Tropical Tuna Commission. 1982. Annual report of the Inter-American Tropical Tuna Commission, 1981. *Annu.Rep.I-ATTC*, (1981):303 p.

Inter-American Tropical Tuna Commission. 1983. Annual report of the Inter-American Tropical Tuna Commission, 1982. *Annu.Rep.I-ATTC*, (1982):294 p.

Inter-American Tropical Tuna Commission. 1984. Annual report of the Inter-American Tropical Tuna Commission, 1983. *Annu.Rep.I-ATTC*, (1983):272 p.

Inter-American Tropical Tuna Commission. 1987a. Annual report of the Inter-American Tropical Tuna Commission, 1986. *Annu.Rep.I-ATTC*, (1986):264 p.

Inter-American Tropical Tuna Commission. 1987b. Quarterly report of the Inter-American Tropical Tuna Commission, 1987. *Annu.Rep.I-ATTC*, (1987) 4:4-6.

Inter-American Tropical Tuna Commission. 1989. Annual report of the Inter-American Tropical Tuna Commission, 1988. *Annu.Rep.I-ATTC*, (1988):288 p.

Inter-American Tropical Tuna Commission. 1991. Annual report of the Inter-American Tropical Tuna Commission, 1989. *Annu.Rep.I-ATTC*, (1989):270 p.

Joseph, J. 1963. The fecundity of yellowfin tuna (*Thunnus albacares*) and skipjack (*Katsuwonus pelamis*) from the eastern Pacific Ocean. *Bull.I-ATTC*, 7(4):255-92.

Joseph, J., and T.P. Calkins. 1969. Population dynamics of the skipjack tuna (*Katsuwonus pelamis*) of the eastern Pacific Ocean. *Bull.I-ATTC*, 13(1):1-273.

Joseph, J., and F.R. Miller. 1988. El Niño and the surface fishery for tunas in the eastern Pacific. *In* Proceedings of the Tuna Fisheries Research Conference, Jap.Fish.Agency Far Seas Fish.Res.Lab. *Maguro Gyogyo Kyogikai Gijiroku Suisancho-Enyo Suisan Kenkyusho*:199-207.

Josse, E., J.C. Le Guen, R. Kearney, A. Lewis, A. Smith, L. Marec and P.K. Tomlinson. 1979. Growth of skipjack. *Occas.Pap.S.Pac.Comm.*, (11):83 p.

Kawasaki, T. 1952. On the populations of the skipjack, *Katsuwonus pelamis* (Linnaeus), migrating to the north-eastern sea area along the Pacific coast of Japan. *Bull.Tohoku Reg.Fish.Res.Lab.*, (1):1-14.

Kimura, K., M. Iwashita, and T. Hattori. 1952. Image of skipjack and tuna recorded on echo sounding machine. *Bull.Tohoku Reg.Fish.Res.Lab.*, (1):15-9.

Klawe, W.L. 1963. Observations on the spawning of four species of tuna, *Neothunnus macropterus*, *Katsuwonus pelamis*, *Auxis thazard*, and *Euthynnus lineatus*, in the eastern Pacific Ocean, based on the distribution of their larvae and juveniles. *Bull.I-ATTC*, 6(9):447-540.

Kleiber, P., A.W. Argue, and R.E. Kearney. 1983. Assessment of skipjack (*Katsuwonus pelamis*) resources in the central and western Pacific by estimating standing stock and components of population turnover from tagging data. *Tech.Rep.Tuna Billfish Assess.Programme, S.Pac.Comm.*, (8):38 p.

Kleiber, P., A.W. Argue, J.R. Sibert, and L.S. Hammond. 1984. A parameter for estimating potential interaction between fisheries for skipjack tuna (*Katsuwonus pelamis*) in the western Pacific. *Tech.Rep.Tuna Billfish Assess.Programme S.Pac.Comm.*, (12): 11 p.

Kleiber, P., A.W. Argue, and R.E. Kearney. 1987. Assessment of Pacific skipjack tuna (*Katsuwonus pelamis*) resources by estimating standing stock and components of population turnover from tagging data. *Can.J.Fish.Aquat.Sci.*, 44(6):1122-34.

Lasker, R. 1975. Field criteria for survival of anchovy between inshore chlorophyl maximum layers and successful first feeding. *Fish.Bull.NOAA-NMFS*, 73(3):453-62.

Laurs, R.M., R. Nishimoto, and J.A. Wetherall. 1985. Frequency of increment formation on sagittae of north Pacific albacore (*Thunnus alalunga*). *Can.J.Fish.Aquat.Sci.*, 42(9):1552-55.

Leis, J.M., T. Trnski, M. Harmelin-Vivien, J.-P. Renon, V. Dufour, M.K. El Moudni, and R. Gazlin. 1991. High concentrations of tuna larvae (Pisces: Scombridae) in near-reef waters of French Polynesia (Society and Tuamotu Islands). *Bull.Mar.Sci.*, 48:150-8.

Lenarz, W.H. 1974. Length-weight relations for five eastern tropical Atlantic scombrids. *Fish.Bull.NOAA-NMFS*, 72(3):848-51.

Lewis, A.D. 1981. Population genetics, ecology and systematics of Indo-Australian scombrid fishes, with particular reference to skipjack tuna (*Katsuwonus pelamis*). Ph.D. Thesis. Australian National University.

Lindberg, G.U. 1971. Fishes of the world. A key to families and a checklist. New York, John Wiley and Sons, 545 p.

Marcille, J., and B. Stequert. 1976. Etude préliminaire de la croissance du listao (Katsuwonus pelamis) dans l'ouest de l'Océan Indien tropical. *Cah.ORSTOM(Ser.Oceanogr.)* 14(2):139-51

Marr, J.C. 1948. Observations on the spawning of oceanic skipjack (*Katsuwonus pelamis*) and yellowfin tuna (*Neothunnus macropterus*) in the northern Marshall Islands. *Fish.Bull.U.S.Fish. WildL.Serv.*, 51(44):201-206.

Matsumoto, W.M. 1975. Distribution, relative abundance, and movement of skipjack tuna, *Katsuwonus pelamis*, in the Pacific Ocean based on Japanese tuna longline catches, 1964-67. *NOAA Tech.Rep.NMFS(Spec.Sci.Rep.-Fish.Ser.)*, (695):30 p.

Matsumoto, W.M., R.A. Skillman and A.E. Dizon. 1984. Synopsis of biological data on skipjack tuna, *Katsuwonus pelamis*. *NOAA Tech.Rep.NMFS Circ.*, (451):92 p.

Miller F.R., and R.M. Laurs. 1975. The El Niño of 1972-1973 in the eastern tropical Pacific Ocean. *Bull.I-ATTC*, 16(5):403-48.

Mori, K. 1972. Geographical distribution and relative apparent abundance of some scombrid fishes based on the occurrences in the stomachs of apex predators caught on tuna longline-I. Juvenile and young of skipjack tuna (*Katsuwonus pelamis*). *Bull.Far Seas Fish.Res.Lab.*, (6):111-57.

Mullen, A.J. 1989. Mobility of tuna. *Mar.Policy*, 13(1):77-8.

Naganuma, A. 1979. On spawning activities of skipjack tuna in the western Pacific Ocean. *Bull.Tohoku Reg.Fish.Res.Lab.*, (40):1-13.

Nakamura, E.L., and J.H. Uchiyama. 1966. Length-weight relations of Pacific tunas. *In* Proceedings of the Governor's Conference on Central Pacific Fishery Resources, edited by T.A. Manar, 28 February - 12 March1966, Honolulu and Hilo, Hawaii, pp. 197-201.

Nishikawa, Y., M. Honma, S. Ueyanagi, and S. Kikawa. 1985. Average distribution of larvae of oceanic species of scombroid fishes., 1956-1981. *S Ser.Far Seas Fish. Res.Lab.*, (12):99 p.

Orange, C.J. 1961. Spawning of yellowfin and skipjack in the eastern tropical Pacific, as inferred from studies of gonad development. *Bull.I-ATTC*, 5(6):459-526.

Orange, C.J., M.B. Schaefer, and F.M. Larmie. 1957. Schooling habits of yellowfin tuna (*Neothunnus macropterus*) and skipjack (*Katsuwonus pelamis*) in the eastern Pacific Ocean as indicated by purse seine catch records, 1946-1955. *Bull.I-ATTC*, 2(3):83-126.

Owen, R.W. 1981a. Microscale plankton patchiness in the larval anchovy environment. *Rapp.P.-V.Reun.CIEM*, 178:364-8.

Owen, R.W. 1981b. Microscale patchiness of small plankton on the Cimbote Shelf, Peru. *Vol.extr.Inst.Mar.Per., Bol., (ICANE)*: 274-9.

Peterson, C.L., and W.H. Bayliff. 1985. Organization, functions, and achievements of the Inter-American Tropical Tuna Commission. *Spec.Rep.I-ATTC*, (5):56 p.

Philander, S.G. 1990. El Niño, La Niña, and the Southern Oscillation. *Ser.Internat. Geophys.*, Academic Press, Vol.46:293 p.

Pianet, R. 1974. Relations poids-longueur des listaos (*Katsuwonus pelamis*) pêches dans le secteur de Pointe-Noire. *Collect.Vol.Sci.Pap.ICCAT*, 9:275-81.

Pickard, G.L. 1968. Descriptive physical oceanography. London, Permagon Press, 200 p.

Polacheck, T. 1990. Another perspective on the need for international skipjack and yellowfin tuna management. *Mar.Policy*, 14(6):526-9.

Pope, J.G. 1972. An investigation of the accuracy of virtual population analysis using cohort analysis. *Res.Bull.Int.Comm.Northwest Atl.Fish.*, 9:65-74.

Radtke, R.L. 1983. Otolith formation and increment deposition in laboratory-reared skipjack tuna, *Euthynnus pelamis*, larvae. *In* Proceedings of the Workshop on Age Determination of Oceanic Pelagic Fishes: Tunas, Billfishes, and Sharks, edited by E.D. Prince, and L.M. Pulos. *NOAA Tech.Rep.NMFS* 8:99-103.

Raju, G. 1964a. Studies on the spawning of the oceanic skipjack *Katsuwonus pelamis* (Linnaeus) in Minicoy waters. <u>In</u> Proceedings of the Symposium on Scombrid Fishes. *Symp.Ser.Mar.Biol.Assoc.India*, 1:744-68.

Raju, G. 1964b. Fecundity of the oceanic skipjack *Katsuwonus pelamis* (Linnaeus) of Minicoy. *In* Proceedings of the Symposium on Scombrid Fishes. *Symp.Ser.Mar.Biol.Assoc.India* 1:725-32.

Richards, W.J., and G.R. Dove. 1971. Internal development of young tunas in the genera *Katsuwonus, Euthynnus, Auxis,* and *Thunnus* (Pisces, Scombridae). *Copeia* 1971:72-8.

Richards, W.J., and D.C. Simmons. 1971. Distribution of tuna larvae (Pisces, Scombridae) in the northwestern Gulf of Guinea and off Sierra Leone. *Fish.Bull.NOAA-NMFS*, 69(3):555-68.

Richardson, B.J. 1978. Skipjack tuna stock identification. *Occas.Publ.Fish.Res. Div.New Zealand Minist.Agri.Fish.* (15):63-4.

Richardson, B.J. 1983. Distribution of protein variation in skipjack tuna (*Katsuwonus pelamis*) from the central and south-western Pacific. *Aust.J.Mar.Freshwat.Res.*, 34(2):231-51.

Richardson, B.J., and G. Habib. 1987. A genetic study of the origins and structuring of the skipjack tuna population exploited by the New Zealand fishery. *New Zealand J.Mar.Freshwat.Res.*, 21(1):109-16.

Robins, J.P. 1952. Further observations on the distribution of striped tuna, *Katsuwonus pelamis* L., in eastern Australian waters, and its relation to surface temperature. *Aust.J.Mar.Freshwat.Res.*, 3(2):101-10.

Robson, D.S., and D.G. Chapman. 1961. Catch curves and mortality rates. *Trans.Am. Fish.Soc.*, 90(2):181-9.

Ronquillo, I.A. 1963. A contribution to the biology of Philippine tunas. *FAO Fish. Rep.*, 6(3):1683-752.

Rothschild, B.J. 1963. Skipjack ecology. *In* Progress in 1961-62, edited by W.G. Van Campen, *Circ.U.S.Fish.Wildl.Serv.*, 163:35 p.

Rothschild, B.J. 1965. Hypothesis on the origin of exploited skipjack tuna (*Katsuwonus pelamis*) in the eastern and central Pacific Ocean. *Spec.Sci.Rep.U.S.Fish Wild.Serv. (Fish.)*, (512):20 p.

Rothschild, B.J. 1967. Estimates of growth of skipjack tuna (*Katsuwonus pelamis*) in the Hawaiian Islands. *Proc.Indo-Pac.Fish.Coun.*, 12(2):100-11.

Seckel, G.R. 1972. Hawaiian-caught skipjack tuna and their physical environment. *Fish.Bull.NOAA-NMFS*, 72(3):763-87.

Shabotiniets, E.I. 1968. Opreedelenie vozraste tuntsov Indiiskogo Okeana (Age determination of Indian Ocean tunas). *Trudy VNIRO*, 64; *Trudy AzcherNIRO*, 28: 374-376 [Engl.transl. by W.L. Klawe, *I-ATTC*:5 p.].

Schaefer, K.M. 1987. Reproductive biology of black skipjack, *Euthynnus lineatus*, an eastern Pacific tuna. *Bull.I-ATTC*, 19(2):169-260.

Schaefer, M.B. 1961. Report on the investigations of the Inter-American Tropical Tuna Commission for the year 1960. *Annu.Rep.I-ATTC*, (1960):40-183.

Schaefer, M.B. 1963. Report on the investigations of the Inter-American Tropical Tuna Commission for the year 1962. *Annu.Rep.I-ATTC*, (1962):35-149.

Schaefer, M.B., and C.J. Orange. 1956. Studies on the sexual development and spawning of yellowfin tuna (*Neothunnus macropterus*) and skipjack (*Katsuwonus pelamis*) in three areas of the eastern Pacific Ocean, by examination of gonads. *Bull.I-ATTC*, 1(6):281-349.

Schaefer, M.B., B.M. Chatwin, and G.C. Broadhead. 1961. Tagging and recovery of tropical tunas, 1955-1959. *Bull.I-ATTC*, 5(5):341-455.

Sharp, G.D. 1978. Behavioral and physiological properties of tuna and their effects on vulnerability to fishing gear. *In* The physiological ecology of tunas, edited by G.D. Sharp and A.E. Dizon. New York, Academic Press, pp. 397-449.

Sibert, J.R. 1984. A two-fishery tag attrition model for the analysis of mortality, recruitment and fishery interaction. *Tech.Rep.Tuna Billfish Assess.Programme, S.Pac.Comm.*, (13):27 p.

Sibert, J.R., R.E. Kearney, and T.A. Lawson. 1983. Variation in growth increments of tagged skipjack (*Katsuwonus pelamis*). *Tech.Rep.Tuna Billfish Assess.Programme, S.Pac.Comm.*, (10):22 p.

Simmons, D.C. 1969. Maturity and spawning of skipjack tuna (*Katsuwonus pelamis*) in the Atlantic Ocean, with comments on nematode infestation of the ovaries. *Spec.Sci.Rep.U.S.Fish Wildl.Serv. (Fish.)*, (580):17 p.

South Pacific Commission. 1980. Review of preliminary results from genetic analysis of skipjack blood samples collected by the Skipjack and Assessment Programme. *Tech.Rep.Tuna Billfish Assess.Programme, S.Pac.Comm.*, (1):22 p.

South Pacific Commission. 1981a. Report of the second Skipjack Survey and Assessment Programme workshop to review results from genetic analysis of skipjack blood samples. *Tech.Rep.Tuna Billfish Assess.Programme, S.Pac.Comm.*, (6):39 p.

South Pacific Commission. 1981b. Skipjack migration, mortality and fishery interactions. Paper presented at the 13th Regional Technical Meeting on Fisheries, S. Pac. Comm., 24-28 August 1981, W.P.9:35 p.

South Pacific Commission. 1983. An assessment of the tuna and billfish resources of Papua New Guinea. *Final Ctry.Rep.Skipjack Surv.Assess.Programme, S.Pac.Comm.*, Number 12.

South Pacific Commission. 1988. Interaction among South Pacific skipjack fisheries. Paper presented at the 20th Regional Technical Meeting on Fisheries, S. Pac. Comm., 1-5 August 1988, Noumea, New Caledonia, I.P.4.

South Pacific Commission. 1990. Collaborative study between SPC and Japan National Research Institute for Far Seas Fisheries: General Report. Paper presented at the Third Standing Committee Meeting on Tuna and Billfish, S. Pac. Comm., Noumea, New Caledonia, W.P.9.

South Pacific Commission. 1991. Status of tuna fisheries in the SPC area during 1990, with annual catches since 1952. Paper presented at the Fourth Standing Committee Meeting on Tuna and Billfish, S. Pac. Comm., 17-19 June 1991, Port Vila, W.P.3.

South Pacific Commission. 1992. Solomon Islands country report 1992. *Ctry.Rep.Tuna Billfish Assess.Programme, S.Pac.Comm.*, (5):81 p.

Sosa-Nishizaki, O., M. Shimizu, and Y. Nose. 1989. The potential use of second dorsal fin rays of skipjack tuna *Katsuwonus pelamis* as an aging character. *Nippon Suisan Gakkaishi*, 55(9):1559-64.

Stequert, B. 1976. Etude de la maturité sexualle, de la ponte et de la fécondité du listao (*Katsuwonus pelamis*) de la côté nord-ouest de Madagascar. *Cah.ORSTOM(Sér.Océanogr.)*, 14:227-47.

Strasburg, D.W. 1960. Estimates of larval tuna abundance in the central Pacific. *Fish.Bull.U.S.Fish Wildl.Serv.*, (167):231-55.

Strasburg, D.W., E.C. Jones, and R.T.B. Iversen. 1968. Use of a small submarine for biological and oceanographic research. *J.Cons.CIEM*, 31(3):410-26.

Suzuki, Z., P.K. Tomlinson and M. Honma. 1978. Population structure of Pacific yellowfin tuna. *Bull.I-ATTC*, 17(5):273-441.

Tester, A.L., and E.L. Nakamura. 1957. Catch rate, size, and sex of tunas and other pelagic fishes taken by trolling off Oahu, Hawaii, 1951-55. *Spec.Sci.Rep.U.S.Fish Wildl.Serv. (Fish.)*, (250):25 p.

Toole, J.M., E. Zou and R.C. Millard. 1988. On the circulation of the upper layers in the western equatorial Pacific Ocean. *Deep-Sea Res.*, 35(7A):1451-82.

Uchiyama, J.H., and P. Struhsaker. 1981. Age and growth of skipjack tuna, *Katsuwonus pelamis*, and yellowfin tuna, *Thunnus albacares*, as indicated by daily growth increments of sagittae. *Fish.Bull.NOAA-NMFS*, 79(1):151-62.

Ueyanagi, S. 1969. Observations on the distribution of tuna larvae in the Indo-Pacific Ocean with emphasis on the delineation of the spawning areas of albacore, *Thunnus alalunga*. *Bull.Far Seas Fish.Res.Lab.*, (2):177-256.

Ueyanagi, S., K. Mori, and Y. Nishikawa. 1969. Research on distribution of larvae. *Ser.Far Seas Fish.Res.Lab.*, 1:12-7.

Ueyanagi, S., K. Mori, Y. Nishikawa, and A. Suda. 1973. Report on experiments on the development of tuna culturing techiques (April, 1970-March, 1973). *S Ser.Far Seas Fish.Res.Lab.*, 8:165 p.

de Vlaming, V.L. 1982. On the use of the gonosomatic index. *Comp.Biochem.Physiol.*, 73A (1):31-9.

Vooren, C.M. 1976. Biological data on skipjack in New Zealand waters, 1973-76. *In* Proceedings of the Skipjack Tuna Conference, July 1976. *Occas.Publ.Fish.Res.Div.New Zealand Minist.Agri.Fish.*, 11:12-6.

Wade, C.B. 1950. Observations on the spawning of Philippine tuna. *Fish.Bull.U.S.Fish Wildl.Serv.*, 51(55):409-23.

Wild, A. 1986. Growth of Yellowfin tuna, *Thunnus albacares*, in the eastern Pacific Ocean based on otolith increments. *Bull.I-ATTC*, 18(6):423-82.

Wild, A., and T.J. Foreman. 1980. The relationship between otolith increments and time for yellowfin and skipjack tunas marked with tetracycline. *Bull.I-ATTC*, 17(7):507-60.

Williams, F. 1970. Sea surface temperature and the distribution and apparent abundance of skipjack (*Katsuwonus pelamis*) in the eastern Pacific Ocean, 1951-1968. *Bull.I-ATTC*, 15(2):229-81.

Williams, F. 1972. Consideration of three proposed models of the migration of young skipjack tuna (*Katsuwonus pelamis*) into the eastern Pacific Ocean. *Fish.Bull.NOAA-NMFS*, 70(3):741-62.

Wyrtki, K. 1965. Surface currents of the eastern tropical Pacific Ocean. *Bull.I-ATTC*, 9(5):269-304

Wyrtki, K. 1967. Circulation and water masses in the eastern equatorial Pacific Ocean. *J.Int.Oceanol.Limnol.*, 1(2):117-47.

Wyrtki, K., E. Firing, D. Halpern, R. Knox, G.J. McNally, W.C. Patzert, E.D. Stroup, B.A. Taft, and R. Williams. 1981. The Hawaii to Tahiti shuttle experiment. *Science* 211(4477):22-8.

Yabe, H. 1954. A study on spawning of skipjack in the Satsunan Sea area. *In* General view of fishery science, Tokyo. *Jap.Assoc.Adv.Sci.*: 182-199 [Engl. transl. by G.Y. Beard, 1959: 9p].

Yamanaka, H., Y. Kurohiji, and J. Morita. 1966. General results of the investigation in the south western Pacific Ocean by the fish-finder. *Rep.Nankai Reg.Fish.Res.Lab.*, (24):115-27.

Yamanaka, K.L. 1990. Age, growth and spawning of yellowfin tuna in the southern Philippines. *Indo-Pac.Tuna Dev.Mgt. Programme,* IPTP/90/WP/21:87 p.

Yao, M. 1981. Growth of skipjack tuna in the western Pacific Ocean. *Bull.Tohoku Reg.Fish.Res.Lab.*, (43):71-82.

Yoshida, H.O. 1960. Marquesas area fishery and environmental data, January -March 1959. *Spec.Sci.Rep.U.S.Fish Wildl.Serv. (Fish.)*, (348):37 p.

Yoshida, H.O. 1966. Skipjack tuna spawning in the Marquesas Islands and Tuamotu Archipelago. *Fish.Bull.U.S.Fish Wildl.Serv.*, 65(2):479-88.

Yoshida, H.O. 1971. The early life history of skipjack tuna, *Katsuwonus pelamis*, in the Pacific Ocean. *Fish.Bull.NOAA-NMFS*, 69(3):545-54.

Yuen, H.S.H. 1963. Schooling behavior within aggregations composed of yellowfin and skipjack tuna. *FAO Fish.Rep.*, 6(3):1419-29.

A REVIEW OF THE BIOLOGY AND FISHERIES FOR YELLOWFIN TUNA, *THUNNUS ALBACARES*, IN THE EASTERN PACIFIC OCEAN

Alex Wild
Inter-American Tropical Tuna Commission
La Jolla, California

1.   INTRODUCTION

This paper is a review of the biology, resource and fisheries associated with yellowfin tuna, *Thunnus albacares*, in the eastern Pacific Ocean. Under certain headings, such as Early Life History and Age and Growth, information from the central and western Pacific and the eastern Atlantic has also been included to offer a wider perspective. The material in the section on Gear Interaction is intended to be both a general review and an interaction-specific review for the purse-seine and longline gears operating in the eastern Pacific. The task of compiling the material was simplified considerably through the extensive use of two major sources of information: the synopsis of biological data on yellowfin by Cole (1980), and the many anonymous contributions made to the Inter-American Tropical Tuna Commission's (IATTC's) Annual Reports.

2.   CLASSIFICATION

Among fisheries biologists it is generally accepted that the expression "tunas and related species" applies to all genera of the family Scombridae, except *Rastrelliger* and *Scomber*, and to all genera of the families Istiophoridae and Xiphiidae (Klawe, 1980). The mackerels, *Rastrelliger* and *Scomber*, are excluded because of their mode of life. The nature of the fisheries directed toward mackerels, and the marketing of their catches, differ considerably from those of the tunas and related species. The listing below identifies the position of yellowfin tuna within the family Scombridae, following Lindberg's (1971) scheme to the family level:

Phylum Chordata
   Subphylum Vertebrata
      Superclass Gnathostomata
         Class Teleostomi
            Subclass Actinopterygii
               Order Perciformes
                  Suborder Scombroidei
                     Family Scombridae
                        Sub-family Scombrinae
                           Tribe Thunnini
                              Genus *Thunnus*
                                 Species *albacares*

The classification of yellowfin and other tunas is discussed by Gibbs and Collette (1967), Sharp and Pirages (1978), and Collette and Nauen (1983).

3. EARLY LIFE HISTORY

The distribution of yellowfin eggs has not been determined because it is not possible to differentiate them from those of many other scombrids (Cole, 1980).

The development from the pre-larval to the post-larval phase of yellowfin hatched from artificially fertilized eggs has been described by Harada *et al.* (1971), Mori *et al.* (1971) and Harada *et al.* (1980). Descriptions of larval and post-larval yellowfin collected at sea are given by Mead (1951), Wade (1951), Matsumoto (1958, 1962), Sun' (1960), Ueyanagi (1966, 1969) and Matsumoto *et al.* (1972). The internal anatomy of yellowfin larvae has been described by Richards and Dove (1971). The juvenile forms have been described by Schaefer and Marr (1948), Wade (1950a), Yabe *et al.* (1958) and Matsumoto (1961). Guides to the identification of early life history stages are offered by Fahay (1983), Collette *et al.* (1984), Nishikawa and Rimmer (1987) and Richards (1989). Yellowfin larvae in the total length (TL) range of 3-10 mm can be distinguished from those of bigeye tuna (*T. obesus*) by the presence in bigeye of a single melanophore on the ventral tail (Richards *et al.*, 1990). The distinction becomes obscured during transformation from larvae to juveniles because both species may then display the melanophore. Graves *et al.* (1988), however, successfully applied an electrophoretic technique to an apparently mixed-species sample and identified all of the small (10-12 mm) juveniles as yellowfin.

Yellowfin larvae distribution has been examined by Wade (1951), Sun' (1960), Yabe and Ueyanagi (1962), Ueyanagi *et al.* (1969) and Mori (1970) for the western Pacific; by Matsumoto (1958), Strasburg (1960), Sun' (1960) and Nakamura and Matsumoto (1967) for the central Pacific; and by Mead (1951), Klawe (1963) and Klawe *et al.* (1970) for the eastern Pacific. Much of the information given by these investigators has been incorporated into the works of Yabe *et al.* (1963), Matsumoto (1966), Ueyanagi (1969), Nishikawa *et al.* (1985) and Suzuki *et al.* (1978) in investigations of the distribution of yellowfin larvae in the entire Pacific Ocean. These studies indicate that the larvae are trans-Pacific in occurrence, although their distribution is limited latitudinally to tropical and subtropical waters. Larvae occur year-round in equatorial waters, but there is a seasonal change in density in the subtropical waters of the central and western Pacific. Seasonal peaks in density of larvae occur in the Kuroshio Current area during May to June and in the East Australian Current during November to December. In the eastern Pacific the range of yellowfin larvae is compressed somewhat during the northern winter by cold water converging toward the equator from the north and south. Although data are not available for an entire year, on a quarterly basis there appears to be a peak in larval density from April to June off Central America (Figure 1).

Klawe (1963) examined the vertical distribution of yellowfin larvae in the eastern Pacific. He found no evidence of occurrence of larvae below the thermocline. Information on the vertical distribution of yellowfin larvae in the central Pacific has been provided by Matsumoto (1958) and Strasburg (1960), and in the Indo-Pacific Ocean by Ueyanagi (1969). These investigators indicated that yellowfin larvae are probably restricted to the upper 50 or 60 metres of the ocean. Richards and Simmons (1971) found that in the northwestern Gulf of Guinea and off Sierra Leone, yellowfin and bigeye tuna larvae migrate to the surface during the day, while the larvae of skipjack tuna (*Katsuwonus pelamis*) migrate to the surface at night. As an aside, Davis *et al.* (1990)

reported that skipjack larvae behave similarly in the east Indian Ocean, whereas southern bluefin (*T. maccoyii*) and albacore (*T. alalunga*) larvae migrate to the surface layer in the daytime.

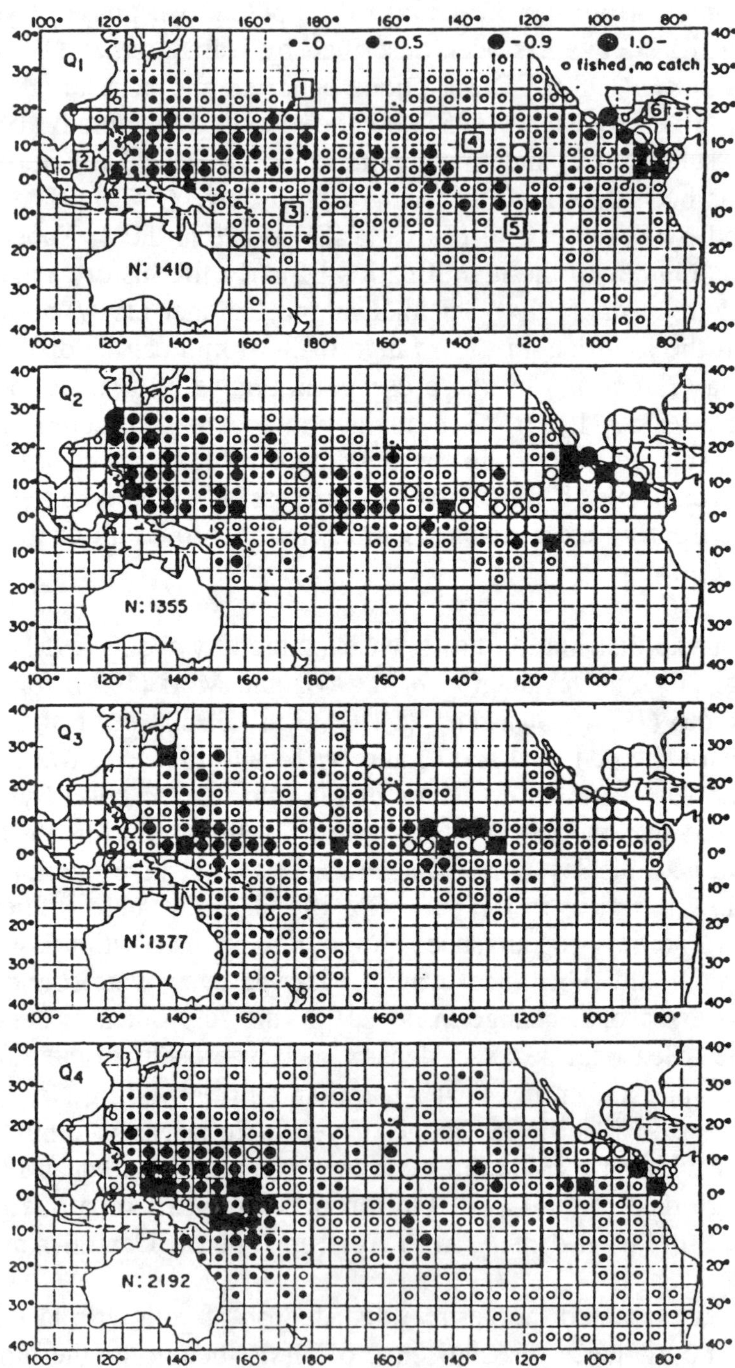

Figure 1. Quarterly density distributions of larval yellowfin sampled by surface horizontal tows. Solid and open circles denote the densities in 5-degree areas calculated from five or more tows and less than five tows respectively. N indicates the number of nominal tows and the numerals represent areas (shown surrounded by heavy lines) for examining seasonal changes of density in them. (Reproduced from Suzuki *et al.*, 1978).

Higgins (1967) summarized the published accounts on the capture of juvenile yellowfin. He reported that they have been collected in the western Pacific as far north as approximately 31°N, near the coast of Japan, and as far south as 23°S. In the central Pacific the northernmost record is 23°N, near the Hawaiian Islands, and the southernmost record is 23°S. In the eastern Pacific juvenile yellowfin have been recorded from approximately 24°N, off Baja California, to approximately 2°S, off the coast of Ecuador. Higgins (1967) found no records of juvenile yellowfin in the area bounded by 150°W and 112°W, but he attributed this hiatus to insufficient sampling in this area. Higgins (1970) also examined the distribution of juvenile tunas in Hawaiian waters from July to September, 1967. Although skipjack was the dominant species encountered, he found that juvenile yellowfin were more abundant offshore than inshore, and that they exhibited no extensive vertical migrations.

4. AGE AND GROWTH

4.1 Age Determination and Growth

The history of age determination of Pacific tunas has been summarized by Hayashi (1957), Bell (1964), Shomura (1966) and Suzuki (1971). The benefits and limitations associated with the scale-reading method were reviewed by Suzuki (1974). At first, the principal methods of estimating age relied on the growth marks in hard parts, such as vertebrae (Aikawa and Kato, 1938) and scales (Nose *et al.*, 1957; Yabuta *et al.*, 1960; Yang *et al.*, 1969), but a consistent and accurate interpretation of age was hampered by the lack of an independent method of verification. Emphasis was therefore directed to the assessment of growth by means of size-frequency modal analysis.

In the eastern Pacific (Hennemuth, 1961a; Davidoff, 1963; Díaz, 1963), eastern Atlantic (Le Guen and Sakagawa, 1973) and western Pacific (Wankowski, 1981) the analyses were based on fork-length frequencies, while Moore (1951) in the central Pacific utilized weight frequencies. The resulting growth curves were affected to some degree by the uncertainties generated by spawning periods of extensive duration, and by the need to anchor the growth curves to the time axis by a suitable choice of age for a particular size. In the eastern Pacific the results were also influenced by the relative lack of large fish (> 100 cm fork length) in the catches during the late 1950s and early 1960s, and by the suspicion of gear selectivity for sizes < 70 cm. Despite these difficulties, there was a gradual convergence of size-at-age estimates prepared by the scale and modal progression methods (Suzuki, 1971).

Mark-recapture experiments involving tetracycline injection at sea demonstrated that eastern Pacific yellowfin north of the equator deposit growth increments daily on their otoliths in the thawed fork-length range from 40 to 110 cm (Wild and Foreman, 1980). Based on this result, the following growth equation was developed for combined sexes over the length (l_t) range of 30 to 168 cm,

$$l_t = 188.2 \ (1.0 + 0.434.\exp[-0.724\{t(\mathrm{yr})-1.825\}])^{-2.30} \tag{1}$$

by estimating the ages in days of 196 yellowfin sampled from 1977 through 1979 (Wild, 1986). To avoid the effects of size selection and differential mortality, each 10-cm interval of the growth curve contained about 15 items. Additional tetracycline, mark-recapture experiments have since extended the validated range of daily deposition to 148 cm (IATTC, 1988) and confirmed about 65 percent of the growth curve from 0-180 cm, *i.e.* from 30-148 cm. Circumstantial evidence suggests that the ratio of increments to days is also 1:1 in the range from 0 to 30 cm and from 148 to 168 cm, and the estimates of ages are therefore believed to be accurate. In the above equation, weighted regressions identified the Richards (1959) function as providing the best statistical fit to the data from the following choice of models:

$$l_t = L_\infty(1\text{-exp}[-K(t\text{-}t_o)])$$ von Bertalanffy (1938)

$$l_t = L_\infty\text{exp}(\text{exp}[-K(t\text{-}t_*)])$$ Gompertz (Ricker, 1979)

$$l_t = L_\infty(1\text{-}(1\text{-}m)\text{exp}[-K(t\text{-}t_*)])^{1/(1\text{-}m)}$$ Richards (1959; $m \neq 0$ or 1)

where l_t is the fork length at time t; L_∞ is the asymptotic size; K is a growth parameter; t_o is the time-axis intercept; t_* is the inflection point of the curve; and m is a shape parameter related to the ratio of l_t/L_∞ at t_*. Either the Gompertz or the Richards model provided a significantly improved fit to the growth data of Moore (1951), Hennemuth (1961a), Davidoff (1963) and Le Guen and Sakagawa (1973), once the constraint of the non-inflective von Bertalanffy model was removed. The comparative growth curves shown in Figure 2 include these adjustments. The growth equation parameters and estimates of sizes at different ages appear in Table 1. Counts of otolith increments were also used by Uchiyama and Struhsaker (1981) to estimate the ages of 14 central Pacific yellowfin. Over the length range from 6 to 93 cm, growth was best described by two linear stanzas. Draganik and Pelczarski (1984) also estimated the growth of central Atlantic yellowfin by assuming that the annuli in sectioned dorsal fin rays were equivalent to years.

Sexual dimorphism has been detected in eastern Pacific yellowfin in terms of growth in length, weight and the length of the counting path on the otolith (Wild, 1986). After about 95 cm in length, the growth rate of females becomes progressively slower than that of males. Although these sexual differences are statistically significant, they are small and may be unimportant in a practical sense. Growth for individual sexes was best described by the Gompertz model, and the equation parameters for length appear in Table 1, together with Le Guen and Sakagawa's (1973) estimates for western Pacific males and females calculated from the data of Yabuta *et al.* (1960). For combined sexes the annual variation in growth in length was significant (Wild, 1986), and confirmed earlier reports of annual variation by Hennemuth (1961a), Davidoff (1963), and Le Guen and Sakagawa (1973). In the length range from 30 to 110 cm, yellowfin caught inshore were significantly heavier than those caught offshore, but the situation was reversed for fish longer than 110 cm (Wild, 1986). This result suggests that there may be size-related changes in the diet of yellowfin that are associated with feeding opportunities in the two areas. In the same study, the following equations describe growth in weight (w) for combined sexes of thawed, whole fish;

$$w\ [\text{kg}] = 178.4\ \text{exp}(\text{-exp}[-0.555(t\ [\text{yr}]\text{-}3.638)])\ (\text{range: } 0.45\text{-}4.76\ \text{yr}) \tag{2}$$

and the relationship between length and weight for combined sexes in the inshore-offshore regions;

$$\ln(w) \text{ [kg]} = -11.1830 + 3.086 \ln(l) \text{ [cm]} \quad \text{(range: 30-168 cm)} \tag{3}$$

Although this last equation is applicable over a broad length range, it is based only on 196 fish. In addition, the weights of small fish (30-40 cm) tend to be overestimated somewhat due to a slight, but insignificant, degree of nonlinearity. On the other hand, the equation developed by Chatwin (1959) for pooled data in the eastern Pacific,

$$\ln(w) \text{ [kg]} = -10.8980 + 3.020 \ln(l) \text{ [cm]} \quad \text{(range: 48-115 cm)} \tag{4}$$

is applicable over a much narrower length range, and it is also weighted toward the dominant sizes of yellowfin (50-85 cm) available to the baitboat fishery during 1956-57. Although significant differences existed among relationships for different areas, they were not considered biologically important at the time. After a conversion to natural logarithms, the length-weight equation reported by Nakamura and Uchiyama (1966) for central Pacific yellowfin, *i.e.*,

$$\ln(w) \text{ [kg]} = -11.1230 + 3.058 \ln(l) \text{ [cm]} \quad \text{(range: 70-180 cm)} \tag{5}$$

underestimates the weight in the eastern Pacific for a given length. Therefore the relationships for the two regions are probably distinct.

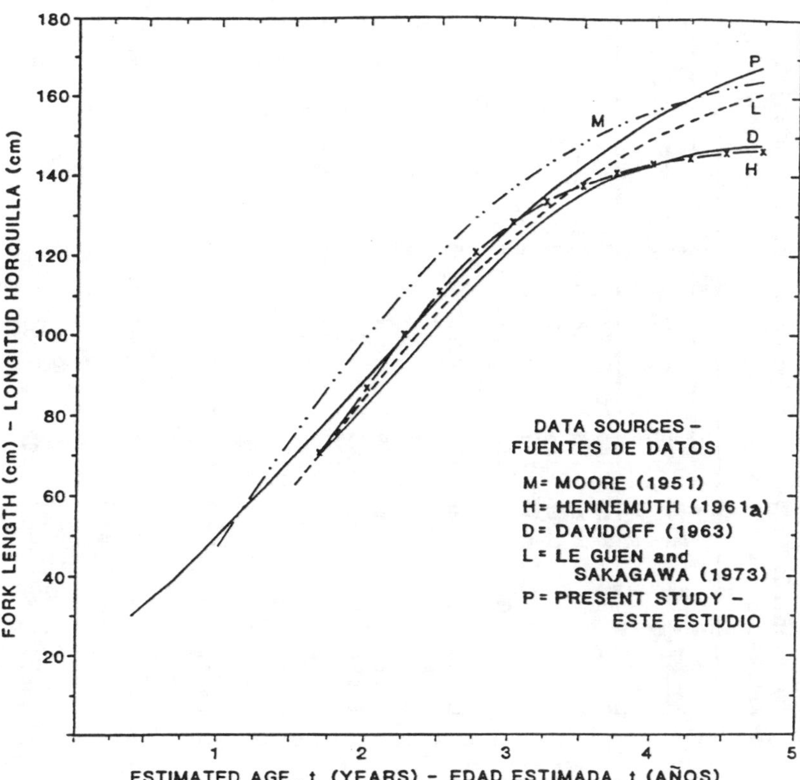

Figure 2.   The fork-length growth curve (P) of yellowfin tuna in the eastern Pacific based on otolith-increment age estimates compared to previous curves derived from size-frequency analysis. (Reproduced from Wild, 1986.)

TABLE 1. Comparison of growth equation parameters and estimated sizes at age for yellowfin tuna for different oceanic regions and by different investigators. Legend: Eq. = equation (see section on Age and Growth in text for growth models); R = Richards (1959); G = Gompertz (Ricker, 1979); B = von Bertalanffy (1938); L = linear segments; f = fixed parameter. Bracketed quantities represent extrapolations slightly beyond the range of the data.

Region	Eq.	Sex	Parameters					Estimated size at age					Range	Data type	Sources
			L_∞ cm	K annual	t* year	m	t_o year	1	2	3	4	5	cm		
East.	R	both	148.0f	1.720	2.000	2.903		-	85	127	143	(147)	70-146	Length modes	Hennemuth (1961a)
Pacif.	R	both	149.0f	1.888	2.294	4.111	-	-	82	123	143	148	70-148	" "	Davidoff (1963)
	R	both	188.2	0.724	1.825	1.434		49	89	127	154	(171)	30-168	Otoliths	Wild (1986)
	G	male	194.7	0.617	1.585			(46)	90	128	155	(172)	50-168	" "	" "
	G	female	184.2	0.591	1.427			(51)	90	124	(148)		54-142	" "	" "
Cent.	G	both	172.7	0.857	1.308			47	99	137	156	165	47-168	Length modes	Moore (1951)
Pacif.	L	both						53	90				52- 93	Otoliths	Uchiyama and Struhsaker (1981)
West.	B	both	190.0	0.33			0.	-	84	111	131	(146)	70-140	Scales	Yabuta et al. (1960)
Pacif.	B	both	195.2	0.36			0.27	-	90	122	(144)		60-139	Scales	Yang et al. (1969)
	B	both	180.9	0.292			0.	46	80	(105)			30- 96	Length modes	Wankowski (1981)
	B	male	202.1	0.276			0.	-	86	114			58-119	Scales	From Yabuta et al.
	B	female	174.9	0.372			0.	-	92	118			57-119	"	(1960) by Le Guen and Sakagawa (1973)
East.	G	both	176.9	0.733	1.590			-	84	124	149	163	62- 165	Length modes	Le Guen and Sakagawa (1973)
Atl.	L	both						49	67				6-24 mo.	" "	Fonteneau (1980)
	B	"	166.4	0.864			1.292			128	150	160	30-84 mo.		
	L	both						53					35-65	" "	Bard (1983)
	B	"	196.5	0.474			0.847		83	126	152	169	65-180		
Cent.	B	both	192.4	0.37			-0.003	-	(101)	129	149	162	96-185	Dorsal fin rays	Draganik and Pelczarski (1984)
Atl.															

4.2 Growth Rates

The growth rates of yellowfin in the eastern Pacific have also been estimated by means of tag-recapture experiments conducted from 1955 through 1981 (Bayliff, 1988).

The rates for particular length groups appear to have annual and areal components, but they are not persistent, and it is therefore more informative to mention general trends. Overall, the growth rate ($\pm$ standard error) of tagged yellowfin in the length range from 25 to 100 cm is about 0.85 $\pm$ .01 mm/day. Growth appears to be most rapid near the Revillagigedo Islands (1.11 $\pm$ .03 mm/day) and slower off Baja California (0.69 $\pm$ .02 mm/day), Central America-Colombia (0.69 $\pm$ .02 mm/day) and Ecuador-Peru (0.52 $\pm$ .10 mm/day). Over the length range from 25 to 100 cm, growth for tagged fish was nearly linear and parameters for curvilinear models could not be calculated. In contrast to growth rates estimated from modal analysis for small yellowfin in the eastern Atlantic (40-70 cm: Fonteneau, 1980; 35-65 cm: Bard, 1984), the southwestern Pacific (30-50 cm; Brouard *et al.*, 1984) and the western Indian Ocean (Marsac and Lablanche, 1985), juvenile yellowfin in the eastern Pacific do not exhibit a period of slow or reduced growth rate, *i.e.* $\leq$ 0.5 mm/day. By means of otolith increments Yamanaka (1990) also demonstrated that yellowfin near the southern Philippines grow rapidly, 2.5 mm/day, in the length range of 15-55 cm, and 0.96 mm/day in the interval from 55-79 cm.

To facilitate comparisons of growth rates obtained from tagging, otolith increment counts and modal progressions, the equation (1) describing growth of eastern Pacific yellowfin for combined sexes (Wild, 1986) was differentiated to yield instantaneous rates and plotted versus age and length (Figure 3). In general, for fish caught north of the equator, the average rate for the untagged yellowfin used in the otolith study was $\geq$ 1 mm/day, or $\geq$ 3.04 cm/mo., during growth in the length range from 50 to 115 cm. This rate compares favourably with the average value of 1.00 mm/day derived from the measured change in otolith sizes of tagged and tetracycline-injected fish (Wild and Foreman, 1980). Bayliff (1988) reports that for yellowfin $\geq$ 50 cm tagged north of the equator, the average growth rate was 0.85 mm/day based on the change in fork length.

5.   MATURATION, SPAWNING AND SEX RATIO

5.1   <u>Maturation</u>

Wade (1950b) found a ripe male 525 mm in length near the Philippine Islands. Buñag (1956) encountered a mature female of a similar size (567 mm) in the same area, and also described seven stages of maturation, including the resting stage, based on the diameter of ova preserved in formalin. On the other hand, by means of gonado-somatic indices, Yuen and June (1957) found that in the central equatorial Pacific a few yellowfin reach maturity at about 70-80 cm, but the majority do not mature until they reach 120 cm. Similarly, Kikawa (1962) reported that female yellowfin attain first maturity at more than 110 cm in the western and central Pacific longline grounds, although a few individuals were found to be mature at lengths between 80 to 110 cm. In the eastern Pacific longline grounds east of 130°W, Shingu *et al.* (1974) found the minimum size at first spawning for both sexes to be between 91 and 100 cm; however, their data also indicated that a high percentage of fish reached first maturity at lengths above 120 cm. Orange (1961) found the minimum size of mature yellowfin along the coast of Central America to be 50 cm; in this area 20 percent of the females in the 50-60 cm size class were found to be mature. More recently, however, the length at sexual maturity in two regions of the eastern Pacific was assessed using histological criteria, the mean diameter of oöcytes in the most advanced mode, and microscopic detection of residual hyaline oöcytes (IATTC, 1990). In the areas bound by 20°-30°N, 110°-120°W, and by

0°-10°N, 80°-90°W, the smallest female found with mature ovaries was 84 cm, and the estimated length at 50-percent maturity was 95 cm.

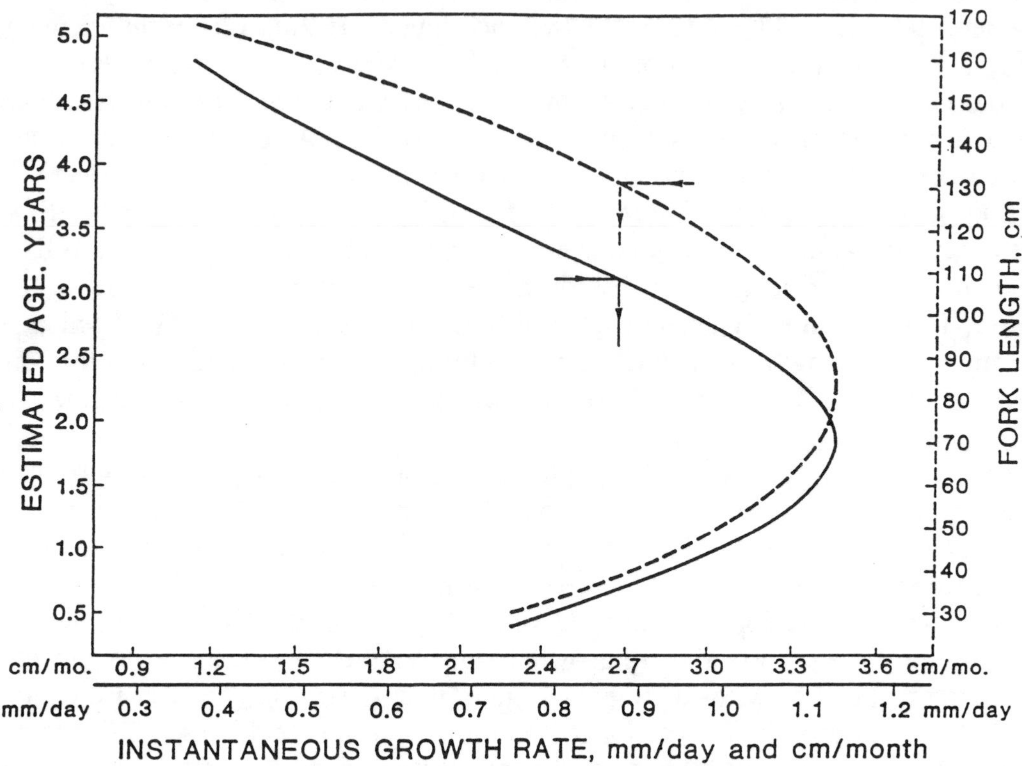

Figure 3. The instantaneous growth rate of yellowfin tuna in the eastern Pacific relative to the fork length or estimated age of the fish, derived from the growth curve based on otolith-increment age estimates by Wild (1986). The solid curve is used to estimate the instantaneous growth rate from the estimated-age axis, and the dashed curve is used to estimate the growth rate from the fork-length axis.

Kikawa (1966) postulated that estimates of the spawning potential of small (< 100 cm) yellowfin derived from longline-caught fish in open areas of the western Pacific were likely to be low. Suzuki *et al.* (1978) pointed out that such may also be the case in the eastern Pacific. Significantly higher percentages of sexually mature yellowfin under 120 cm in length were found in samples taken from purse-seine catches than in samples gathered from longline catches made in the same area and month. Hisada (1973) also reported differences in sexual maturity of yellowfin sampled from surface hand-line and longline catches in the Coral Sea. These results seem to imply that each gear type by itself may be an inappropriate sampling device for maturation studies. It is possible that a re-evaluation of the spawning potential of smaller size groups, as suggested by Kikawa (1966) and Suzuki *et al.* (1978), will show that a greater proportion of fish under 120 cm in open waters of the Pacific are mature than was previously thought.

Kikawa (1966) calculated the following equations for the relationships between body length and ovary weight (both ovaries combined) for yellowfin in the tropical central and western Pacific:

$$< 160 \text{ cm} \quad Y = (9.596 \times 10\text{-}5)X^{3.436}$$
$$> 160 \text{ cm} \quad Y = 2760 + 5.318X$$

where Y is ovary weight (gm), and X is fork length (cm). Fecundity studies have been carried out by June (1953) and Joseph (1963) for yellowfin in the central and eastern Pacific, respectively. In the latter area, the relationship between the length and the number of maturing ova is expressed by:

$$Y = (8.955 \times 10\text{-}9)X^{2.791}$$

where Y is millions of ova in the most advanced mode, and X is the fork length (mm). For example, yellowfin that were 1,000, 1,300 or 1,500 mm long would produce approximately 2.1, 4.4, or 6.6 x 10^6 ova, respectively.

With the exception of Buñag's (1956) work cited above, the gonado-somatic index, or the ovary weight expressed as a function or percentage of body weight, has been used as the primary indicator of the state of reproductive activity in yellowfin. This index is biased, however, because larger females develop larger ovaries in proportion to body weight than smaller females (de Vlaming, 1982). In his analysis of black skipjack's (*Euthynnus lineatus*) reproductive biology, Schaefer (1987) stated that if the gonadal index was to be interpreted correctly with respect to reproductive activity, it should be validated with histology and/or oöcyte diameter and adjusted for the size of individuals. Such work is in progress by the IATTC regarding yellowfin in the eastern Pacific.

5.2   Spawning

Ueyanagi (1969, 1978) stated that 26°C is probably the lower limit for spawning of yellowfin. Spawning in the western and central regions of the Pacific Ocean takes place in northern latitudes during the spring and summer of the Northern Hemisphere (Table 2). Spawning takes place year-round in the northern equatorial waters of the western and central Pacific. In southern equatorial waters, however, yellowfin spawn mostly during the first half of the year, with minimal spawning during the second half because of the intrusion of water cooler than 26°C into the area. Off the coasts of Mexico and Central America yellowfin spawn throughout the year, but the peaks of spawning occur at different times in different areas. Spawning appears to be more sporadic and shorter in duration in the coastal spawning areas than in northern equatorial waters.

By identifying the recruitment cohorts present on the spawning grounds, Knudsen (1977) was able to show that yellowfin spawn at least twice a year off the coast of southern Mexico and Central America. Although samples were not available for all months, further offshore the spawning period is at least seven months and may occur all year. The periods of spawning vary in length and time of occurrence from year to year. In preparation for a more intensive study on yellowfin reproduction in the eastern Pacific, the details of which appear in IATTC (1989), Schaefer (1988) estimated that spawning near Clipperton Island occurs between 2000-2400 h. The interval between spawnings was about 1.27 days in samples collected in 1986 and 1.25 days in the 1987 samples.

5.3   Sex Ratio

Everett and Punsly (IATTC; pers. commun.) re-examined sex-ratio data collected in the eastern Pacific from the longline fishery during 1958-62, and from the surface fishery in 1953-62 and 1970-73. Departures from the expected ratio of 1:1 occurred as a result of year, area, gear and length-class effects, but on a quarterly basis within areas and years the data were homogeneous. Despite the complexity of the results, the rapid decline in the percentage of females around 140 cm was a consistent feature of all data sets. Possible explanations for this result are that females may experience differential mortality, growth and vulnerability to the fishery. During development of the sexually-dimorphic growth curves, Wild (1986) did not detect an accumulation of large females in the length range from 130-140 cm, a situation that would result from a cessation of female growth. He therefore attributed the virtual disappearance of females beyond 140 cm to an increase in their natural mortality rate.

TABLE 2. Spawning seasons of yellowfin in the Pacific Ocean. (Reproduced from Cole, 1980).

General ocean region	Approximate months spawning occurs	Investigators
Western Pacific		
North of 10°N to southern coast of Japan and 120°-170°E	April-July	Kikawa, 1962; Matsumoto, 1966; Mori, 1970; Suzuki et al., 1978
10°N-15°S and 120°E-180°	Year-round; peak months July-November	Wade, 1950a,; Shimada, 1951; Kikawa, 1959
Northeast coast of Australia and 10°-30°S to 180°	October-March; peak months November-February	Kikawa, 1959 and 1962; Legand 1960; Suzuki et al., 1978
Central Pacific		
Hawaiian Islands	May-September; peak June-August	June, 1953; Matsumoto, 1966
10°N-10°S and 180°-120°W	Year-round; peak months March-September	Yuen and June, 1957; Matsumoto 1966; Suzuki et al., 1978
15°-25°S and 150°-130°W	December-March	Kikawa, 1959
Eastern Pacific		
Revillagigedo Is., coast of Mexico and Central America nearshore	Year-round; each area with different peak months	Orange, 1961; Klawe, 1963; Knudsen, 1977
0°-10°N and 130°-90°W	Year-round; greatest activity during first half of year	Shingu et al., 1974; Knudsen 1977
0°-10°S and 130°-90°W	Principally January-June	Shingu et al., 1974

6. STOCK STRUCTURE, DISTRIBUTION, AND MIGRATION

6.1 Subpopulations and Stocks

Most of the inferences about the subpopulation structure, or self-sustaining genetic units, of yellowfin in the Pacific Ocean have come from indirect sources such as morphometric comparisons (Godsil, 1948; Godsil and Greenhood, 1951; Schaefer, 1952, 1955; Royce, 1953, 1964; Kurogane and Hiyama, 1957; Broadhead, 1959; Yang, 1971; Schaefer, 1989, 1991), length-frequency and catch-and-effort analysis (Yabuta *et al.*, 1960; Kamimura and Honma, 1963), tagging experiments (see below), spawning studies (Section 5.2), and other studies dealing with various aspects of the life history of yellowfin. A considerable amount of genetic research has been carried out in an attempt to develop a direct method to discriminate yellowfin subpopulations (Suzuki, 1962; Barrett and Tsuyuki, 1967; Sprague, 1967; Fujino and Kang, 1968a and 1968b; Fujino,

1970; IATTC, 1971-74 and 1976-80). Some of these studies have produced inferences contradictory to those obtained from the indirect sources; however, due to sampling difficulties, limited geographical coverage and the need for further systematic follow-up investigations, they are still considered to be preliminary in nature (Suzuki *et al.*, 1978; IATTC, 1980). Therefore, determination of the existence or non-existence of subpopulations of Pacific yellowfin has yet to be made.

On the basis of the available indirect data, however, Suzuki *et al.* (1978) drew the following conclusions with respect to population structure:

1) Although no definite geographical break in the continuity of yellowfin distribution across the Pacific is indicated, the extent of movement appears to be insufficient for there to be much intermingling among yellowfin of the eastern and central Pacific, and probably of the western and central Pacific as well.

2) Three stocks, which are more or less independent, are envisioned: a western Pacific stock which inhabits the area between about 120°E and 170°W; an eastern Pacific stock inhabiting an area corresponding to the IATTC's yellowfin regulatory area (CYRA) (Figure 4); and a central Pacific stock inhabiting the area between the western and eastern stocks. (A stock was defined to mean "... an exploitable subset of the population existing in a particular area and having some uniqueness relative to exploitation.")

3) The three stocks may well be composed of subpopulations which will have to be discriminated by more direct methods. Should the existence of subpopulations be established, however, it is possible that even direct genetic studies will not demarcate their spatio-temporal boundaries.

In his discussion of the contribution of life history data to the analysis of population structure, Bayliff (1983) noted that the average size of yellowfin within the CYRA was reduced by fishing, whereas this effect was hardly apparent in the area between the CYRA and 150°W. This observation suggested that yellowfin in the two areas may be relatively more independent than those in the central and western Pacific. However, more recent results (IATTC, 1991a) indicate that the apparent difference in averages may have been caused by the scarcity of data in each area due to the effects of regulated and unregulated periods of fishing. In the absence of regulations since 1980, the trends in the average sizes have become more similar as fishing effort in the two areas has become more evenly distributed in space and time. This subject is explored more fully in Section 12.

In a recent study involving discriminant analyses of morphometric characters, Schaefer (1989) concluded that eastern Pacific yellowfin sampled from north of 15°N-20°N were different from those sampled from south of 15°N-20°N. Somewhat distinct regional groups were suggested by the absence of any clinal relationship between the characters and latitude or longitude. On a larger oceanic scale, yellowfin samples from Mexico, Ecuador, Australia, Japan and Hawaii showed significant differences in gill-raker counts and morphometrics, suggesting that these different regions represent separate groups (Schaefer, 1991).

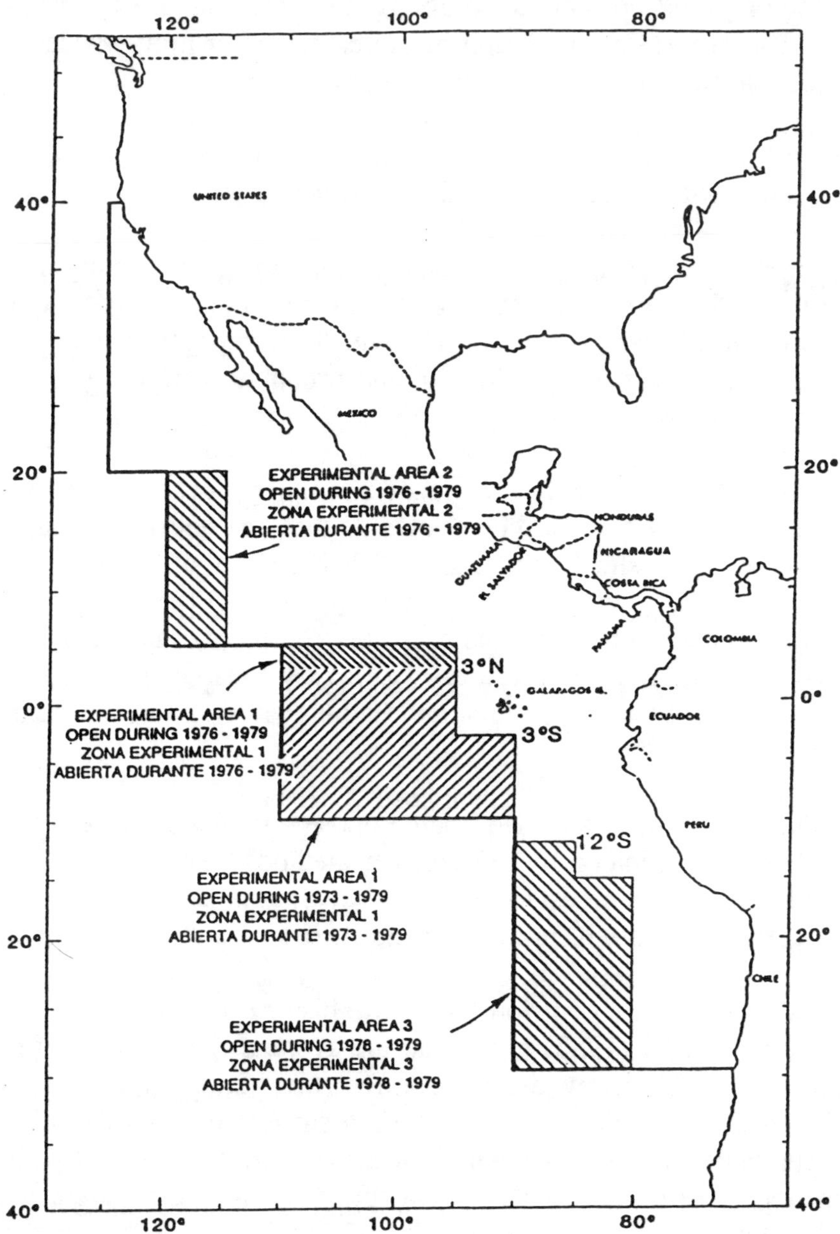

Figure 4. The Inter-American Tropical Tuna Commission's Yellowfin Regulatory Area (CYRA). (Reproduced from IATTC, 1991a).

6.2   Distribution

Yellowfin are distributed worldwide, occurring in the tropical and subtropical waters of the Indian, Pacific, and Atlantic Oceans, and in all the warm seas of the world except the Mediterranean Sea. The coastal limits of their distribution in the Pacific were documented by Rosa (1950). He reported that in the eastern Pacific, yellowfin range from Point Conception, California, to San Antonio or Talcahuano, Chile, and in the western Pacific they are distributed from Hokkaido, Japan, through the Indonesian

Archipelago, to Cape Howe, Australia, and New Zealand. Blackburn (1965) suggested that the latitudinal limits of yellowfin are about 35°N and 33°S in the eastern Pacific, and 40°N and 35°S in the western Pacific. Kamimura and Honma (1963) stated that yellowfin are distributed in practically every part of the Japanese longline fishing grounds in the Pacific, which at the time extended from about 40°N to about 40°S, and from the Asian coast eastward along the equator to about 100°W. The longline fishery has since expanded further into the eastern Pacific (Figure 5). Catch records from this fishery and the purse-seine fishery indicate the relatively continuous distribution of yellowfin within its range (Figure 6), which extends across the Pacific and is roughly enclosed by the 40°N and 40°S latitudes (Calkins and Chatwin, 1967, 1971; Shingu *et al.*, 1974; Calkins, 1975; Fisheries Agency of Japan, 1974-77; IATTC, 1980, 1989; Suzuki *et al.*, 1978; Miyabe and Bayliff, 1987).

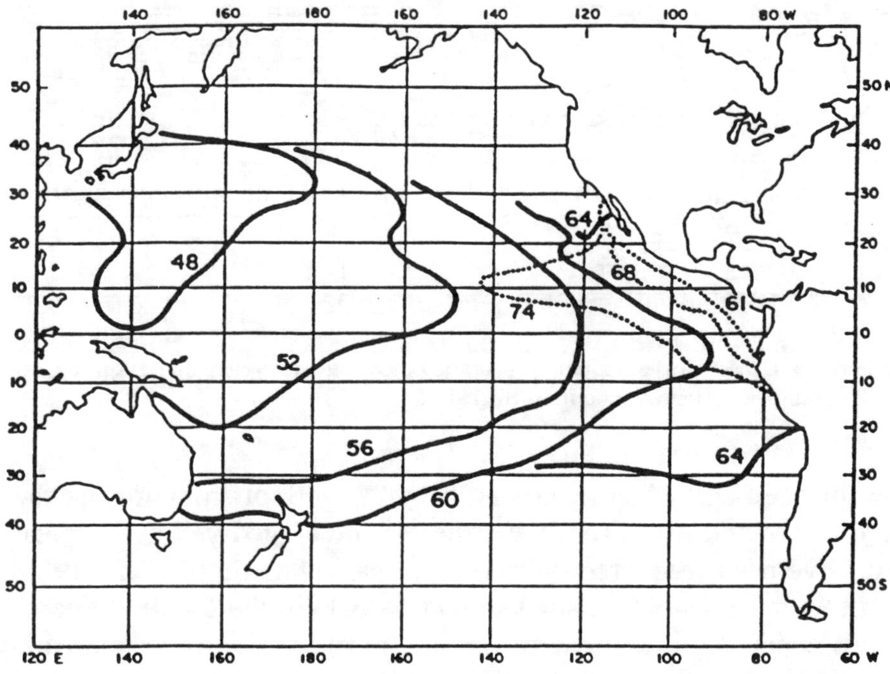

Figure 5. Geographical expansion of the Japanese longline fishery (solid curves) and the surface fishery in the eastern Pacific Ocean (dotted curves). Numerals denote calendar years. (Reproduced from Suzuki *et al.*, 1978).

6.3 <u>Movement</u>

In the eastern Pacific, yellowfin movement was initially studied by Blunt and Messersmith (1960) and Schaefer *et al.* (1961). A later report by Fink and Bayliff (1970) included the data on all tagged fish prior to 1965. They concluded that from 1952-1964 the fishery operated on two main groups of yellowfin: a northern group off the west coast of Baja California, in the Gulf of California, and around the Revillagigedo Islands; and a southern group from the Tres Marias Islands to northern Chile. The seasonal movements of the two groups appeared to be largely in a coastal direction with some mixing taking place between them. The lack of tag recoveries by the Japanese longline fleet operating outside the CYRA and on grounds partially overlapping the nearshore surface fishery, suggests that there was no large-scale offshore-inshore movement of yellowfin during the

1952-1964 period (Joseph *et al.*, 1964). Schaefer *et al.* (1961) also pointed out that fish tagged near fishing banks tended to disperse very slowly.

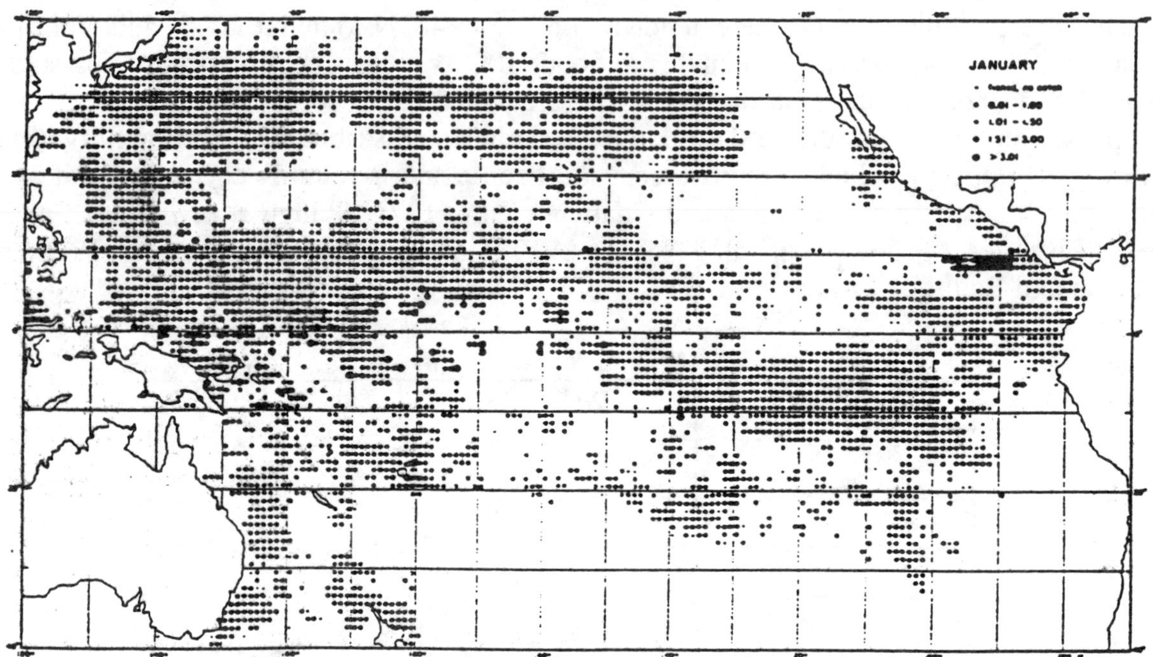

Figure 6. An example of a monthly distribution of average catch rates for yellowfin caught by Japanese longline boats, for 1967-72 combined. (Reproduced from Suzuki *et al.*, 1978).

As the surface fishery began moving progressively offshore during the late 1960s, more fish were recaptured offshore. Nevertheless, no tagged yellowfin released in the eastern Pacific have been recovered outside the eastern Pacific (Bayliff, 1984), and the concept of three, semi-independent stocks (Section 6.1) in the Pacific Ocean has not been refuted. Studies conducted off southern Mexico, northern Central America and offshore (Bayliff and Rothschild, 1974; Bayliff, 1979, 1984) indicate that differences in the catches of smaller fish in the inshore and offshore areas were due mainly to differences in vulnerability. In addition there were no pronounced tendencies for movements in an east-west or north-south direction even though the movements were not random. Although the overall implication is that yellowfin do not move in excess of several hundred miles (see Table 2, Miyabe and Bayliff, 1987), exceptions exist and a few interesting examples appear in Table 3.

7. NATURAL MORTALITY

Hennemuth (1961b) analysed length-frequency data to obtain an estimate of 1.72 for the total annual instantaneous mortality rate (Z) of yellowfin in the eastern Pacific for 1954-59. Using the coefficient of catchability estimated by Schaefer (1957) and an estimate of fishing effort, he estimated the annual instantaneous fishing mortality rate (F) to be 0.95 and the annual instantaneous natural mortality rate (M) to be 0.77, with 95-percent confidence limits of 0.64 to 0.90. Hennemuth regarded 0.80 as the most probable value of M and 0.60 to 1.00 as the "extreme possible" values. Schaefer (1967), however, considered the lowest and highest probable values to be 0.55 and 1.05,

TABLE 3. Unusual returns of yellowfin tagged in the Pacific Ocean (IATTC, 1980, 1982, 1983).

| | Release | | | Recapture | | | |
Date	Area	Length (cm)	Date	Area	Length (cm)	Days free	Net distance (nm)
Mar. 1, 1978	7°58'S– 139°58'W	52	Jul. 12, 1979	11°47'N 130°25'W	?	498	1,315
Apr. 2, 1979	12°58'N– 92°19'W	46	Aug. 11, 1982	10°21'N– 109°20'W	?	1,228	1,012
Apr. 26, 1979	6°13'N– 85°02'W	?	Sep. 16, 1982	10°46'N– 107°36'W	?	1,240	1,366
Feb. 4, 1980	24°54'S– 130°03'W	75	Dec. 16, 1980	14°37'S– 90°50'W	120	317	2,290
Feb. 4, 1980	25°00'S– 130°03'W	76	Mar. 5, 1981	7°45'S– 121°50'W	?	457	1,139
Feb. 12, 1980	16°21'S 146°57'W	?	Jul. 13, 1981	8°14'N 139°25'W	?	518	1,541
Feb. 16, 1980	17°46'S– 150°32'W	85	Aug. 31, 1982	2°13'N– 120°06'W	?	928	2,162
Apr. 21, 1980	16°01'S– 179°48'E	?	Aug. 31, 1982	2°54'N– 118°55'W	?	862	3,806
Oct. 18, 1981	10°18'N– 109°13'W	63	May 18, 1982	10°19'S– 84°42'W	91.8	213	1,916

respectively. Murphy and Sakagawa (1977) evaluated the published estimates of M and concluded that the values around 0.80 were the "best" estimates. Bayliff's (1971) result of $M < 2.0$ does not dispute this conclusion although the estimate, derived from tagging, was apparently affected by the temporal variation in vulnerability and the failure of tagged and untagged fish to mix completely. Francis (1977), however, based on a simulation model of the dynamics of the eastern Pacific surface fishery, suggested that the value of 0.80 was too high. He used this rate and Hennemuth's (1961b) lowest value of $M = 0.60$ as input parameters to modified versions of an earlier model (Francis, 1974) and found that 0.60 was a better estimate in validating the versions which best mimicked the fishery. Nevertheless, until recently, the value of $M = 0.80$ was used by the IATTC (IATTC, 1989) in age-structured assessment models because the interpretation of data from the fishery was most compatible with this value. As a result of the virtual disappearance from the fishery of females larger than 140 cm, different values of M have been applied to the sexes beginning with stock assessments in 1989. This subject is discussed more fully in Section 12.

8.   OCEANOGRAPHIC FEATURES ASSOCIATED WITH THE SPECIES

The principal circulation pattern in the upper waters of the Pacific Ocean (Figure 7) is driven by the easterly trade winds and is not symmetric with respect to the equator (Pickard, 1968). The well-developed surface components of the system include a westward-flowing South Equatorial Current (SEC) between 8° to 10°S and 3°N, a westward-flowing North Equatorial Current from about 8° to 20°N, and between them a narrow North Equatorial Counter Current flowing to the east. An additional easterly component that normally lies beneath the SEC (IATTC, 1984) consists of the generally weaker South Equatorial Counter Current (SECC). This system can be traced from the Gulf of Panama almost to the Philippines, a distance of some 12,000 km. The relationship of this circulatory pattern to other, eastwardly-flowing, subsurface currents is demonstrated in Figure 8 by means of a transect from Hawaii to Tahiti (Wyrtki and Kilonsky, 1984).

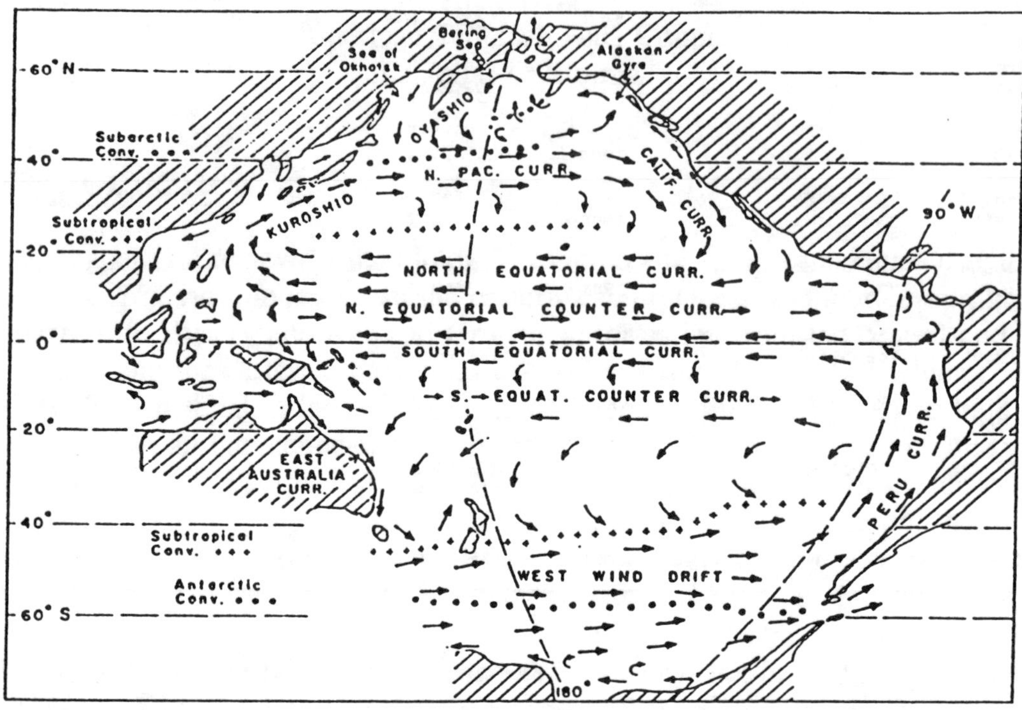

Figure 7. Pacific Ocean surface circulation. (Reproduced from Pickard, 1968).

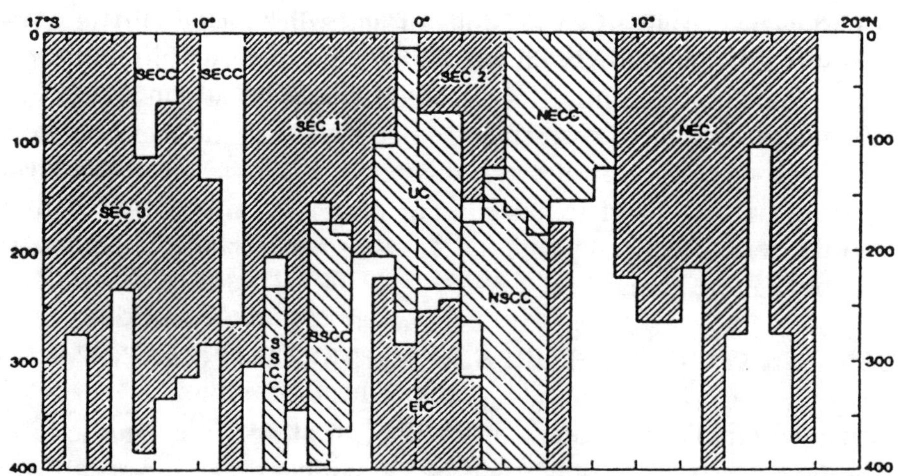

Figure 8. Areas occupied by the main zonal currents between Hawaii and Tahiti in the upper 400 m. Dark shading is westward flow, light shading eastward, and blank areas have zonal speeds less than 2 cm/s. Abbreviations: NEC - North Equatorial Current; SEC - South Equatorial Current; NECC - North Equatorial Counter Current; SECC - South Equatorial Counter Current; NSCC - North Sub Counter Current; SSCC - South Sub Counter Current; UC - Under Current; EIC - Equatorial Intermediate Current. (Reproduced from Wyrtki and Kilonsky, 1984).

Of the three major subsurface currents, the core of the Equatorial Undercurrent, or the Cromwell Current, is located in the thermocline during normal, or non-El Niño periods, and extends approximately 10,000 km westward from the Galapagos Islands. During these periods the decrease in the sea-surface temperature from west to east is associated with a shoaling of isotherms (Figure 9), and a gradual rise in the depth of the thermocline from about 150 m in the western Pacific to 40 m or less in the eastern Pacific (Philander, 1990). Figure 9 also illustrates that the greatest change in the thermocline depth occurs in the central and eastern Pacific regions.

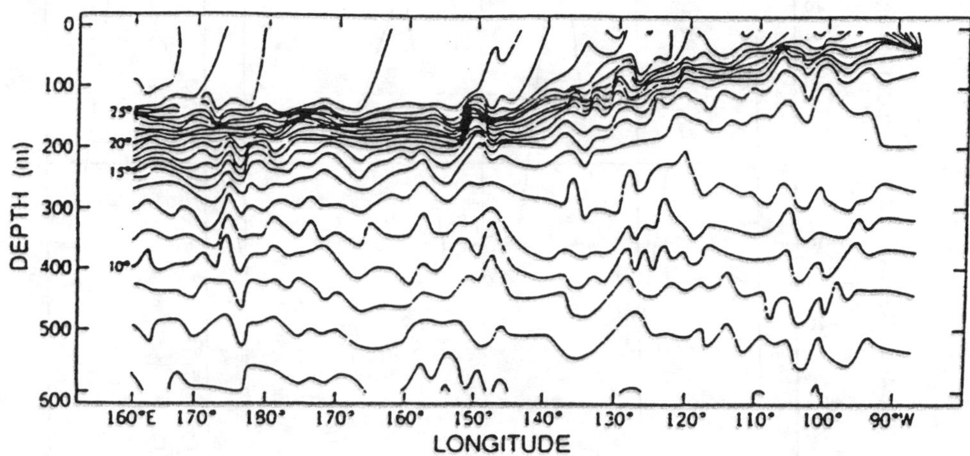

Figure 9. Temperature (°C) as a function of depth and longitude along the equator in the Pacific Ocean, as measured in 1963 by Colin et al., (1971). (Reproduced from Philander, 1990).

During El Niño episodes all of the tropical currents in Figure 7, including the California and Peru (Humboldt) currents in the eastern boundary, undergo changes in their normal positions and strengths (IATTC, 1984). Following a weakening of the southeasterly trade winds in the eastern part of the south Pacific, some of the principal changes that take place in the eastern Pacific include a rise in the sea surface due to a deceleration of the hydrospheric circulation in the south Pacific Ocean, elevated coastal and equatorial temperatures together with a movement of warmer surface water over the normally upwelling regions off Ecuador and Peru, and a deepening of the thermocline (Miller and Laurs, 1975; Joseph and Miller, 1988). In addition, while the SEC weakens, the volume of the SECC may increase and the current is then frequently found at the surface between 5°-10°S. The Cromwell Current also weakens, and during a particularly strong El Niño, such as in the last quarter of 1982 and into 1983, the current is difficult to locate in transect data at the equator. Collectively, these events are thought to have an effect on the recruitment of yellowfin and the vulnerability of yellowfin and skipjack to capture.

As Figure 10 illustrates, strong, positive recruitment anomalies in 1971, 1974, 1978 and 1985 were preceded by pronounced El Niño conditions in 1969, 1972, 1976, and 1983 (Joseph and Miller, 1988). Weaker El Niño events also occurred during the third quarter in 1982 and throughout 1987. Consequently from 1982 through 1986, waters in the eastern Pacific were warmer than usual and recruitment during the 1984-1988 period was the greatest on record for a five-year interval (IATTC, 1989).

Although the causal relationship between enhanced recruitment and El Niño conditions...if it exists...is not understood, the survival of young fish and the rates at which eggs, larvae or postlarvae are retained in the eastern Pacific are probably increased due to the diminished flow of the altered current patterns.

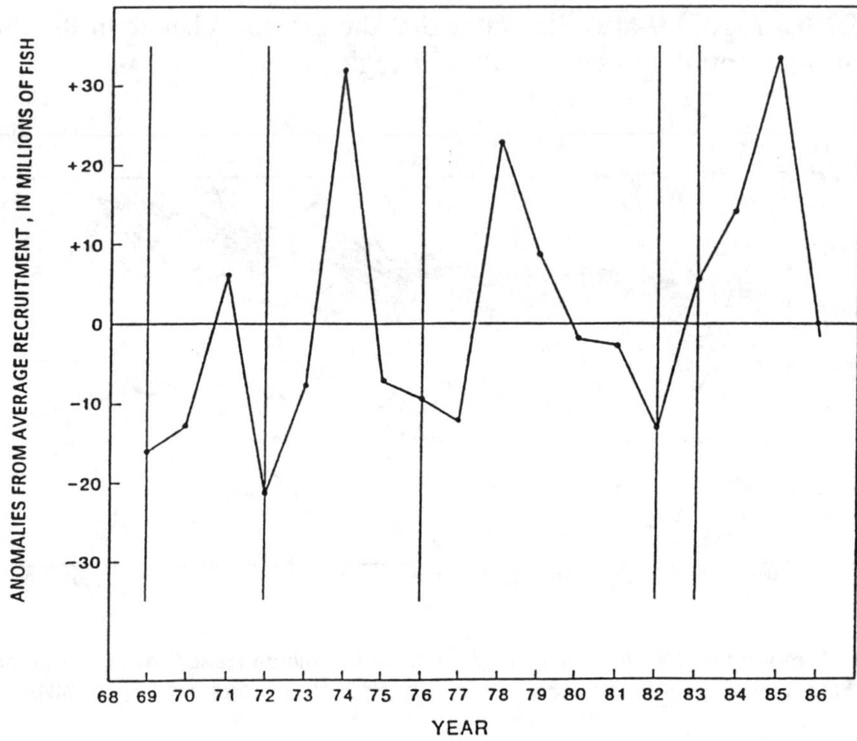

Figure 10. Anomalies from average recruitment of yellowfin, in millions of fish, in the eastern Pacific Ocean. (Reproduced from Joseph and Miller, 1988).

El Niño episodes also affect the vulnerability of yellowfin and skipjack tunas to purse-seining operations. This has been studied in three heavily-fished areas: the inshore coastal area off Ecuador and northern Peru; the Costa Rica Dome (CRD), about 200 miles west of Nicaragua and Costa Rica; and the triangle between Cape San Lucas, Cape Corrientes (Mexico) and the Revillagigedo Islands (IATTC, 1989). In particular, for yellowfin within a 300-mile radius of the CRD, the trends in catch per day's fishing (Section 11), or CPUE, the successful-set ratio (SSR), *i.e.* the ratio of successful sets to all sets, and the availability-vulnerability (AV) index are shown in Figure 11. The index is calculated as CPUE/(1 - SSR) so that an increase in either CPUE or SSR will also increase the index. The 1983 data reflect conditions during the most extensive and severe El Niño in this century (IATTC, 1988). Reduced upwelling, that persisted from November 1982 to May 1983 in the CRD region, reduced the vulnerability of the fish to capture, and depressed the CPUE and hence the AV index. During the last half of the year the El Niño began to weaken, but not sufficiently to allow the fishery to recover. At the same time a major portion of the fleet departed for the western Pacific. During 1984 the average CPUE recovered to 13.5 mt/day compared to 2.5 in the previous year. In contrast to 1983, the mild El Niño episode of 1987 affected only the region south of 5°N. The thermocline was 15 m above normal at the CRD, but 20 m deeper south of 5°N and east of 135°W. This situation apparently increased the CPUE to above-normal levels throughout most of the year and, combined with an elevated SSR, these conditions produced an AV index at least twice the long-term 1975-88 average. The oceanic conditions of 1987 persisted through 1988, with additional upwelling and a shallower thermocline. In response to these favourable conditions, the AV index maintained a high level through increases in both the CPUE and SSR.

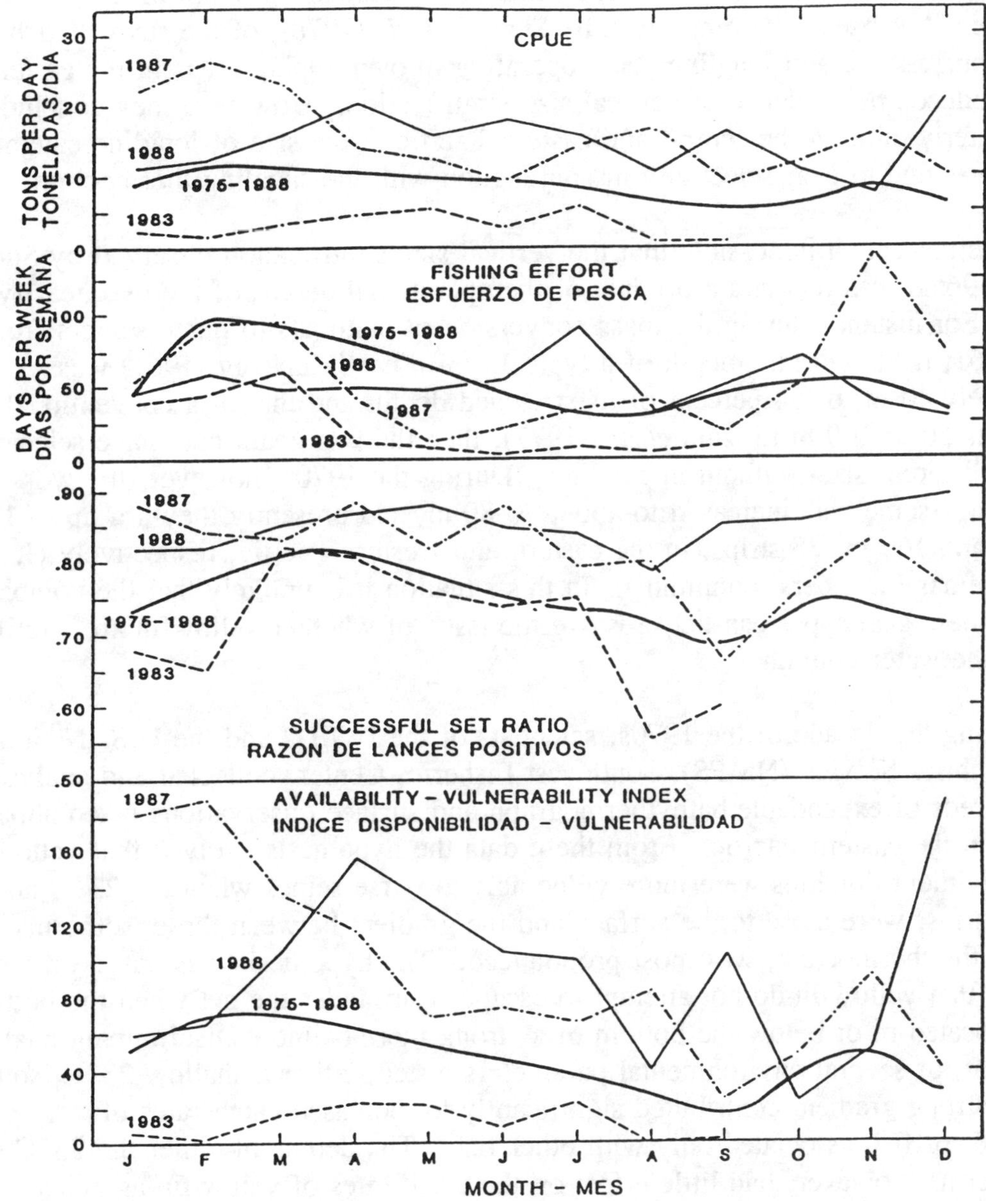

Figure 11. Monthly mean values of CPUE, effort, successful set ratio, and availability-vulnerability index for yellowfin at the Costa Rica Dome. (Reproduced from IATTC, 1989).

There is little information describing the effect of current patterns and other hydrodynamic activity on the fishing success of yellowfin other than in association with the thermocline. For example, Suda and Schaefer (1965a) confirmed earlier observations that in the equatorial zone there is a gradual increase in the average weight of longline-caught yellowfin in the direction from the western to the eastern Pacific. They speculated that the size gradient was a product of gear selectivity, size stratification within the depth habitat of yellowfin, and the easterly shoaling of the thermocline (see above). They recognized that while yellowfin of all sizes occurred throughout the mixed layer and the upper part of the thermocline, larger fish may be more abundant in the lower mixed layer and the upper thermocline. In addition, since conventional longlines fished at a maximum depth of about 160 m during the early 1960s, the gear became increasingly effective in catching larger fish in an easterly direction due to the decreasing depth of the thermocline. A subsequent comparison by Suzuki *et al.* (1978), of the sizes of fish caught by purse-seine and longline gears operating in overlapping areas in the eastern Pacific, produced no evidence of vertical size stratification. However, they concluded that the easterly cline in the central and eastern Pacific in the size of longline-caught yellowfin was due to gear selectivity in conjunction with the shoaling thermocline.

In retrospect, it is possible that the vertical size stratification observed by Suda and Schaefer (1965a) was real and a product of the operational depths of the two gear types at that time. For instance during the mass conversion of baitboats to purse seiners from 1959 to 1961, if the working depth of a typical, 73-m net (McNeely, 1961) was approximately 40 m, or 54 percent of its stretched depth, and that of a conventional longline was 50 to 100 m (Suzuki *et al.*, 1977), then the two gears may have selected somewhat different sizes without interacting. During the 1970s, however, the working depth of purse-seine nets increased to about 70-80 m, and presently they are up to 110 m (19 strips) and 160 m (28 strips) in the eastern and western Pacific, respectively (J. Lewis, Casamar Inc.; pers. commun.). In this situation it is unlikely that the comparative catches of these gear types can help resolve the issue of whether yellowfin are stratified by size in the water column.

During the decade of the 1970s, scientists of the IATTC and the U.S. National Marine Fisheries Service (NMFS), Southwest Fisheries Center, collected and analysed many hundreds of expendable bathythermograph and surface observations taken aboard tuna boats in the eastern Pacific. From these data the hypothesis evolved that yellowfin in schools without dolphins were more vulnerable to purse seines when the 23°C and 15°C isotherms, were close to the surface and the gradient between these isotherms, or strength of the thermocline, was most pronounced. The hypothesis was based on the assumption that yellowfin do not attempt to escape from under the net when the bottom of the net is located in or below the bottom of a strong thermocline. Discriminant analysis revealed that, of several environmental parameters tested, either a shallow 23°C isotherm depth or a strong gradient contributed significantly to increased catch rates of "schoolfish", or fish associated only with other fish. The depths of either the 23°C or 15°C isotherms, however, had little effect on the catch rates of yellowfin associated with dolphins. In this case a strong gradient led to the greatest catch rates. Collectively, these results suggest that the depth of the yellowfin habitat may be determined to a large extent by the vertical temperature structure, or related properties, such as oxygen content (IATTC, 1982). Acceptance of these results must also be tempered by the fact that changes in market needs and in fishing strategies can mask the environmental effects on fishing effort. There have also been numerous occasions when the presence of strong gradients and a shallow 23°C isotherm have not resulted in improved catches. The apparent abundance of schoolfish may therefore be more closely related to other factors, such as ocean currents which aggregate or disperse food (IATTC, 1981; Sund *et al.*, 1981).

The use of sonic tagging devices has augmented, but not changed, the earlier belief that yellowfin occupy the mixed layer and the upper portion of the thermocline. Carey and Olson (1982) attached acoustic transmitters to three, 87-98 cm yellowfin in the Clipperton Island-Gulf of Panama region and found that their primary orientation was to the thermocline, with frequent vertical excursions above and below this reference. Solitary fish also tended to stay near the thermocline, while those which joined schools moved up into the mixed layer. During daytime the fish were found at greater depths than at night, but collectively they spent little time at the surface. Similar experiments by Holland *et al.* (1990) near the Hawaiian Islands indicate that the mixed layer and the first degree (°C) of reduced temperature at the top of the thermocline accounted for 68 percent of the daytime distribution of four, 54-74 cm yellowfin, and 77 percent at night. The introduction of deep longlines, whose maximum hook depth can penetrate to 160 m or 250 m, depending on the area fished, increased the hook rate of bigeye tuna, but not that of yellowfin (Hanamoto, 1974; Suzuki *et al.*, 1977). This finding supports the view that the principal habitat of yellowfin is in the upper boundary of the thermocline and mixed layer. Deep forays to 168 m and 464 m (Carey and Olson, 1982), where the oxygen partial pressures were estimated to be 40 to 60 percent and 20 percent of air saturation, respectively, suggest that yellowfin are tolerant of low oxygen levels. However, Green (1967) noted that, based on typical data from the EASTROPIC cruises, the oxygen content in the mixed layer was 4 to 6 ml/l followed by a sharp decline to 1 ml/l just below the thermocline. Bushnell *et. al.* (1990) found that yellowfin responded significantly in terms of gape, ventilation volume and heart rate when hypoxic oxygen levels reached 2.7 to 3.3 ml/l. Consequently, it is not clear whether the orientation of yellowfin to the top of the thermocline is a preferential response to temperature, or if it represents the maximum depth at which a minimal amount of oxygen is consistently available.

9. INTERACTION WITH OTHER SPECIES

9.1 <u>Dolphins</u>

In the eastern Pacific a situation exists in which predominantly moderate-to-large sizes of yellowfin frequently associate with one or more species of dolphins. Fishermen have long recognized this relationship and purposely search for the surface activity of dolphins because it may indicate the presence of sub-surface yellowfin. Seabirds, such as the sooty terns (*Sterna fuscuta*), wedge-tailed shearwaters (*Puffinus pacificus*) and boobies (*Sula* spp.), are frequently found over dolphin schools, and extend the fishermen's horizon of detection (Au *et al.*, 1979). Exploitation of the yellowfin-dolphin association began in earnest by purse seiners following the conversion of most of the baitboat fleet to purse seiners during 1959-1961. At that time about 62 percent of the catch was taken in association with dolphins (Perrin, 1969). Concern for dolphin mortality incidental to the fishing process has subsequently led to legislation in the USA to monitor their abundance and mortality, and to reduce mortality through application of the Marine Mammal Protection Act (MMPA) of 1972 (U.S. Dept. of Commerce, 1977). The NMFS implemented these provisions, including the establishment of annual quotas for the different dolphin species, enforcement of the regulations, and placement of observers aboard USA vessels. The research programme conducted by the NMFS also focussed on the essential factors of population estimates, reproductive rates, and gear and procedural modifications to reduce mortality (U.S. Dept. of Commerce, 1978). The MMPA was strengthened in 1984 by recommendations that: 1) the maximum dolphin mortality rate inflicted by vessels of a foreign nation conform to that of USA vessels, under penalty of embargo of commercial fish or fish products, and 2) that these restrictions also apply to intermediary nations. In 1977 the IATTC initiated a programme that involved the co-operation of the international purse-seine fleets operating in the eastern Pacific. The plan is similar to that conducted by the NMFS, but lacks enforcement provisions. Attributes of the programme that are oriented toward conservation include gear research, the use of educational workshops to disseminate information on safe practices, and behavioural research surrounding the interaction of dolphins and yellowfin tuna (IATTC, 1978). In 1990 all of the principal tuna canners in the USA announced that they would no longer purchase tunas caught in association with dolphins. The consequences of this action are presently unknown.

Yellowfin are found together primarily with spotted dolphin (*Stenella attenuata*), to a lesser extent with spinner (*S. longirostris*) and common (*Delphinus delphis*) dolphins, and relatively seldom with striped dolphin (*S. coeruleoalba*) and other species (Hammond, 1981). These degrees of involvement may indicate that the basis of the association is partly due to a shared or preferred food source. For example, Perrin *et. al.* (1973) found that the food items in the stomachs of yellowfin and spotted dolphin caught in the same nets overlapped considerably, with ommastrephid squids representing the most common forage item. The diets of the spinner dolphin and yellowfin, on the other hand, appeared to be quite different. With common dolphins the association with yellowfin seems to have seasonal, areal and yearly components. If a common food preference is one ingredient in the linkage, the disadvantages of the implied competition must be offset by one or more beneficial elements. Hammond (1981) explored the interactive possibilities between two populations, including competition, parasitism and commensalism, as well as the implications of tunas seeking dolphins or the reverse. While there is an overriding need for observational or experimental evidence to evaluate these possibilities, there is nevertheless some indication that, in addition to the benefits of schooling, the association may also be based on enhanced protection from mutual predators such as sharks and small whales.

9.2   Skipjack and Bigeye Tunas

In studying the size composition of tuna schools in the eastern Pacific, Orange *et. al.* (1957) and Broadhead and Orange (1960) noted that yellowfin and skipjack occurred in mixed schools. In general, about 10 percent of the purse-seine catch and 35 percent of the baitboat catch resulted from mixed schools, with the remainder in each case being made up of fish from pure schools of either species. The relative occurrence of mixed schools appeared to be greatest when the proportion of pure yellowfin and skipjack schools were evenly divided. On average, the yellowfin and skipjack in mixed schools were similar in size, but the sizes of yellowfin in pure schools appeared to be larger and more variable. The skipjack from pure and mixed schools tended to be more alike.

Yuen (1963) challenged the implied species interaction in mixed schools with evidence from underwater observations and photography of baitboat fishing. The yellowfin and skipjack in seemingly composite aggregations actually maintained a significant degree of separation into schools or species groupings, and the random response of individuals to the proffered food created the illusion of complete mixing. The conclusion was drawn that while yellowfin and skipjack prefer to school by species, their co-occurrence is probably the congregation of separate schools responding to the common attraction of an external stimulus, such as food.

The amount of bigeye tuna caught annually by longline gear in the eastern Pacific has varied by a factor of two, from about 36,300 to 72,600 mt, in the interval from 1961 to 1980 (IATTC, 1991a). Over a slightly longer period, from 1961 to 1988, the modal catch of this species by the surface fleet was less than 907 mt, with a range from 68 to 15,400 mt. Differences of this magnitude between gear types suggest that the habitat preference of the two species preclude any large degree of interaction.

A recent study (IATTC, 1991b) indicated that the possibility of misidentifying bigeye tuna as yellowfin does not occur to an appreciable degree in the eastern Pacific surface fishery. During the interval from April, 1987, to September, 1989, observers aboard USA purse seiners recorded that, in floating-object and school sets, 46 mt of bigeye were caught and loaded compared to 17,155 mt of other tunas, or a tonnage ratio of 0.0027. The observers had received special training to identify bigeye ($> 30\text{-}35$ cm) in the commercial catch. Based on logbook entries, cannery unloading weights and length-frequency samples, the comparable ratio for all other USA purse seiners in the same set-type, time-area strata was 0.0026. Since the two ratios are essentially equal, it does not appear that the bigeye catch by USA vessels is misrepresented by the current methods of evaluation. Non-USA and USA purse seiners fish in the same overall area in the eastern Pacific, but the non-USA vessels tend to capture greater amounts of bigeye near the Galapagos Islands and south of the equator. Consequently, the tonnage ratio of bigeye to all other tunas for the non-USA vessels was 0.0066. However, since the same sources of information are used to estimate the landings for both segments of the fishery, there is no reason to suspect that bigeye landings are misrepresented for the purse-seine fleet as a whole.

10. GENERAL DESCRIPTION OF THE FISHERY

Prior to the late 1950s the surface fishery for yellowfin in the eastern Pacific was dominated by baitboats operating in coastal waters and in the vicinity of offshore islands. The conversion of most of the baitboats to purse seiners in the late 1950s and early 1960s fostered and accelerated the seaward expansion of the fishery. In 1968 the purse-seine fleet began fishing in the area outside the CYRA (Figure 4), and by 1974 fishing was conducted as far west as 150°W (Calkins, 1975; Peterson and Bayliff, 1985). As a result of these changes the number of baitboats declined from 204 to 36 during the 1950-88 period, while their capacity decreased from about 35,380 to 2,810 mt. During the same interval the number of purse seiners increased from about 67 to 182, with an increase in capacity from approximately 7,160 to 134,300 mt (IATTC, 1989, Table 4). Fishing pressure on yellowfin also increased due to the eastward expansion of the Japanese longline fishery that began in 1948. Catches to the east of 130°W were first reported in 1956, and trans-Pacific longlining was essentially completed by 1962 (Figure 5) (Suda and Schaefer, 1965a). With the exception of the region surrounding French Polynesia, the fishery in the eastern Pacific Ocean (EPO) now comprises the area to the east of 150°W between the 39°N and 35°S latitudes. In addition to the gear types just mentioned, lesser amounts of yellowfin are also caught by coastal dayboat operations involving smaller vessels (bolicheras and jigboats). By 1990 vessels of at least 10 nations had participated in the yellowfin surface fishery and three in the longline fishery.

During the 1979-87 period the average catch in the CYRA by the surface fleet and the Japanese longline fleet was 162.6 x 10^3 mt (range: 82.8-247.6), and the preliminary estimate for 1988 was 267.6 x 10^3 mt. For only the surface fleet in the region between the boundary of the CYRA and 150°W, the average catches for the 1979-87 period and in 1988 were 22.2 x 10^3 mt (range: 12.2-38.7) and 20.7 x 10^3 mt, respectively. Total catches in the EPO in 1988 were therefore at least 288.3 x 10^3 mt, the greatest amount on record, and 15.8 x 10^3 mt in excess of the previous record in 1987 (IATTC, 1989). The distribution of catches for the 1979-87 and 1988 surface segments of the fishery appear in Figures 12a and b, respectively.

"The need for regulation of the fishery for yellowfin was first apparent in 1961, but the concerned governments participating in the IATTC were not able to implement regulations until 1966. Thereafter regulations were in force through 1979. For 1980 and subsequent years [except for 1987] the IATTC has recommended quotas, but the countries which participate in the fishery have not been able to agree on their implementation" (Peterson and Bayliff, 1985).

As an oversimplification, in years in which regulations were in effect, the fleet operated throughout the EPO during the initial part of the year until established quotas were met, and then outside the CYRA for the balance of the year. Under these conditions it is difficult to disentangle what may be the seasonal component of fish movement from the obligatory movement of the fleet. Even in the absence of quotas after 1979, it is also possible that the observed distribution of CPUE is the result of the confounding effect of fish and fleet movements. Nevertheless, Punsly (1987) noted certain trends in the distribution of sets on schoolfish, and fish associated with dolphins or logs in several season-area (SA) strata (Figure 13). For example, during the 1970-1985 period, which included both regulated and non-regulated years, the schoolfish catch rate was greatest in February through April near the Gulf of Panama (SA 8), second in rank during November through March near the tip of Baja California (SA 20), and third during September and October off Ecuador (SA 6). The catch rates on fish associated with dolphins on the other hand, were greatest during November and December in the area offshore south of the equator (SA 3), and second off the tip of Baja California (SA 20). Catch rates on fish associated with floating objects were greatest during September and October off Central America and southern Mexico (SA 13), and second near the Gulf of Panama (SA 8).

11. TRENDS IN FISHING EFFORT, CATCH AND CATCH PER UNIT OF FISHING EFFORT

The catch per day's fishing (CPDF), as an index of yellowfin abundance, has been revised several times during the history of the fishery. When baitboats dominated the surface fleet in the 1950s, the catch and CPDF data for vessels of different size classes were standardized (S) to calculate the CPSDF for Class-4 (182-272 mt capacity) baitboats. Following the conversion of the fleet to purse seining by the early 1960s, Class-3 (93-181 mt) seiners became the new standard that persisted until 1967. As early as 1962, however, the CPSDF was adjusted by factors related to the SSR to take increases in fleet

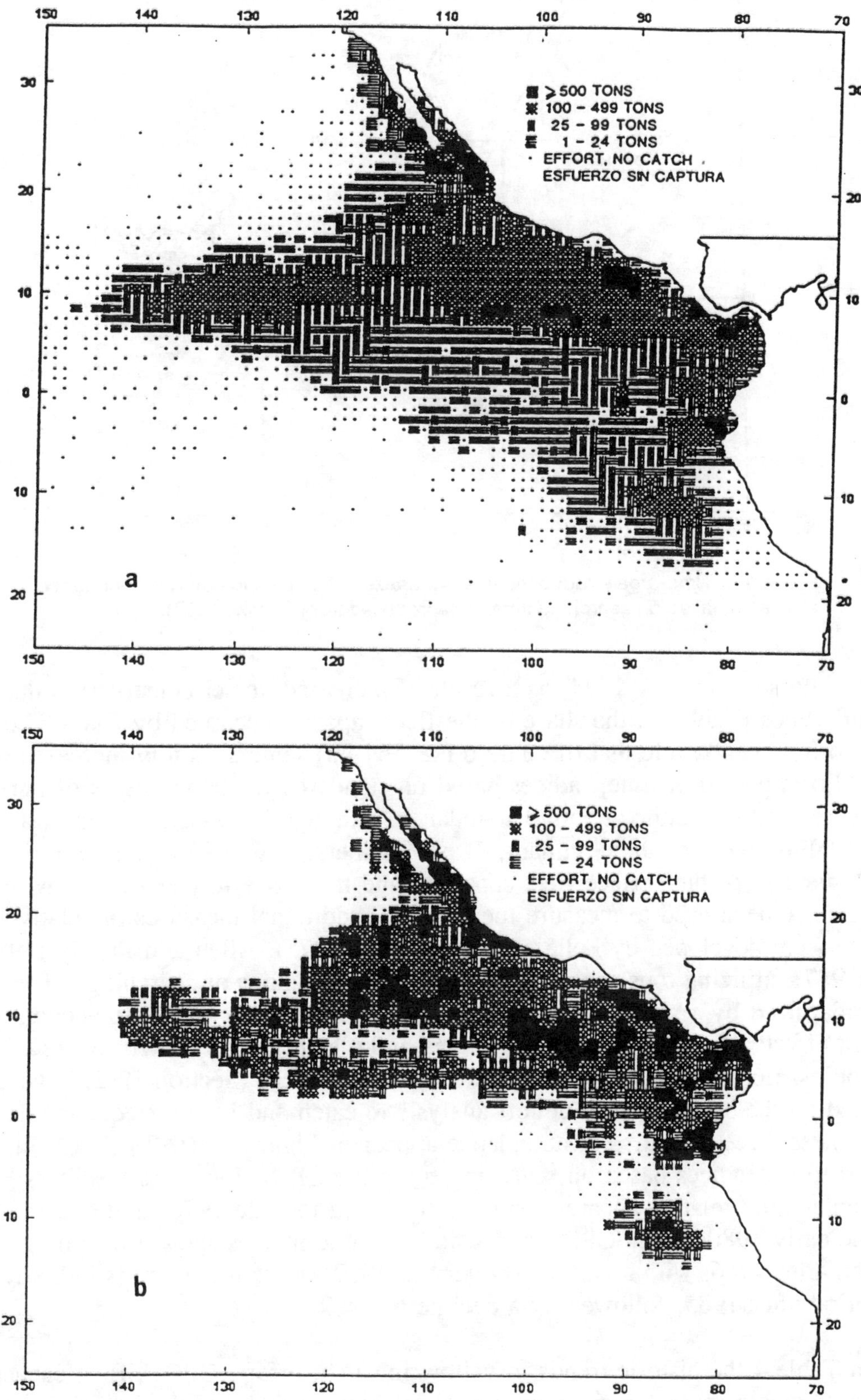

Figure 12. Average annual catches of yellowfin in the EPO during 1979-87 (a), and catches in the EPO in 1988 (b), for all purse-seine trips for which usable logbook data were obtained. (Reproduced from IATTC, 1989).

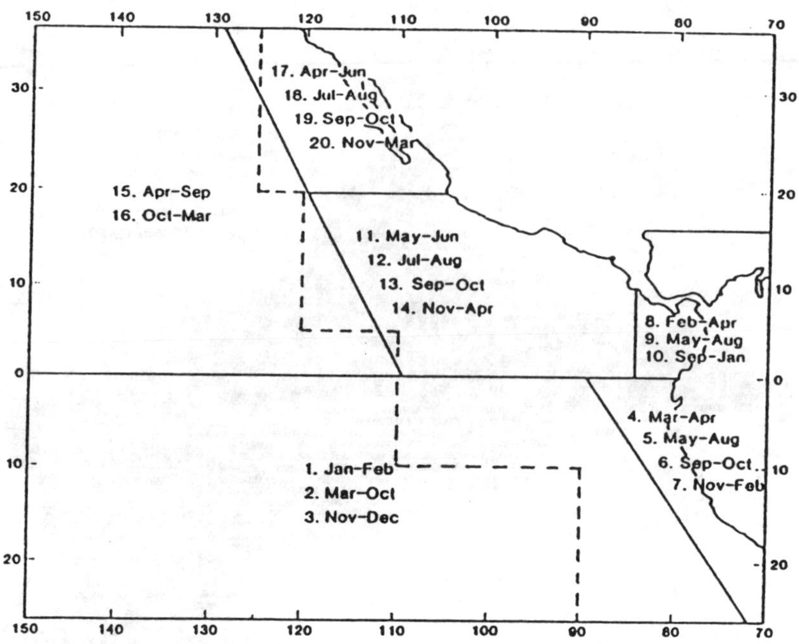

Figure 13. Time-area strata used to standardize catch rates to develop abundance estimates based on searching time. (Reproduced from Punsly, 1987).

efficiency into account. By 1974, as a result of increased vessel construction that began in the mid 1960s (Table 4), the share of the fleet capacity occupied by Class-3 seiners declined to 3 percent while that of Class-6 ($\geq$ 364 mt) seiners rapidly increased to 81 percent. For a period of time, indices based on standards for both classes of purse-seine vessels were used to monitor relative abundance, although increased reliance was placed on the CPDF derived for the dominant, Class-6 seiners. By 1983 this class represented almost 90 percent of the surface fleet capacity, and the raw effort expended by this class alone has since been used to measure the CPDF. Additional measures of relative abundance were developed by Pella and Psaropulos (1975), Allen and Punsly (1984) and Punsly (1987), utilizing tons caught by purse seiners per hour of searching. The rates were standardized by a weighted, generalized linear model that took into account the significant effects of vessel speed, season-area and whether the yellowfin were caught with dolphins, floating objects or skipjack. Biomass indices (Section 12.2) have also been constructed by the application of cohort analysis to catch and length-frequency data. The trends in these three measures of abundance appear in Figure 14 (IATTC, 1991a). Compared to the indices based on searching time, the CPDF for Class-6 seiners underestimated the relative apparent abundance during the late 1970s and overestimated it during the early 1980s. The CPDF and searching-time indices show a declining population after 1986, with a partial recovery in 1989, while the biomass index reflects a stable period after 1985, followed by a decline in 1989.

In Table 4 the historic trends in yellowfin catch, effort, CPDF, fleet capacities and recruitment contain several periods of special interest. Following the introduction of fishing regulations in 1966 and the first ventures by purse seiners outside the CYRA in 1968, the CPDF recovered to its highest level (15.5 mt/day) in 1969 since the earliest days of the fishery. This was due in part to the fishery's concentration on larger fish, such as 2-, 3- and 4-year olds, that allowed the fish of the average recruitments of 1966

TABLE 4. Catches of yellowfin, effort and catch per days fishing (CPDF) by the surface fleet and longline fleets in the eastern Pacific Ocean. The historic carrying capacities and numbers of vessels (No.) in the surface fleet, and the initial number of fish in the X and Y cohorts recruited at 30 cm in length are also included. Weight units are short tons (st). (Adapted from IATTC 1991a).

| | Catches, st x 10^{-3} | | | | | | Effort* | | Number and carrying capacity | | | | | | |
| | Surface Fleet | | | Long-line | Miscel- | | 10^{-3} days | CPDF** | Purse seiner | | Baitboat | | Total | Initial recruitment |
Year	Inside CYRA	Outside CYRA	Total	CYRA	laneous	Total	fishing	st	No.	st x 10^{-3}	No.	st x 10^{-3}	st x 10^{-3}	x 10^{-6}
1961	113.1	0	113.1	2.6	–	115.7	–	–	124	30.1	93	10.5	40.6	–
2	81.6	0	81.6	5.6	–	87.2	–	–	130	33.9	89	6.7	40.6	–
3	68.4	0	68.4	3.7	–	72.1	–	–	141	39.8	108	6.0	45.8	–
4	97.7	0	97.7	3.7	–	101.4	–	–	134	40.3	88	4.7	45.0	–
5	87.0	0	87.0	3.2	–	90.2	–	–	146	42.3	109	5.8	48.1	–
6	88.9	0	88.9	2.6	–	91.5	–	–	126	39.9	113	6.2	46.1	–
7	88.1	0	88.1	2.0	9.8	99.9	13.0	6.8	122	40.2	108	5.9	46.1	60.4
8	111.2	1.2	112.4	3.3	14.1	129.8	6.9	16.3	139	50.6	89	5.7	56.3	50.5
9	122.8	19.2	143.0	4.1	14.4	161.6	8.4	17.1	149	57.0	69	5.0	62.0	55.9
1970	140.9	30.7	171.6	1.6	13.4	186.6	12.3	14.0	162	67.5	49	4.3	71.8	63.3
1	112.6	22.8	135.4	1.3	6.8	143.5	13.3	10.2	185	88.8	102	5.6	94.4	48.7
2	150.5	44.8	195.3	2.4	14.6	212.3	13.5	14.5	206	112.4	108	6.7	119.1	55.1
3	176.7	49.5	226.2	1.3	12.3	239.8	18.0	12.6	216	131.9	106	6.9	138.8	119.0
4	190.9	41.0	231.9	0.7	9.7	242.3	23.0	10.1	230	147.0	111	7.8	154.8	67.5
5	175.1	47.7	222.8	0.9	13.0	236.7	24.5	9.1	249	163.8	102	7.4	171.2	56.2
6	209.7	50.8	260.5	0.8	15.9	277.2	25.8	10.1	250	176.5	99	7.1	183.6	48.7
7	201.4	17.8	219.2	1.0	11.4	231.6	27.4	8.0	250	178.8	79	5.4	184.2	107.3
8	183.0	16.0	199.0	1.0	9.4	209.4	29.3	6.8	262	180.8	68	5.0	185.8	85.0
9	193.9	15.2	209.1	1.1	9.5	219.7	34.8	6.0	268	183.7	45	4.0	187.7	72.5
1980@	145.3	29.6	174.9	1.2	11.0	187.1	32.4	5.4	258	184.6	46	3.8	188.4	66.8
1	173.9	26.5	200.4	0.8	7.4	208.6	32.3	6.2	247	183.7	39	3.1	186.8	56.9
2	117.8	20.1	137.9	0.9	8.6	147.4	26.0	5.3	221	167.8	36	2.7	170.5	72.5
3	90.4	13.4	103.8	1.0	8.8	113.6	18.2	5.7	199	137.8	52	3.5	141.3	83.9
4	141.7	18.2	159.9	0.5	9.6	170.0	16.0	10.0	165	113.2	40	3.1	116.3	82.7
5	212.2	25.0	237.2	3.1	10.0	250.3	17.4	13.6	175	127.3	25	2.4	129.7	82.8
6	251.5	42.8	294.3	1.4	19.2	314.9	16.8	17.5	165	122.6	17	1.9	124.5	94.2
7	273.8	26.6	300.4	0.8	13.2	314.4	21.3	14.1	177	143.8	29	2.2	146.0	106.8
8	294.9	22.8	317.7	1.1	13.8#	332.6#	23.7#	13.5	185	148.2	36	3.1	151.3	97.6
9##	267.6	49.3	316.9	1.2	8.2	326.3	22.0	14.4	172	133.5	30	3.0	136.5	93.8

* effort for all surface vessels in class-6 (>400 st) purse-seine units.
** for class-6 purse seiners.
# preliminary data for 1988.
all data for 1989 are preliminary.
@ Quotas established but not implemented from 1980 through to present.

and 1967 to grow and contribute to that part of the population (IATTC, 1981). However, the interaction of three factors: below-average recruitment from 1969 to 1972 and in 1976 and 1977, accelerated construction of vessels in the early 1970s, and exploitation of both large and small fish, ultimately reduced abundance to its lowest levels by 1981-82. The extremely large recruitment of 1973 temporarily raised the abundance of small fish and thereafter contributed to the biomass of larger fish in 1975 through 1977. In fact, the total catch in 1976, 251.4 x 10^3 mt, was the largest on record at that time, but the pulse of this strong recruitment moving through the fishery was not sufficient to halt the downward trend in abundance. This condition, coupled with the onset of the severe El Niño of 1982-83, led to reduced catches and an exodus of fishing vessels to the western Pacific. During the ensuing period from 1984-86, several factors were thought to contribute to the recovery of the fishery, including: (1) above-average recruitment following the El Niño event; (2) declining effort from 1982-86, although some vessels began returning from the western Pacific in 1984; (3) reduced effort on small yellowfin due to a reduction in prices for skipjack and small yellowfin; and (4) an increase in the effort directed toward larger-than-average fish (IATTC, 1989). The elevated sea-surface temperatures resulting from the 1982-83 El Niño were extended by the weaker El Niño of 1987 and may have contributed to the exceptional recruitment of 1987. In addition, as in 1985-87, it appears that the fishery began to concentrate on large fish again in 1989, following a brief period in which small fish were also captured in 1988. These favourable conditions no doubt contributed to the succession of increased catches that culminated in records of 301.7 x 10^3 mt in 1988 and an estimated 296.0 x 10^3 mt in 1989, but such large catches may have also initiated a decline in abundance (Figure 14).

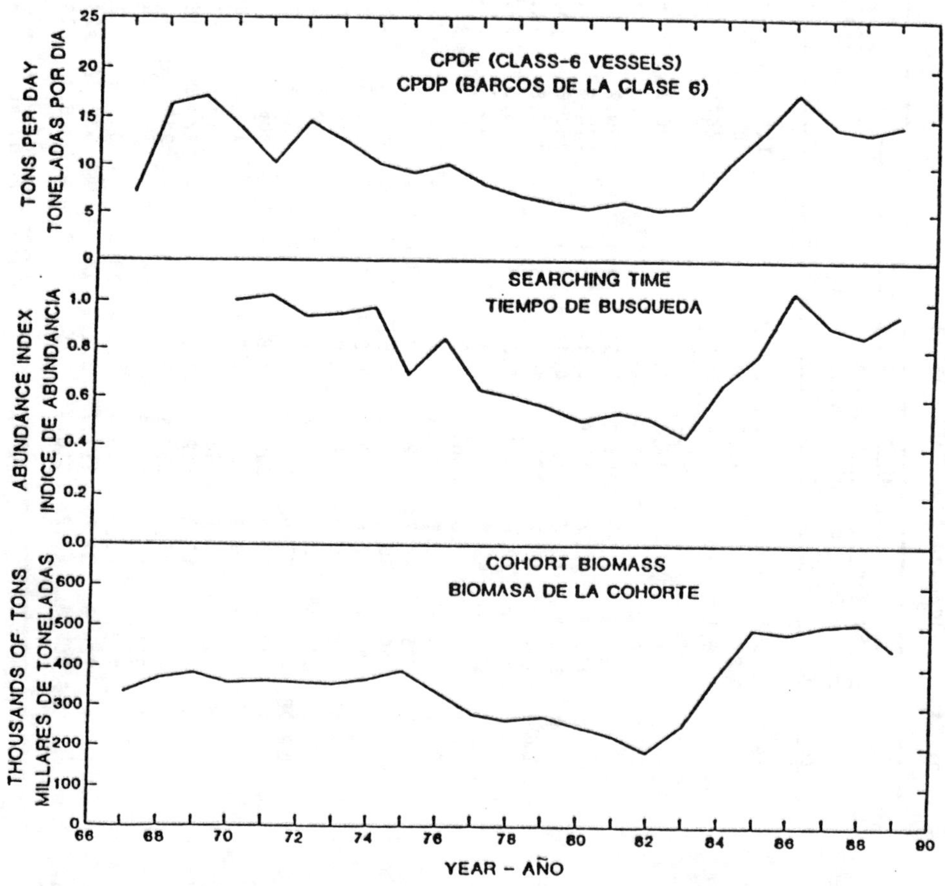

Figure 14. Three indices of abundance of yellowfin in the eastern Pacific. (Reproduced from IATTC, 1991a).

12.   POPULATION DYNAMICS

12.1   Recent Adjustments to the Basic Data

In 1989 the IATTC staff introduced several analytical modifications to its assessments of the yellowfin stock in the EPO (IATTC, 1991a). One of them was the separate treatment of yellowfin inside and outside the CYRA. In the past, tagging experiments (Section 6.3) and studies on population structure (Suzuki *et al.*, 1978) supported the view that the fish inside the CYRA represented the eastern Pacific stock while those outside the CYRA, but east of 150°W, belonged to the stock of the central Pacific. This separation was also upheld by historic differences in the relationship between catch and effort in the two regions. Inside the CYRA catch versus effort generated a curvilinear production model (see below) with the implied need for regulation, but outside the CYRA the linear relationship suggested that no constraints were necessary, at least during the 1968-88 period (IATTC, 1989). However, the fishery has been unregulated since 1979, and in the interim the distribution of effort has become more continuous over the adjacent areas. As a result, the increased amount of data has revealed that since the early 1970s the average sizes of fish in the catches (Figure 15, upper panel) and the CPDFs (Figure 15, lower panel) in both areas displayed similar trends. Analyses carried out in 1990 therefore treated yellowfin to the east of 150°W as a single stock, and this practice will probably be continued in the future. The growth relationships based on otolith increments (Wild, 1986) have also been incorporated in the analyses, and they indicate that the growth rate in length and weight is more rapid than previously thought. As mentioned earlier (Section 5.3), studies on sex ratios point to the

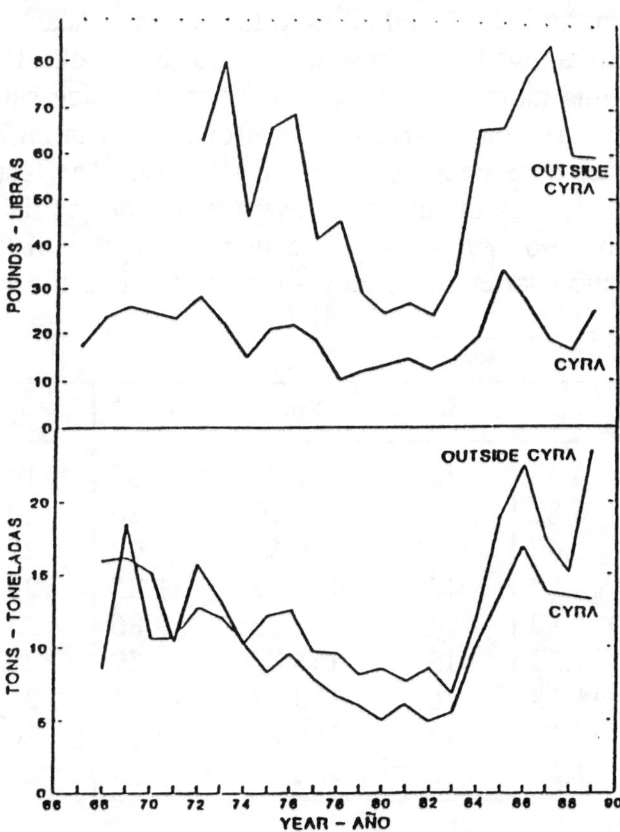

Figure 15. Average weights of yellowfin caught (upper panel) and average catches of yellowfin per day's fishing (lower panel) by Class-6 purse seiners in the eastern Pacific Ocean. (Reproduced from IATTC, 1991a).

virtual disappearance from the fishery of females larger than 140 cm, and this has prompted the use of different natural mortality rates for the sexes. Additionally, the length at recruitment has been adjusted downward from 40 to 30 cm, and the number of recruits has therefore increased as a result of their entering the fishery two months earlier. Finally, the speed of convergence and accuracy of fishing mortality estimates prepared through cohort analysis have been improved by using monthly, rather than quarterly, calculations. Despite all of these adjustments the biomass estimates (next Section) have not changed to a large extent, and the trends in the fishery remain intact (IATTC, 1991a).

12.2 Biomass Analyses

The IATTC staff has collected length-frequency samples from the yellowfin fishery in the EPO since 1954, and it has access to the corresponding information for 1951-54 collected by the California Department of Fish and Game. As a first step in the development of biomass estimates, the samples and total landings for each year are stratified by month, by 13 areas, and by gear type. Samples within strata are weighted and combined and, by procedures involving the growth-in-length relationship, the numbers of fish caught of each age group of the X and Y cohorts are estimated. Total monthly estimates of these fish within age groups and cohorts are obtained by summation across areas and gear types. When these monthly estimates are arranged chronologically in ascending order of years, the data are amenable to the application of cohort analysis, or the back-calculation to recruitment of all the numbers of fish in a cohort that contributed to its catch history. For this purpose, an annual, instantaneous natural mortality rate (M) of 0.8 is used for males of all ages, while for females M is set at 0.8 for the first 30 months in the fishery and allowed to increase linearly to 6.5 at 80 months. In practice these rates are allowed to vary somewhat for each cohort. Values of the annual instantaneous fishing mortality (F) used to initiate the back calculations are chosen on the basis of experience, seasonal variations in effort, and data on completely-fished age groups whose F values have converged toward stability. The results of these calculations appear in Table 5 as the quarterly average number of fish in each age group for each cohort during the 1967-89 period. The average weights of the fish during each quarter, beginning with the quarter (qtr.) they are recruited into the fishery were also estimated to be:

Qtr.	lb	kg	Qtr.	lb	kg	Qtr.	lb	kg
1	0.84	0.38	10	61.24	27.80	19	224.11	101.75
2	1.80	0.82	11	77.19	35.80	20	240.35	109.12
3	3.52	1.60	12	94.54	42.92	21	255.52	116.01
4	6.33	2.87	13	112.89	51.25	22	269.60	122.40
5	10.58	4.80	14	131.86	59.86	23	282.55	128.28
6	16.60	7.54	15	151.07	68.59	24	294.40	133.66
7	24.62	11.18	16	170.17	77.26	25	305.18	138.55
8	34.76	15.78	17	188.86	85.74	26	314.95	142.99
9	47.01	21.34	18	206.90	93.93	27	323.75	146.98

83

The average annual biomass is obtained by multiplying the data in Table 5 by the corresponding quarterly weight and averaging over four quarters. These biomass estimates appear in the bottom panel of Figure 14.

Compared to the abundance indices derived from CPDF or searching time, those calculated by the above method appear to be smoothed, or less sensitive to smaller fluctuations. Trends are preserved, however, including the decline during the mid 1970s to 1982 that resulted from escalating effort and the capture of both small and large fish. The recovery of the fishery also began in 1982, according to the cohort biomass estimate, one year earlier than the other indices began to respond. This may be because the cohort biomass is more responsive to the contribution of recruits. By 1985 the biomass appears to have stabilized at more than twice the level in 1982. While this may be partially the result of increasing trends in average recruitment (Figure 16 and Table 4), the tendency of the fishery to concentrate on larger fish must also have had a beneficial and substantial impact.

TABLE 5. Estimated average numbers of yellowfin in the eastern Pacific Ocean, in thousands, for the X and Y cohorts during the 1967 through 1989 seasons. (Reproduced from IATTC, 1991a).

Year Año	Quarter Trimestre	X Cohort--Cohorte X							Y Cohort--Cohorte Y						
		0	I	II	III	IV	V	VI	0	I	II	III	IV	V	VI
1967	1	0	14437	8315	1200	401	106	50	0	17991	2795	759	200	67	34
	2	0	11820	5742	923	276	83	41	37341	12777	1719	523	142	54	27
	3	0	8996	3998	712	196	66	34	30572	8167	1343	364	108	44	22
	4	21081	7217	3167	521	142	53	27	25029	6486	1064	250	84	36	18
1968	1	0	17259	5696	2379	369	108	43	0	20459	5057	810	175	67	29
	2	0	14124	3884	1218	243	84	35	25304	15220	3459	556	123	54	24
	3	0	11518	2838	872	169	67	29	20718	9635	2450	375	93	44	20
	4	23568	9214	2262	636	124	54	24	16942	7589	1947	259	73	36	16
1969	1	0	19296	7053	1797	423	93	44	0	13643	5408	1293	175	58	29
	2	0	15693	4949	1403	248	72	36	27047	9005	3226	631	121	47	24
	3	0	12476	3887	1083	170	58	29	22145	6245	2457	396	92	38	20
	4	27041	9358	3079	752	122	47	24	18128	4637	1883	242	72	31	16
1970	1	0	22139	6265	2206	478	93	38	0	14561	2975	1314	158	57	25
	2	0	18119	3727	1310	233	72	31	26627	10425	1719	768	102	46	20
	3	0	14692	2816	917	151	58	25	21800	7091	1326	475	77	38	17
	4	34572	11365	2132	595	105	46	21	17832	5255	1010	289	60	31	14
1971	1	0	28302	8231	1562	375	79	38	0	14297	3774	679	153	48	25
	2	0	22996	5654	1186	232	60	31	27402	9885	2283	427	103	39	21
	3	0	18455	4411	871	159	48	25	22435	7351	1706	280	78	31	17
	4	19739	14168	3481	547	109	39	21	18365	5790	1271	152	61	26	14
1972	1	0	16160	10615	2196	314	78	32	0	14530	4023	694	80	48	21
	2	0	13213	6550	1070	175	59	26	36708	10739	2811	324	46	38	17
	3	0	10727	4878	744	83	48	21	30054	7769	2136	161	31	31	14
	4	16617	8632	3485	466	44	38	17	24603	6127	1592	71	24	25	11
1973	1	0	13605	6882	2419	227	31	31	0	19366	4202	1082	41	19	21
	2	0	10847	5147	1321	106	24	25	68568	9270	2549	592	26	15	17
	3	0	8733	4097	942	56	19	21	56139	5202	1893	263	17	13	14
	4	46548	6953	3203	466	30	16	17	45629	3624	1456	84	13	10	11
1974	1	0	38110	5313	1982	238	17	13	0	33755	2518	990	23	10	8
	2	0	30958	4029	1131	142	13	10	28913	18225	1705	646	13	8	7
	3	0	24779	3242	783	55	10	8	23669	10006	1301	316	9	6	6
	4	36388	19319	2517	463	22	8	7	18476	7139	976	148	7	5	5
1975	1	0	29780	12768	1709	278	11	7	0	14517	5028	683	79	5	4
	2	0	23695	8546	1002	135	7	6	36914	10683	3484	387	28	4	3
	3	0	17789	6092	618	49	5	5	30134	8027	2371	197	11	3	3
	4	17412	13237	4516	356	19	4	4	24059	6428	1588	85	6	3	2
1976	1	0	14251	9798	2930	212	11	3	0	16120	4938	1111	33	5	2
	2	0	11537	5824	1988	91	6	3	28482	9309	3722	640	13	4	2
	3	0	8911	4110	1265	31	5	2	23285	5616	2327	243	9	3	1
	4	18656	6427	3133	702	15	4	2	18572	3802	1445	78	6	2	1
1977	1	0	15194	4864	2266	266	8	3	0	13899	2675	1056	35	5	2
	2	0	12028	3194	1504	155	6	3	69564	7988	1871	714	23	4	2
	3	0	9254	2219	680	58	5	2	56610	4594	1087	258	12	3	1
	4	34272	6679	1561	312	15	4	2	43514	3197	780	76	7	3	1

Table 5. (continued).

Year Año	Quarter Trimestre	X Cohort--Cohorte X							Y Cohort--Cohorte Y						
		0	I	II	III	IV	V	VI	0	I	II	III	IV	V	VI
1978	1	0	27913	4875	1170	180	9	3	0	28988	2324	570	32	5	2
	2	0	19549	3313	767	115	6	3	42647	13602	1574	388	18	4	2
	3	0	12173	2464	528	55	5	2	34214	7955	1169	231	10	3	1
	4	39511	8890	1760	298	14	4	2	25833	5742	855	80	4	3	1
1979	1	0	32319	5938	1268	157	7	3	0	16779	4014	534	39	2	2
	2	0	24285	3918	770	89	5	3	39983	9765	2224	347	23	2	2
	3	0	16476	3003	504	35	4	2	31792	6057	1525	181	13	2	1
	4	30104	9376	2246	322	22	3	2	20479	4164	1047	116	9	1	1
1980	1	0	24140	6211	1619	165	9	2	0	13553	3109	603	52	6	1
	2	0	16850	3176	1205	76	3	2	42579	7343	2319	383	16	2	1
	3	0	9968	2255	848	32	2	2	34400	4628	1599	189	7	1	1
	4	21997	6650	1659	563	16	2	1	26573	3376	1135	102	4	1	1
1981	1	0	17731	4726	1162	309	10	1	0	19186	2571	838	51	2	1
	2	0	12690	3039	817	136	4	1	31099	11151	1797	544	14	2	1
	3	0	7698	2039	549	64	3	1	25163	5716	1295	235	8	1	0
	4	23899	4121	1359	320	42	3	1	19207	3360	828	142	5	1	0
1982	1	0	19146	2537	968	180	10	2	0	12952	2121	571	40	3	1
	2	0	12545	1720	669	102	5	2	38886	8156	946	380	20	3	1
	3	0	7612	1307	446	50	4	2	30700	5114	678	185	12	2	1
	4	31253	5313	1022	254	24	2	1	21003	3951	448	94	7	2	0
1983	1	0	25405	4021	802	79	6	2	0	14884	3201	275	20	2	1
	2	0	19329	2933	612	28	3	1	41428	10188	2478	152	7	2	1
	3	0	13295	2232	411	13	2	1	33766	7542	1741	83	5	1	1
	4	39752	9868	1710	270	8	1	1	26865	6016	976	43	3	1	1
1984	1	0	32240	7960	1315	126	5	1	0	21168	4739	597	19	2	1
	2	0	25281	6059	985	56	3	1	38674	16308	3272	351	8	2	1
	3	0	17805	4770	666	18	2	1	31420	12568	2299	151	2	1	1
	4	41355	12279	3587	360	6	2	1	25586	9439	1507	59	1	1	0
1985	1	0	33858	9417	2628	143	2	1	0	20839	7013	933	20	1	1
	2	0	27078	7310	1759	57	1	1	34462	16225	4780	456	7	1	1
	3	0	21756	5277	1075	17	1	1	27976	12736	3127	174	2	1	1
	4	45636	16805	3998	589	4	1	1	22743	9976	2038	56	1	1	0
1986	1	0	37175	10302	2927	301	2	1	0	17305	7584	1279	25	1	0
	2	0	30113	7317	1827	96	1	0	42169	13302	5003	602	7	1	0
	3	0	23962	5848	1042	21	1	0	34321	9650	3266	169	3	1	0
	4	48943	18666	4629	544	7	1	0	26159	6334	2206	55	2	0	0
1987	1	0	39804	14361	3494	288	4	1	0	19754	4657	1463	16	1	0
	2	0	31679	10114	2141	158	2	1	62542	14358	2996	856	5	1	0
	3	0	23753	7548	1222	58	1	0	50984	11067	2039	327	2	1	0
	4	40731	15641	5091	633	19	1	0	40875	8705	1427	145	1	1	0
1988	1	0	33236	11269	3762	336	5	1	0	30323	6665	1023	44	1	0
	2	0	25862	7488	2729	119	2	1	55012	17089	4945	604	6	1	0
	3	0	17448	5365	1925	38	1	0	44877	11070	3711	309	1	1	0
	4	39380	11936	3651	1266	14	1	0	35771	7258	2513	154	0	1	0
1989	1	0	32214	8054	2593	776	5	1	0	26364	4774	1753	73	0	0
	2	0	24681	4839	1568	439	2	1	45781	19377	3015	1067	25	0	0
	3	0	18275	2675	940	253	1	0	37398	14794	1751	525	16	0	0
	4	45008	12751	1487	502	159	1	0	29920	11153	980	267	11	0	0

12.3 Yield-Per-Recruit Analysis

There does not appear to be a discernible relationship between the biomass of large fish in the EPO in any given year and the number of recruits to the fishery one year later (Figure 17; IATTC, 1991a). Constant recruitment has therefore usually been assumed in yield-per-recruit analysis. Recent changes in the distribution of effort though, have helped to identify two major periods in which differences in age-specific fishing

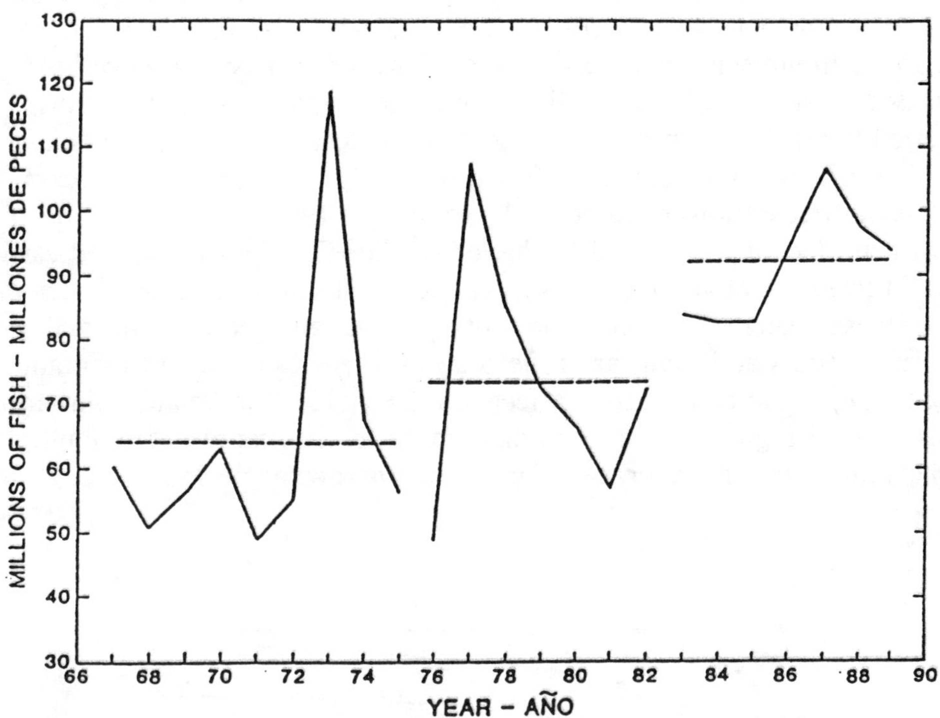

Figure 16. Estimated initial recruitments of yellowfin in the eastern Pacific Ocean, with the average recruitments for 1967-75, 1976-82, and 1983-89. (Reproduced from IATTC, 1991a).

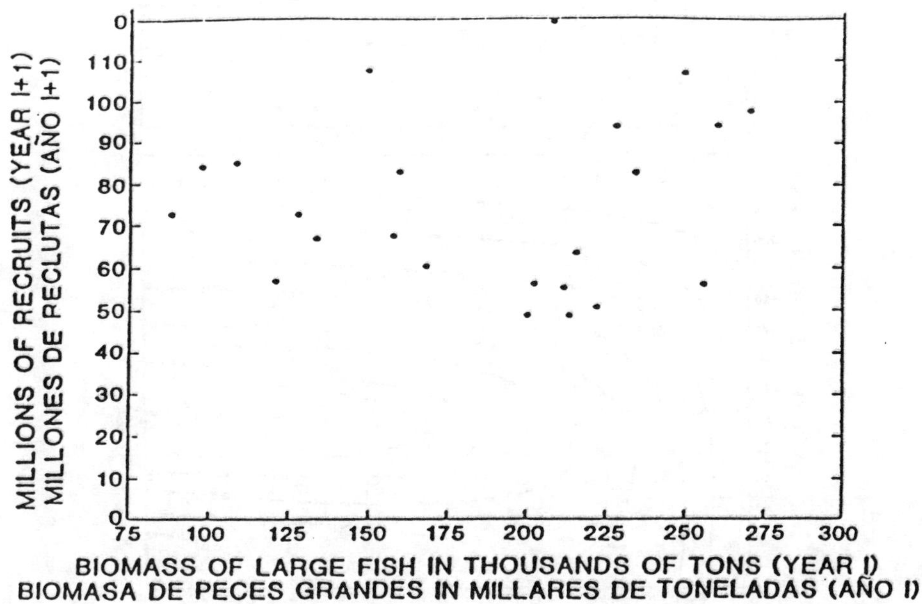

Figure 17. Relationship between stock and recruitment for yellowfin in the eastern Pacific Ocean for 1967 through 1989. Each dot represents the average recruitment of the X and Y cohorts. (Reproduced from IATTC, 1991a).

mortality rates have caused a significant change in the yield per recruit. For example, beginning in the last few months of 1973, the purse-seine fleet began to focus increasingly on the smaller fish associated with floating objects. This re-direction of effort was partly a response to the diminishing numbers of large fish, but it may also have been a reaction to the increased awareness of and public response to dolphin mortality. During the 1978-82 period the distribution of effort had evolved to the point that the age-specific mortalities of small, medium and large fish were similar. In subsequent years, however, effort was again directed toward larger fish, so that by 1985-89 their age-specific mortality was substantially greater than that of smaller fish. As mentioned earlier, a rising trend in recruitment (Figure 16) was also evident during this latter period. Collectively, these changes in fishing mortality patterns and recruitment, and recognition that recruitment occurs at 30 cm, are reflected in the yield-per-recruit diagrams for each period (Figure 18). The comparative effects on the yield-effort relationship are also more clearly illustrated in Figure 19. In each diagram the x-axis represents a multiple of effort in which the value 1.0 is the average effort for the particular period.

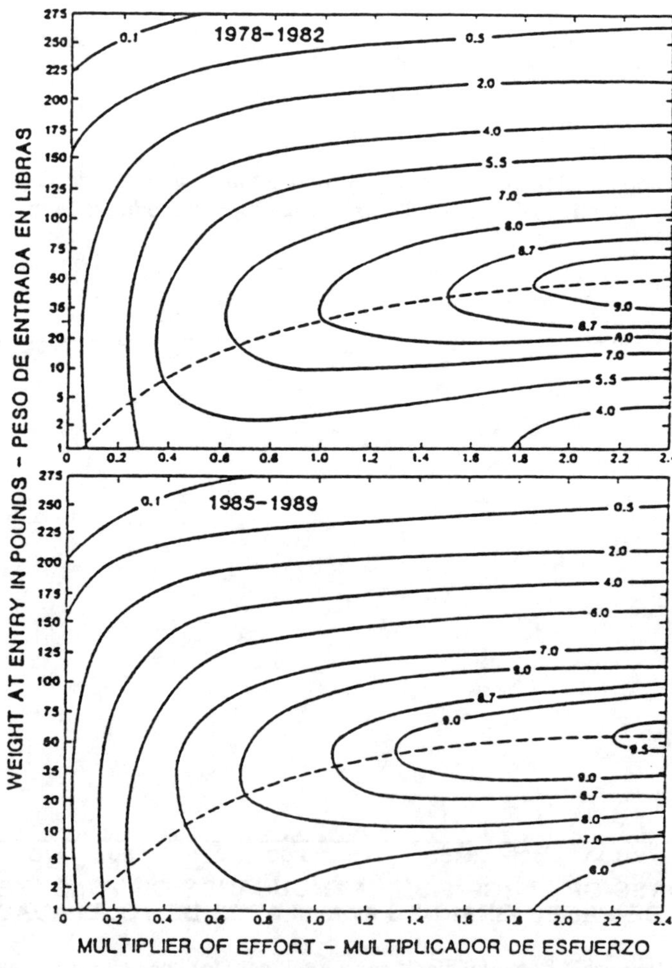

Figure 18. Relationships among size at entry, fishing effort, and yield per recruit for yellowfin for two different patterns of age-specific fishing mortality. (Reproduced from IATTC, 1991a).

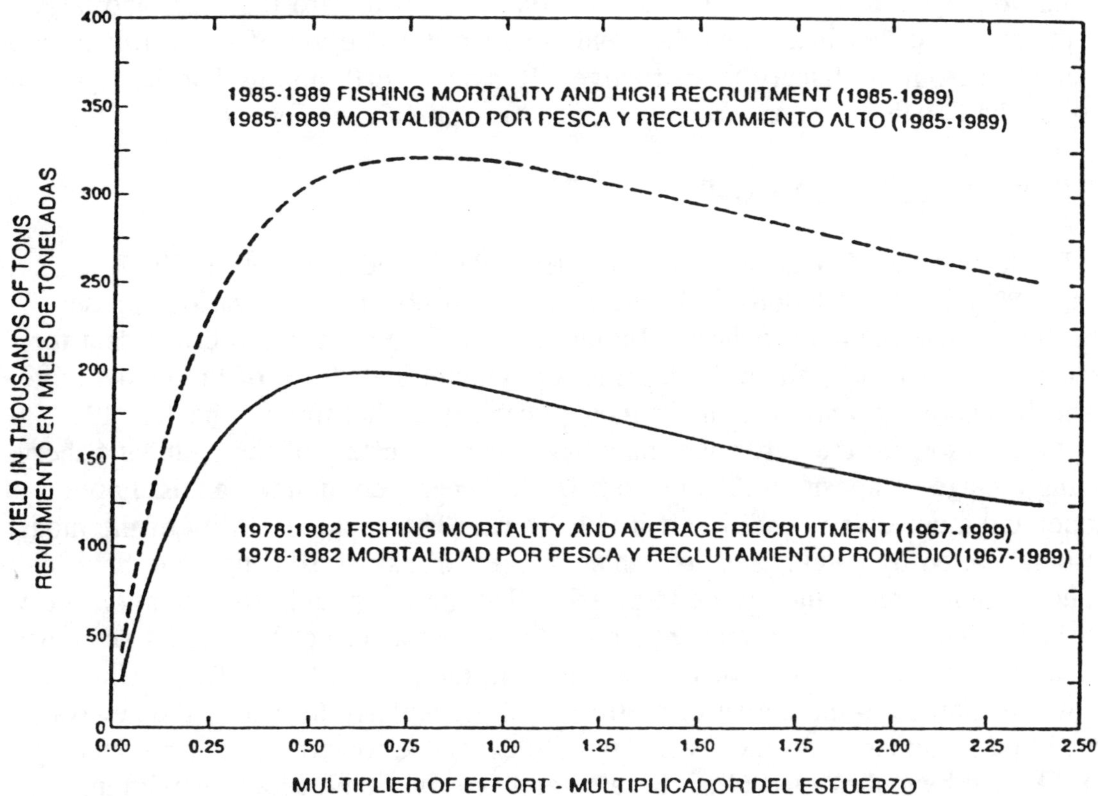

Figure 19. Yields for average recruitment and age-specific fishing mortality similar to that of 1978-1982 and for high recruitment and age-specific mortality similar to that of 1985-89. (Reproduced from IATTC, 1991a).

The major differences between the two periods for the surface fleet are summarized below:

Average	1978-1982	1985-1989	Percent Change
Effort (x 10^{-3} days)	30.96	20.24	-35
CPDF (mt)	5.39	13.26	+146
Catch (x 10^{-3} mt)	167.2	266.1	+59
Recruitment (x 10^{-3})	70.7	95.0	+34
Y/R (kg)	2.26	3.04	+34
Biomass (x 10^{-3} mt)	243	471	+94

It is reasonable to assume that the increased catch during the 1985-89 period stems from the increase in biomass, and in turn, that the biomass had benefitted from the reduction in effort and the increase in recruitment. These factors alone, however, do not explain the disproportionately large increase (146 percent) in CPDF. As a rough estimate, only 64 percent of this increase can be accounted for by changes in effort and recruitment. The remaining 36 percent stems from the reallocation of effort toward the older and larger fish in the stock, further increasing the yield per recruit. The use of less effort to obtain a greater yield from the fishery is instructive. It is also worth noting that the substantial increase in the CPDF has been clearly beneficial to some fishermen.

12.4   Production Model Analysis

Production model analysis (Schaefer, 1954; Pella and Tomlinson, 1969) has been useful in describing the relationship between catch and effort during various phases of development of the yellowfin fishery. Initially, the 1934-55 catch and effort data for the near-shore baitboat fishery fitted the logistic, or symmetrical, form of the model quite well, and there was no reason to doubt its applicability at that time (Schaefer, 1957; 1967). One reason for the agreement may have been the effect of the baitboat fishery on stabilizing the age composition (Tomlinson, IATTC; pers. commun.), an assumption of the model and a necessary condition if it is to work. The apparent utility of the model was also supported by tagging results, which suggested a slow exchange rate between the near- and offshore areas, and by the impression that the catch and effort relationship was not masked by changes in environmental conditions. As a further sign of its usefulness, the model detected the first overfishing condition in the early 1960s. The decline in CPDF was attributed to the increase in effort, and particularly to that of the newly-introduced purse seiners. As the fishery rapidly expanded toward the outer boundary of the CYRA, and beyond it in 1968, it became apparent that the average maximum sustainable yield (AMSY) exceeded that predicted for the inshore area. A series of experiments was therefore undertaken to permit the use of successively larger quotas as a means of empirically determining the AMSY, while simultaneously monitoring the effect on the apparent abundance (Peterson and Bayliff, 1985). The experimental programme remained in effect until 1979.

The production model intentionally describes average conditions over the time series of the available data and it is insensitive to rapid, short-term changes in recruitment or age-specific mortality (IATTC, 1991a). In addition, the effect of changes in age composition on the yield are not incorporated into the model. Partly as a result of these limitations, the increasingly large catches of the mid 1980s, generated by the rising trends in recruitment and the capture of larger fish after 1982 (see above), began to exceed the AMSY considerably. The declines in stock size predicted by the model, however, did not materialize and contrasted with the opposite trends indicated by CPUE, cohort-biomass and yield-per-recruit analyses. To adjust for these imbalances, the stock size in 1985, a year in which all three indices were relatively large (Figure 14), was re-estimated by dividing the CPDF (for Class-6 purse seiners) by the catchability coefficient (q). Although the parameter estimates of the model were not particularly affected by this process, the 1987 biomass estimate increased substantially and the fishery was effectively divided into two time periods. The early portion corresponded to the interval 1968-1983 with $q = 2.7 \times 10^{-5}$ and an AMSY of 181×10^3 mt, and the late stage included 1984-1989 with $q = 3.9 \times 10^{-5}$ and AMSY = 298×10^3 mt. The general production models (Pella and Tomlinson, 1969) fitted to the catch and effort data for these two distinct phases (Figure 20, upper panel) are skewed to the left, as indicated by the shape parameter, $m = 0.8$. For the latest phase the distribution of points indicates that, with the current level of recruitment and concentration on larger fish, the fishery is operating in the optimum region of the AMSY. In relation to the observed CPDF, the values predicted by the model (Figure 20, lower panel) are smoothed by the averaging process. The differences between the two types of values in the late period are also exaggerated because the fit is based on only five data points.

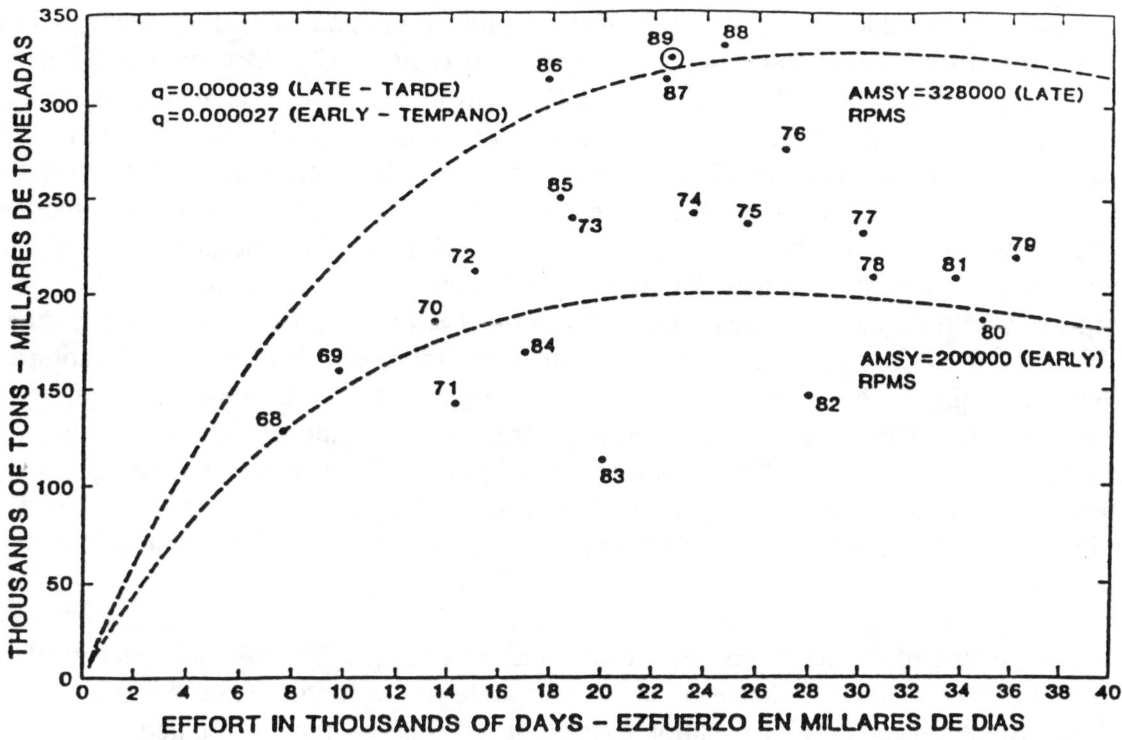

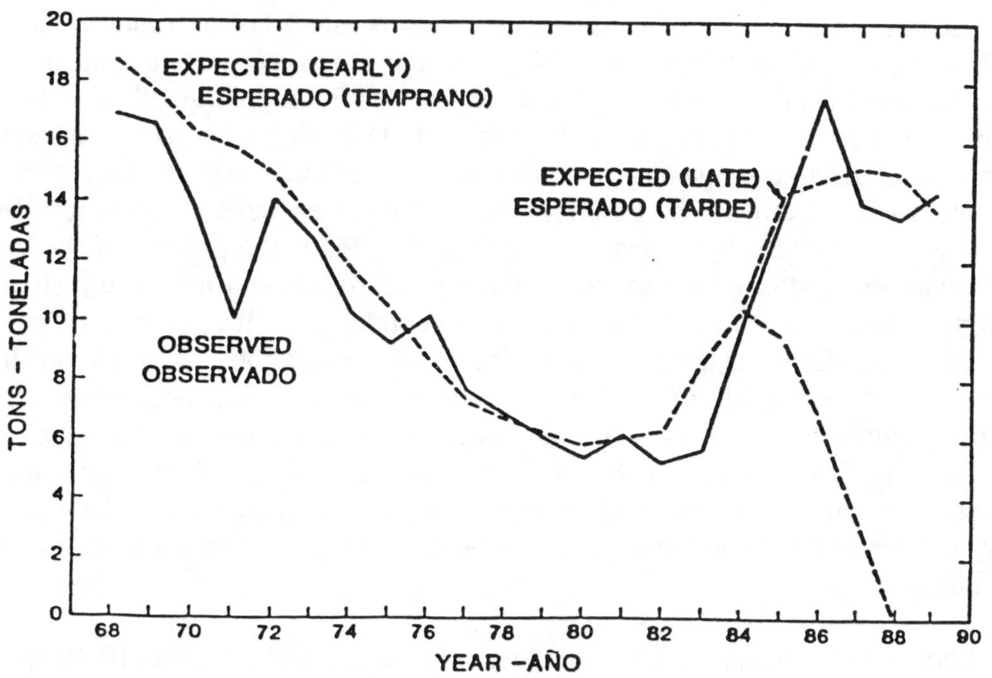

Figure 20. Relationships between effort and catch for the yellowfin fishery in the eastern Pacific Ocean during 1968-89 for the generalized stock production (asymmetrical) model with m set at 0.8 (upper panel), and observed and predicted values of CPDF (lower panel). (Reproduced from IATTC, 1991a).

13. GEAR INTERACTIONS

The interaction between longline and purse-seine gears involving yellowfin in the EPO began to materialize through encounters in the early 1960s. Since then the subject has appeared in longline-fishery reviews of the region (Suda and Schaefer, 1965b; Kume and Schaefer, 1966; Kume and Joseph, 1969; Shingu *et al.*, 1974; Miyabe and Bayliff, 1987), in population studies (Suzuki *et al.*, 1978) and in exploratory analyses (Lenarz and Zweifel, 1979; Hilborn, 1989). The topic has also been reviewed by the IATTC (1987) and Suzuki (1988). A rather consistent theme throughout this literature is that the fish caught by longlines are almost entirely large, whereas the surface fleet captures smaller yellowfin. While this may be historically correct on average, the emergence of the purse seine as the dominant surface gear has modified this viewpoint. A more accurate description would be that both gears catch the same sizes of large fish, mostly between 90-150 cm, but additionally, because purse seines are more efficient, they also capture smaller fish in the 30-90 cm range. This overall range of sizes developed through evolution of the purse-seine fishery in seeking larger fish. In the process the tonnage caught by longlining in the EPO fell below that of purse seining in all size categories for the first time beginning in the early 1970s (Miyabe and Bayliff, 1987), and since then the quantity of longline-caught fish has declined progressively to a small fraction of that of the surface fleet (Table 4).

One of the topics examined by Lenarz and Zweifel (1979) was the effect on the Y/R of the degree of independence of stocks available to the longline and surface fisheries. Since no reliable information existed at the time on the simultaneous availability of yellowfin to both gears, they constructed stocks for each fishery based on the average size composition caught by the different gears in the eastern Atlantic during 1967-71 (Lenarz *et al.*, 1974). The stocks also incorporated the sex ratios found in the central Pacific (Murphy and Shomura, 1972) for each gear type, and realistically portrayed the rapid decline in the proportion of females for fish larger than 135-140 cm. Simulation was then used to determine the effect of: 1) fishing a homogeneous stock by both gears, 2) allowing each gear type to fish a separate stock, and 3) allowing access of purse seiners to 50 percent of the longline stock. They drew several inferences from the study that were considered to be generally applicable. First, the portion of the longline catch that came from stock(s) exploited by the surface fishery did not have a significant effect on the relative values of Y/R within each hypothetical fishing situation. Consequently, "... decisions concerning minimum size regulations ..." were not likely to be affected by the availability of stock(s) to the two fisheries. Second, the effect of one gear on the yield of the other was less pronounced for the realistic sex ratio than for the usual, 1:1 assumption. Finally, if the fish are equally available to both gears, the surface fishery depressed the Y/R of the longline gear to a greater extent than in the reverse situation, and the Y/R for combined gears was somewhat greater than the sum of each gear operating alone.

Hilborn (1989) reanalyzed the data used by Lenarz and Zweifel (1979) in the context of a spatially-structured population model in which the separate stocks fished by longlines and purse seiners were connected only by movements of individual fish. With constant recruitment, the Y/R analysis agreed with that of Lenarz and Zweifel (1979) in that the maximum amount was obtained when both fisheries were operating. In a stock-recruit situation, the total yield was still maximized by a combined fishery even though the overall yield was reduced for both gear types. Hilborn (1989) emphasized that the results were conditional upon the size distribution in the catch and the assumed growth curve. It should be mentioned that the data used in the above studies represents an historic situation because the catch in numbers for both gears were roughly the same. Today, as a consequence of mounting pressure from the purse-seine fleet, the longline catch in the eastern Atlantic and Pacific fisheries is about an order of magnitude less than that of purse seiners. Suzuki (1988) also pointed out that the study by Lenarz and Zweifel (1979) implied that the Y/R for both fisheries were about equal, and masked the impact of purse seining on the longline fishery. By assuming that in the eastern Atlantic the purse-seine and longline gears caught fish between 1-8 and 3-8 years, respectively, a Y/R analysis demonstrated that longlining was relatively much more sensitive to competition than purse seining.

Tagging results have indicated that there is a relatively low rate of mixing in an east-west direction within the EPO (Bayliff, 1979; 1984), and therefore it is reasonable to assume that the longline and purse-seine fisheries operate on the same stock. This has been difficult to demonstrate, however, given that depth appears to be the only major obstacle that separates the different gears. For example, tags from only six, large fish originally caught by surface gear have been returned by longline fishermen (IATTC, 1973, 1975, 1982), and one fish in the reverse direction (IATTC, 1982). In search of other linkages between the fisheries, Suzuki *et al.* (1978) noted that in the same area and similar period of time, female yellowfin caught by purse seines demonstrated a substantially higher degree of sexual maturity (gonad index) than those caught on longlines. Females captured by surface gear also displayed seasonal trends in reproductive activity in different size categories, a feature that was not as apparent for the longline-caught fish. Similar results were reported by Koido and Suzuki (1989), but for non-overlapping time periods. The amount of spatial overlap between the fisheries has also changed over time and obscured their relationship to a common stock. During 1971-80, for example, the longline fleet was largely absent from the region between $10°-20°N$ in the EPO, an area that was previously shared to a degree by both fisheries. At the same time, longlining operations expanded into the region of $20°-30°S$, well beyond the southern limit of the purse-seine fishery.

Longline effort in the CYRA increased from 70×10^6 hooks during 1963-75 to 160×10^6 hooks in 1983 (Suzuki, 1988, Figure 30), but despite this substantial change the tonnage of yellowfin caught by this fishery has remained relatively stable since 1970 (Table 4, column 5). If the purse-seine and longline fisheries operate on the same stock, this stability is difficult to explain considering that over the same period the trends in the catch of the surface fleet reached high (288.2×10^3 mt) and low (94.1×10^3 mt) extremes (Table 4). In part, the stability may be due to the increased use of deep longlining and other selective practices to capture bigeye tuna, so that yellowfin has become an incidental catch. The fact that the longline yield of bigeye in the EPO has increasingly surpassed that of yellowfin since the early 1960s tends to support this view (IATTC, 1989). The stability may have also arisen from a dynamic balance achieved through increasing longline effort and the long-term effect of the purse-seine fishery's interception of a portion of the longline recruits. In this situation the change in the age composition of the longline catch would be reflected in a reduction in average weight, and indeed this quantity has gradually declined from approximately 59 kg (130 lb) during 1958-62 to 43 kg (95 lb) in 1967-80 (Miyabe and Bayliff, 1987). Further evidence that the two gear types act on the same stock is derived from a comparison of the CPUE of Class-6 seiners and longlines specifically for large fish (Figure 21, upper panel). The trends in these variables are visibly reinforced with appropriate scaling (Figure 21, lower panel) such that 73,800 hooks appear to be equivalent to one day of Class-6 effort (IATTC, 1987). In Section 11 it was mentioned that the escalation of effort and harvest of smaller fish by the purse-seine fleet during the 1970s and early 1980s eventually drove the biomass estimate to its lowest level in 1982. Since the CPUEs for both gear types paralleled each other during this period, and since the decline in abundance is reflected in the CPUE for the purse-seine fleet (Figure 14), there is little doubt of the impact that the purse-seine fishery has had on the longline catch.

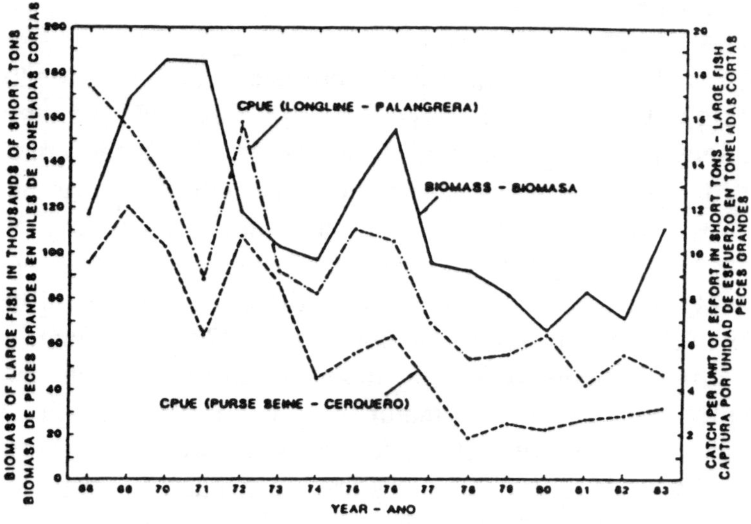

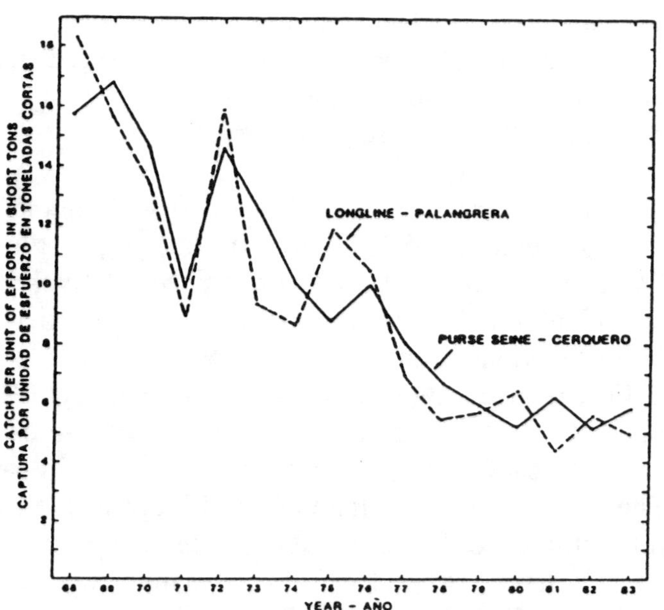

Figure 21. CPUEs of large yellowfin by Class-6 purse seiners and by longliners, compared to biomass estimates of large yellowfin (upper panel); and catch per day's fishing by Class-6 purse seiners, compared to catch of yellowfin per 73,800 hooks by longliners (lower panel). All comparisons are for the eastern Pacific Ocean east of 150°W. (Reproduced from IATTC, 1987).

To help clarify the effect of the longline and purse-seine fisheries on each other and on the stock of yellowfin they exploit, research in a few key areas is needed. First, the timing and behavioural movement of fish from one fishery to the other will probably be best illuminated by tagging experiments. However, steps should be taken first to discover or circumvent the reasons why past experiments have produced such poor recovery rates. If tagged purse-seine- and longline-caught fish have little prospect of survival, the use of other gear, such as baitboats, to seek out and tag large yellowfin may be more successful. Since the two types of fisheries may interact in the Pacific, Atlantic and Indian Oceans, the location with the greatest potential for producing information should also be chosen. Second, the sex ratio and stages of female sexual maturity in the same time-area strata should be explored more fully in both fisheries. This is important for clarifying seasonal reproductive patterns and determining if changes in sex ratio are indicative of movement between the fishing zones. Third, an effort should be made to update and maintain a current position on the catch and CPUE of longline operations in the EPO. The information is needed to identify the response of the fishery relative to the increased recruitment and post-1983 recovery of the purse-seine fishery. The data are also necessary to monitor and detect changes or similarities in the CPUE between the two types of gear.

14.   ACKNOWLEDGEMENTS

I would like to thank several members of the Commission's staff, and particularly Wm. Bayliff, R. Deriso and Pat. Tomlinson, for generously providing their time and counsel during the preparation of the manuscript.

15.   REFERENCES CITED

Aikawa, H., and M. Kato. 1938. Age determination of fish-I [In Jpn.; Engl.summary]. *Bull.Jap.Soc.Sci.Fish.*, 7(2):79-88. English translation *In* W.G. Van Campen, 1950, *Spec.Sci.Rep.U.S.Fish Wildl.Serv. (Fish.)*, (21):22 p.

Allen, R., and R. Punsly. 1984. Catch rates as indices of abundance of yellowfin tuna, *Thunnus albacares*, in the eastern Pacific Ocean. *Bull.I-ATTC*, 18(4):303-79.

Au, D.W.K., W.L. Perryman, and W.F. Perrin. 1979. Dolphin distribution and the relationship to environmental features in the eastern tropical Pacific. *Admin.Rep.NMFS-SWFC La Jolla*, LJ-79-43:59 p.

Bard, F.X. 1983. Croissance de l'albacore (*Thunnus albacares*) Atlantique, d'apres les donnes des marquages. *Collect.Vol.Sci.Pap.ICCAT*, 20(1):104-16.

Bard, F.X. 1984. Aspects de la croissance de l'albacore est Atlantique (*Thunnus albacares*) a partir des marquages. *Collect.Vol.Sci.Pap.ICCAT*, 21:108-14.

Barrett, I., and H. Tsuyuki. 1967. Serum transferrin polymorphism in some scombroid fishes. *Copeia*, (3):551-57.

Bayliff, W.H. 1971. Estimates of the rates of mortality of yellowfin tuna in the eastern Pacific Ocean derived from tagging experiments. *Bull.I-ATTC*, 15(4):379-436.

Bayliff, W.H. 1979. Migrations of yellowfin tuna in the eastern Pacific Ocean as determined from tagging experiments initiated during 1968-1974. *Bull.I-ATTC*, 17(6):447-506.

Bayliff, W.H. 1983. Analysis of population structure of fishes from life-history data. *Collect.Vol.Sci.Pap.ICCAT*, 18:776-91.

Bayliff, W.H. 1984. Migrations of yellowfin and skipjack tuna released in the central portion of the eastern Pacific Ocean, as determined by tagging experiments. *Intern.Rep.I-ATTC*, (18):107 p.

Bayliff, W.H. 1988. Growth of skipjack, *Katsuwonus pelamis*, and yellowfin, *Thunnus albacares*, tunas in the eastern Pacific Ocean, as estimated from tagging data. *Bull.I-ATTC*, 19(4):311-85.

Bayliff, W.H., and B.J. Rothschild. 1974. Migrations of yellowfin tuna tagged off the southern coast of Mexico in 1960 and 1969. *Bull.I-ATTC*, 16(1):1-64.

Bell, R.R. 1964. A history of tuna age determinations. Proc. Symp.Scombroid Fishes; *J.Mar.Biol.Assoc.India*, 2:693-706.

von Bertalanffy, L. 1938. A quantitative theory of organic growth. *Hum.Biol.*, 10:181-213.

Blackburn, M. 1965. Oceanography and the ecology of tunas. *Oceanogr. Mar.Biol.Ann.Rev.*, 3:299-322.

Blunt, C.E., Jr., and J.D. Messersmith. 1960. Tuna tagging in the eastern tropical Pacific, 1952-1959. *Calif.Fish Game*, 46(3):301-69

Broadhead, G.C. 1959. Morphometric comparisons among yellowfin tuna, *Neothunnus macropterus*, the eastern tropical Pacific Ocean. *Bull.I-ATTC*, 3(8):353-91.

Broadhead, G.C., and C.J. Orange. 1960. Species and size relationships within schools of yellowfin and skipjack tuna, as indicated by catches in the eastern tropical Pacific Ocean. *Bull.I-ATTC*, 4(7):449-92.

Brouard, F., R. Grandperrin and E. Cillaurren. 1984. Croissance des juenes thons jaunes (*Thunnus albacares*) et de bonites (*Katuwonus pelamis*) dans de Pacifique tropical occidental. [In Fr.; Engl. summary]. *Doc.d'Ocean.ORSTOM Port-Vila*, 10:23 p.

Buñag, D.M. 1956. Spawning habits of some Philippine tuna based on diameter measurements of the ovarian ova. *J.Philipp.Fish.*, 4(2):145-77.

Bushnell, P.G., R.W. Brill and R.E.Bourke. 1990. Cardiorespiratory responses of skipjack tuna, *Katsuwonus pelamis*; yellowfin tuna, *Thunnus albacares*; and bigeye tuna, *T. obesus*, to acute reductions of ambient oxygen. *Can.J.Zool.*, :1857-65.

Calkins, T.P. 1975. Geographical distribution of yellowfin and skipjack tuna catches in the eastern Pacific Ocean and total catch statistics, 1971-1974. *Bull.I-ATTC*, 17(1):1-116.

Calkins, T.P., and B.M. Chatwin. 1967. Geographical distribution of yellowfin tuna and skipjack catches in the eastern Pacific Ocean, by quarters of the year, 1963-1966. *Bull.I-ATTC*, 12(6): 433-508.

Calkins, T.P., and B.M. Chatwin. 1971. Geographical catch distribution of yellowfin and skipjack tuna in the eastern Pacific Ocean, 1967-1970, and fleet and total catch statistics, 1962-1970. *Bull.I-ATTC*, 15(3):283-377.

Carey, F.G., and R.J. Olson. 1982. Sonic tracking experiments with tunas. *Coll.Vol. Sci.Pap.ICCAT*, 17(2):458-66.

Chatwin, B.M. 1959. The relationships between length and weight of yellowfin tuna (*Neothunnus macropterus*) and skipjack tuna (*Katsuwonus pelamis*) from the eastern tropical Pacific ocean. *Bull.I-ATTC*, 3(7):307-52.

Cole, J.S. 1980. Synopsis of biological data on the yellowfin tuna, *Thunnus albacares* (Bonnaterre, 1788), in the Pacific Ocean. *Spec.Rep.I-ATTC*, (2):75-150.

Collette, B.B., and C.E. Nauen. 1983. FAO species catalogue. Vol. 2. Scombrids of the world. An annotated and illustrated catalogue of tunas, mackerels, bonitos and related species known to date. *FAO Fish.Synop.*, 2(125):137 p.

Collette, B.B., T. Potthoff, W.J. Richards, S. Ueyanagi, J.L. Russo and Y. Nishikawa. 1984. Scombroidei: development and relationships. *In* Ontogony and systematics of fishes, edited by H.G. Moser. *Spec.Publ.Am.Soc.Ichthyol.Herpetol.*, (1): 591-620.

Davidoff, E.B. 1963. Size and year class composition of catch, age and growth of yellowfin tuna in the eastern tropical Pacific Ocean, 1951-1961. *Bull.I-ATTC*, 8(4):199-251.

Davis, T.L.O., G.P. Jenkins and J.W. Young. 1990. Diel patterns of vertical distribution in larvae of southern bluefin *Thunnus maccoyii*, and other tuna in the east Indian Ocean. *Prog.Ser.Mar.Ecol.*, 59:63-74.

Díaz, E.L. 1963. An increment technique for estimating growth parameters of tropical tunas, as applied to yellowfin tuna (*Thunnus albacares*). *Bull.I-ATTC*, 8(7):381-416.

Draganik, B., and W. Pelczarski. 1984. Growth and age of bigeye and yellowfin tuna in the central Atlantic as per data gathered by R/V "Wieczno." *Collect.Vol.Sci.Pap.ICCAT*, 20:96-103.

Fahay, M.P. 1983. Guide to the early stages of marine fishes occurring in the western North Atlantic Ocean, Cape Hatteras to the southern Scotian Shelf. *J.Northwest.Atl.Fish.Sci.*, 423 p.

Fink, B.D., and W.H. Bayliff. 1970. Migrations of yellowfin and skipjack tuna in the eastern Pacific Ocean as determined by tagging experiments, 1952-1964. *Bull.I-ATTC*, 15(1):1-227.

Fishery Agency of Japan. 1974. Annual report of catch and effort statistics by area on Japanese longline fishery 1972. *Annu.Rep.Effort Catch Stat.Area Jap.Tuna Longline Fish.*, (1972): 279 p.

Fishery Agency of Japan. 1975. Annual report of catch and effort statistics by area on Japanese longline fishery 1973. *Annu.Rep.Effort Catch Stat.Area Jap.Tuna Longline Fish.*, (1973):265 p.

Fishery Agency of Japan. 1976. Annual report of catch and effort statistics by area on Japanese longline fishery 1974. *Annu.Rep.Effort Catch Stat.Area Jap.Tuna Longline Fish.*, (1974):267 p.

Fishery Agency of Japan. 1977. Annual report of catch and effort statistics by area on Japanese longline fishery 1975. *Annu.Rep.Effort Catch Stat.Area Jap.Tuna Longline Fish.*, (1975):269 p.

Fonteneau, A. 1980. La croissance de l'albacore de l'Atlantique est. *Collect.Vol.Sci. Pap.ICCAT*, 9:152-68.

Francis, R.C. 1974. TUNPOP, a computer simulation model of the yellowfin tuna population and the surface tuna fishery of the eastern Pacific Ocean. *Bull.I-ATTC*, 16(3):233-79.

Francis, R.C. 1977. TUNPOP: a simulation of the dynamics and structure of the yellowfin tuna stock and surface fishery of the eastern Pacific Ocean. *Bull.I-ATTC*, 17(4):233-79.

Fujino, K. 1970. Immunological and biochemical genetics of tunas. *Trans.Am.Fish. Soc.*, 99(1):152-78.

Fujino, K., and T. Kang. 1968a. Transferrin groups of tunas. *Genetics*, 59:79-91.

Fujino, K, and T. Kang. 1968b. Serum esterase groups of Pacific and Atlantic tunas. *Copeia*, (1):56-63.

Gibbs, R.H. Jr., and B.B. Collette. 1967. Comparative anatomy and systematics of the tunas, genus *Thunnus*. *Fish.Bull.U.S.Fish Wildl.Serv.*, 66(1):65-130.

Godsil, H.C. 1948. A preliminary population study of the yellowfin tuna and the albacore. *Fish Bull.Calif.Dep.Fish Game*, (70):90 p.

Godsil, H.C., and E.E. Greenhood. 1951. A comparison of the populations of the yellowfin tuna, *Neothunnus macropterus*, from the eastern and central Pacific. *Fish Bull.Calif.Dep.Fish Game*, (82):33 p.

Graves. J.E., M.A. Samovich, and K.M. Schaefer. 1988. Electrophoretic identification of early juvenile yellowfin tuna, *Thunnus albacares*. *Fish.Bull.NOAA-NMFS*, 86(4):835-38.

Green, R.E. 1967. Relationships of the thermocline to success of purse seining for tuna. *Trans.Am.Fish.Soc.*, 96(2):126-30.

Hammond, P.S. (editor). 1981. Report on the workshop on tuna-dolphin interactions. *Spec.Rep.I-ATTC*, 4:259 p.

Hanamoto, E. 1974. Fishery oceanography of bigeye tuna-I. *La Mer (Bulletin de la Société Franco-Japonaise d'Océanographie)*, 12(3):128-36.

Harada, T., O. Murata, and S. Oda. 1980. Rearing of and morphological changes in larvae and juveniles of yellowfin tuna. *Bull.Fac.Agric.Kinki Univ.*, (13):33-6.

Harada, T., K. Mizuno, O. Murata, S. Miyashita, and H. Furutani. 1971. On the artificial fertilization and rearing of larvae in yellowfin tuna. *Bull.Fac.Agric.Kinki Univ.*, (4):145-51.

Hayashi, S. 1957. A review on age determination of the Pacific tunas. *Indo-Pac. Fish.Coun.*, 7(II-III):53-64.

Hennemuth, R.C. 1961a. Size and year class composition of catch, age and growth of yellowfin tuna in the eastern tropical Pacific Ocean. *Bull.I-ATTC*, 5(1):1-112.

Hennemuth, R.C. 1961b. Year class abundance, mortality and yield-per-recruit of yellowfin in the eastern Pacific Ocean, 1954-1959. *Bull.I-ATTC*, 6(1):1-51.

Higgins, B.E. 1967. The distribution of juveniles of four species of tunas in the Pacific Ocean. *Proc.Indo-Pac.Fish.Coun.*, 12(2):79-99.

Higgins, B.E. 1970. Juvenile tunas collected by midwater trawling in Hawaiian waters, July-September 1967. *Trans.Am.Fish.Soc.*, 99(1):60-73.

Hilborn, R. 1989. Yield estimation for spatially connected populations: an example of surface and longline fisheries for yellowfin tuna. *North Amer.J.Fish.Mgmt.* 9(4):402-10.

Hisada, K. 1973. Investigation on tuna hand-line fishing ground and some biological observations on yellowfin and bigeye tunas in the northwestern Coral Sea. *Bull.Far Seas Fish.Res.Lab.*, 8:35-69.

Holland, K.N., R.W. Brill and R.K.C. Chang. 1990. Horizontal and vertical movements of yellowfin and bigeye tuna associated with fish aggregating devices. *Fish.Bull.NOAA-NMFS*, 88(3):483-507.

Inter-American Tropical Tuna Commission. 1971. Annual report of the Inter-American Tropical Tuna Commission, 1970. *Annu.Rep.I-ATTC*, (1970):127 p.

Inter-American Tropical Tuna Commission. 1972. Annual report of the Inter-American Tropical Tuna Commission, 1971. *Annu.Rep.I-ATTC*, (1971):129 p.

Inter-American Tropical Tuna Commission. 1973. Annual report of the Inter-American Tropical Tuna Commission, 1972. *Annu.Rep.I-ATTC*, (1972):166 p.

Inter-American Tropical Tuna Commission. 1974. Annual report of the Inter-American Tropical Tuna Commission, 1973. *Annu.Rep.I-ATTC*, (1973):150 p.

Inter-American Tropical Tuna Commission. 1975. Annual report of the Inter-American Tropical Tuna Commission, 1974. *Annu.Rep.I-ATTC*, (1974):169 p.

Inter-American Tropical Tuna Commission. 1976. Annual report of the Inter-American Tropical Tuna Commission, 1975. *Annu.Rep.I-ATTC*, (1975):176 p.

Inter-American Tropical Tuna Commission. 1977. Annual report of the Inter-American Tropical Tuna Commission, 1976. *Annu.Rep.I-ATTC*, (1976):180 p.

Inter-American Tropical Tuna Commission. 1978. Annual report of the Inter-American Tropical Tuna Commission, 1977. *Annu.Rep.I-ATTC*, (1977):155 p.

Inter-American Tropical Tuna Commission. 1979. Annual report of the Inter-American Tropical Tuna Commission, 1978. *Annu.Rep.I-ATTC*, (1978):163 p.

Inter-American Tropical Tuna Commission. 1980. Annual report of the Inter-American Tropical Tuna Commission, 1979. *Annu.Rep.I-ATTC*, (1979):227 p.

Inter-American Tropical Tuna Commission. 1981. Annual report of the Inter-American Tropical Tuna Commission, 1980. *Annu.Rep.I-ATTC*, (1980):234 p.

Inter-American Tropical Tuna Commission. 1982. Annual report of the Inter-American Tropical Tuna Commission, 1981. *Annu.Rep.I-ATTC*, (1981):303 p.

Inter-American Tropical Tuna Commission. 1983. Annual report of the Inter-American Tropical Tuna Commission, 1982. *Annu.Rep.I-ATTC*, (1982):272 p.

Inter-American Tropical Tuna Commission. 1984. Annual report of the Inter-American Tropical Tuna Commission, 1983. *Annu.Rep.I-ATTC*, (1983):272 p.

Inter-American Tropical Tuna Commission. 1987. Annual report of the Inter-American Tropical Tuna Commission, 1986. *Annu.Rep.I-ATTC*, (1986):264 p.

Inter-American Tropical Tuna Commission. 1988. Annual report of the Inter-American Tropical Tuna Commission, 1987. *Annu.Rep.I-ATTC*, (1987):222 p.

Inter-American Tropical Tuna Commission. 1989. Annual report of the Inter-American Tropical Tuna Commission, 1988. *Annu.Rep.I-ATTC*, (1988):288 p.

Inter-American Tropical Tuna Commission. 1990. Quarterly report of the Inter-American Tropical Tuna Commission. *Quart.Rep.I-ATTC*, 2:5-6.

Inter-American Tropical Tuna Commission. 1991a. Annual report of the Inter-American Tropical Tuna Commission, 1989. *Annu.Rep.I-ATTC*, (1989):270 p.

Inter-American Tropical Tuna Commission. 1991b. Quarterly report of the Inter-American Tropical Tuna Commission. *Quart.Rep.I-ATTC*, 2:(in press).

Joseph, J. 1963. The fecundity of yellowfin tuna (*Thunnus albacares*) and skipjack (*Katsuwonus pelamis*) from the eastern Pacific Ocean. *Bull.I-ATTC*, 7(4):255-92.

Joseph, J., and F.R. Miller. 1988. El Ni±o and the surface fishery for tunas in the eastern Pacific. *In* Proceedings of the Tuna Fishery Research Conference, Far Seas Fish.Res.Lab. *Maguro Gyogyo Kyogikai Gijiroku, Suisancho-Enyo Suisan Kenkyusho*:199-207.

Joseph, J., F.G. Alverson, B.D. Fink, and E.B. Davidoff. 1964. A review of the population structure of yellowfin tuna, *Thunnus albacares*, in the eastern Pacific Ocean. *Bull.I-ATTC*, 9(2):53-112.

June, F.C. 1953. Spawning of yellowfin tuna in Hawaiian waters. *Fish.Bull.U.S.Fish Wildl.Serv.*, 54(77):47-64.

Kamimura, T., and M. Honma. 1963. Distribution of the yellowfin tuna *Neothunnus macropterus* (Temminck and Schlegel) in the tuna longline fishing grounds of the Pacific Ocean. *FAO Fish.Rep.*, 6(3):1299-328.

Kikawa, S. 1959. Notes on the regional difference of spawning season of Pacific yellowfin tuna. *Rep.Nankai Reg.Fish.Res.Lab.*, (11):59-76.

Kikawa, S. 1962. Studies on the spawning activity of Pacific tunas, *Parathunnus mebachi* and *Neothunnus macropterus*, by the gonad index examination. *Rep.Nankai Reg.Fish.Res.Lab.*, (1):43-56.

Kikawa, S. 1966. The distribution of maturing bigeye and yellowfin and an evaluation of their spawning potential in different areas in the tuna longline grounds in the Pacific. *Rep.Nankai Reg.Fish.Res.Lab.*, (23):131-208.

Klawe, W.L. 1963. Observations on the spawning of four species of tuna, *Neothunnus macropterus, Katsuwonus pelamis, Auxis thazard*, and *Euthynnus lineatus*, in the eastern Pacific Ocean, based on the distribution of their larvae and juveniles. *Bull.I-ATTC*, 6(9):447-540.

Klawe, W.L. 1980. Classification of tunas, mackerels, billfishes, and related species and their geographical distribution. *Spec.Rep.I-ATTC*, (2):7-16.

Klawe, W.L., J.J. Pella, and W.S. Leet. 1970. The distribution, abundance and ecology of larval tunas from the entrance to the Gulf of California. *Bull.I-ATTC*, 14(4):505-44.

Knudsen, P.F. 1977. Spawning of yellowfin tuna and the discrimination of subpopulations. *Bull.I-ATTC*, 17(2):117-69.

Koido, T.V., and Z. Suzuki. 1989. Main spawning of yellowfin, *Thunnus albacares*, in the western tropical Pacific Ocean based on the gonad index. *Bull.Far Seas Fish.Res.Lab.*, (26):153-63.

Kume, S., and J. Joseph. 1969. The Japanese longline fishery for tunas and billfish in the eastern Pacific Ocean east of 130°W, 1964-1966. *Bull.I-ATTC*, 13(2):275-418.

Kume, S., and M.B. Schaefer. 1966. Studies of the Japanese long-line fishery for tuna and marlin in the eastern tropical Pacific Ocean during 1963. *Bull.I-ATTC*, 11(3):101-70.

Kurogane, K., and Y. Hiyama. 1957. Morphometric comparison of the yellowfin taken from the equatorial Pacific. *Bull.Jap.Soc.Sci.Fish.*, 23(7-8):388-93.

Le Guen, J.C., and G.T. Sakagawa. 1973. Apparent growth of yellowfin tuna from the eastern Atlantic Ocean. *Fish.Bull.NOAA-NMFS*, 71(1):175-87.

Legand, M. 1960. Longuer, repartition des sexes et maturation sexuelle des thons a nageoires jaunes de Nouvelle-Caledonie. *Rapp.Sci.ORSTOM Nouvelle-Caledonie*, (11):6-20.

Lenarz, W.H., and J.R. Zweifel. 1979. A theoretical examination of some aspects of the interaction between longline and surface fisheries for yellowfin tuna, *Thunnus albacares*. *Fish.Bull.NOAA-NMFS*, 76(4):807-25.

Lenarz, W.H., W.W. Fox, Jr., G.T.Sakagawa, and B.J. Rothschild. 1974. An examination of the yield per recruit basis for a minimum size regulation for Atlantic yellowfin tuna, *Thunnus albacares*. *Fish.Bull.NOAA-NMFS*, 72(1):37-61.

Lindberg, G.U. 1971. Fishes of the world. A key to families and a checklist. New York, John Wiley and Sons, 545 p.

Marsac, F., and G. Lablanche. 1985. Preliminary study of the growth of yellowfin (*Thunnus albacares*) estimated from purse seine data in the western Indian Ocean. *In* Expert Consultation on the Stock Assessment of Tunas in the Indian Ocean, Nov. 28-Dec.2, 1985. *Indo-Pac.Tuna Dev.Mgmt.Programme*, Doc. 31:13 p.

Matsumoto, W.M. 1958. Description and distribution of larvae of four species of tuna in central Pacific waters. *Fish.Bull.U.S.Fish Wild.Serv.*, 58(128):31-72.

Matsumoto, W.M. 1961. Collections and descriptions of juvenile tunas from the central Pacific. *Deep-Sea Res.*, 8(3-4):279-86.

Matsumoto, W.M. 1962. Identification of larvae of four species of tuna from the Indo-Pacific region I. *Dana Report*, 55:16 p.

Matsumoto, W.M. 1966. Distribution and abundance of tuna larvae in the Pacific Ocean. *In* Proceedings of the Governor's Conference on Central Pacific Fishery Resources, edited by T.A. Manar, :221-30.

Matsumoto, W.M., E.H. Ahlstrom, S. Jones, W.L. Klawe, W.J. Richards, and S. Ueyanagi. 1972. On the clarification of larval tuna identification particularly the genus *Thunnus*. *Fish.Bull.NOAA-NMFS*, 70(1):1-12.

McNeeley, R.L. 1961. The purse seine revolution in tuna fishing. *Pac.Fish.*, 59(7):27-58.

Mead, G.M. 1951. Postlarval *Neothunnus macropterus*, *Auxis thazard*, and *Euthynnus lineatus* from the Pacific coast of Central America. *Fish.Bull.U.S.Fish Wildl.Serv.*, 52(63):121-27.

Miller, F.R., and R.M. Laurs. 1975. The El Niño of 1972-1973 in the eastern tropical Pacific Ocean. *Bull.I-ATTC*, 16(5):403-48.

Miyabe, N., and W.H. Bayliff. 1987. A review of the Japanese longline fishery for tunas and billfishes in the eastern Pacific Ocean, 1971-1980. *Bull.I-ATTC*, 19(1):1-163.

Moore, H.L. 1951. Estimation of age and growth of yellowfin tuna (*Neothunnus macropterus*) in Hawaiian waters by size frequencies. *Fish.Bull.U.S.Fish Wildl.Serv.*, 52(65):132-49.

Mori, K. 1970. A consideration on the spawning of the tunas, especially of the yellowfin tuna (*Thunnus albacares*) in the adjacent sea of the Pacific coast of Japan. *Bull.Far Seas Fish.Res.Lab.*, (3):215-28.

Mori, K., S. Ueyanagi, and Y. Nishigawa. 1971. The development of artificially fertilized reared larvae of the yellowfin tuna, *Thunnus albacares*. [In Jpn.; Engl. synop.]. *Bull.Far Seas Fish.Res.Lab.*, (5):219-32.

Murphy, G.I, and R.S. Shomura. 1972. Pre-exploitation abundance of tunas in the equatorial central Pacific. *Fish.Bull.U.S.Fish Wildl.Serv.*, (70):875-913.

Murphy, T.C., and G.T. Sakagawa. 1977. A review and evaluation of estimates of natural mortality rates of tunas. *Collect.Vol.Sci.Pap.ICCAT*, 6(1):117-23.

Nakamura, E.L., and W.M. Matsumoto. 1967. Distributions of larval tunas in Marquesan waters. *Fish.Bull.U.S.Fish Wildl.Serv.*, 66(1):1-12.

Nakamura, E.L., and J.H. Uchiyama. 1966. Length weight relations of Pacific tunas. *In* Proceedings of the Governor's Conference on Central Pacific Fishery Resources, Hawaii, :197-201.

Nishikawa, Y., and D.R. Rimmer. 1987. Identification of larval tunas, billfishes, and other scombroid fishes (Suborder Scombroidei): an illustrated guide. *CSIRO Rep.Aust.*, (186):20 p.

Nishikawa, Y., M. Honma, S. Ueyanagi, and S. Kikawa. 1985. Average distribution of larvae of oceanic species of scombroid fishes, 1956-1981. *S Ser.Far Seas Fish.Res.Lab.*, (12):99 p.

Nose, Y., H. Kawatsu, and Y. Hiyama. 1957. Age and growth of Pacific tunas by scale reading [In Jpn.; Engl. summary]. *Suisan Gaku Shusei*, Tokyo University Press, pp. 701-16.

Orange, C.J. 1961. Spawning of yellowfin tuna and skipjack in the eastern tropical Pacific, as inferred from studies of gonad development. *Bull.I-ATTC*, 5(6):457-526.

Orange, C.J., M.B. Schaefer, and F.M. Larmie. 1957. Schooling habits of yellowfin tuna (*Neothunnus macropterus*) and skipjack (*Katsuwonus pelamis*) in the eastern Pacific ocean as indicated by purse seine catch records, 1946-1955. *Bull.I-ATTC*, 2(3):83-126.

Pella, J.J., and C.T. Psaropulos. 1975. Measures of tuna abundance from purse-seine operations in the eastern Pacific Ocean adjusted for fleet-wide evolution of increased fishing power, 1960-1971. *Bull.I-ATTC*, 16(4):283-400.

Pella, J.J., and P.K. Tomlinson. 1969. A generalized stock production model. *Bull.I-ATTC*, 13(3):421-96.

Perrin, W.F. 1969. Using porpoise to catch tuna. *World Fishing*, 18(6):42-5.

Perrin, W.F., R.R. Warner, C.H. Fiscus, and D.B. Holts. 1973. Stomach contents of porpoise, *Stenella* spp., and yellowfin tuna, *Thunnus albacares*, in mixed-species aggregations. *Fish.Bull.NOAA-NMFS*, 71(4):1077-92.

Peterson, C.L. and W.H. Bayliff. 1985. Organization, functions, and achievements of the Inter-American Tropical Tuna Commission. *Spec.Rep.I-ATTC*, (5):56 p.

Philander, S.G. 1990. El Niño, La Niña, and the Southern Oscillation. Vol. 46, *Ser.Internat.Geophys.*, Academic Press, 293 p.

Pickard, G.L. 1968. Descriptive Physical Oceanography. London, Permagon Press, 200 p.

Punsly, R. 1987. Estimation of the relative annual abundance of yellowfin tuna, *Thunnus albacares*, in the eastern Pacific Ocean during 1970-1985. *Bull.I-ATTC*, 19(3):265-306.

Richards, F.J. 1959. A flexible growth function for empirical use. *J.Exp.Bot.*, 10(29):290-300.

Richards, W.J. 1989. Preliminary guide to the identification of early life history stages of scombroid fishes of the western central Atlantic. *NOAA Tech.Memo.NMFS-SWFC, La Jolla*, (240):101 p.

Richards, F.J., and G.R. Dove. 1971. Internal development of young tunas of the genera *Katsuwonus, Euthynnus, Auxis* and *Thunnus* (Pisces, Scomberidae). *Copeia*, (1):72-8.

Richards, F.J., and D.C. Simmons. 1971. Distribution of tuna larvae (Pisces, Scombridae) in the northwestern Gulf of Guinea and off Sierra Leone. *Fish.Bull.NOAA-NMFS*, 69(3):555-68.

Richards, F.J., T. Potthoff, and J.-M. Kim. 1990. Problems identifying tuna larvae species (Pisces: Scombridae: *Thunnus*) from the Gulf of Mexico. *Fish.Bull.NOAA-NMFS*, 88(3): 607-609.

Ricker, W.E. 1979. Growth rates and models. *In* Fish Physiology, edited by S. Hoar, D.J. Randall and J.R. Brett, Vol. 8., New York, Academic Press, 786 p.

Rosa, H., Jr. 1950. Scientific and common names applied to tunas, mackerels and spearfishes of the world with notes on their geographical distribution. Washington, *FAO*, 235 p.

Royce, W.F. 1953. Preliminary report on a comparison of the stocks of yellowfin tuna. *Proc.Indo-Pac.Fish.Coun.*, 4(2):130-45.

Royce, W.F. 1964. A morphometric study of yellowfin tuna *Thunnus albacares* (Bonnaterre). *Fish.Bull.U.S.Fish Wildl.Serv.*, 63(2):395-443.

Schaefer, K.M. 1987. Reproductive biology of black skipjack, *Euthynnus lineatus*, an eastern Pacific tuna. *Bull.I-ATTC*, 19(2):169-260.

Schaefer, K.M. 1988. Time and frequency of spawning of yellowfin tuna at Clipperton Island, and plans for future studies. *In* Proceedings of Tuna Fishery Research Conference, Far Seas Fishery Research Laboratory. *Maguro Giyiroku, Suisancho-Enyo Suisan Kendyusho*, :118-26.

Schaefer, K.M. 1989. Morphometric analysis of yellowfin tuna, *Thunnus albacares*, from the eastern Pacific Ocean. *Bull.I-ATTC*, 19(5):389-427.

Schaefer, K.M. 1991. Geographic variation in morphometric characters and gill-raker counts of yellowfin tuna *Thunnus albacares* from the Pacific Ocean. *Fish.Bull.NOAA-NMFS*, 89(2):289-97.

Schaefer, M.B. 1952. A comparison of yellowfin tuna of Hawaiian waters and of the American west coast. *Fish.Bull.U.S.Fish Wildl.Serv.*, 52(72):353-73.

Schaefer, M.B. 1954. Some aspects of the dynamics of populations important to the management of the commercial marine fisheries. *Bull.I-ATTC*, 1(2):27-56.

Schaefer, M.B. 1955. Morphometrics comparison of yellowfin tuna from southeast Polynesia, Central America and Hawaii. *Bull.I-ATTC*, 1(4):89-136.

Schaefer, M.B. 1957. A study of the dynamics of the fishery for yellowfin tuna in the eastern tropical Pacific Ocean. *Bull.I-ATTC*, 2(6):245-85.

Schaefer, M.B. 1967. Fishery dynamics and present status of the yellowfin tuna population of the eastern Pacific Ocean. *Bull.I-ATTC*, 12(3):87-136.

Schaefer, M.B., and J.C. Marr. 1948. Juvenile *Euthynnus lineatus* and *Auxis thazard* from the Pacific Ocean and Central America. *Pac.Sci.*, 2(4):262-71.

Schaefer, M.B., B.M. Chatwin, and G.C. Broadhead. 1961. Tagging and recovery of tropical tunas, 1955-1959. *Bull.I-ATTC*, 5(5):341-455.

Sharp, G.D., and S. Pirages. 1978. The distribution of red and white swimming muscles, their biochemistry, and the biochemical phylogeny of selected scombrid fishes. *In* The physiological ecology of tunas, edited by G.D. Sharp and A.E. Dizon. New York, Academic Press, 485 p.

Shimada, B.M. 1951. Contributions to the biology of tunas from the western equatorial Pacific. *Fish.Bull.U.S.Fish Wildl.Serv.*, 52(62):111-9.

Shingu, C., P.K. Tomlinson, and C.L. Peterson. 1974. A review of the Japanese longline fishery for tunas and billfishes in the eastern Pacific Ocean, 1967-1970. *Bull.I-ATTC*, 16(2):65-230.

Shomura, R.S. 1966. Age and growth studies of four species of tunas in the Pacific Ocean. *In* Proceedings of the Governor's Conference on Central Pacific Fishery Resources, edited by T.A. Manar. Hawaii, pp. 203-219.

Sprague, L.M. 1967. Multiple molecular forms of serum esterase in three tuna species from the Pacific Ocean. *Hereditas*, 57:198-204.

Strasburg, D.W. 1960. Estimates of larval tuna abundance in the central Pacific. *Fish.Bull.U.S.Fish Wildl.Serv.*, 60(167):231-55.

Suda, A., and M.B. Schaefer. 1965a. Size-composition of catches of yellowfin in the Japanese long-line fishery in the eastern tropical Pacific east of 130°W. *Bull.I-ATTC*, 10(4):267-331.

Suda, A., and M.B. Schaefer. 1965b. General review of the Japanese tuna long-line fishery in the eastern tropical Pacific Ocean 1956-1962. *Bull.I-ATTC*, 9(6):307-462..

Sun', Tszi-Dzen'. 1960. Lichiniki i mal'ki tuntsov, parunsnikov i mech-ruby (Thunnidae, Istiophoridae, Xiphiidae) tsentral 'noi i zapadnoi chasti Tikhogo okeana (Larvae and juveniles of tunas, sailfishes and swordfish (Thunnidae, Istiophoridae, Xiphiidae) from the central and western part of the Pacific Ocean. *Trudy Inst.Okeanol.*, 41:175-91.

Sund, P.N., M. Blackburn, and F. Williams. 1981. Tunas and their environment in the Pacific Ocean: a review. *Oceanogr.Mar.Biol.Ann.Rev.*, 19:443-512.

Suzuki, A. 1962. On the blood types of yellowfin and bigeye. *Amer.Natur.*, 96(889): 239-46.

Suzuki, Z. 1971. Comparison of growth parameters estimated for the yellowfin tuna in the Pacific Ocean [In Jpn.; Engl. synopsis]. *Bull.Far Seas Fish.Res.Lab.*, (5):89-105.

Suzuki, Z. 1974. Re-examination of scale reading method of yellowfin tuna taken in the western and central Pacific Ocean. [In Jpn.; Engl. synopsis]. *Bull.Far Seas Fish.Res.Lab.*, (5):89-105.

Suzuki, Z. 1988. Study of interaction between longline and purse seine fisheries on yellowfin tuna, *Thunnus albacares* (Bonnaterre). *Bull.Far Seas Fish.Res.Lab.*, (25): 73-143.

Suzuki, Z., P.K. Tomlinson, and M. Honma. 1978. Population structure of Pacific yellowfin tuna. *Bull.I-ATTC*, 17(5):273-441.

Suzuki, Z., Y. Warashina, and M. Kishida. 1977. The comparison of catches by regular and deep tuna longline gears in the western and central equatorial Pacific. *Bull.Far Seas Fish.Res.Lab.*, (15):51-73.

Uchiyama, J.H., and P. Struhsaker. 1981. Age and growth of skipjack tuna, *Katsuwonus pelamis*, and yellowfin tuna, *Thunnus albacares*, as indicated by daily growth increments of sagittae. *Fish.Bull.NOAA-NMFS*, 79(1):151-62.

Ueyanagi, S. 1966. On the pigmentation of larval tuna and its usefulness in species identification. *Rep.Nankai Reg.Fish.Res.Lab.*, (24):41-8.

Ueyanagi, S. 1969. Observations on the distribution of tuna larvae in the Indo-Pacific Ocean with emphasis on the delineation of the spawning areas of albacore, *Thunnus alalunga*. *Bull.Far Seas Fish.Res.Lab.*, (2):177-256.

Ueyanagi, S. 1978. Recent tuna culture research in Japan. Paper presented at the International Ocean Development Conference, 5(C1):23-39 (preprint).

Ueyanagi, S., K. Mori, and Y. Nishikawa. 1969. Research on distribution of larvae. *S Ser.Far Seas Fish.Res.Lab.*, (1):12-7.

U.S. Dept. of Commerce. 1977. Administration of the Marine Mammal Protection Act of 1972. *NOAA-NMFS*. Reprinted from Fed. Register, 42(147):38982-9030.

U.S. Dept. of Commerce. 1978. The Marine Mammal Protection Act of 1972. Annual Rep., April 1, 1977, to March 31, 1978 : 202 p.

de Vlaming, V.L. 1982. On the use of the gonosomatic index. *Comp.Biochem.Physiol.*, 73A (1):31-9.

Wade, C.B. 1950a. Juvenile forms of *Neothunnus macropterus*, *Katsuwonus pelamis* and *Euthynnus yaito* from Philippine seas. *Fish.Bull.U.S.Fish Wildl.Serv.*, 51(53):395-404.

Wade, C.B. 1950b. Observations on the spawning of Philippine tuna. *Fish.Bull.U.S. Fish Wildl.Serv.*, 51(55):409-23.

Wade, C.B. 1951. Larvae of tuna and tuna-like fishes from Philippine waters. *Fish. Bull.U.S.Fish Wildl.Serv.*, 51(57):445-85.

Wankowski, J.W.J. 1981. Estimated growth of surface-schooling skipjack tuna, *Katsuwonus pelamis*, and yellowfin tuna, *Thunnus albacares*, from the Papua New Guinea region. *Fish.Bull.NOAA-NMFS*, 79(3):517-45.

Wild, A. 1986. Growth of yellowfin tuna, *Thunnus albacares*, in the eastern Pacific Ocean based on otolith increments. *Bull.I-ATTC*, 18(6):423-82.

Wild, A., and T.J. Foreman. 1980. The relationship between otolith increments and time for yellowfin and skipjack tunas marked with tetracycline. *Bull.I-ATTC*, 17(7):507-60.

Wyrtki, K., and B. Kilonsky. 1984. Mean water and current structure during the Hawaii-to-Tahiti shuttle experiment. *J.Phys.Oceanogr.*, 14:242-54.

Yabe, H., and S. Ueyanagi. 1962. Contributions to the study of the early life history of tunas. *Occas.Rep.Nankai Reg.Fish.Res.Lab.*, (1):57-72.

Yabe, H., N. Anraku, and M. Yukinawa. 1958. Studies on the yellowfin tuna-III. Annual variations of the size composition and the hooked-rates of yellowfin tuna distributing in the equatorial Pacific. *Rep.Nankai Reg.Fish.Res.Lab.*, (7):88-104.

Yabe, H., Y. Yabuta, and S. Ueyanagi. 1963. Comparative distribution of eggs, larvae, and adults in relation to biotic and abiotic environmental factors. *FAO Fish.Rep.*, 6(3):979-1,009.

Yabe, H., S. Ueyanagi, S. Kikawa, and H. Watanabe. 1958. Young tunas found in the stomach contents. *Rep.Nankai Reg.Fish.Res.Lab.*, (8):31-48.

Yabuta,Y., M. Yukinawa, and Y. Warashina. 1960. Growth and age of yellowfin tuna II. Age determination (scale method). [In Jpn.; Engl. summary]. *Rep.Nankai Reg.Fish.Res.Lab.*, (12):63-74.

Yamanaka, K.L. 1990. Age, growth and spawning of yellowfin tuna in the southern Philippines. Colombo, Sri lanka, *Indo-Pac.Tuna Dev.Mgt.Programme*, IPTP/90/WP/21:87 p.

Yang, R-T. 1971. Population study of yellowfin tuna in the waters adjacent to Taiwan. *Nat.Taiwan Univ.Sci.Rep.Acta Oceanogr.Taiwanica*, (1):137-55.

Yang, R-T., Y. Nose, and Y. Hiyama. 1969. A comparative study on the growth of yellowfin tunas from the Atlantic and Pacific Oceans. *Bull.Far Seas Fish.Res.Lab.*, (2):1-21.

Yuen, H.S.H. 1963. Schooling behavior within aggregations composed of yellowfin and skipjack tuna. *FAO Fish.Rep.*, 6(3):1419-29.

Yuen, H.S.H., and F.C. June. 1957. Yellowfin tuna spawning in the central equatorial Pacific. *Fish.Bull.U.S.Fish Wildl.Serv.*, 57(112):251-64.

A REVIEW OF THE BIOLOGY AND FISHERIES FOR YELLOWFIN TUNA (*THUNNUS ALBACARES*) IN THE WESTERN AND CENTRAL PACIFIC OCEAN

Ziro Suzuki
National Research Institute of Far Seas Fisheries
Shimizu-shi, Japan

1. INTRODUCTION

The species synopsis of yellowfin tuna by Cole (1980) was referred to extensively for preparation of this review paper. The layout of this paper follows a similar review paper on the yellowfin tuna in the eastern Pacific (Wild, 1993).

2. IDENTITY

The species name for the yellowfin tuna is *Thunnus albacares* (Bonnaterre, 1788) (Collette and Nauen, 1983).

3. EARLY LIFE HISTORY

The morphological description of larval yellowfin tuna has been reported by Wade (1951), Yabe and Ueyanagi (1961), Matsumoto (1962) and Mori *et al.* (1971) *etc.* Detailed observations on rearing and morphological change of artificially fertilized yellowfin were made available by Mori *et al.* (1971), Harada *et al.* (1971), Far Seas Fisheries Research Laboratory (FSFRL, 1973), and Harada *et al.* (1980a). Harada *et al.* (1980a) reported that they could rear yellowfin from eggs to a maximum of 51 mm in total length (1.35 g in body weight) 38 days after hatching. It appears from these artificial fertilization experiments that there are two critical periods for larval mortality, one at 4 to 5 days and the other about 11 days after hatching (FSFRL, 1973). In this regard, Nishikawa (in FSFRL, 1973) pointed out that the second critical period, at about 7-10 mm in total length, corresponded to the period of changing prey foods from crustaceans to fish larvae which was observed in the stomach contents of the wild yellowfin larvae.

General aspects of geographical distribution of larval yellowfin tuna in the western and central Pacific are found in several works (*e.g.*, Wade, 1951; Matsumoto, 1958; Strasburg, 1960 ; Ueyanagi, 1969; and Nishikawa *et al.*, 1985). Among them the studies by Ueyanagi (1969) and Nishikawa *et al.* (1985) are comprehensive papers covering extensive areas. Figure 1 shows the occurrence of larval yellowfin tuna. Ueyanagi (1969) stated that larval yellowfin tuna occurred all the year round in tropical waters and the range of their distribution expanded to the higher latitudes in the respective summer months in the northern and southern hemispheres. He further mentioned that the lower limiting temperature of larval distribution of yellowfin tuna was presumed to be about 26°C although some larvae occurred in waters lower than 26°C, as low as 24°C according to his data. Mori (1970) reported that the minimum sea-surface temperature where larval yellowfin were found was about 24°C in the waters adjacent to Japan. Harada *et al.* (1980b) showed in the experiments on artificially-fertilized yellowfin eggs

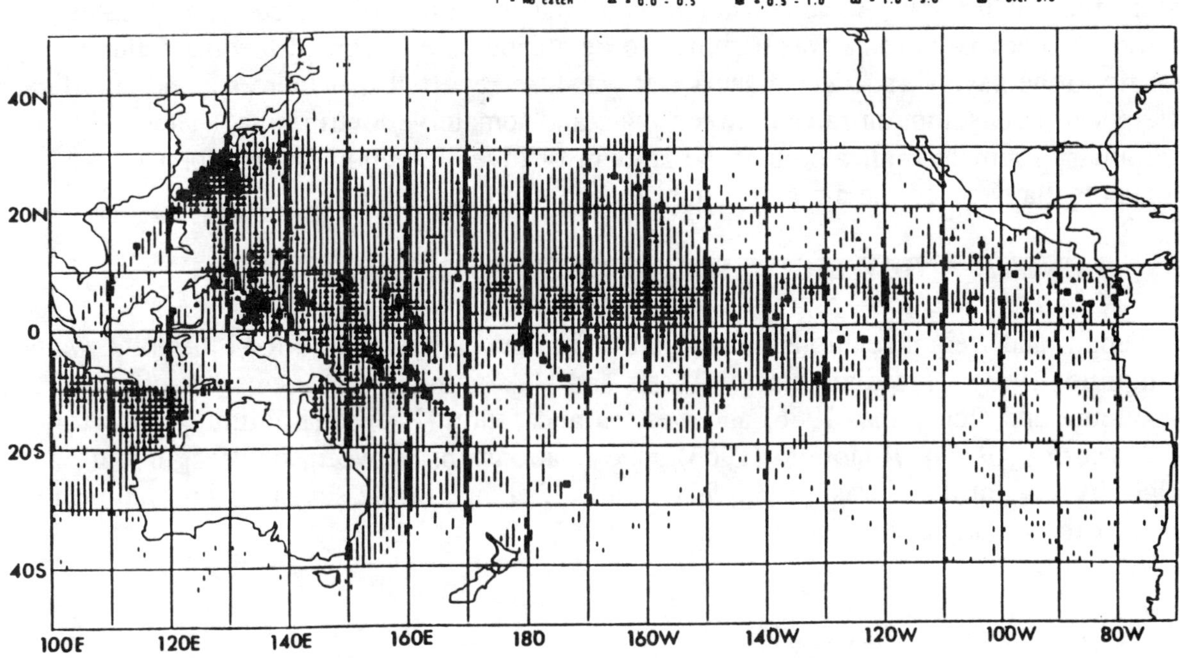

Figure 1.   Distribution of larval yellowfin tuna by plankton net survey; after Nishikawa *et al.* (1985).

that the rates of occurrence of normally hatched larvae were highest for an average water temperature range from 26.4°C to 27.8°C, but no normal larvae were found in the temperature range below 18.7°C or over 31.9°C.

Comparisons of occurrence of yellowfin larvae at the surface and the subsurface layer (FSFRL, 1978) appear to indicate that the larvae in the western and central Pacific were more abundant at the surface, especially within the layer 50 meters from the surface, than in the subsurface layers, and more abundant at night than during the day. However, a detailed statistical analysis with more precise vertical-sampling methods is required to define the vertical distribution of the yellowfin larvae since the studies by Strasburg (1960), Klawe (1963), and Ueyanagi (1969) indicated no diurnal difference while Richards and Simmons (1971) indicated that yellowfin larvae in the equatorial eastern Atlantic migrate to the surface during the day.

4.   FEEDING AND FOOD

The feeding of yellowfin appears to take place mainly during day time and they accept a great variety of prey species, including fish, crustaceans, and cephalopods (Reintjes and King, 1953; Watanabe, 1958). Areal variation is noted in occurrence of empty stomachs and volume of stomach contents (Watanabe, 1960). There was a significant difference in prey species composition for the specimens between those around fish aggregating devices (FAD) and non-FAD associated yellowfin in the Hawaiian waters (Brock, 1985).

Yesaki (1983) and Barut (1988) reported the food and feeding habits of yellowfin tuna caught by handline around payaos in the Moro Gulf, Philippines. They found significantly higher daily-meal consumption, about twice that estimated for the eastern Pacific counterpart and this was assumed to be attributable to forced feeding (chumming) specific to the payao-associated fishery. It was also reported that there was no significant difference in consumption rates between male and female yellowfin tuna. Their preliminary estimates indicated that the amount of juvenile yellowfin cannibalized was high, potentially twice the amount caught by the handline fishery.

5. LENGTH AND WEIGHT RELATIONSHIP

Length and weight relationships of yellowfin tuna are available for Philippine waters (Ronquillo, 1963), Hawaiian waters (Tester and Nakamura, 1957), central Pacific (Nakamura and Uchiyama, 1966) and western and central Pacific (Kamimura and Honma, 1959; Morita, 1973). Ronquillo (1963) gave relationships between total length and weight by sex but there was only a very small difference in the sex-specific relationships. Some of these relationships are shown in Table 1.

Table 1. Length and weight relationships ($W=aL_b$) of yellowfin tuna in the western and central Pacific Ocean. Fork length (L) is in cm and weight (W)1[1] in kg.

Author	Fishing gear	sample size	size range (cm)	Fishing ground	a	b
Kamimura and Honma (1959)[2]	Longline	6785	100-150	Mainly central and western Pacific	6.44×10^{-6}	3.1878

L	50	70	90	110*	130*	150*	170
W(1.15)	2.0	5.6	13	24	41	64	96
W(r)	2.4	6.5	14	24	40	61	88

Author	Fishing gear	sample size	size range (cm)	Fishing ground	a	b
Nakamura and Uchiyama (1966)	-	4822	70-180	Central Pacific	1.4769×10^{-5}	3.0583

L	50	70*	90*	110*	130*	150*	170*
W	2.3	6.5	14	26	43	67	98

Author	Fishing gear	sample size	size range (cm)	Fishing ground	a	b
Morita(1973)	Mainly Longline	2043	26-157	western and central Pacific	2.512×10^{-5}	2.9396

L	50*	70*	90*	110*	130*	150*	170
W	2.5	6.7	14	25	41	63	91

1) Whole weight except Kamimura and Honma (1966) who measured gilled-and gutted weight.

Kamimura and Honma (1959) estimated the length-weight relationship between the fork length and gilled-and-gutted weight for yellowfin tuna taken by the Japanese longline boats in the western and central tropical Pacific.

Length and live-weight relationship for the western and central Pacific for yellowfin tuna taken mainly by the Japanese longline boats was formulated by Morita (1973). Morita (1973) also gave a conversion factor from the gilled-and-gutted (*GG*) weight to the live weight in the following expression:

$$log\ r\ =\ 0.5707\text{-}0.2445\ log\ L$$

where r = conversion factor at the fork length L(cm).

The r value is multiplied by the GG weight to obtain the corresponding live weight. The samples used cover fish larger than about 70 cm in fork length caught by the longline boats in the western tropical Pacific with a small number of samples from the tropical Atlantic.

At present, the value of 1.15 is being used by the Japanese scientists in the conversion from the *GG* to the live weight regardless of the area and size of fish for the yellowfin taken by the Japanese longline boats. However, as is obvious in the conversion equation shown previously, the factor changes with the length of the fish, smaller fish have larger values than larger fish (*e.g.*, r=1.32 for 70-80 cm class, 1.18 for 110-120 cm class and 1.09 for 150-160 cm class). Length and weight values in selected sizes among the three papers [*W(r)*for Kamimura and Honma (1959)]in Table 1 are relatively similar.

White (1982) shows a length-weight relationship which is apparently derived from the Philippine fisheries. However, the range of the data and the method used to fit the curve were not mentioned.

It is noted that there appears to be no length and weight relationship available for yellowfin caught by the purse seine fishery in the equatorial western Pacific. Since the purse seine fishery in that area is the most dominant yellowfin fishery, the relationship specific to the fishery should be established urgently.

6.   AGE AND GROWTH

Age and growth of this species in the western and central Pacific were studied through hardpart analysis (Aikawa and Kato, 1938; Nose *et al.*, 1957; Yabuta *et al.*, 1960; Yokota *et al.*, 1961 (including length composition data); Tan *et al.*, 1965; Yang *et al.*, 1969; Uchiyama and Struhsaker, 1981; Yamanaka, 1990), weight frequency analysis (Kimura, 1932; Moore, 1951) and length frequency analysis (Yabuta and Yukinawa, 1957,1959; Wankowski, 1981; White, 1982; Yesaki, 1983; Ingles and Pauly, 1984; Yamanaka, 1990). Suzuki (1971) reviewed the age and growth studies in the Pacific and suggested the superiority of the methods using the hard parts relative to those using length or weight frequencies. The growth parameters in the major works and sizes at age in the selected papers for the western and central Pacific are shown in Table 2 and Table 3, respectively. Except for the early studies based on small numbers of samples, there are some significant differences in age and growth among the works (Table 3).

Recent work by Yamanaka (1990) based on counting the daily rings on otoliths of juvenile yellowfin tuna in the Philippine waters showed that the growth was best

Table 2. Comparison of growth equation parameters (von Bertalanffy) for yellowfin tuna in the western and central Pacific Ocean.

Author	Fishing gear	Sample size	Method	Range (cm) Fork length	Fishing ground	Growth parameter (annual value, 1 :cm)		
						k	1	t_0
Kimura(1932)	Set net	200	Weight modes	100-120	Japanese waters	-	-	-
Aikawa and Kato (1938)	-	6	Vertebrae	-	-	-	-	-
Moore(1951)	Longline	5000	Weight modes	70-120	Hawaiian waters	0.44	192	0.22
Yabuta and Yukinawa(1957)	Longline and baitboat	50000	Length modes	30-150	Japanese waters	0.55	168	0.35
Yabuta and Yukinawa(1959)	Longline	110000	Length modes	80-150	Western Pacific	0.66	150	0.40
Nose et al (1957)	Longline	300	Scales	100-140	Pacific	-	-	-
Yabuta et al (1960)	Longline	1000	Scales	70-140	Western Pacific	0.33	190	0
Yokota et al (1961)	Baitboat	-	Length modes and scales	35-130	Japanese waters	-	215	-
Tan et al (1965)	Longline	170	Vertebrae	-	Western Pacific	-	-	-
Yang et al (1969)	Longline	200	Scales	60-140	Western Pacific	0.36	195	0.27
Uchiyama and Struhsaker(1981)	Mainly troll	14	Otoliths	52-93	Hawaiian waters	Linear segments		
Wankowski(1981)	Baitboat	-	Length modes	30-96	Papua New Guinea waters	0.29	181	-
White(1982)	Various	-	Length modes		Philippine waters:			
				20-60	Santa Cruz	0.25	189	-
				20-70	Opol	0.20	169	-
				20-60,90-150	General-Santos	0.43	182	-
				20-60	Labuan	0.29	179	-
Yesaki(1983)	Various	-	Length modes	20-60, 120-160	Philippine waters:			
					Male	0.30	175	-
					Female	0.32	173	-
Ingles and Pauly (1984)(from Bayliff, 1988)	-	-	Length modes	-	Philippine waters?	-	148	0.42
Yamanaka(1990)	Ringnet	207	Otoliths	20-80	Philippine waters	Linear segments		

After Suzuki (1971) for the works published till 1969 except Yokota et al. (1961)

Table 3.   Comparison of size (Fork length in cm) at age in selected growth studies for yellowfin tuna in the western and central Pacific.

Age	Moore (1951)	Yabuta & Yukinava (1957)	Yabuta et al. (1960)	White (1982) (General-Santos)	Yamanaka (1990)
1	56.	50*	53.	64*	57*
2	104*	100*	92*	105*	
3	135*	129*	119*	132*	
4	156.	145*	139*	149*	
5	169.	155.	154.	161.	
6	177.	160.	164.	168	

* Estimates within the size range of the sample.

described by two linear stanzas. The two stanzas cover the range from 15 to 35 cm and 35 to 79 cm in fork length, and from 50 to 2,350 g and 2,350 to 9,200 g in whole weight. The Philippine yellowfin tuna attains 57 cm at age one according to the otolith study which is close to the age one estimate of previous studies. However, the size reached in one year derived by Yamanaka (1990) for Philippine yellowfin was the largest found among the three daily-increment studies, *i.e.*, 53 cm and 49 cm for Hawaiian (Uchiyama and Struhsaker, 1981) and eastern Pacific yellowfin tuna (Wild, 1986). It is not clear whether the largest size at one year in the Philippine waters is due to temperature or other factors, such as the method of otolith preparation or differences in the method of counting increments. Yamanaka (1990) infers from these observations that a significant physiological and ecological change occurs when the fish attains a critical size. This size lies between approximately 35 and 50 cm.

Age and growth studies by sex through the length-frequency analysis of yellowfin tuna in the Philippine waters shows that the growth rate of males is higher than that of females (Yesaki, 1983; Yamanaka, 1990). On the other hand, Wild (1986), using daily ring methods for yellowfin tuna in the eastern Pacific shows that young females are initially larger than males of the same age, the growth curves cross one another at around age 2.0 (about 95 cm in fork length) and thereafter males are larger than females. Wild (1986) assumes that differences in mortality between the sexes may be more attributable to the apparent smaller maximum size reached by females, rather than differences in growth rate or availability to the fisheries.

7.   MATURATION, SPAWNING, FECUNDITY AND SEX RATIO

7.1   Maturation

The size of yellowfin at first spawning (maturation) in the western and central Pacific has been studied through gonadosomatic indices (GI) (Yuen and June, 1957; Kikawa, 1962), external features of the ovaries (Wade, 1950) and microscopic examination of egg diameters (Buñag, 1956). Results ranged from 53 cm for males in the Philippine waters

(Wade, 1950), 57 cm for females also in the Philippine waters (Buñag, 1956), 70-80 cm for females in the central Pacific and 80-110 cm for females in the western and central Pacific (Kikawa, 1962). It appears difficult to judge the size at first spawning without a detailed histological observation of ovaries. The histological examination of ovaries of yellowfin in the eastern tropical Pacific showed that the smallest female found with mature ovaries was 84 cm (IATTC, 1990).

The size at 50 % maturity of yellowfin tuna is estimated to be about 110-120 cm (Yuen and June, 1957; Kikawa, 1962) in the western and central Pacific. However, as in the case of size at first spawning, these studies did not include histological examinations. There is some indication that yellowfin caught by the surface fisheries tend to be more mature than those taken by the longline fishery (Hisada, 1973; Koido and Suzuki, 1989). Hisada (1973) hypothesized that this observation indicates that the fish move to the surface layer when they mature. The review in this section shows the necessity of a detailed histological study on maturity covering both surface and longline data.

7.2 Spawning

Yellowfin tuna in the Pacific spawn over vast areas, throughout the year in the tropical waters (*e.g.*, Kikawa, 1966), and in the respective summers in the higher latitudes where the water temperature is over 24°C. However, there are peak seasons for spawning in specific areas. Kikawa (1966) reported, using the longline data that the peak spawning potential (product of fecundity, rate of group maturity, sex ratio and abundance index) is in December-January in the western tropical Pacific (120°E-180°) and April-May in the central tropical Pacific (140°W-180°). Two spawning peaks in a year were indicated for the yellowfin tuna in the Philippine waters. The major peak, according to GI analysis by Yesaki (1983) covers the period from March-May and a lesser peak during November-December. Yamanaka (1990) showed a similar peak, one in April and the other in October through GI examination.

Some scientists have suggested that yellowfin is a multiple spawner, based on the observation of multiple modes in ova diameter frequency distributions (*e.g.*, June, 1953; Buñag, 1956). Recent preliminary work by Nikaido (1988) based on the observation of post-ovulatory follicles in the ovaries demonstrates that yellowfin caught by purse seine in the western Pacific spawn every 1.7 days. Similar analysis in the eastern Pacific indicates a spawning interval of about 1.3 days (Schaefer, 1988). Schaefer (1988) estimates that spawning occurs between 2000-2400 h.

Yesaki(1983) shows no correlation between the lunar cycle and spawning activity inferred from GI, but Yamanaka (1990) reports that four otoliths back-dated to spawning indicate that spawning occurs during the new moon. Yamanaka (1990) also suggests that the monsoon seasons significantly affect spawning, growth and recruitment of yellowfin tuna in the Philippine waters.

7.3 Fecundity

June (1953) calculated the relationship between weight and batch fecundity of yellowfin tuna taken by the Hawaiian longline fishery. He fitted a straight line to a sample of 11 yellowfin ranging from 47 to 88 kg and number of eggs from 2.4 to 8.6 millions:

$$Y = 125,200\,X - 2,853,000$$

where Y and X denote number of eggs and weight of fish in kg.   Kikawa (1966) proposed a relationship between fork length and ovary weight.   However, it should be noted that Kikawa's relationship is derived from data on bigeye tuna, does not specifically apply to yellowfin tuna.

7.4   Sex Ratio

It is generally agreed that the sex ratio (ratio of female to male) is about 1 until a length of about 120 cm is reached.   Subsequently, the ratio of females decreases steadily for the larger fish in the western and central Pacific (*e.g.*, Kikawa, 1966; Yesaki, 1983; Yamanaka, 1990).   However, in the eastern Pacific, the decrease in number of females is rapid, after reaching the length approximately 140 cm (Wild, 1993).

8.   STOCK STRUCTURE, DISTRIBUTION AND MIGRATION

8.1   Stock Structure

There are many studies on stock structure of yellowfin in the western and central Pacific.   They derive from morphometric methods (Godsil and Greenhood, 1951; Schaefer, 1955; Kurogane and Hiyama, 1957; Royce, 1964; Schaefer, 1991), circumstantial evidence from fisheries-related information (Yabuta *et al.* 1958; Kamimura and Honma, 1963; Honma *et al.*, 1971; Suzuki *et al.* 1978) and from immunological and biochemical methods (Suzuki, 1962; Sprague, 1967; Barrett and Tsuyuki, 1967; Fujino and Kang, 1968).

The morphometric studies tend to show heterogeneity in several areas in the Pacific ranging from very limited areal extent (*e.g.*, Royce, 1964) to ocean-scale such as western, central and eastern Pacific (e.g., Kurogane and Hiyama, 1957; Schaefer, 1991). The inference from the fisheries related information suggests, in most cases, ocean-scale stocks (e.g., Suzuki *et al.*, 1978).   The studies using immunological and biochemical methods indicate no difference in the incidence of serum esterase and transferrin systems between fish from the waters of Hawaii, Line Islands and the eastern Pacific Ocean (Barrett and Tsuyuki, 1967; Fujino and Kang, 1968).

With respect to the stock structure in the western Pacific, it was noticed in the Philippine fisheries, one of the most diverse and substantial yellowfin fisheries in the world, that both very small fish from 15 cm to 60 cm and large fish over 110 cm are common, but that middle-sized fish between 60 cm to 110 cm are relatively scarce (*e.g.*, Yesaki, 1983).   Since the middle-sized fish have been commonly captured in the offshore oceanic western Pacific by the Japanese longline and to some extent by the industrial purse seine fisheries, the observed apparent size segregation by areas deserves to be analyzed quantitatively from the viewpoint of the stock structure.

As described previously for stock structure in the western and central Pacific, there is no commonly-accepted hypothesis at present. However, for practical reasons due to the existence of large fisheries in the region, some preliminary stock assessments were conducted over the whole western Pacific west of 180° including the Philippine fisheries (*e.g.*, Suzuki *et al.*, 1989).

8.2   Distribution

Higgins (1967) reviewed the available information on the distribution of juvenile yellowfin tuna (12 mm to 300 mm). Juveniles appear to be distributed in the higher latitudes in the western Pacific as far north as the coastal areas of southern Japan (about 30°N) and as far south as 23°S along the Australian coast. In the central Pacific, their distribution band seems to be a little bit narrower, especially to the north, than in the western Pacific.

The geographical distribution of adult yellowfin tuna, studied by using longline data, covers a wide area centering on the tropical waters and extending to the temperate waters (*e.g.*, Suzuki *et al.*, 1978). Adults have a much wider distribution than juveniles, covering the whole area between 40°N and 40°S in the western Pacific and becoming narrower latitudinally toward the central Pacific (Figure 2).

It has been inferred from analysis of fisheries-related information (Suda and Schaefer, 1965; Suzuki *et al.*, 1977) that yellowfin tuna inhabit a relatively shallow swimming layer mainly above the thermocline. Direct studies using sonic tags later verified the previous inference but indicated more dynamic aspects of vertical distribution (Yonemori, 1982; Suzuki, 1984; Holland *et al.*, 1990; Koido and Miyabe, 1990). According to the results of sonic tagging (all experiments so far made on juvenile fish less than about 80 cm) the yellowfin show diurnal, vertical movement shallow at night and deep in the daytime. The study around Hawaii by Holland *et al.* (1990) indicates that the fish tend to move away from the FAD at night and return to the same FAD the next morning and stay around the FAD during the day. However, studies by Yonemori (1982), Koido and Miyabe (1990) and Suzuki (1984) do not show such a consistent pattern of diurnal, horizontal movement associated with the FAD.

8.3   Migration

The migratory pattern of yellowfin has been inferred from the change in seasonal fishing grounds together with information on size of fish in the catch in the western and central Pacific. There are several tagging experiments pertinent to the migratory aspects, but they are far less extensive than those conducted in the eastern Pacific. Recent, ongoing large-scale tagging by the SPC is expected to provide valuable information on this subject.

The fisheries-related information indicates a fairly clear seasonal movement along the Kuroshio and the East Australian Currents, moving to the higher latitudes in the warmer seasons and returning to the lower latitudes in the colder seasons (Suzuki *et al.*, 1978). There are several instances of juvenile yellowfin tuna tagged in the equatorial Pacific and recovered in the temperate waters of Japan. These indicate that yellowfin has potential for long-distance migration.

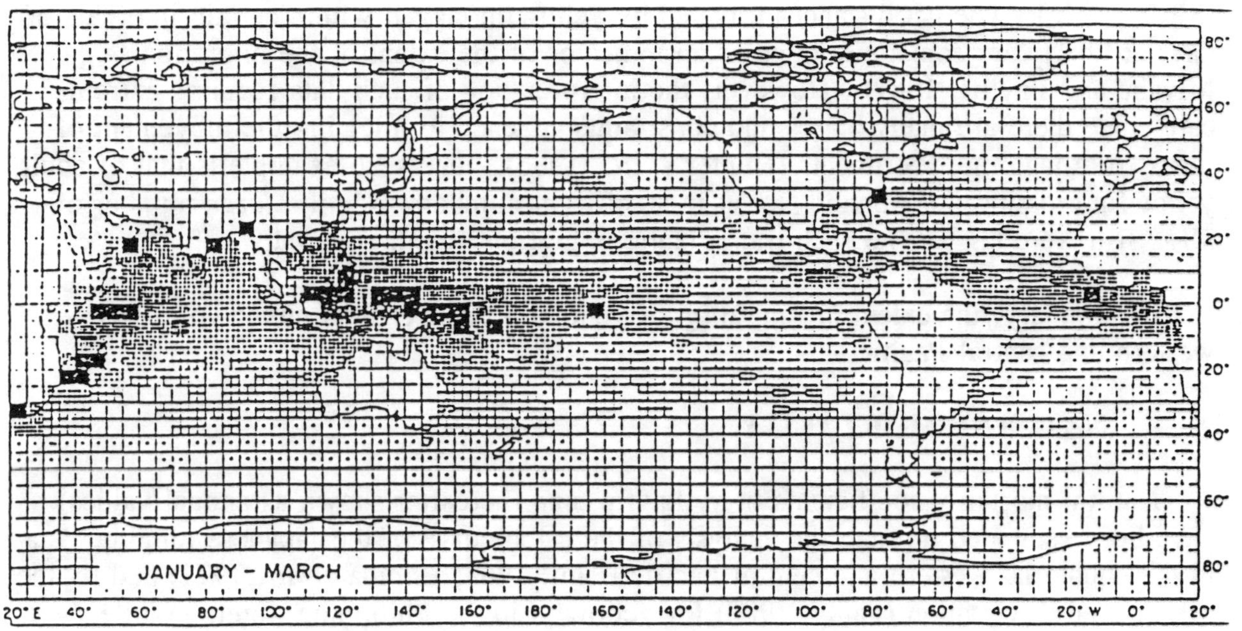

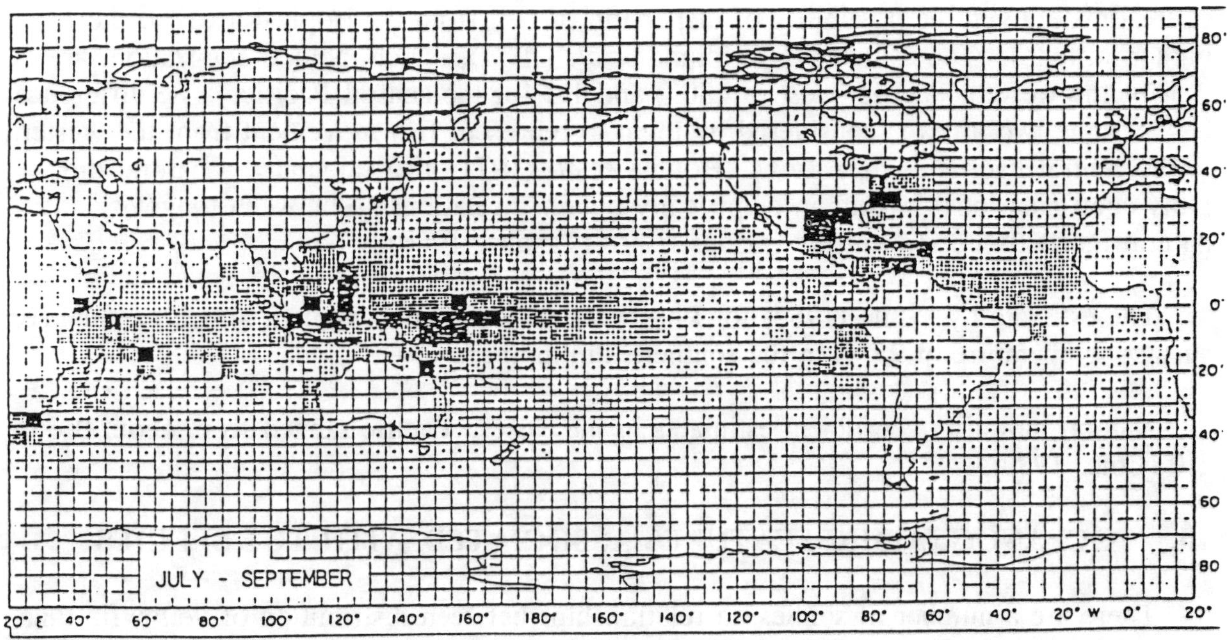

Figure 2.   Average quarterly distribution, January-March and July-September, of yellowfin tuna longline catch rates in the Pacific Ocean, 1966-1975. Areas of darkest shading denote highest catch rates. (After FAO, 1980).

However, little is known about the east-west migration in the western and central tropical Pacific. As shown by Kamimura and Honma (1963), yellowfin tuna may migrate extensively from the western to central Pacific, however, Royce (1964) suggests that most of them may stay within a range of several hundred miles throughout their life span. It should be mentioned that preliminary results from large scale-tagging programs in the western equatorial Pacific for juvenile yellowfin (mostly between 30 cm and 60 cm in fork length) tend to indicate more east-west migrations than north-south migrations

(SPC, 1991a). The net distances traveled by yellowfin tagged within certain intervals of days at liberty are sometimes cited as an indicator of the extent of the migration (*e.g.*, Miyabe and Bayliff, 1987). However, the interpretation of such data requires extreme care since there are many biases and constraints attached to inferring the movement of tunas from tagging experiments (Hunter *et al.*, 1986). One such example is that of adult yellowfin crossing the Atlantic Ocean in significant numbers. Until a *large* number of *large* fish were tagged, no such transatlantic migration was discovered.

In summary, the extent of migration of yellowfin tuna remains unknown in the western and central Pacific.

9.   NATURAL MORTALITY

Natural mortality coefficients of yellowfin tuna (hereafter shown as annual values) have been estimated through the analysis of the catch curve of the Japanese longline data in the equatorial western and central Pacific (Ishii, 1968; Honma *et al.*, 1971) and by the use of Pauly's equation (Pauly, 1980) for yellowfin in the Philippine waters (White, 1982). The study by Ishii (1968), using a sequential-recruitment model with the catch-at-age data, gives an estimate of 0.9 for fish older than 3 years. This might be an overestimate given the possibility of dispersion of fish from the tropical areas to the higher latitudes. The estimates by Honma *et al.* (1971), which include emigration, are 2.5 for the western Pacific (west of 180°) and 1.1 for the central Pacific. The difference between the two areas is considered by them to represent fish movement from the western to the central Pacific as they grow. Honma *et al.* (1971) also calculated a natural mortality coefficient of 0.3 from an empirically-obtained relationship between longevity and natural mortality coefficient for various fishes (Tanaka, 1960).

The estimate by White (1982) for Philippine yellowfin is 0.5 given a mean annual sea surface temperature of 27°C in the southern Philippines.

Natural mortality coefficient estimates in the eastern Pacific ranged from about 0.6 to 1.0 (Cole, 1980).

10.   OCEANOGRAPHIC FEATURES ASSOCIATED WITH THE SPECIES

There are a number of studies on relationships between distribution of yellowfin tuna and various oceanographic factors such as surface water temperature (*e.g.*, Broadhead and Barrett, 1964). While the observed relationships work fairly well in specific areas or for specific developmental (*e.g.*, growth) stages of tunas, they are often not applicable on a global scale. One exception to this is the relatively good-agreement between areas with high basic productivity and those with high abundance of yellowfin tuna (Suda *et al.*, 1969 *etc.*). Nakamura (1965) developed a working hypothesis that each tuna species has its own specific distribution in a specific ocean current with inter-current habitat change depending on the developmental stage. His working hypothesis was based on the observation of longline fishing grounds where the species-specific distribution areas were segregated by longitudinal bands which roughly corresponded to different currents. Yamanaka *et al.* (1969) analyzed water type (T-S diagram) and distribution of tunas in a similar manner to Nakamura (1965). However, as Nakamura (1965) recognized, there were several later findings that could not be explained by his hypothesis.

Kawai(1969) emphasized the importance of all aspects of the water-temperature structure including not only the water temperature itself but also the several other vital factors such as salinity, oxygen content and productivity. Thus, Kawai (1969) defined the conditions which characterized the main fishing ground of Atlantic yellowfin tuna taken by the longliners. Suzuki *et al.* (1977) compared the fishing efficiency of the regular and deep longline gear in the western and central Pacific and hypothesized that the main habitat of yellowfin tuna is above the thermocline.

A series of sonic tag experiments has been made for juvenile yellowfin tuna in the western (Yonemori, 1982) and central Pacific around Hawaii (Holland *et al.*, 1990). These results support the hypothesis that swimming depth of yellowfin tuna is mainly above or in the upper part of the thermocline with a tendency for deep swimming in the daytime and shallower swimming at night. The sonic tag experiments by Holland *et al.* (1990) reveal several behavioural characteristics including diurnal movement pattern away from the FAD at night and towards the FAD in the day. Information was also gained on the effective range of the payao in terms of space, the swimming speed and possible fly-glide behaviour. Further, as mentioned previously, Holland *et al.* (1990) state that "...different phenomena may underlie the association of tuna with natural debris and their association with the FADs," *i.e.*, the tunas associated with natural logs and debris in the western equatorial Pacific tended to aggregate around drifting objects during the night and to leave these in daytime.

Recent studies of El Niño events indicate that these occur on a far more global scale than previously thought and that the perturbation of oceanographic conditions in the western equatorial Pacific is as great as that in the eastern equatorial Pacific. Suzuki (1988) compared the hook rate of yellowfin caught by the Japanese longline boats in the western and eastern Pacific with regard to El Niño events. While it was suggested that the hook rates of yellowfin during or one year after El Niño years tends to be higher than those in other years in the eastern Pacific, there was no apparent relationship between the two factors in the western Pacific. Although Suzuki (1988) did not find any changes in annual hook rates of longline-caught yellowfin related to El Niño and non-El Niño years, analysis based on more detailed time-area stratum appears to indicate a slight inverse relationship in the western equatorial Pacific between the depth of mixed layer (shallower in El Niño years) and the hook rates of yellowfin caught by the longline gear (FSFRL, 1987). In the eastern Pacific, there is an indication that the increases in recruitment follow two years after El Niño conditions since the mid-1960s (IATTC, 1989).

11.   INTERACTIONS WITH OTHER SPECIES

There is no confirmed information as far as the Japanese purse-seine fishery is concerned that yellowfin associate with dolphins in the western and central Pacific. It is generally known in the purse-seine fishery of the western and central tropical Pacific that juvenile yellowfin tuna, usually when associated with drifting objects, are caught mixed with skipjack and bigeye tunas of similar size, whereas adult yellowfin tuna tend to be taken as a pure school. There are several by-catches of marlins, rainbow runners, triggerfish, *etc.*, caught mixed with tunas in the purse-seine fishery of the western Pacific. Yellowfin also associate with whales and sharks (although not so frequently as with drifting objects) in the purse-seine fishery of the western and central equatorial Pacific (Suzuki 1981). In the purse-seine fishery of the Philippine waters, small juvenile yellowfin are frequently caught mixed with a substantial amount of small tunas and small pelagic fishes such as kawakawa, bullet tunas, and round scad.

Juvenile yellowfin and bigeye are often caught together in the purse-seine fishery; bigeye under 1.5 kg are not separately classified in the Japanese unloading sites. This may have a significant implication for estimates of juvenile bigeye catch since a preliminary study by the National Research Institute of Far Seas Fisheries (NRIFSF, unpublished) shows that about 15% of bigeye were included in the same category as small juvenile yellowfin.

12.   GENERAL DESCRIPTION OF THE FISHERIES

The yellowfin fisheries of the central and western Pacific are probably the most substantial and diverse in the world. The fisheries are roughly categorized as purse seine, longline, baitboat, and coastal. The coastal fishery includes various fishing gears such as handline, trap net, trolling, and gillnet. In addition, there are artisanal fisheries throughout the Southeast Asian countries and island countries in the South Pacific. A detailed description of general trends in the tuna fisheries of the western and central Pacific is not available but rough information can be found in a series of publications by the South Pacific Commission (SPC, 1991b *etc.*). The following is a summary description of yellowfin fisheries by country and gear in the western and central Pacific from various sources.

12.1   Longline Fishery

The yellowfin catch by the longline fishery was the largest in the western and central Pacific until the mid 1970s. Suzuki (1988) describes the development of the longline fishery in the Pacific as follows; "The operation of the (Japanese) longline was confined to the northwestern Pacific and did not last throughout the year in and before the 1940s. In that older period, the fishing boats were used as skipjack baitboats in the warmer months and as longline boats in the colder months of the year. After 1952, the longline fishery developed remarkably equipping with advanced navigation and freezing techniques, and expanded the fishing ground, having covered most of the tropical waters by the mid 1950s. In the late 1960s, the longline fishing grounds expanded toward the higher latitudes reflecting the shift of target species from yellowfin tuna and albacore to bigeye and southern bluefin tunas. After Japanese, Taiwanese and Korean tuna fleets started distant water longlining in this area from 1954 and 1964, respectively. Their major fishing grounds are areas south of the equator." For distant-water longlining, the Korean fleet sets its targets on bigeye and yellowfin and the Taiwanese fleet on albacore.

Both Japan and Taiwan have coastal and offshore fishing grounds (approximately corresponding to FAO area 61) for longlining, mostly fished by smaller boats (less than 100 GT) than the distant water boats. A significant amount of yellowfin, about 30 to 40 thousand tons combined for the two countries, has been caught by these fisheries.

12.2   Purse-Seine Fishery

At present, the yellowfin catch by the purse seine fishery is the most dominant in the western and central Pacific. Suzuki (1988) summarizes the development of purse seine fishing in the western and central Pacific, "Before the mid 1970s some Japanese purse seiners fished in the western equatorial Pacific Ocean only during northern winter, offseason for exploiting tuna in the Japanese waters. But there were no extensive operations there. In 1977 the Japanese fishermen discovered it easy to catch tunas associated with such materials as logs, and established year round operations in the western equatorial Pacific Ocean. The discovery resulted in a sharp increase in the amount of fishing effort and catch in waters north of Papua New Guinea (Honma and Suzuki, 1978)." In the 1980s, other countries joined the purse-seine fishery in this area including Korea, Taiwan, USA, Philippines, Indonesia, Solomon Islands, *etc.* This fishery has been the largest yellowfin fishery since the early 1980s. Skipjack is usually dominant in the purse-seine catch, accounting for about two thirds of the total catch. The ratio, however, is often subject to change by country and season. The remainder of the catches is mostly yellowfin with small amount of bigeye as a by-catch.

12.3 Baitboat Fishery

The yellowfin catch from the baitboat fishery has been far smaller than that from other major fisheries. Japan is the only major distant-water baitboat-fishing country operating in the western and central Pacific. Skipjack is predominant in the catch and yellowfin accounts for only a small percentage of the total catch. The Japanese distant-water baitboat fishery is declining due to economic problems caused by such factors as the labour-intensive nature of the fishery.

There are domestic baitboat fisheries in several South Pacific countries including the Solomon Islands, Fiji, Kiribati, Hawaii, *etc.*, all with a relatively small by-catch of yellowfin.

12.4 Philippines and Indonesia

Other than distant-water fishing nations mentioned previously, the Philippines and Indonesia catch a substantial amount of yellowfin. The combined catch of these two countries is the second largest after that by the entire purse-seine fishery. The magnitude of the Philippine and Indonesian yellowfin catch is uncertain since there exists a species separation problem in their reported landings, *i.e.*, other small tuna species are included with yellowfin. The main characteristic of the tuna fisheries in these two countries is the substantial amount of production made by small-scale and artisanal fisheries with diverse fishing gears. Most of the fisheries in these two countries are operated within their respective EEZs but at least 11 and 3 purse seiners have operated beyond the EEZ in 1990 for the Philippines and Indonesia, respectively (SPC, 1991b). The fisheries relating to yellowfin are briefly described below for the two countries.

12.4.1   Philippines

The tunas are exploited by both municipal fishing vessels (less than 3 gross tons) and by the commercial vessels (over 3 gross tons). Fish aggregating devices (FADs), or payaos, characterize almost all the Philippine tuna fisheries. The most important commercial fishing gears are purse seines and ringnets, and for municipal fisheries, hook and line (handlining). However, there is substantial inter-annual variation in catch by gear types both in the municipal and commercial fisheries.

12.4.2   Indonesia

Comprehensive documentation of the Indonesian tuna fisheries, specifying catch by species, fishing gear and area is scarce. Catch data by species, fishing gear and by FAO area are available but most of the catch of yellowfin tuna is listed as unclassified "UNCL" (*e.g.*, IPTP, 1991).

Merta (1985) describes the tuna fishery of Indonesia from 1976 to 1982 and noted that the species category "tuna" in Indonesia includes yellowfin, bigeye, albacore and southern bluefin tuna, but the amount of yellowfin catch appears dominant. The 1982 statistics for "tuna" in FAO area 71 (mostly Pacific side) indicate that the major fishing methods, in order of importance in production in weight, are handlining, trolling and longlining. As in the Philippine tuna fisheries, usage of FADs has assumed major importance.

Recent information on Indonesian tuna fisheries (IPTP, 1991) indicates that the longline, purse-seine and baitboat fisheries all contributed to the substantial, steady increase of yellowfin tuna catch by Indonesia.

13.   TRENDS IN CATCH, FISHING EFFORT, AND CATCH PER UNIT EFFORT

Only approximate trends of overall catch of yellowfin tuna by country are available in the western and central Pacific (*e.g.*, FAO Yearbooks) due to the lack of a coordinating organization responsible for collecting the basic catch and effort statistics, such as those by country, fishing gear and area. Except in the case of some fisheries, effective fishing effort on yellowfin and standardized CPUE have not been estimated.

13.1   Catch

The overall catch of yellowfin tuna in the western and central Pacific is reflected in the statistics of the FAO fishing area 71, and to a lesser degree, area 61. Table 4 shows the catch by country and fishing gear (for Japan) for area 71. As mentioned earlier, the annual catch of yellowfin tuna in area 61 is about 30 to 40 thousand tons, caught predominantly by Japan and Taiwan mostly by longline. The overall catch in area 71 has increased from 24 to 213 thousand tons from 1971 to 1986. Before 1973, most of the yellowfin catch seems to have been made by the Japanese longline fishery. The Japanese catch increased steadily till about 1977 when a catch of 51 thousand tons was landed. After a short period of fluctuating catches, the yellowfin catch remained stable after 1980 at a higher level mostly over 70 thousand tons. The Philippine and Indonesian catches appear in 1973 and 1975 in the statistics, but there must have been substantial yellowfin catch before by those countries and some catch by other island countries in this area as well. The Philippine catch does not show any consistent increasing trend but fluctuated between 40 and 60 thousand tons since 1974. The Indonesian catch has been increasing and reached 34 thousand tons in 1986. The USA catch reached a high level in a short time starting in 1980, and produced about 30 to 50 thousands tons after 1982. Most of the USA catch is from purse seining with small catches by artisanal fisheries mainly around Hawaii by longline, handline and troll gears (Coan, 1993).

Table 4. Catch of yellowfin tuna by countries from the FAO Area 71. (After Suzuki *et al.*, 1989).

Unit:ton

Country (Gear)/Year	1971	1972	1973	1974	1975	1976	1977	1978
Fiji				12	11	74	151	540
Indonesia					11062	8037	10859	10601
Japan Total	23531	24315	29822	31670	33070	37813	50822	73080
Longline	22807	24021	29286		29528		41987	59044
Purse seine	379		481		2176		7159	7036
Baitboat	345	294	55		55		1676	769
Other								6231
Kiribati				25	25	25	2771	2930
Korea Rep.					259	3664	5462	5088
Papua New Guinea			1420	1420	1743	8563	3695	3115
Phillippines			14900	51732	52793	44478	63059	47629
Solomon Is.								
U.S.A.								
Other				4229	4319	2510	2805	1870
Total	23531	24315	46142	89088	92209	105164	138904	144853
Japan/Total	1.00	1.00	0.65	0.36	0.36	0.36	0.37	0.50

Country (Gear)/Year	1979	1980	1981	1982	1983	1984	1985	1986
Fiji	361	240	846	1157	1586	1771	1128	995
Indonesia	14663	17550	21869	24340	20200	26450	31022	34140
Japan, Total	54789	75990	77145	72362	70682	61767	76513	73108
Longline	43488	55888	49003	38162	40193	28433	30786	24872
Purse seine	10528	9918	21827	28054	25567	32057	37523	42388
Baitboat	773	6143	2706	1531	1030	1275	3229	1827
Other		4041	3609	4014	3892	2	4995	4021
Kiribati	3000	3148	3000	3000	2135	4036	4844	1065
Korea Rep.	6881	7424	2712	2528	1156	1373	1893	3251
Papua New Guinea	2881	3019	3516	0	0	372	370	400
Phillippines	49224	48023	56176	51922	62036	58927	64293	59510
Solomon Is.	192	314	1167	2165	3328	2816	3698	2769
U.S.A.		772	12867	14345	51066	41455	28798	36520
Other	2084	2363	1444	1036	433	1031	1062	1009
Total	134075	158843	180762	172855	212622	199998	213621	212767
Japan/Total	0.41	0.48	0.43	0.42	0.33	0.31	0.36	0.34

The overall catch of yellowfin tuna after 1986 remained at a level similar to the early 1980s, except that in 1987 the catch was 260 thousand tons, mostly due to increased USA purse seine catch in this year (FAO, 1990).

It is possible to describe briefly the trend of yellowfin catch by fishing gear type for area 71. The longline catch has been decreasing, especially from the early 1980s both for Japan and Korea. There appears to be a very small yellowfin catch by the Taiwanese longline fishery in area 71 judging from the fishing grounds mostly in the middle latitudes of the South Pacific aimed at albacore. The overall baitboat catch of yellowfin tuna appears to be stable or slightly decreasing although the catch by the Solomon Islands has increased rapidly during the last few years (SPC, 1991b). On the other hand, the purse-seine catch has been increasing notably from the early 1980s. Table 5 shows the increase of the USA purse seine catch, and the recent increase in Korean catch (FAO, 1990) which is also due to increased catch by its purse seiners. Taiwanese catch is not specified in the Table 4.   However, the Taiwanese purse seine catch of yellowfin tuna has increased remarkably in the late 1980s (SPC, 1991b). As mentioned previously for the artisanal fisheries, the Philippine catch appears to be stable whereas the Indonesian catch has been increasing although it is not known which kinds of fishing gear are contributing to the increase. Another source of statistics by SPC gives approximate yellowfin catch by country and fishing gear for the SPC area (SPC, 1991b).   According to SPC estimates, the purse seine catch dominates, followed by artisanal and longline catches (Table 6). The baitboat catch appears to be very small.

Table 5.   USA distant-water purse-seine landings (mt) of tropical tunas in the western and central Pacific. (After Coan, 1993).

YEAR	YELLOWFIN	SKIPJACK	BIGEYE	TOTAL
1976	200	500	-	700
1977	200	700	-	900
1978	200	800	-	800
1979	600	8,000	20	8,620
1980	1,100	9,900	0	11,000
1981	13,000	17,400	170	30,570
1982	22,000	37,900	*	59,900
1983	49,600	104,100	-	153,700
1984	45,100	124,300	60	169,460
1985	29,000	87,700	-	116,700
1986	36,600	93,500	-	130,100
1987	66,400	79,800	-	146,200
1988	25,200	99,400	-	124,600
1989	41,200	90,400	-	131,600

Remarks:
--- "-" indicates that landings are not available, but may be greater than zero.
--- "*" indicates values less than 10 mt.
--- Yellowfin tuna landings include some landings of bigeye tuna.
--- Since trips that start late in one year may land their catch in the next, landings in each calendar year may contain some catches from the previous year.
--- Landings before 1979 are from Pacific Tuna Development Foundation exploratory fishing charters; landings from other USA vessels fishing in 1976 to 1978 are unknown.
--- Values in this table for 1980 to 1985 are different from those in Doulman (1987) due to inclusion here of USA vessels operating out of Guam and direct exports.

Table 6.   Preliminary estimates of catches (mt) of tunas in the western tropical Pacific Ocean in 1989.

COUNTRY	GEAR	SKJ	YFT	BET	ALB	OTH	TOTAL
FIJI	POLE-AND-LINE	5,369	507	7	0	0	5,883
	PURSE SEINE	143	767	7	0	0	917
	TOTAL	5,512	1,274	14	0	0	6,800
INDONESIA	VARIOUS	91,760	34,920	0	0	0	126,680
	PURSE SEINE	2,107	500	0	0	0	2,607
	TOTAL	93,867	35,420	0	0	0	129,287
JAPAN	GILLNET	1,680	0	0	13,161	0	14,841
	LONGLINE	0	16,400	8,800	1,900	0	27,100
	POLE-AND-LINE	118,900	1,100	700	0	0	120,700
	PURSE SEINE	104,000	33,000	1,000	0	0	138,000
	TOTAL	224,580	50,500	10,500	15,061	0	300,641
KIRIBATI	POLE-AND-LINE	857	385	0	0	0	1,242
KOREA	LONGLINE	0	12,950	15,750	2,800	0	31,500
	PURSE SEINE	89,250	15,750	0	0	0	105,000
	TOTAL	89,250	28,700	15,750	2,800	0	136,500
NEW CALEDONIA	LONGLINE	0	248	24	566	310	1,148
NEW ZEALAND	PURSE SEINE	6,974	0	0	0	0	6,974
	TROLLERS	0	0	0	5,205	0	5,205
	TOTAL	6,974	0	0	5,205	0	12,179
PHILIPPINES	VARIOUS	57,906	57,515	0	0	0	115,421
	PURSE SEINE	22,715	15,400	0	0	385	38,500
	TOTAL	80,621	72,915	0	0	385	153,921
SOLOMON ISLANDS	POLE-AND-LINE	24,284	1,475	0	0	109	25,868
	PURSE SEINE	5,892	4,410	0	0	489	10,791
	TOTAL	30,176	5,885	0	0	598	36,659
TAIWAN	GILLNET	0	0	0	12,000	0	12,000
	LONGLINE	0	2,209	2,719	6,119	0	11,047
	PURSE SEINE	76,500	13,500	0	0	0	90,000
	TOTAL	76,500	15,709	2,719	18,119	0	113,047
UNITED STATES	PURSE SEINE	92,179	43,708	0	0	293	136,180
	TROLLERS	0	0	0	5,100	0	5,100
	TOTAL	92,179	43,708	0	5,100	293	141,280
GRAND TOTAL	GILLNET	1,680	0	0	25,161	0	26,841
	LONGLINE	0	31,807	27,293	11,385	310	70,795
	POLE-AND-LINE	149,410	3,467	707	0	109	153,693
	PURSE SEINE	399,760	127,035	1,007	0	1,167	528,969
	TROLLERS	0	0	0	10,305	0	10,305
	SUB-TOTAL	550,850	162,309	29,007	46,851	1,586	790,603
	VARIOUS	149,666	92,435	0	0	0	242,101
	TOTAL	700,516	254,744	29,007	46,851	1,586	1,032,704

13.2 Fishing Effort and Catch per unit Effort

Description of trends in fishing effort and catch per unit effort of yellowfin tuna in the western and central Pacific, is more difficult than describing catch trends due to lesser availability of the fishing effort data. However, in a general sense, a cursory review of available information suggests that the trend of yellowfin catch closely relates to the corresponding trend of fishing effort. These trends in fishing effort and the CPUE are described as follows separated by fisheries by the major countries which capture yellowfin tuna.

Longline fishing for yellowfin is conducted mainly by the Japanese, Korean and Taiwanese boats. Among them the Japanese catch dominates. Suzuki *et al.* (1989) estimated the effective fishing effort and CPUE for yellowfin tuna caught by the Japanese longline fishery for the area west of 180° and north of 40°S and south of 40°N during 1952 to 1986. The effective effort peaked in 1962 at about 170 million hooks and then declined to a low in 1970 of about 75 million hooks, probably reflecting the change in target species from yellowfin to bigeye. The effective effort again began to increase and attained an historically high level in 1981 of about 230 million hooks. It then decreased to about 105 million hooks in 1986. The up-and-down trends in this period are due to the effects of introducing deep longlining aimed at bigeye tuna which resulted in overall increase in fishing effort in the area. This was followed by a reduction in domestic fishing effort implemented in the early 1980s. The trend in CPUE based on effective fishing effort in the early 1980s is about one half the level of that at the beginning of the fishery in the 1950s.

Major purse seining countries in the western and central Pacific are Japan, USA, Korea and Taiwan. Korean and Taiwanese purse seine catches of yellowfin in recent years has increased remarkably while Japanese and USA catches have leveled off with a high variability noted in the USA catches (SPC, 1991b). Information on the effective effort by purse seine fishing on yellowfin is not available due partly to difficulty in accounting for significant differences in the catch rates by countries as well as differences by school types, difficulties in separating the effort reasonably between skipjack and yellowfin catches, and difficulty in taking account of a possible improvement in gear efficiency (Tsuji, 1990). The number of purse-seine boats operating in the region has been limited for the Japanese fleet since 1982 and for the USA since 1988. Table 7 shows the trend of nominal fishing effort and CPUE for the Japanese, USA, Taiwanese and Korean purse-seine fleets (SPC, 1990). Fishing effort by the Japanese purse seiners increased sharply until 1984 and subsequently leveled off. The Japanese CPUE shows an increasing trend until 1986 with some drop in the 1982-1985 period but a sharp drop in 1988. No long-term trend of fishing effort by the USA purse-seine fishery is available and caution must be advised on the use of the series of effort and CPUE data of the USA fleets before 1988 because only part of the data is included. The number of USA purse-seine boats operating peaked in 1983 at 62 then decreased to 34 in 1989 (Coan, in press). Similar to the Japanese CPUE trend, the USA fleet experienced a sharp drop in CPUE in 1988 while no such drop was shown in the catch rate of the Taiwanese and Korean purse seine fisheries.

There are few references on the general trend in fishing effort and CPUE exerted on yellowfin by the various fisheries of the Philippines and Indonesia. In the Philippine tuna fishery, the statistics for the municipal fishery are not precise, but for the commercial fishing boats no trends appear in the number of boats in major fisheries except for an increasing trend of ringnet vessels (Barut and Arce, 1991). In the Indonesian fisheries, the number of longline boats of both domestic and foreign registration started to increase rapidly after the mid 1980s (Naamin and Bahar, 1991). However, most of that longline fleet appears to be operating in the Indian Ocean.

Table 7. Effort (days fishing and searching) and CPUE (tons) for the major purse-seine fleets of the western Pacific. (After SPC, 1990).

Year	Japan		U.S.		Taiwan		Korea	
	Effort	CPUE	Effort	CPUE	Effort	CPUE	Effort	CPUE
1980	973	3.67						
1981	1,732	5.50						
1982	4,433	4.72					177	1.21
1983	4,605	4.06			229	2.91	309	2.44
1984	5,555	4.89	624	11.10	528	1.63	623	1.89
1985	4,657	4.93	774	7.17	1,086	1.82	631	0.94
1986	4,715	6.50	535	9.23	1,136	1.41	546	4.40
1987	4,717	6.51	502	13.06	2,750	0.78	1,520	4.74
1988	4,895	3.26	4,089	3.4	3,706	0.77	1,706	2.91
1989	4,142	5.11	6,514	7.7	4,205	0.87	1,774	1.27

The USA 1988 and 1989 data were substituted by estimates of Coan (1993).

14. POPULATION DYNAMICS

Kamimura *et al.* (1966) assessed the Pacific yellowfin tuna stock, mostly covering the western and central tropical Pacific, based on the Japanese longline data from 1955 to 1964. Their analysis indicated that the longline CPUE in the western Pacific west of 180° did not show any appreciable decreasing trend while that in the central(180°-150°W) and eastern (150°W-120°W) Pacific showed a decreasing trend. The CPUE decreased to one half, and then to one fourth of its original level over the period of the study. They found a significant difference in age composition of the catch between the western and central Pacific, *i.e.*, in the west it was composed mostly of young fish around age 3, while in the central Pacific the fish were older than age 3. During the period of the study, the dominant age in the catch in the central Pacific shifted from age 4 to age 3. These changes were interpreted by the authors to mean that the longline exploitation of yellowfin tuna in the western and central Pacific had reduced the adult stock size, but that the recruitment to the longline fishery was not affected by the increased fishing effort. They concluded that the longline fishery operated around the MSY level in the mid 1960s.

A similar analysis was conducted by Honma *et al.* (1971) using a more comprehensive data series covering the period from 1950 to 1964. Their conclusion was close to that of Kamimura *et al.* (1966) indicating that the longline fishery was exploiting the stock around the MSY (about 30 thousand tons from the tropical Pacific) in the mid 1960s. Furthermore, Honma *et al.* (1971) estimated the recruitment in number of fish at age 1.5 assuming $M=0.3$ and 0.8. It ranged from 1.7 to 3 million and from 4.5 to 7 million for the two values of M. It should be noted that these values are applicable to tropical areas only, as in the case of the MSY previously mentioned. The follow-up studies indicate a much higher MSY for the yellowfin stock available to the longline fishery, from some 60 thousand tons (Honma, 1974) to 70-110 thousand tons (Suzuki *et al.*, 1989).

Preliminary *Y/R* analysis of data from the purse seine (Japan and USA), longline (Japan, Korea and Taiwan) and Philippine fisheries suggests that increasing purse-seine fishing effort beyond the 1980-1982 level will not increase the total catch due to the offset of increased catch of purse-seine fishery by reduction of catch in the longline fishery (Suzuki, 1986). However, the total catch from the western and central Pacific after 1982 increased sharply although the longline catch decreased. Suzuki *et al.*(1989) suggested from the catch trend of the major yellowfin fisheries and catch-at-age information that the yellowfin stock in the western and central Pacific was harvested at a sustained level of 200-210 thousand tons in the 1980s.

White (1982) analyzed the Philippine tuna-fishery data and made several attempts to estimate the effects of limiting the fishing mortality of a specific gear type on various other fisheries and on the total fishery. However, as he stated, it was hard to estimate fishing mortality specific to the Philippine fisheries because yellowfin tuna caught in the Philippines are probably a part of the offshore oceanic stock of the western and central Pacific.

15.   INTERACTIONS AMONG FISHERIES

One of the major concerns of interactions among the fisheries in the western and central Pacific has been the possible impact of the increasing purse seine fishery on either the longline fishery or the coastal fisheries.

At present, no concrete information regarding interaction is available; this shortcoming is largely due to a lack of basic statistics needed for stock assessment. A theoretical study indicates that the total yield from the yellowfin stock is larger in the case of a coexistence of a surface and subsurface fishery than in the case where either one of the fisheries operated solely in the area (Lenarz and Zweifel, 1979). Hilborn (1989) supported the result by Lenarz and Zweifel (1979) and showed more explicitly that the total yield could be maximized if the yellowfin stock fished by the longline fishery was somewhat discrete from that fished by the surface fisheries. If the two fisheries were fishing a homogeneous common stock, a longline fishery would maximize the yield.

A preliminary study of the CPUE for yellowfin tuna in the western and central Pacific based on the SPC data indicates that there is no evidence that the stock available to the surface fisheries has been significantly affected by fishing activity during 1978 and 1988 (Hampton, 1988). Suzuki *et al.* (1989) calculated the trend of CPUE for yellowfin taken by the Japanese longline fishery in the western Pacific west of 180°. The recent trend showed a decrease from a high peak in 1978 to a low in 1986 (Figure 3). During this period the yellowfin catch by the purse seine fishery increased sharply, especially after 1982 (Table 4, Japanese purse seine catch plus the USA catch). However, it should not necessarily be assumed that this reveals the impact of the purse seine fishery on the longline fishery as the decline of the longline CPUE appears to have started before the substantial increase in the purse seine catches. In addition, there is a 1-3 years age gap in longline and purse seine caught yellowfin. When the gap is accounted for, the decline of the longline CPUE started too early if the decline is assumed caused by the purse seine fishery.

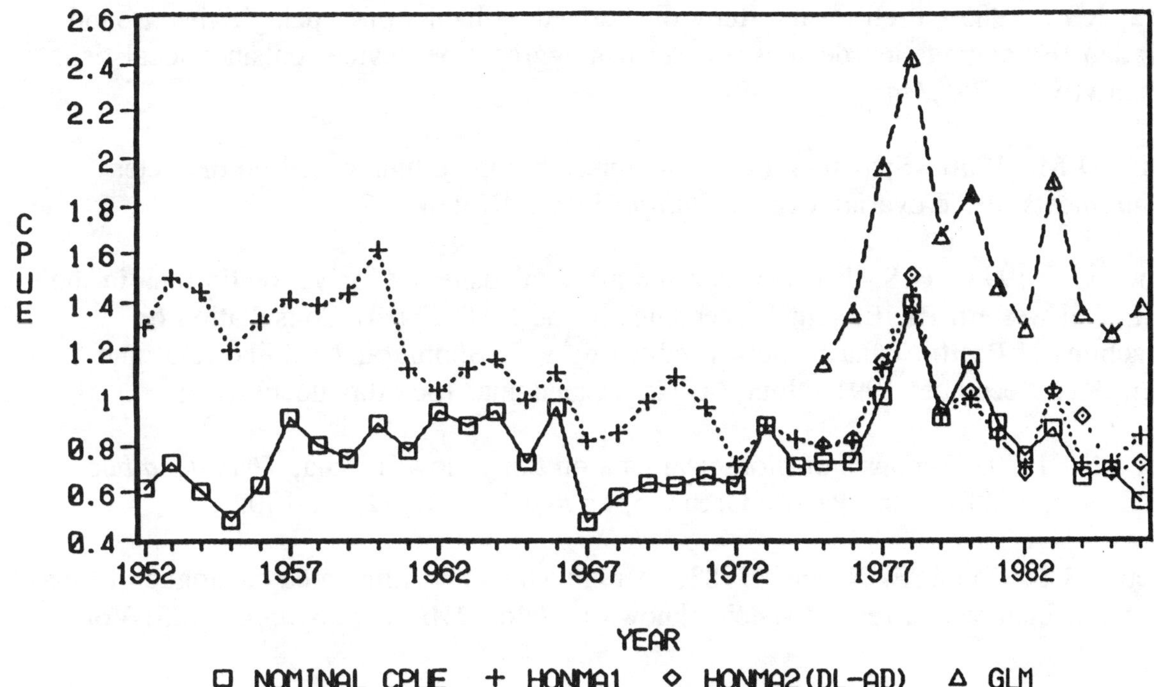

Figure 3.   Trend of CPUE for yellowfin taken by the Japanese longline fishery in the western Pacific. (After Suzuki *et al.*, 1989). The HONMA1, HONMA2(DL-AD), and GLM denote Honma's method without deep and regular longline adjustment, with adjustment, and general linear model, respectively.

16.   REFERENCES CITED

Aikawa, H., and M. Kato. 1938. Age determinations of fish-I. *Bull.Jap.Soc.Sci.Fish.*, 7(1):79-88.

Barrett, I., and H. Tsuyuki. 1967. Serum transferrin polymorphism in some scombroid fishes. *Copeia,* (3):551-7.

Barut, N.C. 1988. Food and feeding habits of yellowfin tuna *Thunnus albacares* (Bonnaterre, 1788), caught by handline around payao in the Moro Gulf. *Indo-Pac.Tuna Dev.Mgt.Programme,* IPTP/88/WP/18:39 p.

Barut, N.C., and F.M. Arce. 1991. The Philippine tuna fisheries: Industry and research. *Indo-Pac.Tuna Dev.Mgt.Programme,* CVWD, SEAC/90/5:12-42.

Bayliff, W.H. 1988. Growth of skipjack, *Katsuwonus pelamis*, and yellowfin, *Thunnus albacares*, tunas in the eastern Pacific Ocean, as estimated from tagging data. *Bull.I-ATTC,* 19(4):311-85.

Broadhead, G.C., and I. Barrett. 1964. Some factors affecting the distribution and apparent abundance of yellowfin and skipjack tuna in the eastern Pacific Ocean. *Bull.I-ATTC,* 8(8):417-73.

Brock, R.E. 1985. Preliminary study of the feeding habits of pelagic fish around Hawaiian fish aggregation devices, or can fish aggregation devices enhance local fish productivity ?. *Bull.Mar.Sci.* 37:40-9.

Buñag, D.M. 1956. Spawning habits of some Philippine tuna based on diameter measurements of the ovarian ova. *J.Philipp.Fish.,* 4(2):145-77.

Coan, A.L. 1993. U.S. distant-water and artisanal fisheries for yellowfin tuna in the central and western Pacific. In Proceedings of the FAO Expert Consultation on Interactions of Pacific Tuna Fisheries, edited by R.S. Shomura, J. Majkowski, and S. Langi, 3-11 December 1991, Noumea, New Caledonia. [See this document.]

Cole, J.S. 1980. Synopsis of biological data on the yellowfin tuna, *Thunnus albacares* (Bonnaterre, 1788), in the Pacific Ocean. *Spec.Rep.I-ATTC,* (2):71-150.

Collette, B.B., and C.E. Nauen. 1983. An annotated and illustrated catalogue of tunas, mackerels, bonitos and related species known to date. *FAO Fish.Synop.,* (125) Vol.2: 137 p.

Doulman, D.J. 1987. Development and expansion of the tuna purse seine fishery. *In* Tuna issues and perspectives in the Pacific islands region, edited by D.J. Doulman. Honolulu, East-West Center, pp. 133-160. [Cited from Coan(1993) without referring to the original paper.]

FAO. 1980. State of selected stocks of tuna and billfish in the Pacific and Indian Oceans. *FAO Fish.Tech.Pap.,* 200:88 p.

FAO. 1990. Catches and landings, 1988. *FAO Yearb.Fish.Statist.,* (66):502 p.

FSFRL. 1973. Report on experiments on the development of tuna culturing techniques (April, 1970-March, 1973). *S Ser.Far Seas Fish.Res.Lab.,* (8):165 p.

FSFRL. 1978. Distribution atlas of larval tunas, billfishes and related species -Results of larval surveys by R/V Shunyo Maru, and Shoyo Maru, 1956-1975. *S Ser.Far Seas Fish.Res.Lab.,* (9):99 p.

FSFRL. 1987. Effect of El Nino events upon reproduction and distribution of tunas. *Far Seas Fish.Res.Lab.,* 13 p.

Fujino, K., and T. Kang. 1968. Transferrin groups of tunas. *Genetics,* 59:79-91.

Godsil, H.C., and E.C. Greenhood. 1951. A comparison of the populations of yellowfin tuna (*Neothunnus macropterus*) from the eastern and central Pacific. *Fish Bull.Calif.Dep.Fish Game,* (82):33 p.

Hampton, J. 1988. Status of tuna fisheries in the western and central Pacific Ocean. Report of the 2nd Southeast Asian tuna conference and 3rd meeting of tuna research groups in the Southeast Asian region, Indo-Pacific Tuna Development and Management Programme, pp. 187-200.

Harada, T., K. Mizuno, O. Murata, S. Miyashita, and H. Furutani. 1971. On the artificial fertilization and rearing of larvae in yellowfin tuna. *Bull.Fac.Agric.Kinki Univ.*, (4):145-51.

Harada, T., O. Murata, and S. Oda. 1980a. Rearing of and morphological changes in larvae and juveniles of yellowfin tuna. *Bull. Fac.Agric.Kinki Univ.*, (13):33-6.

Harada, T., S. Miyashita, and H. Yoneshima. 1980b Effect of water temperature on yellowfin tuna hatching. *Bull.Fac.Agric.Kinki Univ.*, (13):29-32.

Higgins, B.E. 1967. The distribution of juveniles of four species of tunas in the Pacific Ocean. *Proc.Indo-Pac.Fish.Coun.*, 12(2):79-99.

Hilborn, R. 1989. Yield estimation for spatially connected populations: an example of surface and longline fisheries for yellowfin tuna. *North Amer.J.Fish.Mgt.*, 9(4):402-10.

Hisada, K. 1973. Investigation on tuna hand-line fishing ground and some biological observations on yellowfin and bigeye tunas in the northwestern Coral Sea. *Bull.Far Seas Fish.Res.Lab.*, (8):35-69.

Holland, K.N., R.W. Brill, and R.K.C. Chang. 1990. Horizontal and vertical movements of yellowfin and bigeye tuna associated with fish aggregating devices. *Fish.Bull.NOAA-NMFS,* 88:493-507.

Honma, M., T. Kamimura, and S. Hayasi. 1971. Compilation of catch records for assessment of yellowfin tuna stocks in longline fishery in the Pacific Ocean, and application to data for the years of 1950 to 1965. *Bull.Far Seas Fish.Res.Lab.*, (4):1-26.

Honma, M. 1974. Stock assessment of yellowfin tuna in the Pacific. *In* Proceedings of Tuna Conference in Shimizu (Showa 48 nendo), pp. 193-213.

Honma, M., and Z. Suzuki. 1978. Japanese tuna purse seine fishery in the western Pacific. *S Ser.Far Seas Fish.Res.Lab.*, (10):66 p.

Hunter, J.R., A.W. Argue, W. H. Bayliff, A.E. Dizon, A. Fonteneau, D. Goodman, and G.R. Seckel. 1986. The dynamics of tuna movements: an evaluation of past and future research. *FAO Fish.Tech.Pap.*, (227):78 p.

IATTC. 1989. Annual Report of the Inter-American Tropical Tuna Commission, 1988. *Annu.Rep.I-ATTC,* (1978):288 p.

IATTC. 1990. Quarterly Report of the Inter-American Tropical Tuna Commission. *Quart.Rep.I-ATTC,* 2:5-6.

ICCAT. (in press). Report of ICCAT Meeting of the working group on west Atlantic tropical tunas, 17-24 April 1991, Miami.

Ingles, J., and D. Pauly. 1984. An atlas of the growth, mortality and recruitment of Philippine fishes. *ICLARM Tech.Rep.*, 13(3):1278 p. [Cited from Bayliff (1988) without referring to the original paper.]

IPTP. 1991. Indian Ocean and Southeast Asian tuna fisheries data summary for 1989. *Data Summ.Indo-Pac.Tuna Dev.Mgt.Programme*, 11:96 p.

Ishii, T. 1968. On population parameters of yellowfin and bigeye tunas in the Pacific. *In* Proceedings of the Japanese Tuna Conference (Showa 42 nendo), Shimizu, pp. 72-76.

June, F.C. 1953. Spawning of yellowfin tuna in Hawaiian waters. *Fish.Bull.U.S.Fish.Wildl.Serv.*, 54(77):47-64.

Kamimura, T., and M. Honma. 1959. The relationship between length and weight of the landings of yellowfin from the Pacific Ocean. *Rep.Nankai Reg.Fish.Res.Lab.*, (11):88-107.

Kamimura, T., and M. Honma. 1963. Distribution of the yellowfin tuna (*Neothunnus macropterus*) (Temminck and Schlegel) in the tuna longline fishing grounds of the Pacific Ocean. *Rep.Nankai Reg.Fish.Res.Lab.*, (17):31-53.

Kamimura, T., A. Suda, and S. Hayasi. 1966. Symposium on Tuna Fisheries. *Bull.Jap.Soc.Sci.Fish.*, 32(9):756-86 and 829.

Kawai, H. 1969. On the relationship between thermal structure and distribution of longline fishing-grounds of tunas in the intertropical Atlantic-I. Analysis based on isotherms on level surfaces, topographies of thermocline, etc. *Bull.Far Seas Fish.Res.Lab.*, (2):275-303.

Kikawa, S. 1962. Studies on the spawning activity of Pacific tunas, *Parathunnus mebachi* and *Neothunnus macropterus*, by the gonad index examination. *Occas.Rep.Nankai Fish.Reg.Lab.*, (1):43-56.

Kikawa, S. 1966. The distribution of maturing bigeye and yellowfin and an evaluation of their spawning potential in different areas in the tuna longline grounds in the Pacific. *Rep.Nankai Reg.Fish.Res.Lab.*, (23):131-208.

Kimura, K. 1932. Growth curves of bluefin tuna and yellowfin tuna based on the catches near Sigedera on the west coast of Prov. Izu. *Bull.Jap.Soc.Sci.Fish.*, 1(1):1-4.

Klawe, W.L. 1963. Observations on spawning of four species of tuna (*Neothunnus macropterus, Katsuwonus pelamis, Auxis thazard* and *Euthynnus lineatus*) in the eastern Pacific Ocean, based on the distribution of their larvae and juveniles. *Bull.I-ATTC*, 6(9):447-540.

Koido, T., and Z. Suzuki. 1989. Main spawning season of yellowfin tuna, *Thunnus albacares*, in the western tropical Pacific Ocean based on the gonad index. *Bull.Far Seas Fish.Res.Lab.*, (26):153-64.

Koido, T., and N. Miyabe. 1990. II Field observation, 5. Tunas in Application of telemetry to aquatic animal behavior. *Fish.Sci.Series (Suisangaku siriizu)*, 80:55-66.

Kurogane, K., and Y. Hiyama. 1957. Morphometric comparison of the yellowfin tuna taken from the equatorial Pacific. *Bull.Jap.Soc.Sci.Fish.*, 23(7 & 8):388-93.

Lenarz, W., and J.R. Zweifel. 1979. A theoretical examination of some aspects of the interactions between longline and surface fisheries for yellowfin tuna, *Thunnus albacares. Fish.Bull.NOAA-NMFS*, 76(4):807-25.

Matsumoto, W.M. 1958. Description and distribution of larvae of four species of tuna in central Pacific waters. *Fish.Bull.U.S.Fish Wildl.Serv.*, 58(128):30-72.

Matsumoto, W.M. 1962. Identification of larvae of four species of tuna from the Indo-Pacific region I. *Dana-Rep.*, (55):1-16.

Merta, I.G.S. 1985. Tuna fisheries in Indonesia. Paper presented at the 36th Tuna Conference in Lake Arrowhead in 1985, California, 8 p.

Miyabe, N., and W.H. Bayliff. 1987. A review of the Japanese longline fishery for tunas and billfishes in the eastern Pacific Ocean. *Bull.I-ATTC*, 19(1):1-163.

Moore, H.L. 1951. Estimation of age and growth of yellowfin tuna (*Neothunnus macropterus*) in Hawaiian waters by size frequencies. *Fish.Bull.U.S.Fish Wildl.Serv.*, 52(65):133-49.

Mori, K. 1970. A consideration on the spawning of the tunas, especially of the yellowfin tuna (*Thunnus albacares*) in the adjacent sea of the Pacific coast of Japan. *Bull.Far Seas Fish.Res.Lab.*, (3):215-28.

Mori, K., S. Ueyanagi, and Y. Nishikawa. 1971. The development of artificially fertilized and reared larvae of the yellowfin tuna, *Thunnus albacares. Bull.Far Seas Fish.Res.Lab.*, (5):219-32.

Morita, Y. 1973. Conversion factors for estimating live weight from gilled-and-gutted weight of bigeye and yellowfin tunas. *Bull. Far Seas Fish.Res.Lab.*, (9):109-22.

Naamin N., and S. Bahar. 1991. Present status of tuna fisheries in Indonesia. *Indo-Pac.Tuna Dev.Mgt.Programme*, CVWP, SEAC/90/6:43-60.

Nakamura, H. 1965. Tuna resources of the world (1). *Fish.Invest.Ser.Jap.Fish.Res. Conserv.Assoc.*, 10-1:64 p.

Nakamura, E.L., and J.H. Uchiyama. 1966. Length-weight relations of Pacific tunas. *In* Proceedings of the Governor's Conference on Central Pacific Fishery Resources, edited by T.A. Manar, Hawaii, pp. 197-201.

Nikaido, H. 1988. Spawning ecology of yellowfin tuna-investigation of spawning frequency. *In* Proceedings of Japanese Tuna Conference in Shimizu (Showa 62 nendo), pp. 111-4.

Nishikawa, Y., M. Honma, S. Ueyanagi, and S. Kikawa. 1985. Average distribution of larvae of oceanic species of scombroid fishes, 1956-1981. *Far Seas Fish.Res.Lab.*, 99 p.

Nose, Y., S. Tomomatsu, K. Mimura, and Y. Hiyama. 1957. A method determine the time of ring-formation in hard tissues of fishes, especially for the age determination of Pacific tunas. Record of Oceanographic works in Japan, 2(3):9-18.

Pauly, D. 1980. A selection of simple methods for the assessment of the tropical fish stocks. *FAO Fish.Circ.*, (729), FIRM/C729. [Cited from White (1982) without referring to the original paper.]

Reintjes, J.W., and J.E. King. 1953. Food of yellowfin tuna in the central Pacific. *Fish.Bull.U.S.Fish Wildl.Serv.*, 54(81):91-110.

Richards, W.J., and D.C. Simmons. 1971. Distribution of tuna larvae (pisces, scombridae) in the northwestern Gulf of Guinea and off Sierra Leone. *Fish.Bull.NOAA-NMFS*, 69(3):555-68.

Ronquillo, J.A. 1963. A contribution to the biology of Philippine tunas. *In* Proceedings of the World Scientific Meeting on the Biology of Tunas and Related Species. *FAO Fish.Rep.*, 6(3):1683-752.

Royce, W.F. 1964. A morphometric study of yellowfin tuna, *Thunnus albacares* (Bonnaterre). *Fish.Bull.U.S.Fish Wildl.Serv.*, 63(2):395-443.

Schaefer, M.B. 1955. Morphometric comparison of yellowfin tuna from southeast Polynesia, Central America, and Hawaii. *Bull.I-ATTC*, 1(4):89-136.

Schaefer, K.M. 1988. Time and frequency of spawning of yellowfin tuna at Clipperton Island, and plans for future studies. *In* Proceedings of Tuna Conference in Shimizu (Showa 63 nendo), pp. 118-126.

Schaefer, K.M. 1991. Geographic variation in morphometric characters and gill raker counts of yellowfin tuna, *Thunnus albacares* from the Pacific Ocean. *Fish.Bull.NOAA-NMFS,* 89(2):289-97.

SPC. 1990. Collaborative study between SPC and Japan National Research Institute of Far Seas Fisheries: General Report. *Tuna Billfish Assess.Programme, S.Pac.Comm.*, SCTB/WP9:7 p.

SPC. 1991a. Report on the South Pacific Commission's Tuna and Billfish Assessment Programme Activities for 1990-1991. Paper presented at the Twenty-third Regional Technical Meeting on Fisheries, SPC/Fisheries 23/WP2:25 p.

SPC. 1991b. Status of tuna fisheries in the SPC area during 1990, with annual catches since 1952. SPC, SCTB/WP3:75 p.

Sprague, L.M. 1967. Multiple molecular forms of serum esterase in three tuna species from the Pacific Ocean. *Hereditas* 57:198-204.

Strasburg, D.W. 1960. Estimates of larval tuna abundance in the Central Pacific. *Fish.Bull.U.S.Fish Wildl.Serv.*, 60(167):231-55.

Suda, A., and M.B. Schaefer. 1965. Size-composition of catches of yellowfin in the Japanese long-line fishery in the eastern tropical Pacific east of 130° W. *Bull.I-ATTC*, 10(4):267-331.

Suda, A., S. Kume, and T. Shiohama. 1969. An indicative note on a role of permanent thermocline as a factor controlling the longline fishing ground for bigeye tuna. *Bull.Far Seas Fish.Res.Lab.*, (1):99-114.

Suzuki, A. 1962. On the blood types of yellowfin and bigeye tuna. *Am.Natur.*, 96(889):239-46.

Suzuki, Z. 1971. Comparison of growth parameters estimated for the yellowfin tuna in the Pacific Ocean. *Bull.Far Seas Fish.Res.Lab.*, (5):89-105.

Suzuki, Z., Y. Warashina, and M. Kishida. 1977. The comparison of catches by regular and deep tuna longline gears in the western and central equatorial Pacific. *Bull.Far Seas Fish.Res.Lab.*, (15):51-73.

Suzuki, Z., P.K. Tomlinson, and M. Honma. 1978. Population structure of Pacific yellowfin tuna. *Bull.I-ATTC,* 17(5):273-441.

Suzuki, Z. 1981. Recent activities of small and medium sized Japanese purse seine fishery and its characteristics viewed from type of association. *In* Proceedings of Tuna Conference in Shimizu (Showa 55 nendo), pp. 252-61.

Suzuki, Z. 1984. Swimming behavior of tunas by sonic tags. *In* Proceedings of the Japanese Tuna Conference in Shimizu (Showa 58 nendo), pp. 89-95.

Suzuki, Z. 1986. A study of interaction on yellowfin tuna between longline and purse seine fisheries in the western Pacific Ocean. *Indo-Pac.Tuna Dev.Mgt.Programme*, CVWD:175-190.

Suzuki, Z. 1988. Study of interaction between longline and purse seine fisheries on yellowfin tuna, *Thunnus albacares* (Bonnaterre). *Bull.Far Seas Fish.Res.Lab.*, (25):73-143.

Suzuki, Z., N. Miyabe, and S. Tsuji. 1989. Preliminary analysis of fisheries and some inference on stock status for yellowfin tuna in the western Pacific. Paper presented at the Standing Committee on Tuna and Billfish, South Pacific Commission, Suva, Fiji, 1989, SCTB/WP9.

Tan, Hui-Chong, Y. Nose, and Y. Hiyama. 1965. Age determination and growth of yellowfin tuna, *Thunnus albacares* Bonnaterre by vertebrae. *Bull.Jap.Soc.Sci.Fish.*, 31(6):414-22.

Tanaka, S. 1960. Studies on the dynamics and the management of fish populations. *Bull.Tokai Reg.Fish.Res.Lab.*, (28):1-200.

Tester, A.L., and E.L. Nakamura. 1957. Catch rate, size, sex, and food of tunas and other pelagic fishes taken by trolling off Oahu, Hawaii, 1951-55. *Spec.Sci.Rep.U.S.Fish Wildl.Serv.(Fish.)*, (250):25 p. [cited from Nakamura and Uchiyama (1966) without referring to the original paper].

Tsuji, S. 1990. Operation and CPUE of Japanese purse seine in the Western tropical Pacific. Paper presented at the Standing Committee on Tuna and Billfish, South Pacific Commission, SCTB3/WP10:12 p.

Uchiyama, J.H., and P. Struhsaker. 1981. Age and growth of skipjack tuna, *Katsuonus pelamis*, and yellowfin tuna *Thunnus albacares*, as indicated by daily growth increments of sagittae. *Fish.Bull.NOAA-NMFS*, 79(1):151-62.

Ueyanagi, S. 1969. Observations on the distribution of tuna larvae in the Indo-Pacific Ocean with emphasis on the delineation of the spawning areas of albacore, *Thunnus alalunga*. *Bull.Far Seas Fish.Res.Lab.*, (2):177-256.

Wade, C.B. 1950. Observations on the spawning of Philippine tuna. *Fish.Bull.U.S.Fish Wildl.Serv.*, 51(55):409-23.

Wade, C.B. 1951. Larvae of tuna and tuna-like fishes from the Philippine waters. *Fish.Bull.U.S.Fish Wildl.Serv.*, 51(57):445-85.

Wankowski, J.W.J. 1981. Estimated growth of surface-schooling skipjack tuna, *Katsuwonus pelamis*, and yellowfin tuna, *Thunnus albacares*, from the Papua New Guinea region. Fish. Bull., 79(3):517-545.

Watanabe, H. 1958. On the difference of the stomach contents of the yellowfin and bigeye tunas from the western equatorial Pacific. *Rep.Nankai Reg.Fish.Res.Lab.*, (7):72-81.

Watanabe, H. 1960. Regional differences in food composition of the tunas and marlins from several oceanic areas. *Rep.Nankai Reg.Fish.Res.Lab.*, (12):75-84.

White, T.F. 1982. The Philippine tuna fishery and aspects of the population dynamics of tunas in Philippine waters. *Indo-Pac.Tuna Dev.Mgt.Programme,* IPTP/82/WP/5:64 p.

Wild, A. 1986. Growth of yellowfin tuna, *Thunnus albacares*, in the eastern Pacific Ocean based on otolith increments. *Bull.I-ATTC*, 18(6):423-82.

Wild, A. 1993. A review of the biology and fisheries for yellowfin tuna, *Thunnus albacares*, in the eastern Pacific Ocean. *In* Proceedings of the FAO Expert Consultation on Interactions of Pacific Tuna Fisheries, edited by R.S. Shomura, J. Majkowski, and S. Langi, 3-11 December 1991, Noumea, New Caledonia. [See this document.]

Yabe, H., and S. Ueyanagi. 1961. Contributions to the study of the early life history of the tunas. *Occas.Rep.Nankai Reg.Fish.Res.Lab.*, (1):57-72.

Yabuta, Y., and M. Yukinawa. 1957. Age and growth of yellowfin tuna (*Neothunnus macropterus*) in Japanese waters by size frequencies. *Rep.Nankai Reg.Fish.Res.Lab.*, (5):127-33.

Yabuta, Y., N. Anraku, and M. Yukinawa. 1958. Study on the yellowfin tuna-III. Annual variations of the size compositions and the hooked rates of yellowfin tuna distributing in equatorial Pacific. *Rep.Nankai Reg.Fish.Res.Lab.*, (7):88-104.

Yabuta, Y., and M. Yukinawa. 1959. Growth and age of yellowfin tuna (*Neothunnus macropterus*) in the equatorial Pacific. Study of length frequency distribution-I. *Rep.Nankai Reg.Fish.Res.Lab.*, (11):77-87.

Yabuta, Y., M. Yukinawa, and Y. Warashina. 1960. Growth and age of yellowfin tuna-II. Age determination (scale method). *Rep.Nankai Reg.Fish.Res.Lab.*, (12):63-74.

Yamanaka, H., J. Morita, and N. Anraku. 1969. Relation between the distribution of tuna and water types of the North and South Pacific Ocean. *Bull.Far Seas Fish.Res.Lab.*, (2):257-73.

Yamanaka, K.L. 1990. Age, growth and spawning of yellowfin tuna in the southern Philippines. *Indo-Pac.Tuna Dev.Mgt.Programme,* IPTP/90/WP/21:1-87.

Yang, R.T., Y. Nose, and Y. Hiyama. 1969. A comparative study on the age and growth of yellowfin tunas from the Pacific and Atlantic Oceans. *Bull.Far Seas Fish.Res.Lab.*, (2):1-21.

Yesaki, M. 1983. Observations on the biology of yellowfin (*Thunnus albacares*) and skipjack (*Katsuwonus pelamis*) tunas in Philippine waters. *Indo-Pac.Tuna Dev.Mgt.Programme,* IPTP/83/WP/7: 66 p.

Yokota, T., M. Toriyama, F. Kanai, and S. Nomura. 1961. Studies on the feeding habit of fishes. *Rep.Nankai Reg.Fish.Res.Lab.*, (14):234 p.

Yonemori, T. 1982. Study of tuna behavior, particularly their swimming depths, by the use of sonic tags. *Newsl.Far Seas Fish.Res.Lab.*, 44:1-5.

Yuen, H.S.H., and F.C. June. 1957. Yellowfin tuna spawning in the central equatorial Pacific. *Fish.Bull.U.S.Fish. Wildl.Serv.*, 57(112):251-64.

USA DISTANT-WATER AND ARTISANAL FISHERIES FOR YELLOWFIN TUNA IN THE CENTRAL AND WESTERN PACIFIC

Atilio L. Coan, Jr.
La Jolla Laboratory
Southwest Fisheries Science Center
National Marine Fisheries Service, NOAA
La Jolla, California 92038

ABSTRACT

Total landings, catch per days fished and sizes of fish caught are reviewed for two types of USA yellowfin tuna fisheries operating in areas of the central and western Pacific through 1990. The largest of these USA fisheries is the distant-water purse-seine fishery which landed highs of 66,400 mt of yellowfin tuna in 1987, and 57,000 mt in 1990, averaged 9 mt per day fished in 1990, and caught yellowfin tuna between 25 and 186 cm.

1.   INTRODUCTION

A USA distant-water purse seine fishery, a multispecies fishery catching predominantly skipjack with lesser quantities of yellowfin and bigeye tunas, is conducted in the central and western Pacific. The carrying capacities of the majority of the purse seine vessels participating in the fishery are 1,000 metric tons (mt) or more (Doulman 1987). The purse seiners are based in either American Samoa or Guam.

Within the USA Exclusive Economic Zone (EEZ), or 200 miles of the coasts of Hawaii, American Samoa, Northern Marianas and Guam, the U.S. has artisanal fisheries for yellowfin tuna and other large pelagics (*e.g.*, marlins, wahoo, *etc.*). These fisheries are mixtures of commercial, subsistence and recreational fisheries that use a variety of gears. The major gears used to catch yellowfin tuna are longline, handline and troll (Hudgins and Pooley 1987).

The purpose of this paper is to review available data on landings, fishing effort, catch per effort, sizes of fish caught and methods of data collection from these fisheries. Distant-water fisheries are reviewed from their start in 1976 to 1990 and artisanal fisheries from 1954 to 1990.

2.   DISTANT-WATER FISHERIES

2.1   Total Landings

Total landings (metric tons) for the USA distant-water central and western Pacific purse seine fishery for tropical tunas are available since 1976 (Table 1). Landings for 1976 to 1978 are estimates from exploratory fishing of vessels chartered by the Pacific Tuna Development Foundation (Anon., 1977; Anon., 1979; Souter and Broadhead, 1978). Landings for 1979 to 1990 are from cannery receipts of landings or transshipments to USA canneries in American Samoa, Puerto Rico and California (Schug and Galea'i,

Table 1.   U.S. distant-water purse seine landings (metric tons) of tropical tunas in the central and western Pacific. Yellowfin tuna landings include some landings of bigeye tuna. Since trips that start late in one year may land their catch in the next, landings in each calendar year may contain some catches from the previous year.

YEAR	YELLOWFIN	SKIPJACK	BIGEYE	TOTAL
1976	200	500	-	700
1977	200	700	-	900
1978	200	800	-	800
1979	600	8,000	20	8,620
1980	1,100	9,900	0	11,000
1981	13,000	17,400	170	30,570
1982	22,000	37,900	*	59,900
1983	49,600	104,100	-	153,700
1984	45,100	124,300	60	169,460
1985	29,000	87,700	-	116,700
1986	36,600	93,500	-	130,100
1987	66,400	79,800	-	146,200
1988	25,200	99,400	-	124,600
1989	41,200	90,100	-	131,600
1990	57,000	107,400	-	164,400

Remarks:
- indicates that landings are not available but may be greater than zero.
* indicates values less than 10 metric tons.
Landings before 1979 are from Pacific Tuna Development Foundation exploratory fishing charters.
Landings from other U.S. vessels fishing in 1976 to 1978 are unknown.
Values in this table for 1980 to 1985 are different than those in Doulman 1987 due to inclusion here of U.S. vessels operating out of Guam and direct exports.

1987). Some vessels transship their catch directly to foreign ports such as Thailand, Japan, Indonesia, Philippines, Italy and Australia and data are gathered from canneries in those countries whenever possible. Since 1989, landings data for the distant-water fishery are more complete, due to reporting requirements of the South Pacific Regional Tuna Treaty (SPRTT) between the USA and 16 island nations that started in June, 1988 (Coan *et al.*, 1988). There has been 100% coverage of landings since June 15, 1988.

The USA distant-water purse seine landings of yellowfin tuna since 1979 have increased from approximately 600 mt to a high of 66,400 mt in 1987 (Figure 1, Table 1). Skipjack landings increased from 8,000 mt in 1979, to a high of 124,300 mt in 1984. Reported bigeye landings reached a high of 200 mt in 1981, but can be much higher as landings of this species are reported with yellowfin tuna landings. Estimated yellowfin tuna landings from exploratory fishing records in 1976 to 1978 are approximately 200 mt per year and may represent only a portion of the actual landings.

Data in Table 1 have not been corrected for species composition of yellowfin and bigeye tunas or for overlapping of landings between years (trips catching fish late in the year and landing their catches in the next year). Data collected for 1988 to 1990 show that approximately 4% of the 1989 yellowfin tuna landings were actually catches in 1988, and 10% of the 1990 yellowfin landings were actually catches in 1989. Current estimates from species composition samples taken under the SPRTT indicate that yellowfin landings can be between 6% and 9% bigeye tuna.

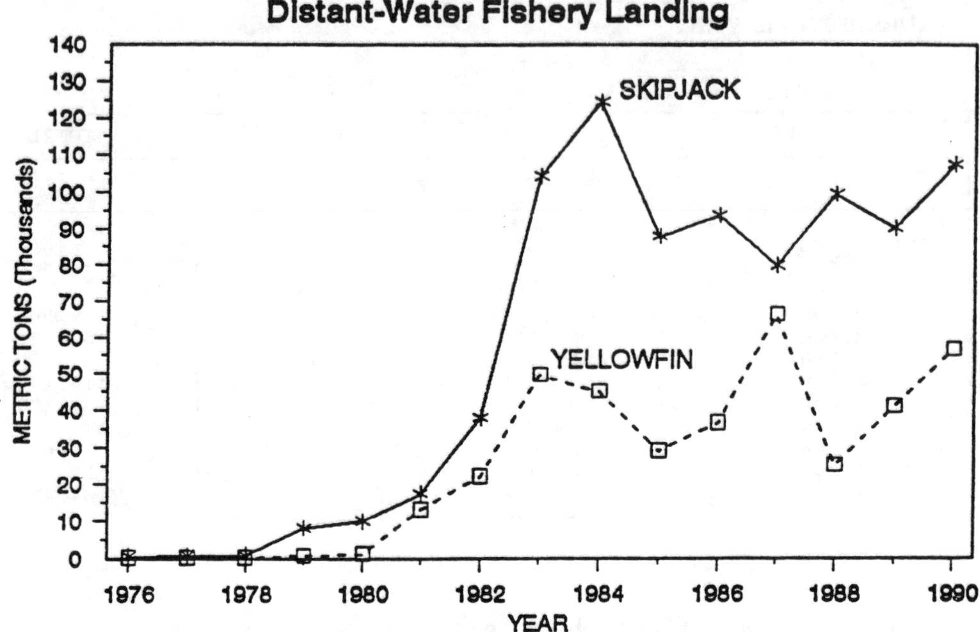

Figure 1.   Landings (metric tons) of yellowfin and skipjack tunas from the U.S. distant-water central and western Pacific purse seine fishery. Bigeye tuna landings are included in yellowfin landings.

2.2   Total Effort

Total effort in number of vessels participating in the USA distant-water central and western Pacific purse seine fishery ranges from 8 to 62 during the period 1979 to 1990 (Table 2). The number of purse seiners participating in the fishery during the period 1976 to 1978 reflects only vessels fishing under an exploratory fishing charter with the PTDF and may not represent the actual number of vessels fishing in those years.

The data for 1979 to 1990 are tabulated from information on landing receipts or from licenses issued under the SPRTT. Since 1979, the number of purse seiners participating in the fishery increased from 8 to 62 in 1983, decreased to the low 30's in 1986 to 1989, and increased to 41 in 1990. Coverage of total effort data has been 100% since June 15, 1988.

2.3   Catch and Effort

The USA distant-water central and western Pacific purse seine fleet fishing logbook data have been collected by the USA tuna industry for 1976 to 1988. Data for 1979 to 1988 exist, but are currently unavailable to the public. Data are available for 1976 to 1978, but are for PTDF charters only and are not included here due to the small sample size. Data for 1987 to 1990, for operations of this fleet within the EEZs of Pacific island countries, have been published in the South Pacific Commission's Regional

Table 2.   Total number of U.S. distant-water purse seiners fishing for tropical tunas in the central and western Pacific. The number of vessels for 1976 to 1978 are Pacific Tuna Development Foundation exploratory fishing charters and do not include an unknown number of other vessels that may have fished in these years.

YEAR	NUMBER OF VESSELS
1976	3
1977	1
1978	2
1979	8
1980	14
1981	14
1982	24
1983	62
1984	61
1985	40
1986	36
1987	35
1988	32
1989	34
1990	41

Tuna bulletin (SPC, 1989; SPC, 1990; SPC, 1991). Since June 15, 1988, the fleet submits logbooks as a requirement of the SPRTT (Coan *et al.*, 1988). The data are collected, computerized and submitted to the Forum Fisheries Agency, the managing agency for the treaty. Coverage since June 15, 1988, has been 100%.

Monthly catch (mt), effort (days fishing) and catch-per-effort (mt/days fishing) are shown in Table 3. The fishery is usually centered north of Papua New Guinea, with occasional fishing off New Zealand (Figure 2). The fishery is primarily for skipjack (67%), with lesser quantities of yellowfin and bigeye tunas (33%). Yellowfin catches are generally highest in June to November, also the months of higher yellowfin catch rates. The average catch rate for yellowfin tuna in 1990 is 9.0 mt per day fished compared to 7.1 in 1989, 3.1 in 1988, and 12.8 in 1987. Catch rates in 1987 and 1988 may be biased due to poor sampling coverage.

2.4   Length Composition

The USA distant-water central and western Pacific purse seine catches are measured for fork length (FL) from landings in American Samoa (Coan *et al.*, 1988; Honda *et al.*, 1988). For data collected before the SPRTT (1981; 1982; and 1984 to 1987) fifty to 100 fish samples were drawn from each well and no species composition samples were taken. For data collected under the SPRTT (1988 to 1990) fifty to 100 fish were randomly selected from each well and unloaded size category. If yellowfin tuna were mixed with skipjack or bigeye tuna in the landings, then 100 fish were drawn, measured and used for species composition sampling (Coan and Yamasaki, 1990).

Yellowfin tuna in landings of USA purse seiners fishing in the central and western Pacific in 1981, 1982 and 1984 to 1990 range from 25 to 186 centimeters (cm) FL (Figure 3). Fish measured are generally in two modes, 40 to 70 cm and 80 to 120 cm or age 1, 2 and 3 year-old fish.

3. ARTISANAL FISHERIES

3.1 Total Landings

Landings data for artisanal fisheries of Hawaii, American Samoa, Northern Marianas and Guam are gathered by local island fisheries agencies. Various techniques are used to gather and process the data (Hamm, 1985; Hamm and Kassman, 1986; Hamm *et al.*, 1986, Hamm and Quach, 1988; Hamm and Quach, 1989; Hamm *et al.*, 1990; Hamm *et al.*, 1991). The data are placed on the Western Pacific Fishery Information Network (WPACFIN).

Yellowfin tuna landings in Hawaii are collected through a commercial fishing license system that requires all commercial fishermen to report their individual landings. Data are available for 1945 to 1953 (Boggs and Pooley, 1987), and for 1954 to 1990 (Table 4). Currently, approximately 2,500 fishermen have licenses. The principal methods of fishing for yellowfin tuna in Hawaii are with handline, longline or troll gears. The number of longline vessels operating in the fishery decreased from 76 in 1950 to a low of 14 in 1979, then increased to 138 by 1990 (TNL, 1990; Ito, 1991). Handline vessels increased from 30 to 40 in 1976 to over 230 in 1980. Approximately 160 troll vessels operated in 1976 (Boggs and Pooley, 1987). Hawaiian commercial landings of yellowfin tuna have increased from less than 200 mt in 1956 to over 1,800 mt in 1988. Yellowfin tuna are also caught by Hawaiian subsistence and recreational fishermen. Catches from these fisheries could be as high as 2,000 mt of skipjack and yellowfin annually, but few data are available (Hudgins and Pooley, 1987).

Yellowfin tuna landings in American Samoa, Northern Marianas and Guam are estimated by a random survey of fishermen or from landing receipts of fish buyers. Yellowfin landings have increased from a low of 10 mt in 1979 to a high of 80 mt in 1985 (Table 4), then decreased to approximately 50 mt in 1990 (Hamm et al., 1992). Fisheries are typically small boat (12 to 48 feet), 1 to 2 day commercial, recreational and subsistence operations using primarily troll fishing gear.

3.2 Length Composition

Yellowfin tuna caught by the Hawaiian artisanal longline fishery have been measured for fork length (FL) or weight at the local fish auction. Data are available for 1948 to 1952 (Otsu, 1954), and for 1960 to 1970 (Figure 4). Fish sampled range between 50 and 190 cm FL and the majority of the fish measured were between 116 and 166 cm FL.

Sizes of yellowfin tuna caught by the Hawaiian artisanal handline fishery are available for 1973-75 (Yuen, 1979), and for 1978 (Figure 5). Fish measured in 1973 to 1975 range between 50 and 180 cm FL, and the majority of the fish were less than 132 cm FL. Fish measured in 1978 tended to be smaller, ranging between 47 and 57 cm FL.

Table 3. Sample catch (metric tons, MT), fishing effort (days fished) and catch per effort (CPE, metric tons per day fished) for the U.S. distant-water central and western Pacific purse seine fishery. Sampling coverage from January, 1987 to June, 1988 is unknown (data from South Pacific Commission). Coverage from July, 1988 to December, 1990 is 100% (data from South Pacific Regional Tuna Treaty sampling).

YEAR	MONTH	DAYS FISHED	YELLOWFIN		SKIPJACK		OTHER	TOTAL	
			MT	CPE	MT	CPE	MT	MT	CPE
1987	JAN	45	243	5.4	1,081	24.0	0	1,324	29.4
1987	FEB	-	-	-	-	-	-	-	-
1987	MAR	-	-	-	-	-	-	-	-
1987	APR	-	-	-	-	-	-	-	-
1987	MAY	-	-	-	-	-	-	-	-
1987	JUN	20	262	13.1	284	14.2	0	546	27.3
1987	JUL	38	510	13.4	528	13.9	0	1,038	27.3
1987	AUG	85	1,594	18.8	186	2.2	18	1,798	21.2
1987	SEP	136	1,379	10.1	1,122	8.3	99	2,600	19.1
1987	OCT	91	1,764	19.4	823	9.0	0	2,587	28.4
1987	NOV	26	159	6.1	1,014	39.0	0	1,173	45.1
1987	DEC	49	351	7.2	1,165	23.8	0	1,516	30.9
	TOTAL	490	6,262	12.8	6,203	12.7	117	12,582	25.7
1988	JAN	18	19	1.1	569	31.6	0	588	32.7
1988	FEB	56	573	10.2	1,004	17.9	0	1,577	28.2
1988	MAR	34	271	8.0	599	17.6	0	870	25.6
1988	APR	133	460	3.5	2,664	20.0	0	3,124	23.5
1988	MAY	143	1,069	7.5	2,064	14.4	0	3,133	21.9
1988	JUN	439	799	1.8	8,470	19.3	2	9,271	21.1
1988	JUL	594	2,563	4.3	9,050	15.2	18	11,631	19.6
1988	AUG	554	1,071	1.9	6,771	12.2	1	7,843	14.2
1988	SEP	638	2,383	3.7	8,118	12.7	1	10,502	16.5
1988	OCT	550	930	1.7	8,062	14.7	0	8,992	16.3
1988	NOV	489	1,343	2.7	7,005	14.3	1	8,349	17.1
1988	DEC	487	1,444	3.0	7,321	15.0	1	8,766	18.0
	TOTAL	4,135	12,925	3.1	61,697	14.9	24	74,646	18.0
1989	JAN	416	3,278	7.9	9,309	22.4	0	12,587	30.3
1989	FEB	441	932	2.1	11,469	26.0	27	12,428	28.2
1989	MAR	355	1,637	4.6	11,028	31.1	0	12,665	35.7
1989	APR	471	1,150	2.4	8,476	18.0	95	9,721	20.6
1989	MAY	672	2,579	3.8	7,749	11.5	286	10,614	15.8
1989	JUN	664	5,771	8.7	9,622	14.5	347	15,740	23.7
1989	JUL	443	3,591	8.1	4,033	9.1	0	7,624	17.2
1989	AUG	755	4,297	5.7	4,026	5.3	0	8,323	11.0
1989	SEP	685	12,231	17.9	5,410	7.9	7	17,648	25.8
1989	OCT	529	4,923	9.3	6,240	11.8	0	11,163	21.1
1989	NOV	618	4,881	7.9	6,144	9.9	54	11,079	17.9
1989	DEC	580	1,524	2.6	8,704	15.0	45	10,273	17.7
	TOTAL	6,629	46,794	7.1	92,210	13.9	861	139,865	21.1
1990	JAN	454	1,255	2.8	8,062	17.8	0	9,317	20.5
1990	FEB	566	3,837	6.8	7,430	13.1	95	11,362	20.1
1990	MAR	572	4,517	7.9	7,241	12.7	30	11,788	20.6
1990	APR	488	5,114	10.5	10,820	22.2	54	15,988	32.8
1990	MAY	427	7,709	18.1	10,483	24.6	87	18,279	42.8
1990	JUN	385	6,407	16.6	5,339	13.9	5	11,751	30.5
1990	JUL	446	2,803	6.3	9,528	21.4	14	12,345	27.7
1990	AUG	584	10,641	18.2	10,023	17.2	14	20,678	35.4
1990	SEP	523	7,503	14.3	6,831	13.1	1	14,335	27.4
1990	OCT	609	4,375	7.2	13,113	21.5	0	17,488	28.7
1990	NOV	644	2,202	3.4	6,860	10.7	0	9,062	14.1
1990	DEC	696	1,343	1.9	10,305	14.8	0	11,648	16.7
	TOTAL	6,394	57,706	9.0	106,035	16.6	300	164,041	25.7

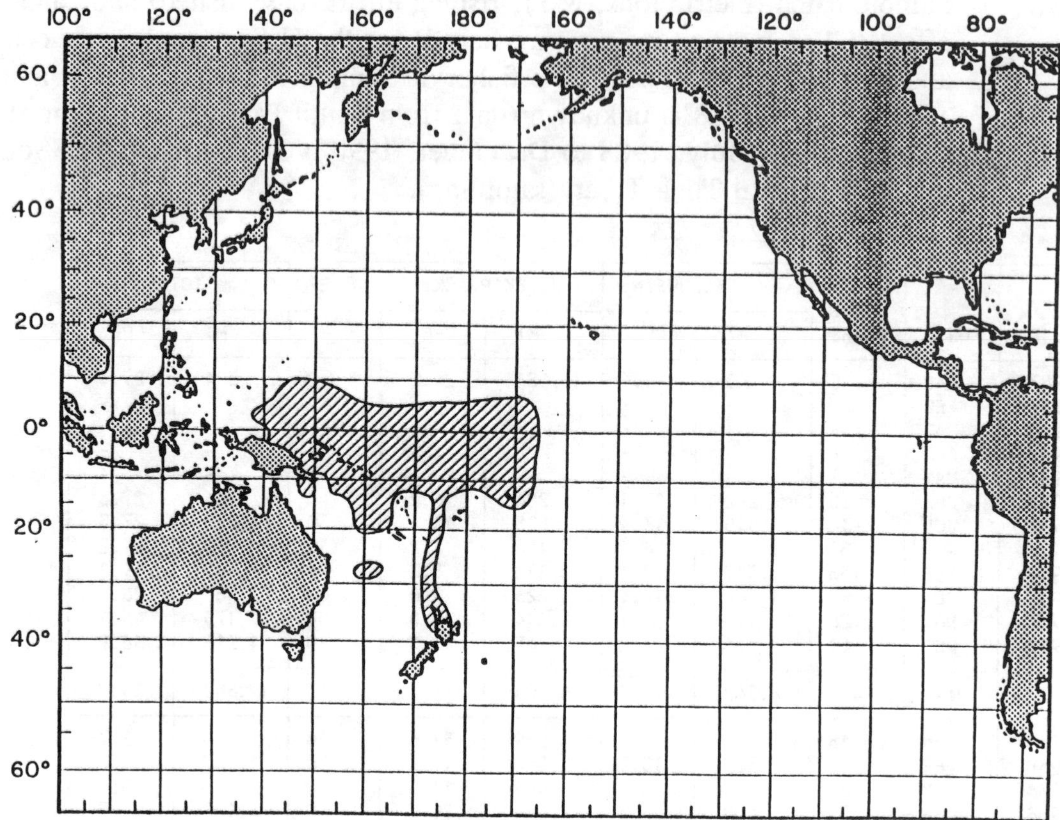

Figure 2.   General area fished by the U.S. central and western Pacific distant-water purse seine fishery, 1988-1990.

4.   SUMMARY

The major USA fishery for yellowfin tuna in the central and western Pacific is the distant-water purse seine fishery. The fishery operates year round, catching mainly skipjack, primarily off Papua New Guinea and has landed as much as 66,400 mt of yellowfin tuna in 1987. Yellowfin tuna catches and catch rates from the fishery are highest from June to November. The average yellowfin tuna catch rate was 9.0 mt per days fishing in 1990. Collection of data from this fishery has improved since the implementation of the SPRTT. Coverage rates for the fleet are currently 100% for logbook and landings data.

The USA artisanal fisheries for yellowfin tuna operate within 200 miles of the coasts of Hawaii, Guam, American Samoa and Northern Marianas. Yellowfin tuna landings from these fisheries are less than 1,800 mt a year. The majority of the landings are from waters around Hawaii and are made with longline, handline or troll fishing gears. Sizes of yellowfin tuna caught by the Hawaiian artisanal longline and handline fisheries ranged from 50 to 190 cm FL. The larger fish (116 to 190 cm FL) are generally caught by longline gears.

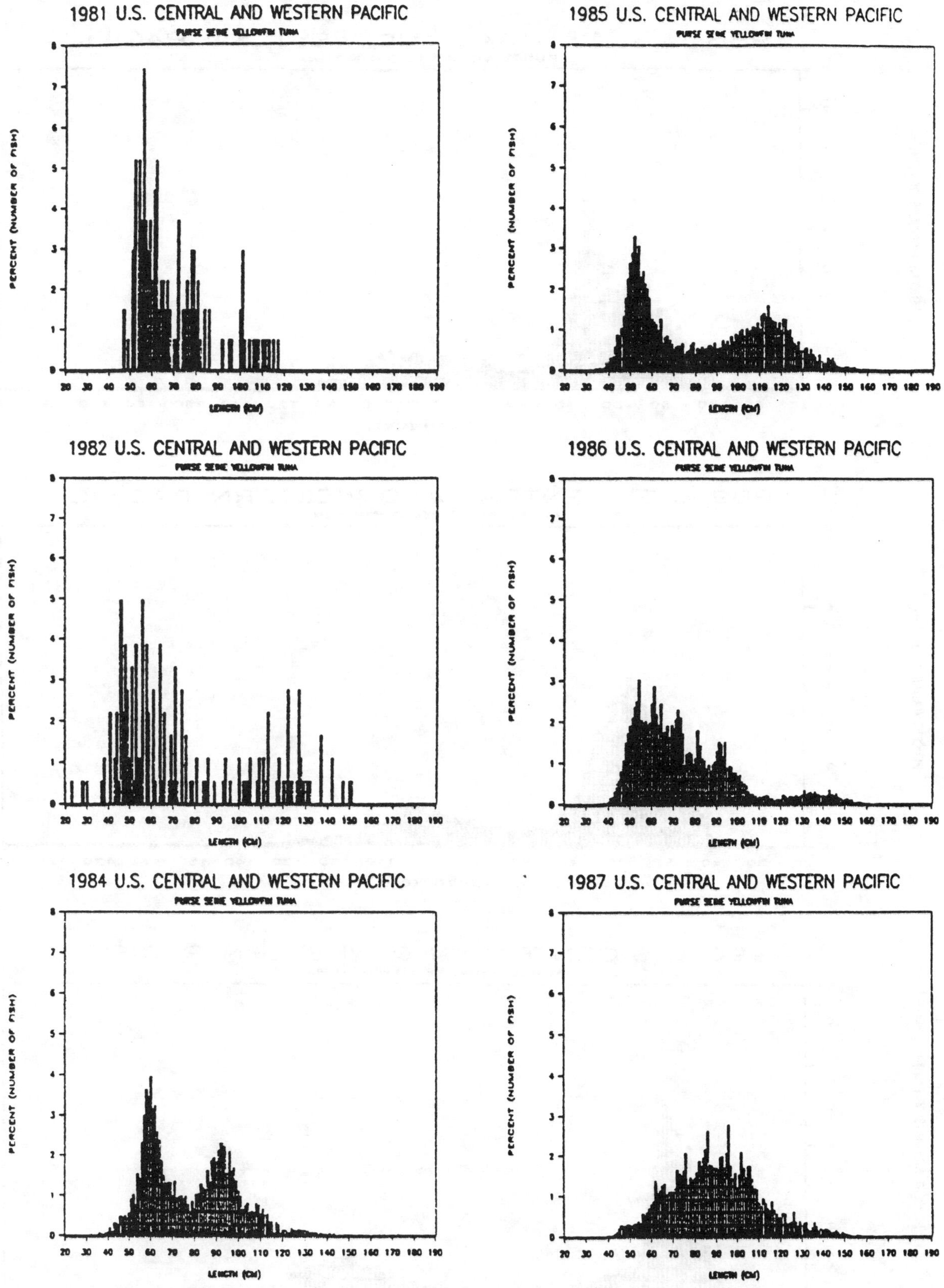

Figure 3. Annual length composition of yellowfin tuna caught by the U.S. central and western Pacific distant-water purse seine fishery, 1981, 1982 and 1984-1990.

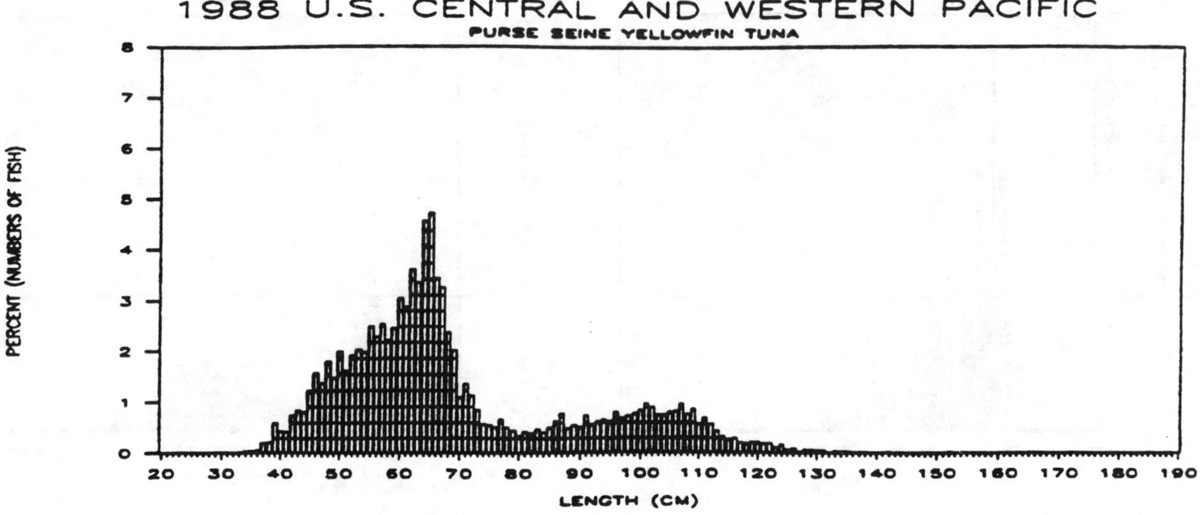

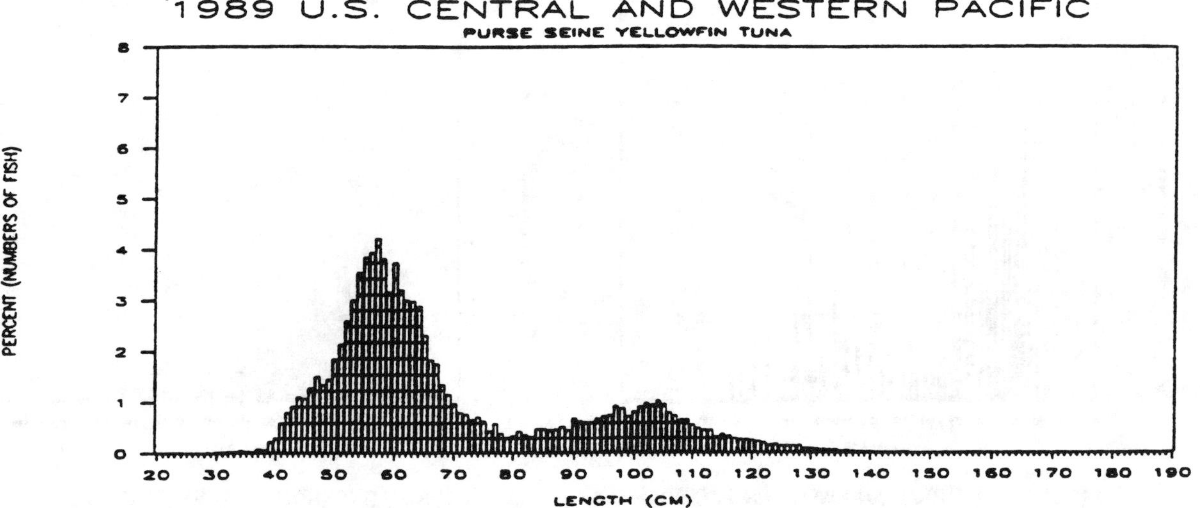

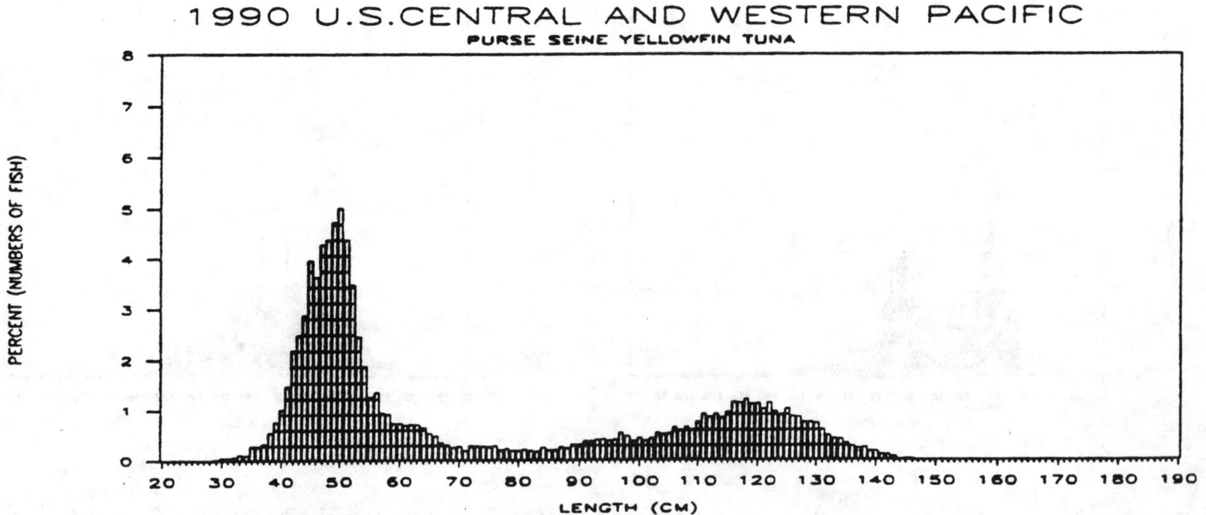

Figure 3.   (Continued)

Table 4.   Yellowfin tuna landings (metric tons) from U.S. artisanal fisheries in the central and western Pacific.

YEAR	HAWAII	AMERICAN SAMOA	NORTHERN MARIANAS	GUAM
1954	240	-	-	-
1955	200	-	-	-
1956	140	-	-	-
1957	170	-	-	-
1958	180	-	-	-
1959	260	-	-	-
1960	160	-	-	-
1961	210	-	-	-
1962	180	-	-	-
1963	180	-	-	-
1964	400	-	-	-
1965	230	-	-	-
1966	230	-	-	-
1967	230	-	-	-
1968	190	-	-	-
1969	190	-	-	-
1970	320	-	-	-
1971	390	-	-	-
1972	380	-	-	-
1973	350	-	-	-
1974	520	-	-	-
1975	780	-	-	-
1976	840	-	-	-
1977	930	-	-	-
1978	960	-	-	-
1979	980	-	*	10
1980	1,200	-	*	20
1981	1,280	-	*	50
1982	820	*	*	60
1983	920	10	10	30
1984	910	30	10	30
1985	1,080	20	*	60
1986	1,720	20	10	20
1987	1,720	10	*	20
1988	1,130	20	10	40
1989	1,350	20	*	20

Remarks:

- indicates that landings are not available but may be greater than zero.

* indicates values less than 10 metric tons.

Hawaiian baitboat and longline fisheries operated before 1954 and as early as the late 1930's, but data are sparse (June 1950, June 1951, Boggs and Pooley 1987)

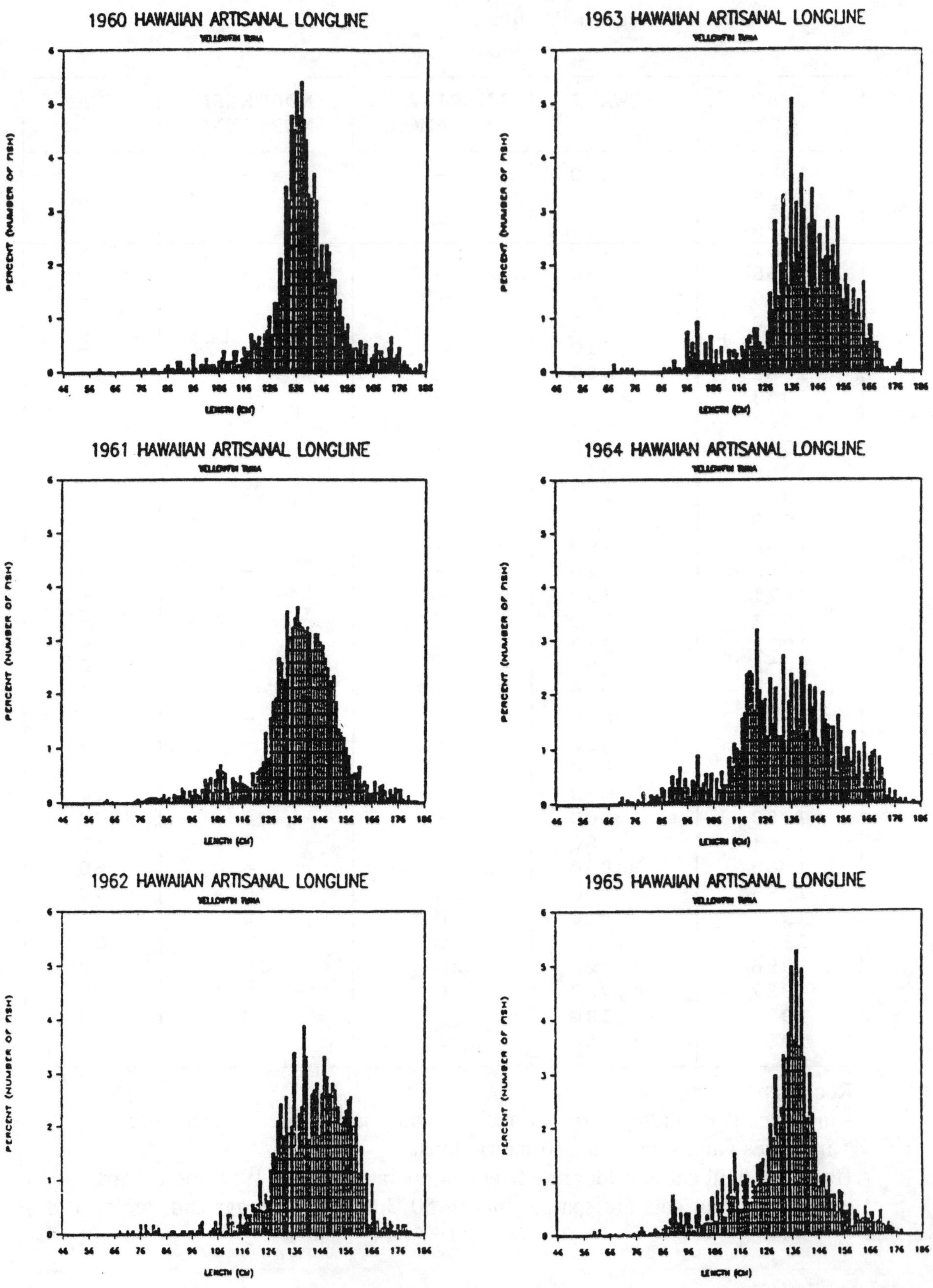

Figure 4.   Annual length composition of yellowfin tuna caught by the Hawaiian artisanal longline fishery.

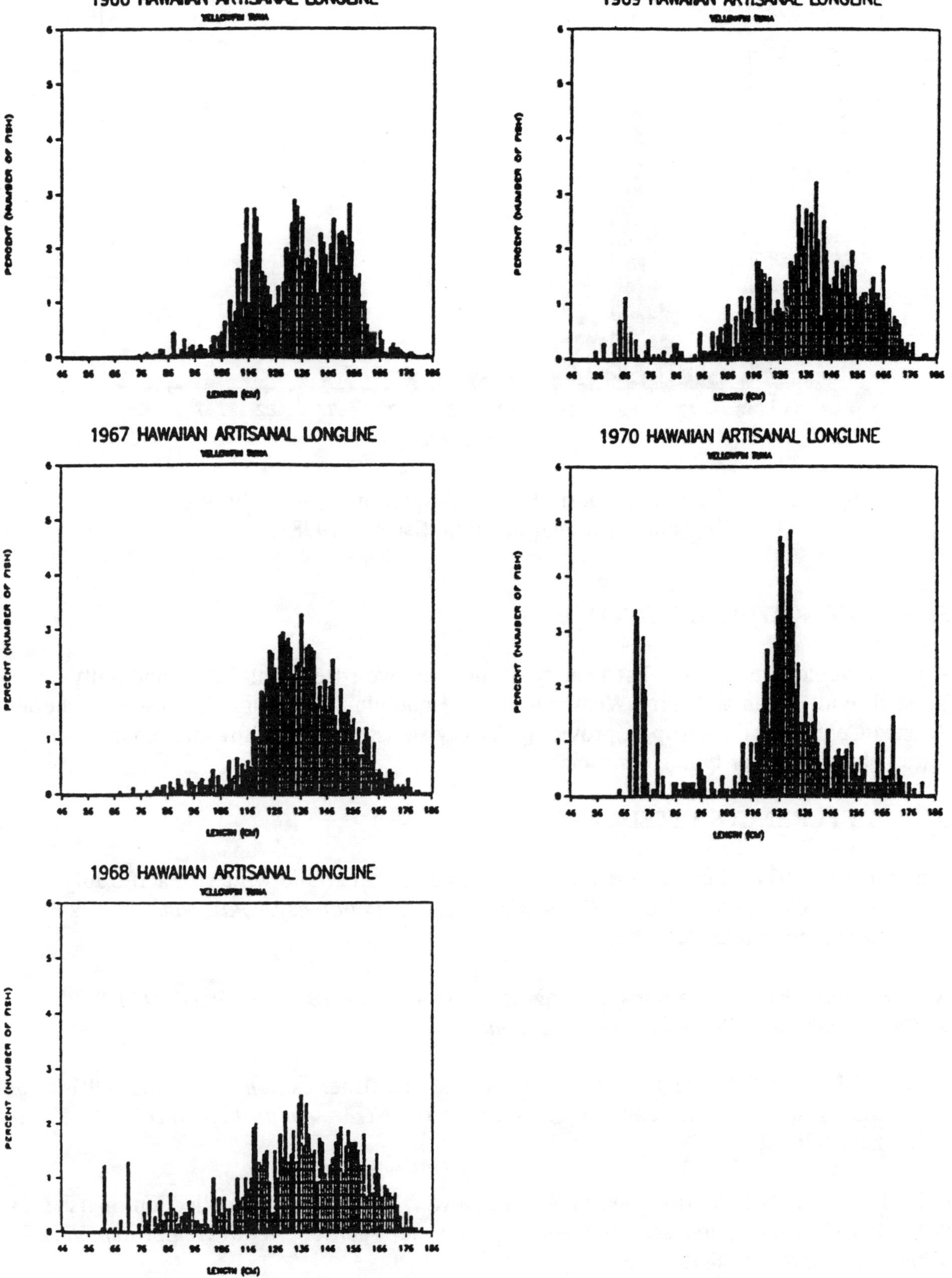

Figure 4. (Continued)

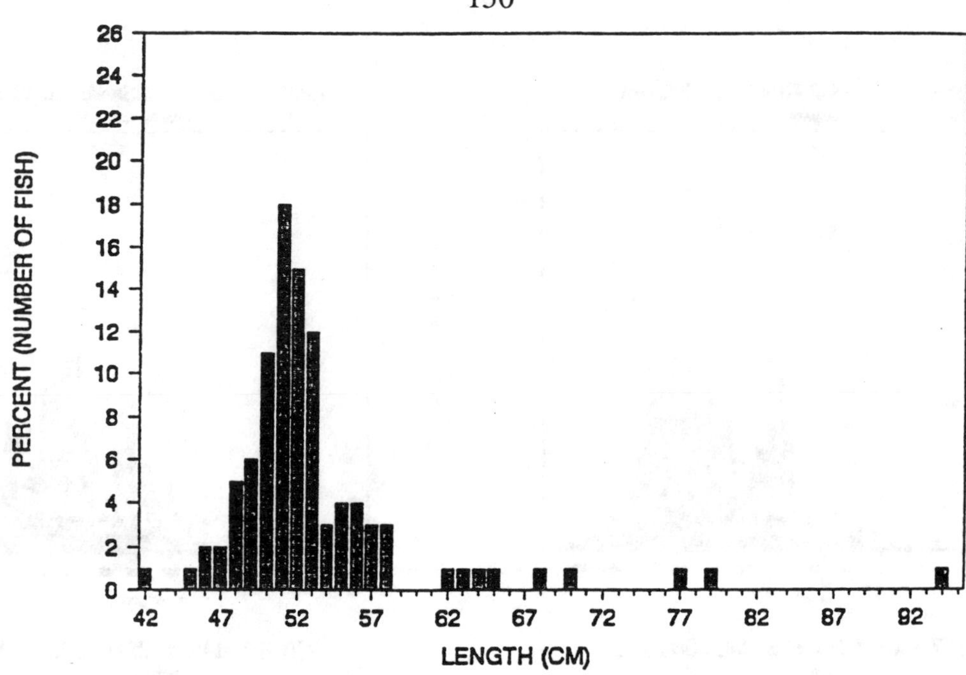

Figure 5.   Length composition of yellowfin tuna caught by the Hawaiian artisanal handline fishery, 1978.

5.   ACKNOWLEDGEMENTS

I would like to thank Pat Donley of the Southwest Regional Office and Sally Kuba, David Hamm and Jerry Wetherall of the Honolulu Laboratory, Southwest Fisheries Science Center for their help in providing fishery statistics needed for this report, and Karen Handschuh for typing assistance.

6.   REFERENCES CITED

Anonymous. 1977. Final report tuna purse-seine charter to the western Pacific July-November, 1976 *Apollo - Mary Elizabeth* and *Zapata Pathfinder. Pac.Tuna Dev.Foundation, Honolulu.*

Anonymous. 1979. Tuna purse seining cruise report July 1978 February 1979 *Bold Venture. Pac.Tuna Dev.Foundation, Honolulu.*

Boggs, C.H., and S.G. Pooley. 1987. Hawaii's tuna fisheries. *In:* Tuna and billfish summaries of major stocks, edited by N.W. Bartoo. *NOAA Admin.Rep.NMFS-SWFC, La Jolla,* LJ-87-26, III-2:1-20.

Coan Jr., A.L., N.B. Bartoo, and G.T. Sakagawa. 1988. Plan for collection of fisheries data from U.S. tuna purse seiners fishing in the south Pacific. *NOAA Admin.Rep.NMFS-SWFC, La Jolla,* LJ-88-19:19 p.

Coan Jr., A.L., and G. Yamasaki. 1990. Sampling manual for collection of biological data from U.S.purse seiners fishing in the south Pacific. *NOAA Admin.Rep.NMFS-SWFC, La Jolla,* LJ-90-02.

Doulman, D.J. 1987. Development and expansion of the tuna purse seine fishery. *In:* Tuna issues and perspectives in the Pacific islands region, edited by D.J. Doulman. Honolulu, East-West Center, pp. 133-160.

Hamm, D.C. 1985. Western Pacific Fishery Information Network organization and design status and issues. *NOAA Admin.Rep.NMFS-SWFC, Honolulu*, H-85-4:35 p.

Hamm, D.C., and T.T. Kassman. 1986. Fishery statistics of the western Pacific, Volume I. *NOAA Admin.Rep.NMFS-SWFC, Honolulu*, H-86-4:206 p.

Hamm, D.C., M.M.C. Quach. and T.T. Kassman. 1986. Fishery statistics of the western Pacific, Volume II. *NOAA Admin.Rep.NMFS-SWFC, Honolulu*, H-86-20:222 p.

Hamm, D.C., and M.M.C. Quach. 1988. Fishery statistics of the western Pacific, Volume III. *NOAA Admin.Rep.NMFS-SWFC, Honolulu*, H-88-4:81 p.

Hamm, D.C., and M.M.C. Quach. 1989. Fishery statistics of the western Pacific, Volume IV. *NOAA Admin.Rep.NMFS-SWFC, Honolulu*, H-89-1:50 p.

Hamm, D.C., M.M.C. Quach, R.S. Antonio, and J.J. Czyz. 1990. Fishery statistics of the western Pacific, Volume V. *NOAA Admin.Rep.NMFS-SWFC, Honolulu*, H-90-09: 50 p.

Hamm, D.C., M.M.C. Quach, and L.K. Timme. 1991. Fishery statistics of the Western Pacific, Volume VI. *NOAA Admin.Rep.NMFS-SWFC, Honolulu*, H-91-01:50 p.

Hamm, D.C., R.S. Antonio, and M.M.C. Quach. 1990. Fishery statistics of the western Pacific, Volume VII. *NOAA Admin.Rep.NMFS-SWFC, Honolulu*, H-92-06.

Honda, V.A., G.S. Yamasaki, and R.Y. Ito. 1988. American Samoa purse seine fishery sampling. *NOAA Admin.Rep.NMFS-SWFC, Honolulu*, H-88-20:35 p.

Hudgins, L.L. and S.G. Pooley. 1987. Growth and contraction of domestic fisheries: Hawaii's tuna industry in the 1980's. *In:* Tuna issues and perspectives in the Pacific islands region, edited by D.J. Doulman. Honolulu, East-West Center, pp. 225-41.

Ito, R.Y. 1991. Western Pacific pelagic fisheries in 1990. *NOAA Admin.Rep.NMFS-SWFC, Honolulu*, H-91-10.

June, F.C. 1950. Preliminary fisheries survey of the Hawaiian Line Islands area: Part 1 - the Hawaiian longline fishery. *Comm.Fish.Rev.*, 12(1):1-23.

June, F.C. 1951. Preliminary fisheries survey of the Hawaiian-Line Islands area: Part 3 - the live bait skipjack fishery of the Hawaiian Islands. *Comm.Fish.Rev.*, 13(2):1-18.

Otsu, T. 1954. Analysis of the Hawaiian longline fishery, 1948-1952. *Comm.Fish.Rev.*, 16(9):1-17.

Schug, D.M., and A.P Galea'i. 1987. American Samoa: The tuna industry and the economy. *In*: Tuna issues and perspectives in the Pacific islands region, edited by D.J. Doulman. Honolulu, East-West Center, pp. 191-202.

Souter, D., and G. Broadhead. 1978. Purse-seine fishing for yellowfin and skipjack in the southern waters of the central and western Pacific *Jeanette C.* charter. *Tech.Bull.Pac.Tuna Dev.Foundation, Honolulu*, 2.

SPC. 1989. Regional Tuna Bulletin. *Reg.Tuna Bull.Tuna Billfish Assess.Programme, S.Pac.Comm.*, (varying pagination).

SPC. 1990. Regional Tuna Bulletin. *Reg.Tuna Bull.Tuna Billfish Assess.Programme, S.Pac.Comm.*, (varying pagination).

SPC. 1991. Regional Tuna Bulletin. *Reg.Tuna Bull.Tuna Billfish Assess.Programme, S.Pac.Comm.*, (varying pagination).

TNL. 1990. Tuna Newsletter, National Marine Fisheries Service, Southwest Region, Southwest Fisheries Center. May, 1990 (97).

Yuen, H.S.H. 1979. A night handline fishery for tunas in Hawaii. *Mar.Fish.Rev.*, August 1979, 41(8):7-14.

STATUS OF KOREAN TUNA LONGLINE AND PURSE-SEINE FISHERIES IN THE PACIFIC OCEAN

Yeong Chull Park, Won Seok Yang and Tae Ik Kim
National Fisheries Research and Development Agency
Republic of Korea

ABSTRACT

This paper reviews the fluctuations in catch and fishing vessels, catch per unit effort (CPUE), fishing grounds, and length composition of the major species based on Korean commercial fishing data in the Pacific Ocean.

1.   INTRODUCTION

The commercial fishing operations of the Korean longline fishery in the Pacific Ocean have been conducted for large-size yellowfin, bigeye, and albacore tunas since the mid-1960s. The Korean purse-seine fishery started experimentally in 1980 and commercial fishing activity began in the western tropical Pacific Ocean in 1982, targetting mostly skipjack tuna.

2.   TRENDS IN FISHING VESSELS AND CATCHES

2.1   <u>Longline Fishery</u>

Nineteen fishing vessels operated in the Pacific Ocean in 1964. The fishing fleet gradually increased year by year to a peak of 270 vessels in 1974, then decreased to a low of 94 in 1985, and increased to 182 in 1990.

Catch of tuna and tuna-like species increased sharply from 2,400 mt in 1964 to 68,800 mt in 1976 with increasing number of vessels. After 1976 the catch declined to a lower level of 25,700 mt in 1984 along with a decrease in the number of vessels. However, it again showed an increasing pattern during the 1985-1990 period. About 44,800 mt of tunas were taken by 182 vessels in 1990 (Figure 1).

The catch of bigeye tuna was higher than that of either yellowfin or albacore during the 1975-1978 and 1987-1990 periods (Figure 2). The catch of yellowfin tuna was predominant during 1978-1980 and that of albacore during 1971-1973, 1981-1982, and 1985-1986. From 1983 to 1984, the catch of these three species showed lower levels than other years due to the decreased number of vessels. However, there seemed to be an increasing trend in catch of bigeye and yellowfin during the 1985-1990 period, except in 1988 and 1989. The catch of albacore showed a distinct decreasing trend from 1986 to 1990 because Korean tuna longliners changed their target species to yellowfin and bigeye tuna in recent years. In 1990, the catches of the major species were 20,800 mt of bigeye, 13,900 mt of yellowfin, and 3,200 mt of albacore.

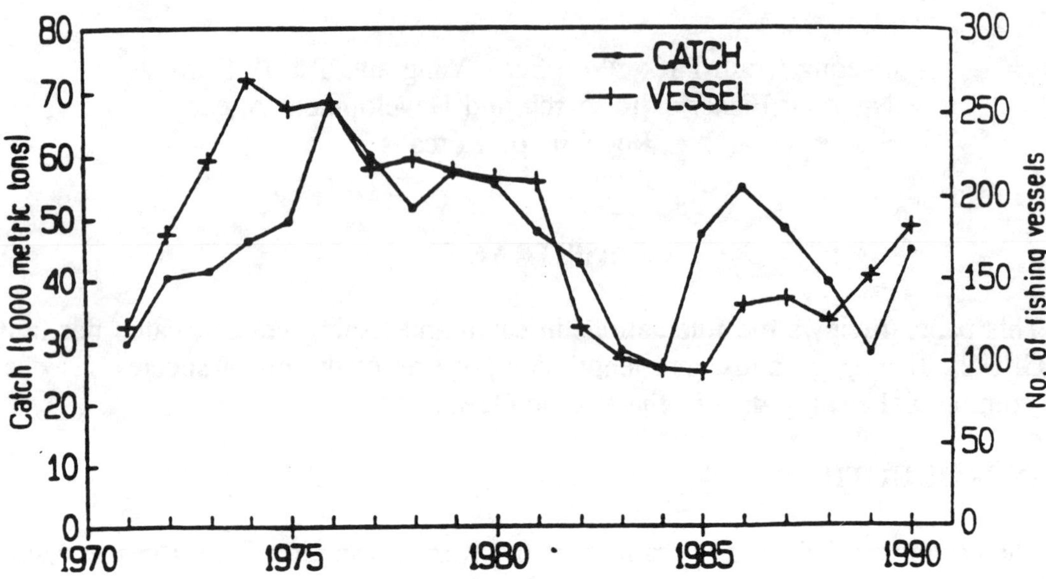

Figure 1. Trend in total catch and fishing vessels for Korean tuna longline fishery in the Pacific Ocean, 1971-1990.

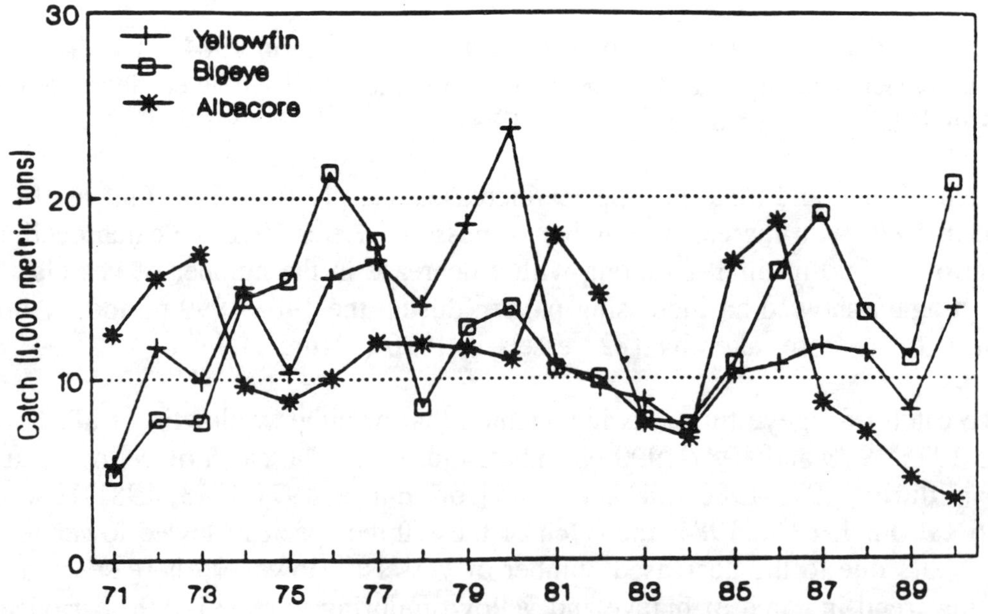

Figure 2. Trend in catch of yellowfin, bigeye, and albacore tuna by Korean tuna longline fishery in the Pacific Ocean, 1971-1990.

2.2   Purse-seine Fishery

Two Korean purse seiners caught 500 mt of tunas in the western tropical Pacific Ocean in 1980. The catch was under 15,000 mt between 1981 and 1985, but it increased considerably from 27,700 mt (13 vessels) in 1986 to 115,800 mt (30 vessels) in 1989. The catch of skipjack and yellowfin tuna also increased sharply from 9,700 mt to 80,900 mt and from 1,600 mt to 34,500 mt, respectively, during the 1985-1989 period. The catch composition was 75.0% skipjack, and 24,8% yellowfin during 1986-1989 (Figure 3 and 4).

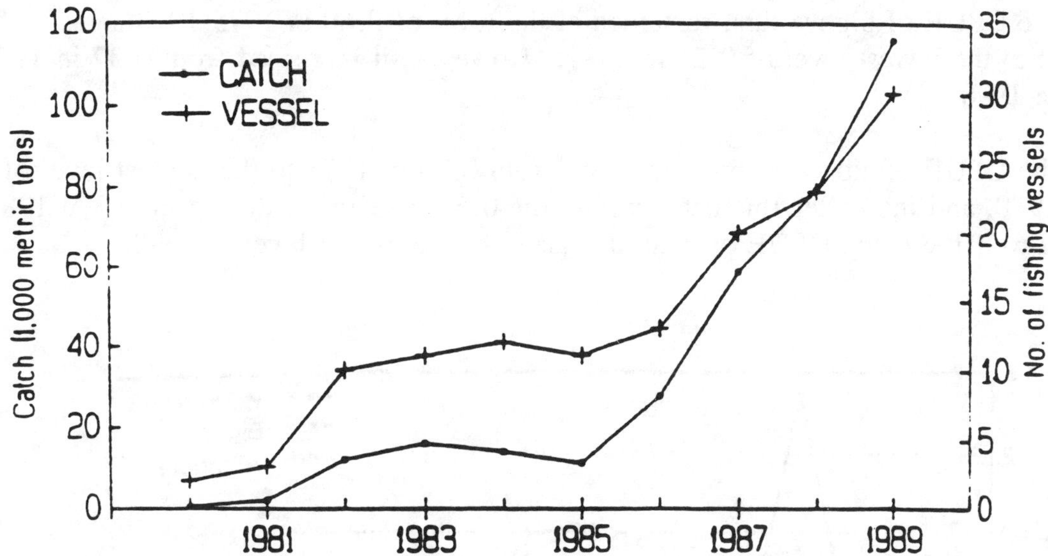

Figure 3.   Trend in total catch and fishing vessels for Korean tuna purse-seine fishery in the western Pacific Ocean, 1980-1989.

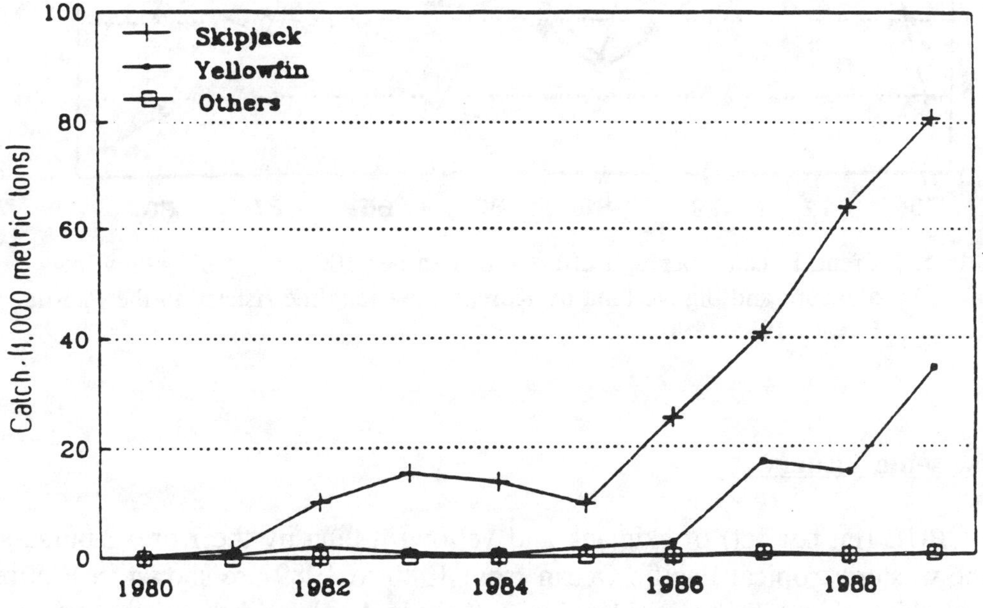

Figure 4   Trend in catch of tunas by Korean tuna purse-seine fishery in the western Pacific Ocean, 1980-1989.

3.   TREND IN CPUE

3.1   <u>Longline</u> <u>Fishery</u>

Catch per unit effort (number of fish per 100 hooks) of yellowfin, bigeye, and albacore tuna taken by Korean longline fishery in the Pacific Ocean from 1975 to 1990 are shown in Figure 5.

The CPUE of yellowfin tuna fluctuated from 0.30 to 0.96 during 1975-1982 and it showed a more stable level of 0.46-0.64 during 1983-1990 period.

The CPUE of bigeye tuna remained at the level of 0.60 in 1975-1978 and decreased to the lowest level of 0.27 in 1981. However, it increased from 0.37 in 1982 to 0.68 in 1990.

The CPUE of albacore tuna increased from 0.20 in 1975 to the highest level of 1.35 in 1978, and thereafter fluctuated within the 0.65-0.99 range during the 1979-1986 period. From 1987, the CPUE decreased sharply to the lowest level in 1990.

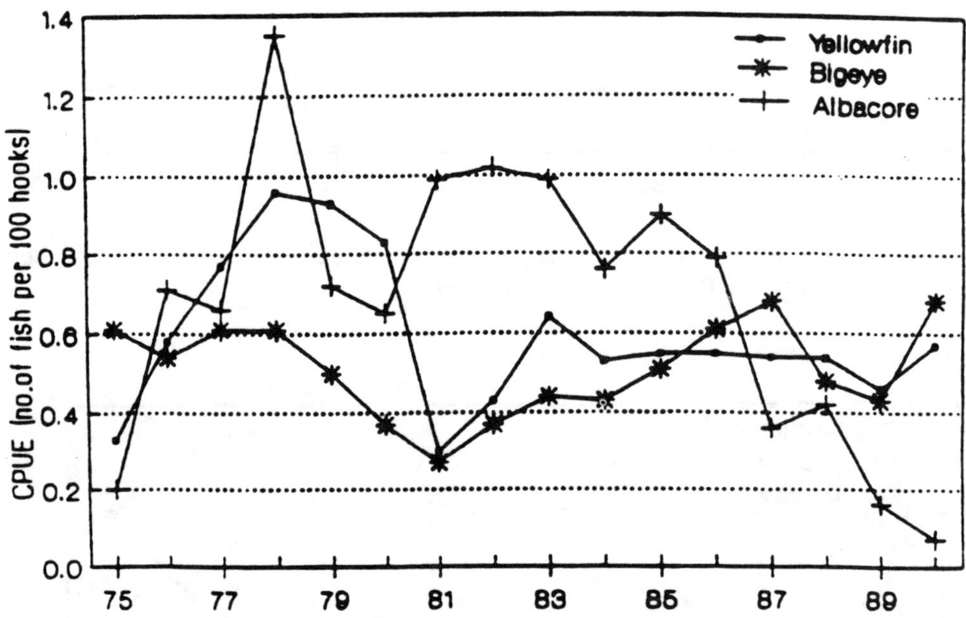

Figure 5.   Trend in catch per unit effort (no. fish per 100 hooks) of yellowfin, albacore and bigeye tuna by Korean tuna longline fishery in the Pacific Ocean, 1975-1990.

3.2   <u>Purse-seine</u> <u>Fishery</u>

The CPUE (mt per set) of skipjack and yellowfin tuna by the Korean purse-seine fishery in the western tropical Pacific Ocean from 1980 to 1989 are shown in Figure 6. The CPUE of skipjack reached a low level of 6.9 mt in 1981, and then it began an increasing trend up to 1989, except for 1984 and 1985. Skipjack tuna CPUE for 1988-

1989 increased more than twice from that of 1984-1985. The CPUE of yellowfin tuna was much lower than that of skipjack. It also increased gradually during the 1981-1989 period.

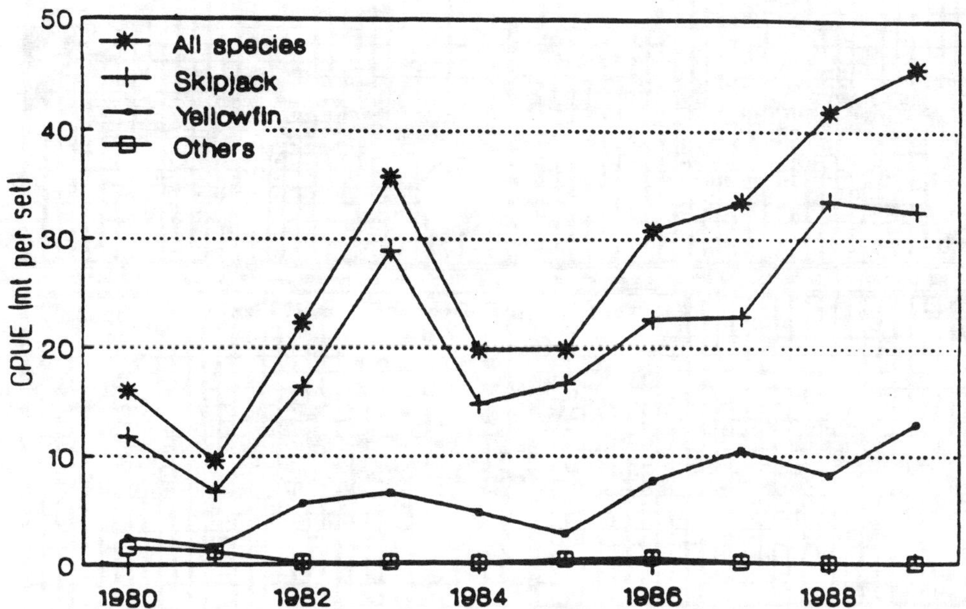

Figure 6.   Trend in catch per unit effort (mt per set) of tuna by Korean tuna purse-seine fishery in the western Pacific Ocean, 1980-1989.

4.   GEOGRAPHICAL DISTRIBUTION OF CPUE

4.1   Longline Fishery

The annual and seasonal geographical distributions of catch per unit effort (number of fish per 100 hooks) of yellowfin and bigeye tuna in the Pacific Ocean from 1988 through 1989 are shown in Figures 7 and 8. The major fishing grounds of yellowfin tuna were generally between 10°N and 15°S for the two years. The areas of high CPUEs were in the western central Pacific between 10°N to 10°S and 150°E to 155°W. From July to September the fishing grounds were more or less extended to the eastern Pacific, but the CPUEs were low.

The fishing grounds of bigeye tuna were mainly in the area between 15°N and 15°S during 1988 and 1989. The areas of high CPUEs were in the region of 15°N-15°S and 155°E-125°W in 1988. In 1989 the areas of high CPUEs were at 10°N-10°S, 160°E-160°W and at 10°N-10°S, 95°-115°W from April to December.

4.2   Purse-seine Fishery

The geographical distributions of CPUE (mt per set) of tunas taken by the Korean tuna purse-seine fishery in the western tropical Pacific Ocean from 1988 to 1989 are shown in Figure 9. The fishing grounds were formed in the area between 7°N-6°S and 140°-165°E. The areas of high CPUEs for skipjack tuna appeared to be uniformly distributed over the fishing grounds without any clear pattern, while high CPUEs for yellowfin appeared to be sparsely distributed.

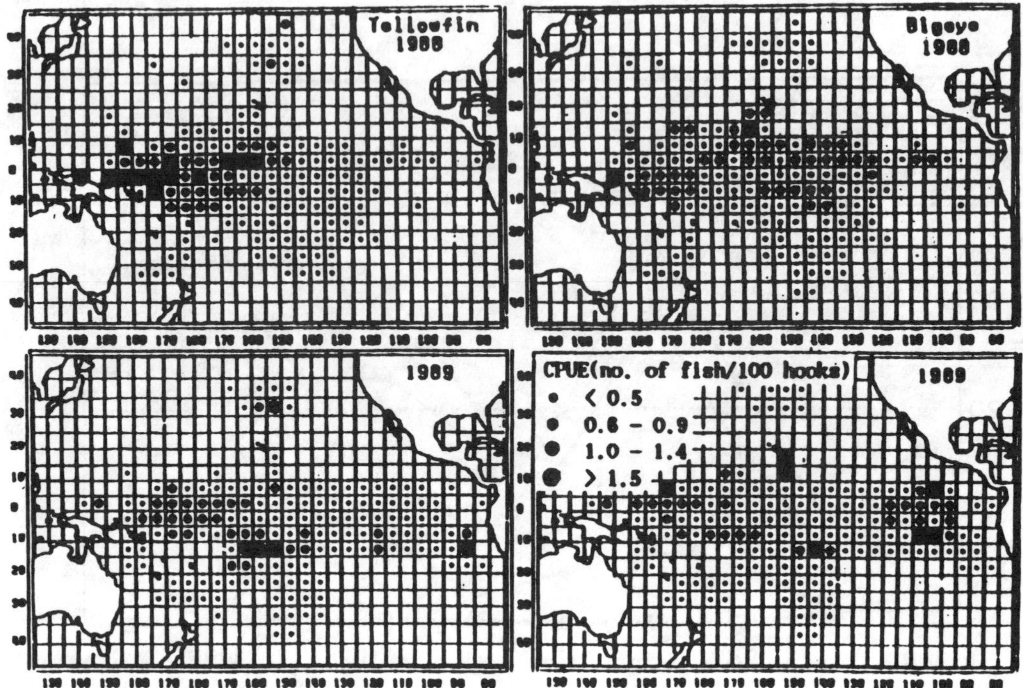

Figure 7.   Annual distribution of CPUE (no. fish per 100 hooks) for yellowfin and bigeye tuna by Korean tuna longline fishery in the Pacific Ocean, 1988-1989.

5.   LENGTH COMPOSITION

5.1   Longline fishery

The fork-length composition of yellowfin and bigeye tuna taken by the Korean longline fishery in the Pacific Ocean from 1988 to 1989 are shown in Figure 10. The range and mean fork length of yellowfin tuna was 62-180 cm and 130.0 cm in 1988, and 70-185 cm and 129.5 cm in 1989, respectively. The length composition of yellowfin tuna for the two years was similar.

The range in fork length for bigeye tuna was 60-200 cm in 1988 and 1989, and the mean size was 132.4 cm in 1988 and 135.0 cm in 1989. The modal size of 1989 was larger than that of 1988.

5.2   Purse-seine Fishery

The fork-length composition of skipjack and yellowfin tuna taken by the Korean purse-seine fishery from 1988 to 1989 are shown in Figure 11. The range in fork length for skipjack was 30-78 cm in both years; the mean sizes were 53.4 cm in 1988 and 54.0 cm in 1989.

The range in fork length for yellowfin tuna was 40-170 cm for 1988 and 198; the mean size was 87.8 cm in 1988 and 88.5 cm in 1989. There seem to be three size groups in the length composition of yellowfin tuna during the two years. The modal lengths were 60-75 cm for small-sized fish, 110-115 cm for middle-sized fish, and 155-160 cm for large-sized fish.

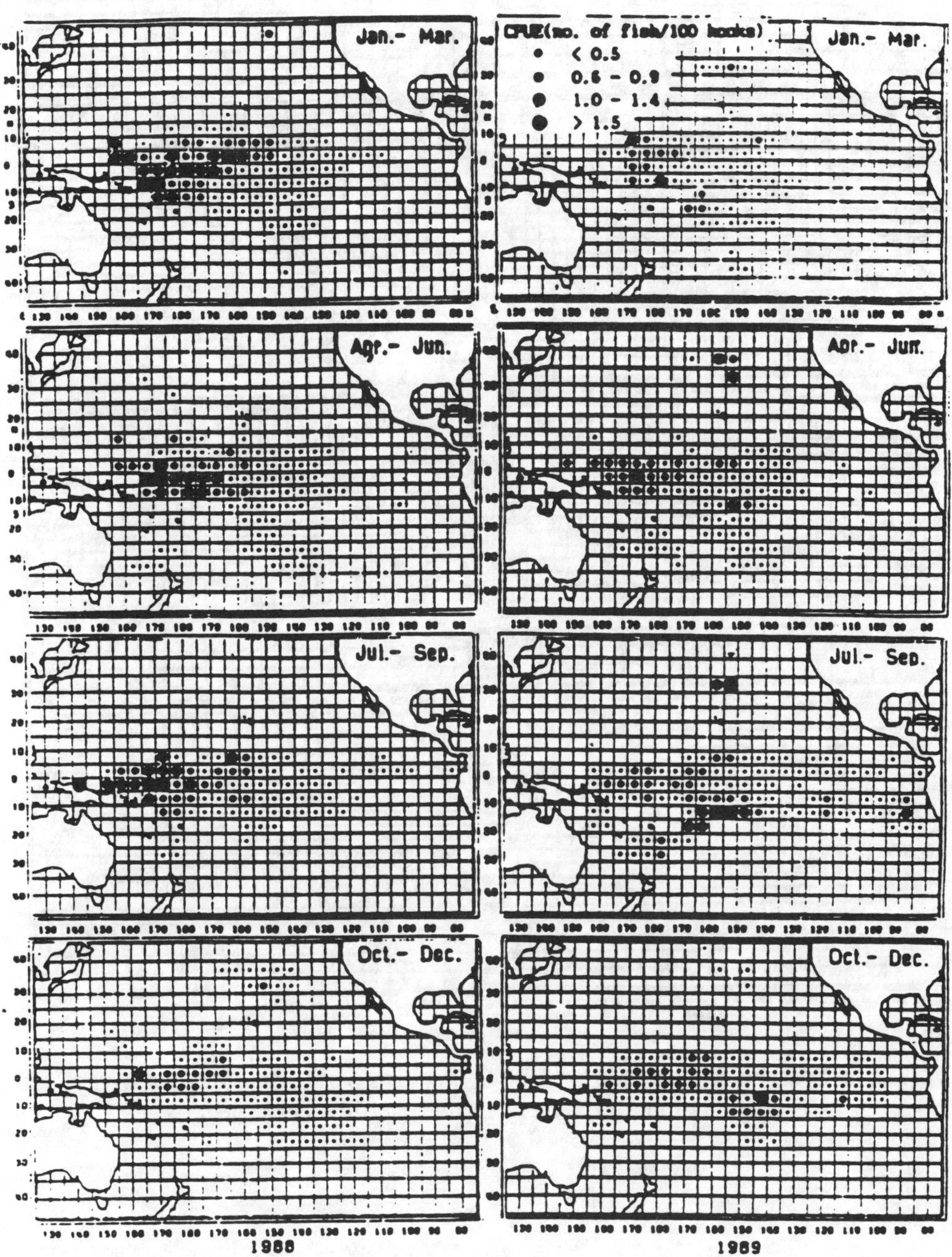

Figure 8-1.   Seasonal distribution of CPUE (no. fish per 100 hooks) for yellowfin tuna by Korean tuna longline fishery in the Pacific Ocean, 1988-1989.

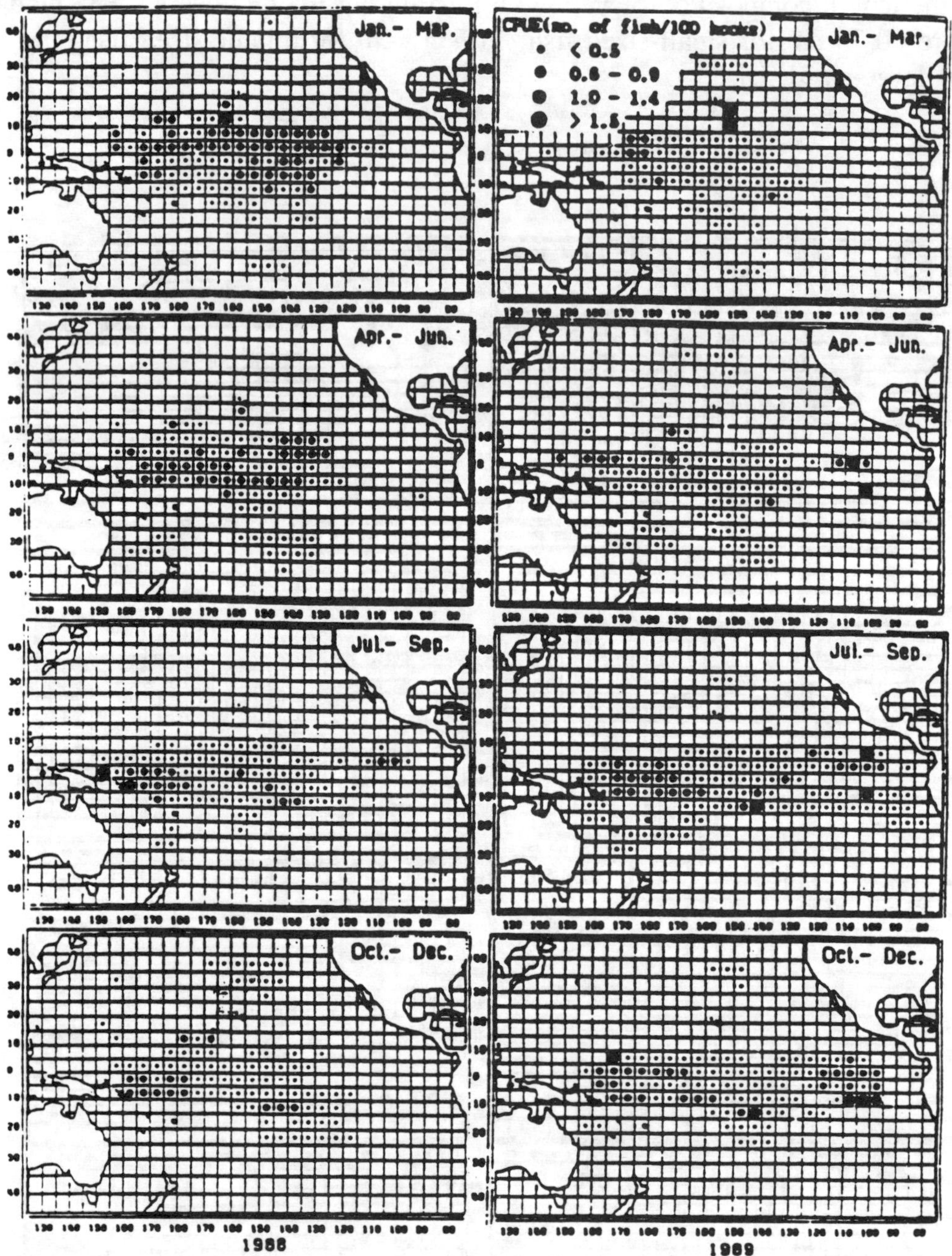

Figure 8-2. Seasonal distribution of CPUE (no. fish per 100 hooks) for bigeye tuna by Korean tuna longline fishery in the Pacific Ocean, 1988-1989.

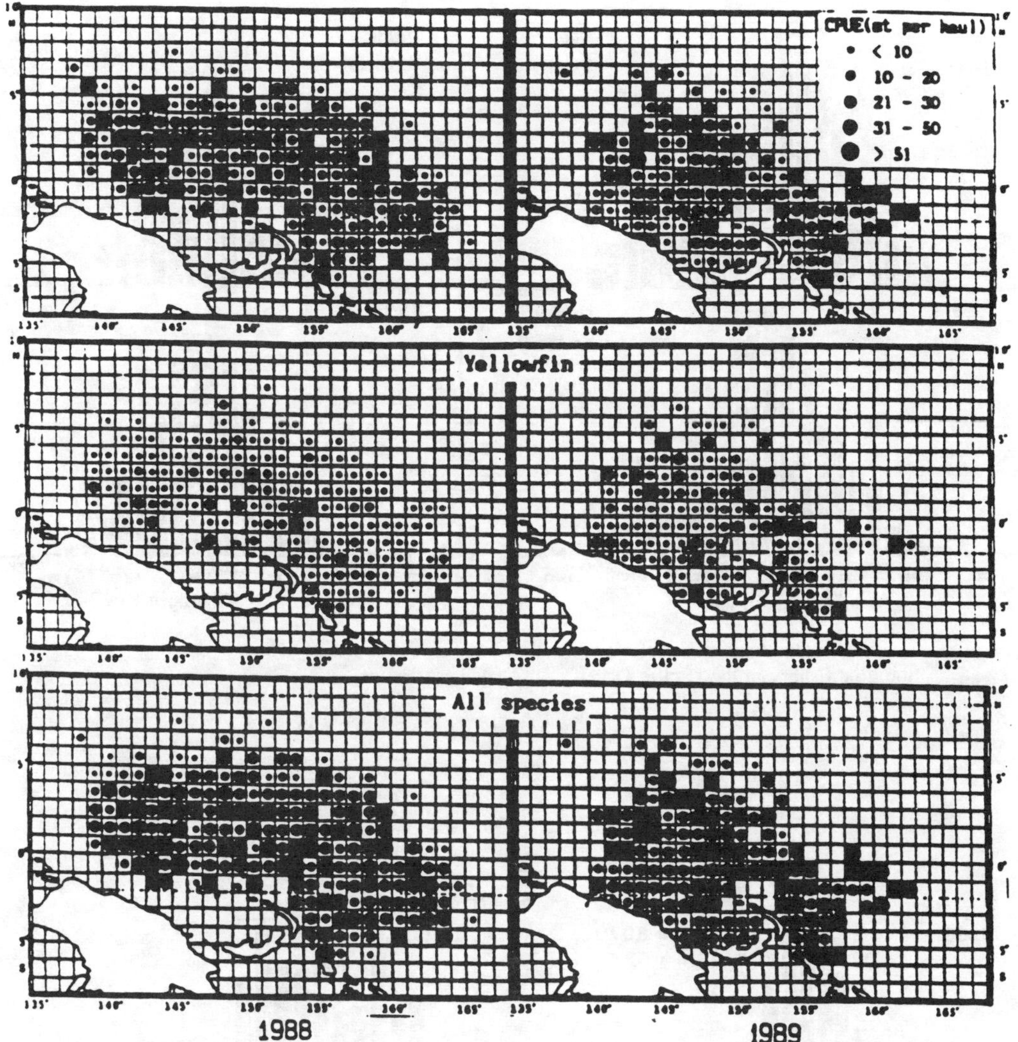

Figure 9.   Annual distribution of CPUE (mt per set) for tunas by Korean tuna purse-seine fishery in the western Pacific Ocean, 1988-1989.

162

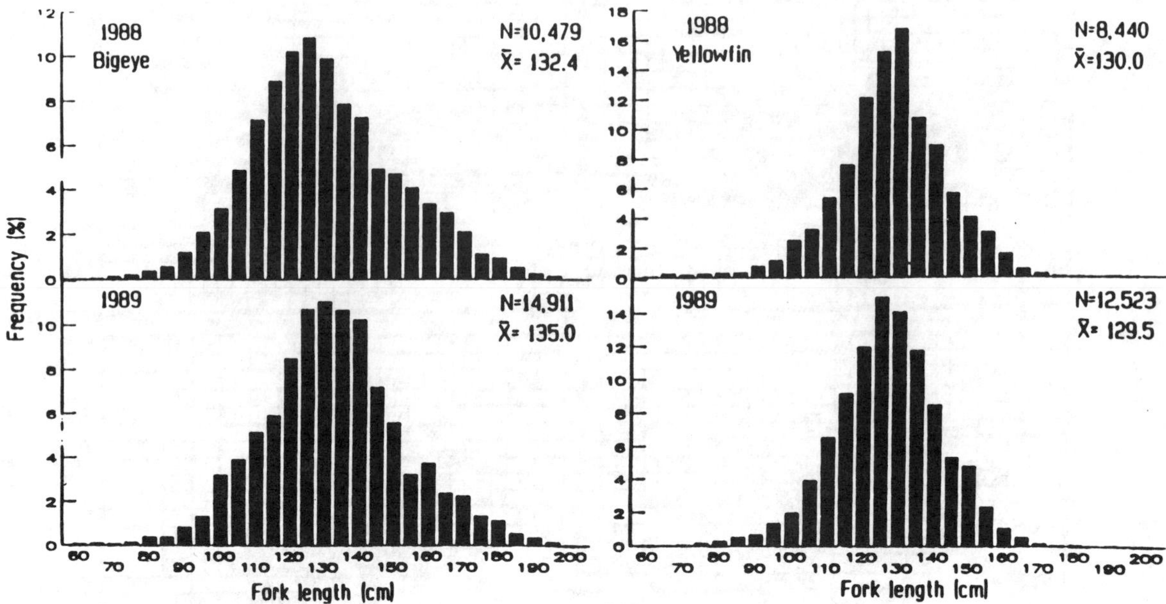

Figure 10. Annual length composition of bigeye and yellowfin tuna taken by Korean tuna longline fishery in the Pacific Ocean, 1988-1989.

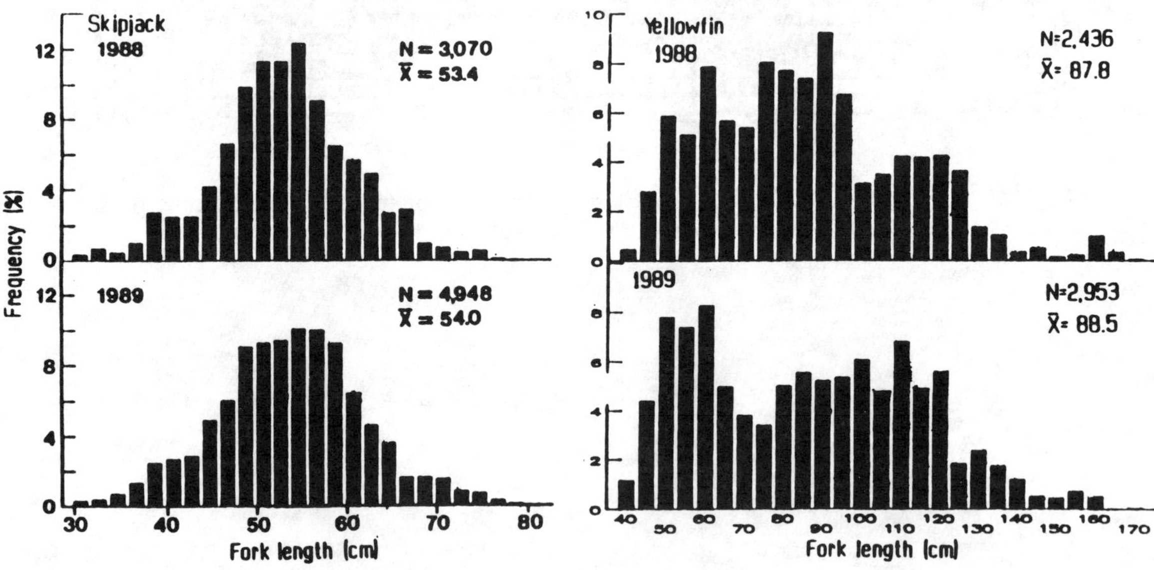

Figure 11. Annual length composition of skipjack and yellowfin tuna taken by Korean tuna purse-seine fishery in the western Pacific Ocean, 1988-1989.

TAIWANESE YELLOWFIN FISHERIES IN THE PACIFIC OCEAN

Chien-Hsiung Wang
Institute of Oceanography
National Taiwan University
Taipei, Taiwan

ABSTRACT

The catch of yellowfin tuna by Taiwan's pelagic fisheries are briefly described in this report. Traditionally, longline gear was the principal gear used to harvest yellowfin; landings in the early 1970s reached a peak of nearly 10,000 mt. Development of the purse-seine fishery began in early 1980s and by 1990 the yellowfin catch exceeded 12,000 mt. In recent years there has been a dramatic shift in base ports and fishing grounds for the inshore longliners, to more distant waters.

1.   INTRODUCTION

Catch statistics of Taiwan are divided into far-seas, inshore, and coastal fisheries. The far-seas fisheries include vessels larger than 50 gross registered tons (GRT), inshore fisheries include all of the power crafts smaller than 50 GRT, and coastal fisheries include nonpowered craft, set net, beach seine, and inland water fisheries.[1]

The Taiwanese catch of yellowfin tuna by all types of gear and from all oceans for the period 1964 to 1990 is given in Table 1. The peak catch occurred in 1990 with a catch of 44,513 metric tons (mt). The far-seas fisheries of the Pacific landed approximately 40% of the total Taiwanese yellowfin catch in 1990 (Table 2).

Yellowfin tuna stocks are mainly exploited by tuna longliners, primarily the powered longliners larger than 20 GRT. Only a few yellowfin are caught by other fishing gear such as small gillnet, trolling, and miscellaneous types of longline.

2.   FAR-SEAS FISHERIES

Longline and purse seine are the two most important types of fishing gear used in exploiting yellowfin stocks.

2.1   Longline

In the Pacific the Taiwanese far-seas longliners mainly fish in the South Pacific (Wang, 1988). The target species of this fishery is albacore. Most of the longliners are of 100-500 GRT (Table 3). In 1977 the total number of far-seas longliners operating in the South Pacific was 178; the number decreased to a low of 44 in 1985 and subsequently

---

[1] From 1990 the definitions for the fisheries were changed to: (a) Far-seas Fisheries -- operating outside of Exclusive Economic Zone (EEZ), (b) Inshore Fisheries -- operating 12 to 200 miles, and (c) Coastal Fisheries -- operating from the coast to 12 miles.

increased to 96 in 1990. The fluctuations in vessel numbers reflect a shift in fishing area and not a change in fleet size.

Table 1. Total Taiwanese yellowfin tuna catch, 1964-1990.

Year	Catch	Year	Catch	Year	Catch
1964	6,372	1973	27,876	1982	23,564
1965	7,170	1974	23,445	1983	27,038
1966	12,221	1975	28,717	1984	26,957
1967	13,784	1976	22,261	1985	24,343
1968	34,717	1977	27,157	1986	27,378
1969	39,638	1978	30,998	1987	40,216
1970	36,343	1979	34,123	1988	43,247
1971	31,531	1980	31,912	1989	35,548
1972	31,000	1981	25,522	1990	44,513

Sources: Taiwan Fisheries Bureau (TFB) annual reports, 1965-1991; Tuna Research Center (TRC) annual reports, 1973-1991.

The peak longline catch in the Pacific occurred in 1972 when the catch was about 10,000 mt (Table 2). Subsequently, the catch declined steadily, reaching a low of 862 mt in 1983. Since 1983 the catch has fluctuated slightly. It reached a post-1983 high of 5,411 mt in 1990.

The catch composition of the Taiwanese far-seas longline fishery for 1970, 1980, and 1990 is given in Table 4. Albacore made up 37%, 47%, and 38% of the catch for 1970, 1980, and 1990, respectively, indicating that it is the target species. Since yellowfin made up a smaller share of the catch, with 32%, 8%, and 12% for the years 1970, 1980, and 1990, it is clearly not the species targeted by Taiwanese far-seas longliners.

2.2   Purse Seine

The catch of yellowfin by the Taiwanese purse-seine fishery was first reported in 1982 (Table 2). The purse-seine catch of yellowfin increased from 298 mt in 1982 to 14,540 mt in 1990. Although details of the catch by oceans are not available, it is estimated that more than 80% of the catch was made in the Pacific. In 1990 Taiwanese purse seiners numbered 44. In the Pacific they fished mainly in waters around Micronesia.

Table 2. Yellowfin tuna catch of Taiwanese far-seas fisheries, 1964-1990

unit: mt

Year	All Oceans				Pacific Ocean			
	LL	PS	GN	Total	LL	PS	GN	Total
1964	3,120	-	-	3,120	-	-	-	-
1965	3,292	-	-	3,292	-	-	-	-
1966	7,434	-	-	7,434	-	-	-	-
1967	9,273	-	-	9,273	2,099	-	-	2,099
1968	28,377	-	-	28,377	5,112	-	-	5,112
1969	33,809	-	-	33,809	4,769	-	-	4,769
1970	29,806	-	-	29,806	3,964	-	-	3,964
1971	26,088	-	-	26,088	8,960	-	-	8,960
1972	27,347	-	-	27,347	9,714	-	-	9,714
1973	17,062	-	-	17,062	8,412	-	-	8,412
1974	15,334	-	-	15,334	4,594	-	-	4,594
1975	14,753	-	-	14,753	2,902	-	-	2,902
1976	8,439	-	19	8,458	3,667	-	7	3,774
1977	10,062	-	196	10,258	2,849	-	69	2,918
1978	9,918	-	398	10,316	3,024	-	139	3,163
1979	9,477	-	274	9,751	3,009	-	96	3,105
1980	9,956	-	-	9,956	5,603	-	-	5,603
1981	6,863	-	247	7,110	2,031	-	86	2,117
1982	5,871	298	323	6,492	1,015	256	97	1,368
1983	9,087	972	401	10,460	862	836	140	1,838
1984	8,953	741	120	9,814	1,430	637	42	2,109
1985	8,317	1,357	290	9,965	1,210	1,167	102	2,479
1986	12,737	2,718	193	15,648	1,213	2,337	68	3,618
1987	17,699	4,599	791	23,090	1,017	3,955	277	5,249
1988	17,927	6,782	618	25,327	3,054	5,833	216	9,103
1989	11,159	11,099	874	23,131	1,270	9,561	305	11,136
1990	18,938	14,540	1,033	34,511	5,411	12,510	0	17,921

Sources: Taiwan Fisheries Bureau (TFB) annual reports, 1965-1991. Tuna Research Center (TRC) annual reports, 1973-1991.

Legend: LL -- Far-seas tuna longliners (1967-1979 data based on logbooks, 1980-1990 data provided by TFB), PS -- Purse seiners (data prior to 1989 based on assumption Pacific PS/All Oceans PS equal to that of 1989); GN -- Gill netters (data prior to 1989 based on assumption Pacific GN/Far-seas GN equal to that of 1989).

Table 3.   Number of Taiwanese far-seas tuna longline vessels in the Pacific Ocean (> 50 GRT), 1977-1990.

Year	Vessel Class (GRT)						
	Total	0-50	50-100	100-200	200-500	500-1,000	Over 1,000
1977	178	2	11	141	24	-	-
1978	169	1	11	133	23	1	-
1979	157	-	7	121	28	1	-
1980	182	-	5	140	36	1	-
1981	140	-	2	114	24	-	-
1982	115	-	1	92	22	-	-
1983	65	-	-	55	10	-	-
1984	61	-	-	51	10	-	-
1985	44	-	-	36	8	-	-
1986	51	-	-	39	12	-	-
1987	60	-	-	37	22	1	-
1988	70	-	-	39	29	2	-
1989	85	-	-	42	37	5	-
1990	96	-	2	37	49	8	-

Source: Data provided by Taiwan Fisheries Bureau (TFB) and Tuna Research Center (TRC).

Table 4.   Catch composition of Taiwanese far-seas longline fishery, 1970, 1980, and 1990.

Species	1970		1980		1990	
	MT	%	MT	%	MT	%
Yellowfin	29,806	31.7	9,956	8.3	18,938	12.1
Bigeye	14,463	15.4	11,982	10.0	12,712	8.2
Albacore	34,745	37.0	56,054	46.9	59,531	38.2
Others	14,918	15.9	41,643	34.8	64,740	41.5
TOTAL	93,932	100.0	119,635	100.0	155,921	100.0

Source: Taiwan Fisheries Bureau (TFB) annual reports, 1971, 1981, and 1991.

3.   INSHORE FISHERIES

The catch of yellowfin tuna by the various inshore fisheries is given in Table 5. Among the several inshore fisheries the dominant gear for yellowfin is the tuna longline.

The yellowfin longline catch increased dramatically from a low of 3,323 mt in 1972 to a high of 22,629 mt in 1979. Much of this increase can be attributed to an increase in effort, as the number of inshore tuna longline vessels during this period increased from 899 to 1,840 vessels (Table 5). The decline in catch since 1979 cannot be easily explained, since fishing effort continued to increase until 1987, when a high of 2,207 vessels was reported (Table 6). It is possible that the decline may reflect discrepancies in reporting and processing of catch statistics. In recent years some of the Taiwanese fishing vessels have based their operations in Singapore, Guam, Palau, and Indonesia. It is conceivable that the catches of these vessels were not reflected in the Taiwanese catch statistics.

Other inshore fisheries are not very important in exploiting yellowfin stocks (Table 5). Target species for the trolling vessels include bonitos and small tunas (less than 50 cm fork length). In 1989 about 15,000 mt of young tuna (including longtail tuna, small yellowfin, and small bigeye) were landed in Taiwan. The reason for the remarkably high catches of the inshore miscellaneous longline fishery in 1987 and 1988 and the drag-net fishery in 1976 is not clear. These unexpectedly high catches may be errors in data recording and processing or the result of changes in fishing gear. In Taiwan, fishing vessels change their gear depending on the season and the species targeted.

3.1 Fishing Effort of Inshore Fisheries

The vessel composition of the Taiwanese inshore fisheries for the period 1964 to 1990 is given in Table 6. With the exception of a recent decline in fleet size of the inshore tuna longline fleet, the general trend for the several fleets of the inshore fisheries has been an increase in number and an increase in average capacity. In 1990 the vessel capacity of the inshore tuna longliners, miscellaneous longliners, and the troll vessels averaged 33, 16, and 8 GRT, respectively.

3.2 Shift of Base Port and Fishing Grounds of Inshore Fishery

A survey carried out from July 1981 to June 1982 (Sun and Yang, 1983) showed that the main fishing grounds for the Taiwan fishery were close to both sides of the Philippine archipelago. Seasonal changes of fishing grounds were very clear. From January to April most of the fishing took place on the western side. From April to June the fishing fleet shifted to the eastern waters of the Philippines and from July to September most of the vessels were operating on the eastern side. From October to December the vessels shifted to the south sector and fished close to the equator.

In 1985 the inshore tuna longliners began seeking new fishing grounds away from Taiwan (personal communication with staff of Tongkang Fisheries Association and captains of fishing vessels). Larger vessels, mainly those larger than 20 GRT, began to use Singapore, Guam, Palau, and Indonesia as bases to land their catch and to obtain necessary supplies. Details of this foreign-based inshore tuna longliner operation in the western Pacific Ocean are not currently available.

The reason for the shift in base ports and fishing grounds is not clear. The shift may reflect exhaustion of traditional fishing grounds or may be the result of the rapid growth in power and capacity of the fishing vessels.

Table 5. Yellowfin tuna catch of Taiwanese inshore fisheries, by gear (1964-1990).

Unit : mt

Year	ILL	IML	IGN	ITR	IOT	Total
1964	2,918	12	2	189	11	3,132
1965	3,459	141	3	154	27	3,784
1966	4,316	93	2	288	7	4,706
1967	3,863	206	1	378	7	4,455
1968	5,805	193	-	252	5	6,255
1969	5,409	182	-	152	7	5,750
1970	6,132	176	-	142	5	6,455
1971	5,080	155	-	113	8	5,356
1972	3,323	85	-	139	-	3,547
1973	10,373	131	-	146	-	10,650
1974	7,778	89	-	84	-	7,951
1975	13,539	196	-	101	2	13,838
1976	12,425	179	-	101	943	13,648
1977	16,471	156	6	135	1	16,769
1978	19,165	58	9	1,304	1	20,537
1979	22,629	749	51	793	9	24,231
1980	20,924	684	42	150	5	21,805
1981	17,778	353	36	83	49	18,299
1982	16,508	328	53	103	2	16,994
1983	16,260	173	33	45	15	16,526
1984	16,107	816	17	155	7	17,102
1985	13,554	631	105	62	1	14,353
1986	10,884	617	106	69	15	11,691
1987	14,061	2,898	35	75	3	17,072
1988	14,337	3,190	20	316	1	17,864
1989	11,933	17	129	284	1	12,364
1990	7,848	143	203	1,662	10	9,866

Data source: Adapted from TFB annual reports, 1965-1991.
Legend: ILL = Inshore tuna longline fishery; ML = Inshore miscellaneous fish longline fishery; ITR = Inshore troll fishery; IOT = Other inshore fisheries.

Table 6. Number of Taiwanese inshore tuna fishing vessels, 1964-1990.

Year	Tuna Longline		Misc. Longline		Troll Line		Total No.
	Number	GRT/V	Number	GRT/V	Number	GRT/V	
1964	935	13.57	1,225	5.13	273	3.11	2,433
1965	499	17.69	1,689	5.94	388	2.85	2,576
1966	561	17.11	1,531	6.57	413	2.85	2,505
1967	607	17.78	1,949	6.62	360	2.42	2,916
1968	552	17.33	1,728	6.06	439	3.08	2,719
1969	420	23.70	2,197	5.91	421	3.27	3,038
1970	829	18.87	1,997	5,83	450	3.02	3,276
1971	863	19.16	1,934	6.07	381	3.26	3,178
1972	899	18.87	1,926	5,88	376	3.26	3,201
1973	1,255	18.26	2,023	6.09	432	3.32	3,710
1974	1,451	16.22	2,130	6.36	406	3.48	3,987
1975	1,411	16.54	1,897	7.77	616	3.71	3,924
1976	1,331	17.56	1,649	7.32	605	3.96	3,585
1977	1,382	17.93	1,596	5.82	428	4.41	3,406
1978	1,670	19.92	1,839	6.88	407	4.61	3,916
1979	1,840	22.55	2,149	9.08	448	4.91	4,437
1980	1,900	22.05	1,863	11.22	486	6.14	4,249
1981	1,846	22.95	1,904	10.69	494	6.23	4,244
1982	1,831	22.64	1,993	9.81	484	6.28	4,308
1983	1,872	22.38	2,125	12.03	534	6.85	4,531
1984	1,944	22.40	2,451	13.41	471	6.68	4,866
1985	2,129	25.23	2,355	17.08	484	6.68	4,968
1986	2,084	26.95	2,818	16.08	478	6.91	5,380
1987	2,207	27.90	2,894	15.05	509	8.81	5,610
1988	1,977	29.37	3,034	14.76	519	8.58	5,530
1989	1,671	33.37	3,742	14.19	515	8.88	5,928
1990	1,139	32.78	4,200	15.77	367	7.83	5,706

Source: Taiwan Fisheries Bureau (TFB) annual reports, 1965-1991.

4.   COASTAL FISHERIES

The yellowfin catch by Taiwanese coastal fisheries is very small and contributes little to the total Taiwanese tuna catch (Table 7).

Table 7. Yellowfin tuna catch of Taiwanese coastal fisheries, by gear (1964-1990).

Year	CST	CGN	CPL	COT	Total
1964	10	-	110	-	120
1965	6	-	88	-	94
1966	13	-	68	-	81
1967	9	-	47	-	56
1968	12	-	73	-	85
1969	6	-	73	-	79
1970	8	-	74	-	82
1971	11	-	76	-	87
1972	5	21	48	32	106
1973	2	60	57	45	164
1974	14	56	46	44	160
1975	4	28	56	38	126
1976	6	34	57	39	136
1977	4	31	57	38	130
1978	8	39	58	38	143
1979	6	45	57	33	141
1980	10	57	58	26	151
1981	10	53	46	7	116
1982	6	34	38	-	78
1983	5	19	28	-	52
1984	10	31	-	-	41
1985	9	12	5	-	26
1986	20	17	2	-	39
1987	26	25	4	-	55
1988	35	16	5	-	56
1989	22	24	2	5	53
1990	5	129	0	1	135

Source: Taiwan Fisheries Bureau (TFB) annual reports, 1964-1990
Legend: CST = Coastal set net fishery; CGN = Coastal gillnet fishery; CPL = Coastal pole-and-line fishery; COT = Other coastal fisheries.

5.   RESEARCH

Prior to 1987, research of western Pacific yellowfin stock was carried out by the Tuna Research Center (Yang *et al.*, 1970a, 1970b; Wang and Yang, 1970; Yang, 1971; Chen and Yang, 1973; Huang and Yang, 1974; Sun and Yang, 1983; and Wang, 1986). In 1991 two new projects were established. One project involves collecting basic catch and effort data of the Taiwanese inshore tuna longline fishery and the other project involves age determination of southwestern Pacific yellowfin stocks. We expect to obtain more information of yellowfin stocks from these two projects.

6.   REFERENCES CITED

Chen, C.J., and R.T. Yang. 1973. Parasites of yellowfin tuna in the waters southwest off Taiwan. *Nat.Taiwan Univ.Sci.Rep.Acta Oceanogr.Taiwanica,* 3:181-98.

Huang, C.C., and R.T. Yang. 1974. Age and growth of yellowfin tuna in the waters around the southern part of Taiwan. *J.Fish.Soc.Taiwan,* 3(2):51-60.

Sun, C.L., and R.T. Yang. 1983. The inshore tuna longline fishery of Taiwan -Fishing grounds, fishing seasons, fishing conditions and a biological study of the major species, yellowfin tuna, 1981-1982. *J.Fish.Soc.Taiwan,* 10(2):11-41.

Taiwan Fisheries Bureau. 1964-1991. Fisheries yearbook, Taiwan Area. *Taiwan Fish.Bur.Dep.Agri.Forestry,* (varying pagination).

Tuna Research Center. 1973-1991. Annual catch statistics of Taiwanese tuna longline fishery. *Tuna Res.Cent.Inst.Oceangr.Nat.Taiwan Univ.,* (varying pagination).

Wang, C.H. 1986. Population dynamics of Taiwanese inshore tuna longline fishery. *China Fish.Monthly,* 400:3-15.

Wang, C.H. 1988. Seasonal changes of the distribution of South Pacific albacore based on Taiwan's tuna longline fisheries, 1975-1985. *Nat.Taiwan Univ.Sci.Rep.Acta Oceanogr.Taiwanica,* 20:13-40.

Wang, C.H., and R.T. Yang. 1970. Preliminary report on study of tuna resources. I. Study of tuna resources of waters adjacent to Taiwan. (3)Fitting non-linear regressions to the relationship between fork length and scale radius of yellowfin tuna. *China Fish.Monthly,* 211:3-8.

Yang, R.T. 1971. Population study of yellowfin tuna in the waters adjacent to Taiwan. *Nat.Taiwan Univ.Sci.Rep.Acta Oceanogr.Taiwanica,* 1:137-56.

Yang, R.T., C.C. Huang, C.H. Wang, and C.L. Tseng. 1970a. Preliminary report on study of tuna resources. I. Study of tuna resources of waters adjacent to Taiwan. (2) Relationship between scale radius and fork length of yellowfin tuna from waters around Shiao-Ryukyu. *China Fish.Monthly,* 209:9-13.

Yang, R.T., C.C. Huang, C.H. Wang, and H.C. Yang. 1970b. Preliminary report on study of tuna resources. I. Study of tuna resources of waters adjacent to Taiwan. (1) Discussion of analytical method and preliminary conclusions of length-weight relationship. *China Fish.Monthly*, 207:2-14.

A REVIEW OF THE BIOLOGY AND FISHERIES FOR NORTH PACIFIC ALBACORE (*THUNNUS ALALUNGA*)

Norman Bartoo
Southwest Fisheries Science Center
National Marine Fisheries Service, NOAA
La Jolla, California 92038

and

Terry J. Foreman
Inter-American Tropical Tuna Commission
La Jolla, California 92038

1. INTRODUCTION

North Pacific albacore have been fished off Japan and North America since at least the turn of the century. These early fisheries were mostly coastal in nature, extending only a few hundred miles offshore. By the early 1950s the longline fishery for albacore extended across the Pacific. Total catches from 1952 to 1960 ranged from 51,000 mt to 94,000 mt. Prices paid (and world demand) for albacore are and have been considerably higher (as much as 2 or 2 1/2 times) than the price paid for tropical tunas used for canning. The albacore is therefore a highly desirable species world-wide.

2. CLASSIFICATION

The North Pacific albacore is the same species as albacore found elsewhere. Specifically, from the suborder level the taxonomy is:

Suborder Scombroidei
 Family Scombridae
  Subfamily Scombrinae
   Tribe Thunnini
    Genus *Thunnus*
     Species *alalunga*

Bonnaterre is credited with the current scientific name.

3. EARLY LIFE HISTORY

Descriptions of larval-phase albacore are found in Matsumoto (1962), Matsumoto *et al.*, (1972), and other literature.

Juvenile albacore are found in tropical and subtropical waters of the North Pacific. Although most documented findings of very small (12-400 mm) juveniles are from the South Pacific, fish in this size range have been found around Hawaii (Yabe *et al.*, 1958;

Yoshida, 1965; and Yoshida, 1968) and in the eastern Pacific near Guadalupe Island (Clemens, 1961).

Small albacore, presumably smaller than 25 cm have been caught in the eastern Pacific in some years. Small albacore (28 - 35 cm) are commonly found in considerable abundance in Japanese coastal waters in spring and summer, with considerable year-to-year variation (Nakamura, 1969).

4.   AGE AND GROWTH

Considerable efforts have been applied to estimating age and growth from hard parts, size composition data and tagging experiments. Foreman (1980) presents a summary of age and growth parameters which shows considerable differences between authors. Laurs *et al.* (1985) found a linear relationship of 0.954 rings per day between detectable otolith-increment counts and days at liberty following tagging and tetracycline injection for fish between 50 and 100 cm. From subsequent (unpublished) whole-otolith, total-ring counts as well as unpublished data from a similar study by one of the authors (Bartoo) it appears that North Pacific albacore have approximately the following size at age: 1 - 35 cm, 2 - 52 cm, 3 - 65 cm, 4 - 76 cm, 5 - 85 cm, 6 - 93 cm. Laurs and Wetherall (1979) demonstrated from tagging data that fish captured south of 40°N had a higher estimated growth rate than fish captured north of 40°N. Yoshida (1968) estimates that juveniles grow at a rate of 3.19 cm per month, which is fairly consistent with the above information.

The von Bertalanffy parameters, $L_\infty = 135.6$ cm, $K = 0.17$, $t_o = -0.87$, fit the sizes at age reasonably well and appear reasonable to use for assessment purposes.

Weight at size estimated from the equation w (kg) $= 2.188$ x 10^{-5} $L^{2.99}$(mm) has been empirically tabulated by Clemens (1961) for sizes between 38 and 100 cm (Table 1). As this relationship was based on fish taken in the North American troll fishery, it is reasonable to assume variability between areas and seasons. A summary of length-weight relationships from various fisheries appears in Foreman (1980).

5.   MATURATION AND SPAWNING

North Pacific albacore mature at approximately 5 years of age, or about 85 cm (Ueyanagi, 1957; Otsu and Uchida, 1963). Fecundity is estimated to be 0.8 to 2.6 million eggs per spawning (Ueyanagi, 1957; Otsu and Uchida, 1959). Peak spawning occurs in subtropical waters from March through July (Foreman, 1980). There is evidence that other tuna species spawn repeatedly throughout their spawning period and evidence for multiple spawning has been found in albacore (Otsu and Uchida, 1959), although no estimates of the spawning frequency exists. Few collections of running ripe fish have been made. Immature albacore (<80 cm) generally have a sex ratio of 1:1, but males predominate in catches of mature fish (Otsu and Uchida, 1959). The sex ratio may vary between years.

6.   STOCK STRUCTURE, DISTRIBUTION, AND MIGRATION

The migration of North Pacific albacore has been described by several authors (Clemens, 1961; Otsu and Uchida, 1963; Laurs and Lynn, 1977). In general, the bulk of the juvenile albacore recruiting into the North Pacific fisheries first enters the Japanese western Pacific fisheries off Japan and then moves eastward. Recovery of tagged juveniles (ages 1 to 5) indicates that fish tagged off Japan appear in the North American fishery; movement is along the North Pacific Transition Zone. Albacore tagged off North America seem to move across the Pacific during the fall and appear in the Japanese late-winter/spring fisheries near Japan. These fish then appear to migrate back to North America. There are few tag returns of mature fish. Based on catch patterns it would seem that adults move to lower latitudes. In addition to this general pattern of movement there may be variations associated with recruitment. It appears that a small portion of the population may spawn further east than the bulk of the population and first enter the fishery off North America (see summary in Foreman, 1980).

Table 1. Relationship between albacore fork length and weight (from Clemens, 1961).

Fork Length (mm)	Weight (lbs.)	Fork Length (in.)	Fork Length (mm)	Weight (lbs.)	Fork Length (in.)	Fork Length (mm)	Weight (lbs.)	Fork Length (in.)
380	2.53	14.96	590	9.41	23.23	800	23.36	31.50
385	2.63	15.16	595	9.65	23.43	805	23.80	31.69
390	2.73	15.35	600	9.89	23.62	810	24.25	31.89
395	2.84	15.55	605	10.14	23.82	815	24.70	32.09
400	2.95	15.75	610	10.39	24.02	820	25.15	32.28
405	3.06	15.94	615	10.65	24.21	825	25.62	32.48
410	3.17	16.14	620	10.91	24.41	830	26.08	32.68
415	3.29	16.34	625	11.17	24.61	835	26.56	32.87
420	3.41	16.54	630	11.44	24.80	840	27.03	33.07
425	3.53	16.73	635	11.72	25.00	845	27.82	33.27
430	3.65	16.93	640	12.00	25.20	850	28.00	33.46
435	3.78	17.13	645	12.28	25.39	855	28.50	33.66
440	3.91	17.32	650	12.56	25.59	860	29.00	33.86
445	4.05	17.52	655	12.85	25.79	865	29.51	34.06
450	4.19	17.72	660	13.15	25.98	870	30.02	34.25
455	4.33	17.91	665	13.45	26.18	875	30.54	34.45
460	4.47	18.11	670	13.75	26.38	880	31.06	34.65
465	4.62	18.31	675	14.06	26.57	885	31.59	34.84
470	4.77	18.50	680	14.38	26.77	890	32.13	35.04
475	4.92	18.70	685	14.70	26.97	895	32.67	35.24
480	5.08	18.90	690	15.02	27.17	900	33.22	35.43
485	5.23	19.09	695	15.34	27.36	905	33.78	35.63
490	5.40	19.29	700	15.68	27.56	910	34.34	35.83
495	5.57	19.49	705	16.01	27.76	915	34.90	36.02
500	5.74	19.69	710	16.36	27.95	920	35.48	36.22
505	5.91	19.88	715	16.70	28.15	925	36.06	36.42
510	6.08	20.08	720	17.05	28.35	930	36.64	36.61
515	6.26	20.28	725	17.41	28.54	935	37.23	36.81
520	6.45	20.47	730	17.77	28.74	940	37.83	37.01
525	6.64	20.67	735	18.14	28.94	945	38.44	37.20
530	6.83	20.87	740	18.51	29.13	950	39.05	37.40
535	7.02	21.06	745	18.89	29.33	955	39.66	37.60
540	7.22	21.26	750	19.27	29.53	960	40.29	37.80
545	7.42	21.46	755	19.65	29.72	965	40.92	37.99
550	7.62	21.65	760	20.04	29.92	970	41.56	38.19
555	7.83	21.85	765	20.44	30.12	975	42.20	38.39
560	8.05	22.05	770	20.84	30.31	980	42.85	38.58
565	8.26	22.25	775	21.23	30.51	985	43.51	38.78
570	8.48	22.44	780	21.66	30.71	990	44.17	38.98
575	8.71	22.64	785	22.08	30.91	995	44.84	39.17
580	8.94	22.84	790	22.50	31.10	1000	45.52	39.37
585	9.17	23.03	795	22.93	31.30			

7.   OCEANOGRAPHIC FEATURES ASSOCIATED WITH THE SPECIES

Water temperature affects albacore distribution (Clemens, 1961). Albacore catches are generally associated with sea-surface temperatures between 15.0°C and

19.4°C, although telemetry experiments have documented that albacore may move vertically into much colder waters for short periods of time. Smaller fish tend to be found in cooler waters (Clemens, 1961).

Albacore concentrate along oceanic fronts (thermal discontinuities) and thus their fisheries extend Pacific-wide (Laurs and Lynn, 1977). Food concentration appears to be the key mechanism (Sund, *et al.*, 1981).

8.   INTERACTION WITH OTHER SPECIES

Recent drift gillnet surveys in the North Pacific indicate that albacore are often found in the same area as pomfret (*Brama* spp.) and blue sharks (*Prionace glauca*). There appears to be no direct species interactions (*i.e.* associations) between albacore and other fish species, although albacore may associate with drifting masses of giant kelp in the eastern Pacific and are sometimes caught mixed with other tunas.

9.   GENERAL DESCRIPTION OF FISHERIES

In the North Pacific, albacore are caught in significant quantities by both surface and subsurface gears.

Longline fisheries (the only ones to use subsurface gear) are operated primarily by Japan (95% of the total catch) and to a lesser extent by Taiwan and Korea. Figure 1 shows the general distribution of the longline fisheries. Longline catches have averaged near 15,000 mt over the last 10 years (Table 2).

Major surface fisheries taking significant amounts of North Pacific albacore include the North American troll fishery, the Japanese pole-and-line (baitboat) fishery, the Taiwanese and Japanese large-mesh drift gillnet fishery, and the Japanese, Taiwanese and Korean small-mesh squid drift gillnet fishery (Figure 1). Catches from the troll and pole-and-line fisheries (Table 2) have shown a marked decline over the last 10 or 15 years, while drift gillnet catches have increased (Table 2) since 1980 (although data are incomplete for this segment of the fishery).

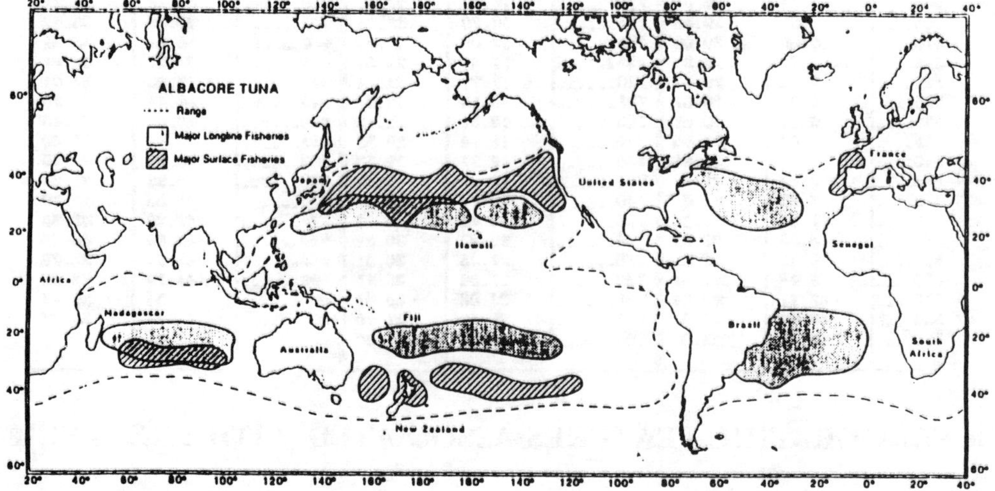

Figure 1.   General distribution of albacore showing distribution of major surface and longline fisheries.

Unit: metric ton

Year	Japan						Taiwan			Korea		USA						Canada	Grand total
	Pole & line	Long-line	Drift-net	Purse seine	Others	Sub total	Long-line	Drift-net	Sub total	Long-line	Drift-net	Pole & line	Troll	Drift-net	Purse seine	Sport	Sub total	Troll	
1952	41,768	28,887	•	151	237	68,861	•	•	•	•	•	•	23,843	•	•	1,373	25,218	71	94,151
1953	32,921	27,777	•	38	132	60,888	•	•	•	•	•	•	15,740	•	•	171	15,911	6	76,784
1954	28,060	20,058	•	23	38	49,088	•	•	•	•	•	•	12,246	•	•	147	12,393		61,481
1955	24,236	18,277	•	8	136	40,657	•	•	•	•	•	•	13,284	•	•	677	13,841		54,498
1956	42,810	14,341	•	•	57	57,208	•	•	•	•	•	•	18,751	•	•	482	19,233	17	76,458
1957	49,500	21,053	•	83	151	70,787	•	•	•	•	•	•	21,185	•	•	304	21,489	8	92,284
1958	22,175	18,432	•	8	124	40,739	•	•	•	•	•	•	14,855	•	•	48	14,903	74	55,716
1959	14,252	16,802	•	•	67	30,121	•	•	•	•	•	•	20,990	•	•	0	20,990	212	51,323
1960	25,158	17,389	•	•	78	42,601	•	•	•	•	•	•	20,100	•	•	557	20,657	6	63,283
1961	18,838	17,437	•	7	268	36,348	•	•	•	•	•	2,037	12,061	•	•	1,355	16,253	4	52,605
1962	8,729	16,784	•	53	191	24,737	•	•	•	•	•	1,085	10,760	•	•	1,681	22,528	1	47,284
1963	26,420	13,484	•	58	210	40,101	•	•	•	•	•	2,432	25,147	•	•	1,181	28,760	6	68,908
1964	23,658	16,458	•	128	310	39,703	28	•	28	•	•	3,411	18,392	•	•	824	22,627	3	62,419
1965	41,401	13,701	•	11	121	55,324	281	•	281	•	•	417	16,545	•	•	731	17,693	15	73,293
1966	22,830	25,050	•	111	685	48,670	271	•	271	•	•	1,600	15,342	•	•	588	17,530	44	66,421
1967	30,481	28,889	•	89	520	59,050	638	•	638	•	•	4,113	17,820	•	•	707	22,640	101	83,404
1968	16,697	23,061	•	287	1,109	41,834	698	•	698	•	•	4,908	20,444	•	•	951	26,303	1,028	69,081
1969	32,107	18,006	•	521	1,480	62,114	634	•	634	•	•	2,098	18,039	•	•	358	20,495	1,385	78,306
1970	24,376	16,372	•	317	794	40,859	1,510	•	1,510	•	•	4,418	21,041	•	•	822	26,281	354	69,008
1971	53,104	11,035	•	902	387	65,502	1,759	•	1,759	•	•	2,071	20,537	•	•	1,175	23,783	1,587	92,631
1972	60,782	12,849	1	277	846	74,335	3,091	•	3,091	•	•	3,750	23,608	•	•	637	27,995	3,558	108,979
1973	69,811	18,059	30	1,363	533	87,705	129	•	129	•	•	2,238	15,667	•	•	84	17,987	1,270	107,181
1974	73,678	13,053	224	181	959	87,973	570	•	570	•	•	4,777	20,187	•	•	94	25,058	1,207	114,808
1975	62,167	10,060	166	159	254	82,798	1,404	•	1,404	310	•	3,243	18,075	•	•	640	21,958	101	87,568
1976	85,338	16,898	1,070	1,109	285	103,898	1,251	•	1,251	871	•	2,700	15,932	•	•	713	19,345	252	125,515
1977	31,834	15,737	868	669	370	49,407	873	•	873	65	•	1,497	10,005	•	•	637	12,139	53	62,437
1978	50,877	13,081	4,029	1,115	2,097	80,170	284	•	284	174	•	950	16,682	•	•	810	18,442	23	99,102
1979	44,682	14,249	2,858	125	1,156	83,050	187	•	187	27	•	303	6,801	•	•	74	7,178	521	70,963
1980	48,743	14,743	2,888	329	1,200	86,010	318	•	318	16	•	382	7,574	•	•	168	8,124	212	74,870
1981	27,428	18,020	10,348	252	904	56,650	330	•	330	600	•	748	12,694	•	•	195	13,637	200	71,728
1982	29,615	18,762	12,511	581	732	60,181	559	•	559	1,070	•	425	6,861	•	•	257	7,343	104	69,257
1983	21,008	16,103	8,862	350	125	43,621	520	•	520	1,233	•	607	9,512	•	•	87	10,206	225	55,712
1984	28,015	16,111	8,848	3,380	518	64,012	471	•	471	1,041	•	1,030	9,378	•	3,728	1,427	15,563	50	71,137
1985	20,714	14,320	11,204	1,633	407	48,178	109	•	109	2,188	•	1,498	6,431	2	•	1,176	9,107	58	69,019
1986	18,008	12,945	7,813	1,542	650	39,048	•	•	•	•	•	432	4,708	3	•	108	5,251	30	44,415
1987	19,001	14,842	8,698	1,205	189	41,825	•	7,700	7,700	•	•	158	2,768	5	•	74	3,005	104	52,632
1988	8,218	13,904	9,074	1,208	177	30,578	38	11,366	11,404	•	•	598	4,212	15	•	64	4,889	155	47,027
1989	8,629	12,899	7,437	2,621	1,421	32,907	504	4,200	4,704	•	•	54	1,860	4	•	160	2,078	200	39,889
1990	12,500	•	2,315	•	•	14,816	•	•	•	•	•	115	2,603	29	71	24	2,842	305	17,963

The following notes apply to this table:

(1) Figure for 1990 is preliminary; the USA troll catches (1984-88) include gillnet.

(2) Japanese longline catches for 1952-60 exclude amount taken by vessels under 20 tons;longline catches in weight are estimated by multiplying annual number of fish caught by average weight statistics.

(3) Japanese pole-and-line catches include fish caught by research vessels.

(4) Japanese longline catches from 1958-68 were readjusted in 1988.

(5) The USA troll catches from 1952-60 include fish caught by baitboats; period from 1961-85 include fish landed in Hawaii.

(6) Korean longline catches calculated from FAO statistics and Korean catch/effort data.

(7) Korean gillnet catches are missing.

(8) Taiwanese gillnet catches are based on personal communications from the Institute of Oceanography, National Taiwan University.

Table 2.   Catch of North Pacific albacore by gear, 1952-1990.

10. TRENDS IN EFFORT, CATCH, AND CATCH-PER-UNIT-EFFORT

10.1 <u>Longline</u>

The longline fishery is best indexed by the Japanese fleet. The Japanese longline fleet has a directed fall-winter fishery for adult albacore although a small proportion of immature fish are taken in some areas. The reduction in catches in the very early 1970s (Table 2) reflects a shift in fishing effort toward bigeye tuna (Sakagawa *et al.*, 1987). After this change the longline catch of albacore stabilized (Figure 2).

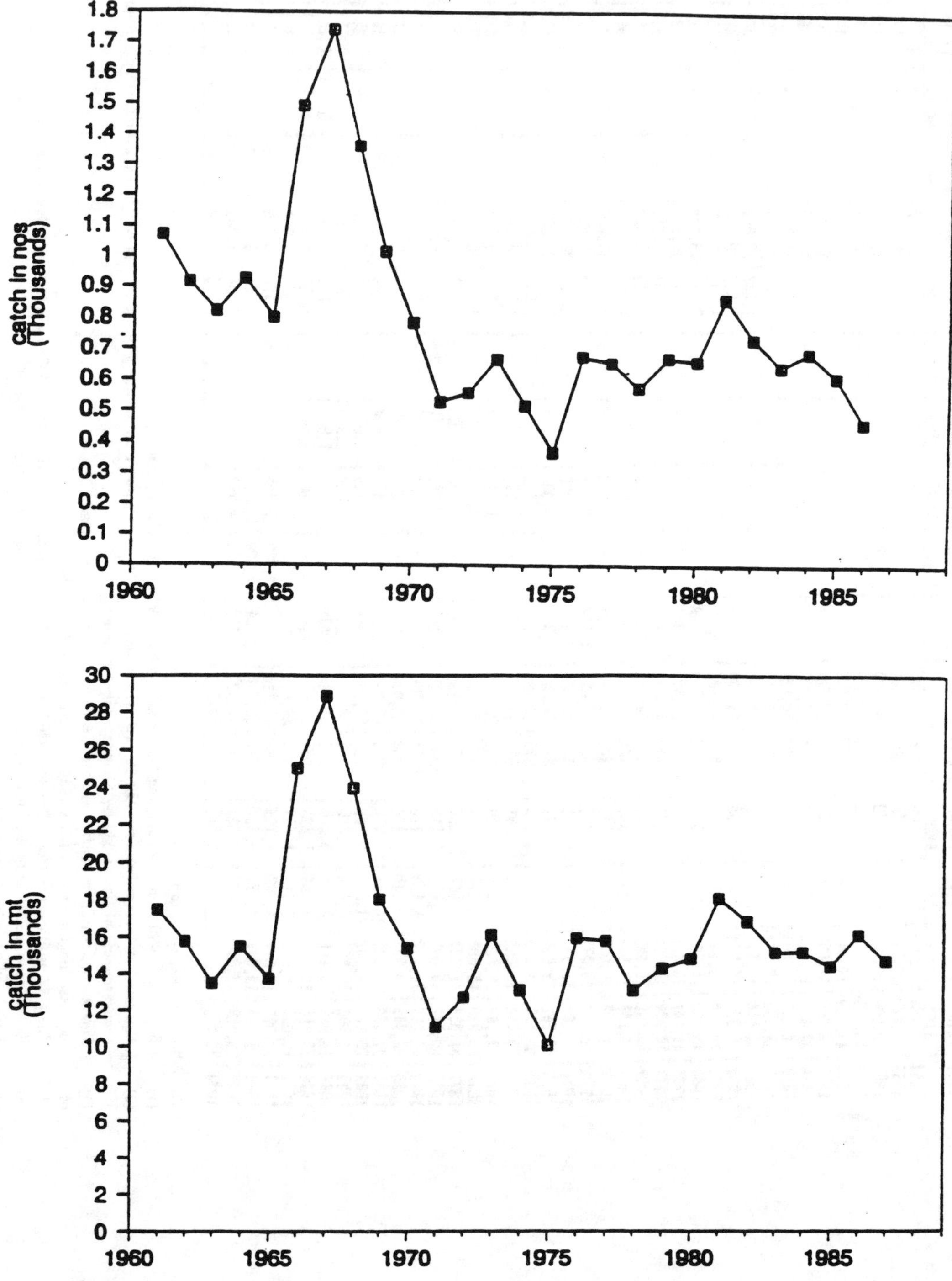

Figure 2.   Japanese fleet longline catch in weight (upper panel) and numbers (lower panel).

Effective effort (Figure 3) has shown great fluctuations during the period associated with the change in fishing operations. After 1975 effective effort increased approximately 24%, with modest fluctuations.

Catch-per-unit-effort (CPUE) in numbers of fish-per-hook has shown a slight down trend from the mid-1970s through 1985 (Figure 4). The CPUE in 1986 was down about 40% from the recent years CPUE average. The CPUE as measured by kg-per-100 effective hooks (Figure 4) showed an erratic trend through 1975 associated with fishery changes. Since 1975 this CPUE index has been relatively stable, declining slightly in the most recent years.

The average weight per fish captured as calculated from total tonnage and total numbers estimated in the catch remained relatively constant through the late 1960s (Figure 5). From 1969 through 1974 (the period of operations change) the average weight increased steadily, and from 1975 through 1986 the average weight remained relatively constant. In 1986 the average weight increased.

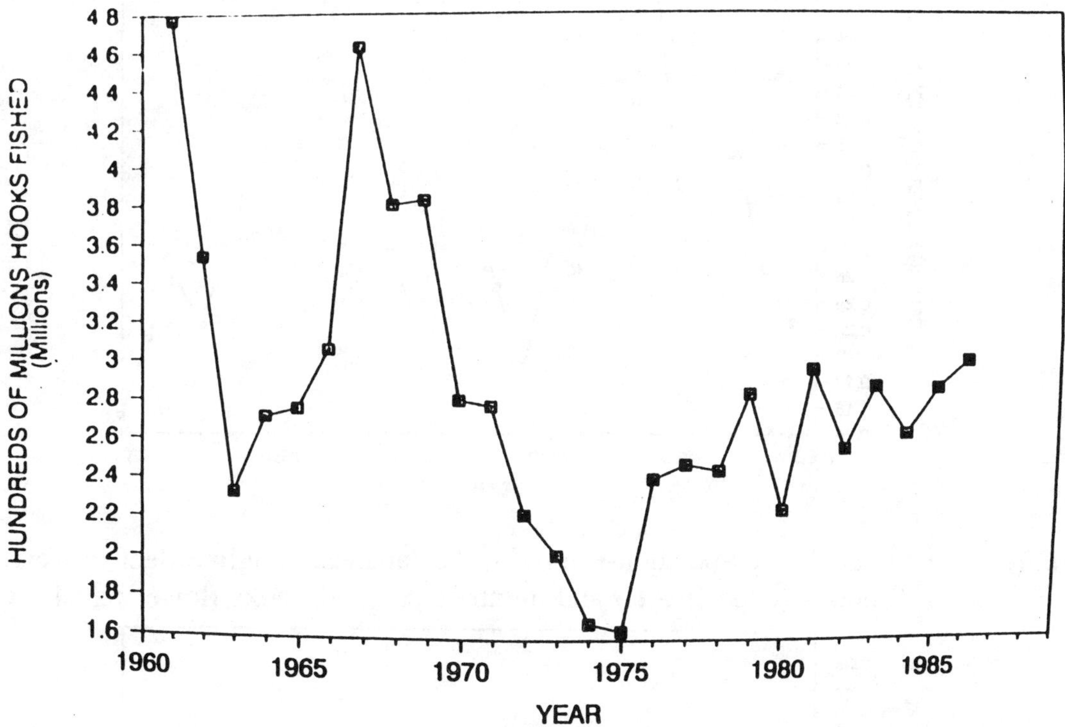

Figure 3. Estimated effective-effort for albacore by the Japanese longline fleet.

10.2 Pole and Line

The pole-and-line (baitboat) fishery is almost exclusively Japanese. The vessels fishing albacore are predominantly 100 gross tons or larger and are part of a multi-species fishery. These vessels fish for skipjack tuna most of the year and as albacore become available, shift to albacore. The fishery for albacore generally begins south of the Japanese home islands in March or April and moves northward and east along the Kuroshio extension waters.

180

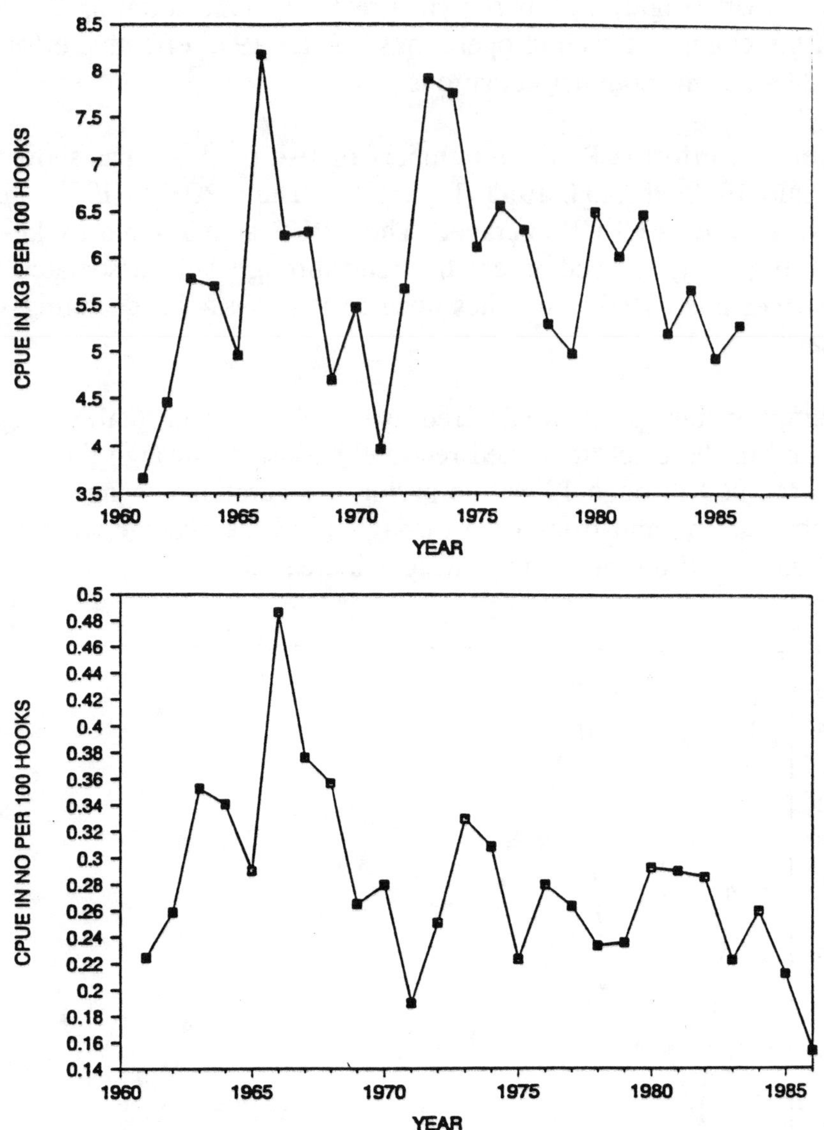

Figure 4.   Estimated catch-per-unit-effort for the Japanese longline fleet in weight per 100 hooks (upper panel) and numbers per 100 hooks (lower panel).

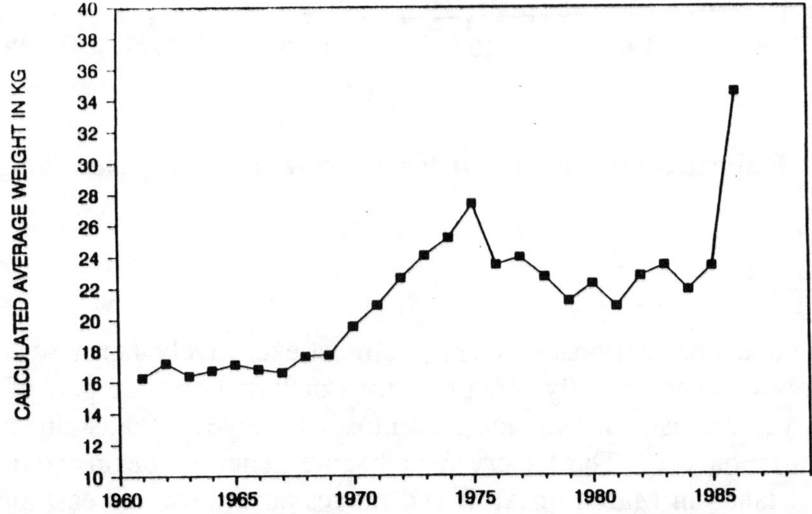

Figure 5. Estimated average weight per fish in the Japanese longline fishery.

Catches for the pole-and-line fleet are shown in Table 2. Peak catches of 70,000 to 85,000 mt were seen in the mid-1970's, but have dropped rather steadily to just over 6,000 mt in 1988. Korea reportedly has had a small live-bait fishery for albacore since the 1960s, although no official statistics are available.

Information on effort in the pole-and-line fishery is available for Japanese boats which exceed 100 gross tons and have licenses to operate over the 1961 to 1987 period (Figure 6). Note that the peak number of boats corresponds to the timing of the peak catches. Recent declines in vessel numbers are due in part to conversion of vessel licenses from pole-and-line to purse seine and reduced economic viability of the fishery.

The CPUE for the pole-and-line fleet (Figure 7), is relatively constant until 1976 when it declined considerably. Since 1976 it appears to have been increasing, although during this period improvements in efficiency may have biased the actual trend upward (see discussion on USA troll fleet below). The effort used in the calculations does not include days with effort and zero catch even if the vessel was searching for fish, due to the way data are reported. Average CPUE per day per boat (mt/day/boat; Figure 8) shows a relation between the number of boats and CPUE. During the period when many boats were in the fleet (the period of peak catches also), CPUE was lower than during the period before and after.

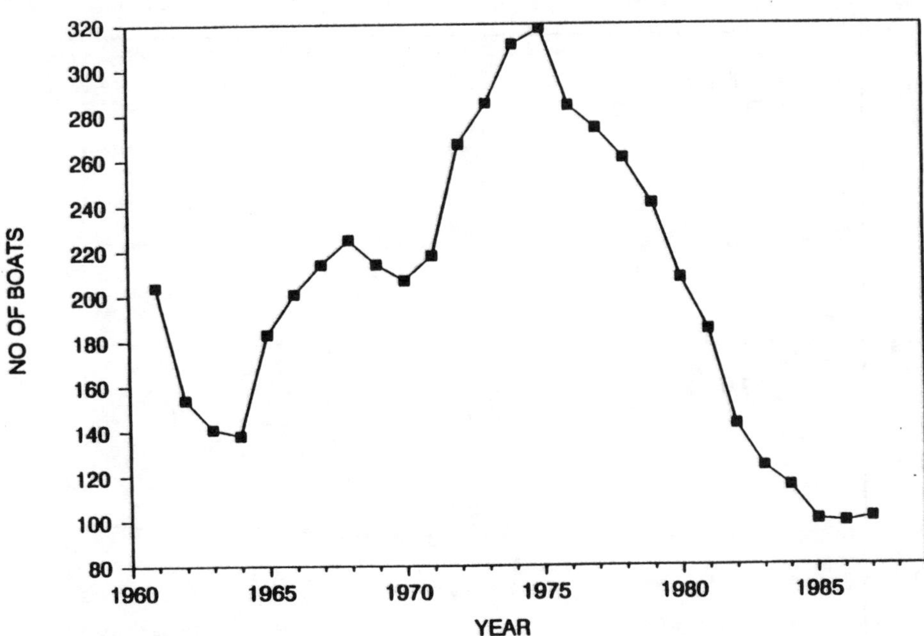

Figure 6.   Number of pole-and-line boats larger than 100 gross tons licensed to fish albacore.

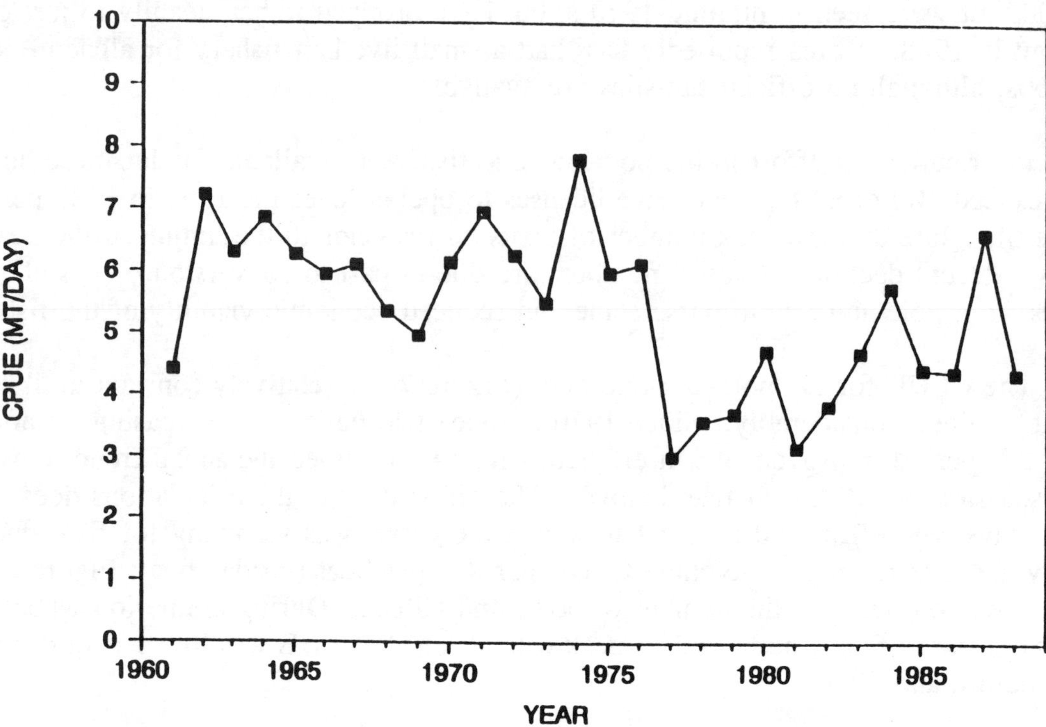

Figure 7.   Estimated catch-per-unit-effort for albacore in the Japanese pole-and-line fleet.

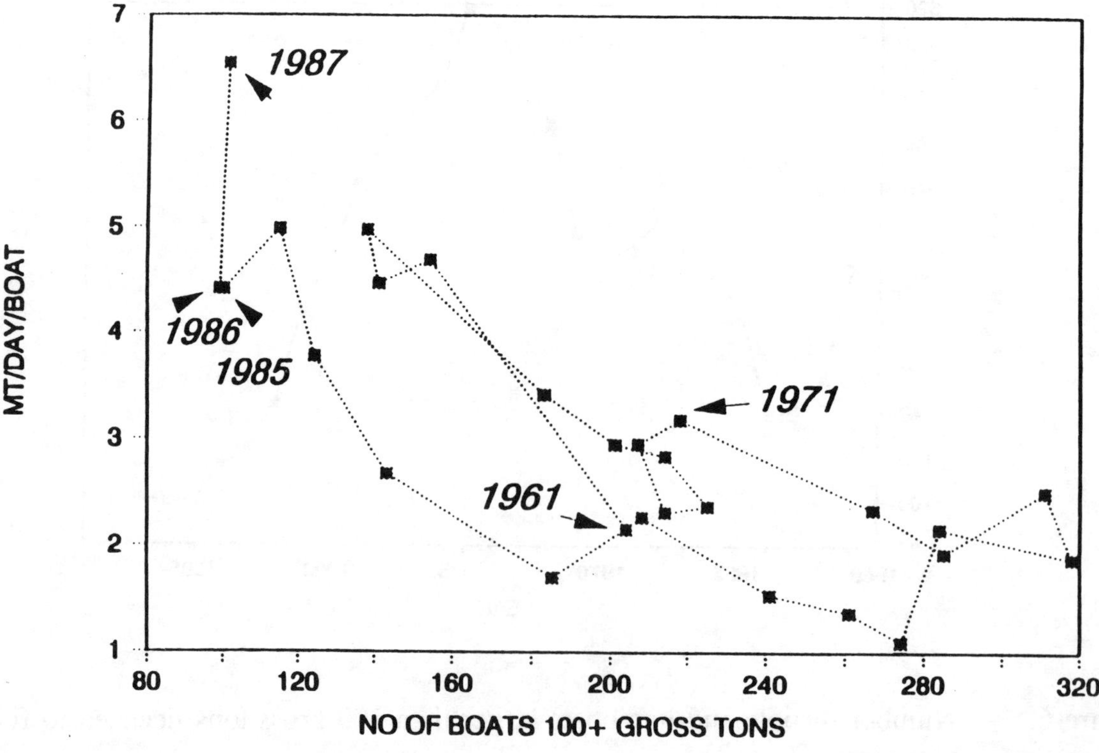

Figure 8.   Number of pole-and-line boats larger than 100 gross tons licensed to fish albacore.

10.3 Troll

The North American troll (jig) fishery is predominantly conducted by USA vessels with a few Canadian vessels. The troll vessels are generally coastal boats averaging 18 m in length, with a maximum size of about 26 m. The fishery begins with larger vessels fishing north of Midway Island in the north-central Pacific in May or June and proceeds eastward toward North America. Smaller boats from coastal bases of the USA begin fishing 200 to 1,000 miles offshore in late July. The fishery moves along the coast of the USA and Canada following the movement of albacore.

Statistics on the troll fishery (Table 2) show a period of high catches between 1959 to 1975 (nearly 20,000 mt in most years) with catches declining to a low of less than 2,000 mt in 1989. Effort in the fishery (Figure 9) estimated from total landings and CPUE, generally increased through the mid-1970s. Effort then began to decline steeply to current levels of about 20% of the average level in the mid-1970s. There is considerable year-to-year variation in effort, due in part to alternative fisheries such as the salmon troll fishery.

The CPUE for the USA fishery measured in numbers of fish-per-days-fishing (Figure 10) shows a continuous decline from the mid-1960s to the present, with year-to-year fluctuations of up to 30%. The CPUE as measured in metric tons per day per boat (Figure 11) is almost identical in trend to CPUE measured in numbers of fish.

The average weight per fish taken in the North American fishery showed no trend in the 1960s and 1970s (Figure 12). In 1979 and 1980 the average size increased considerably but dropped below the long term average in 1981 and, with the exception of a few years, has since been decreasing.

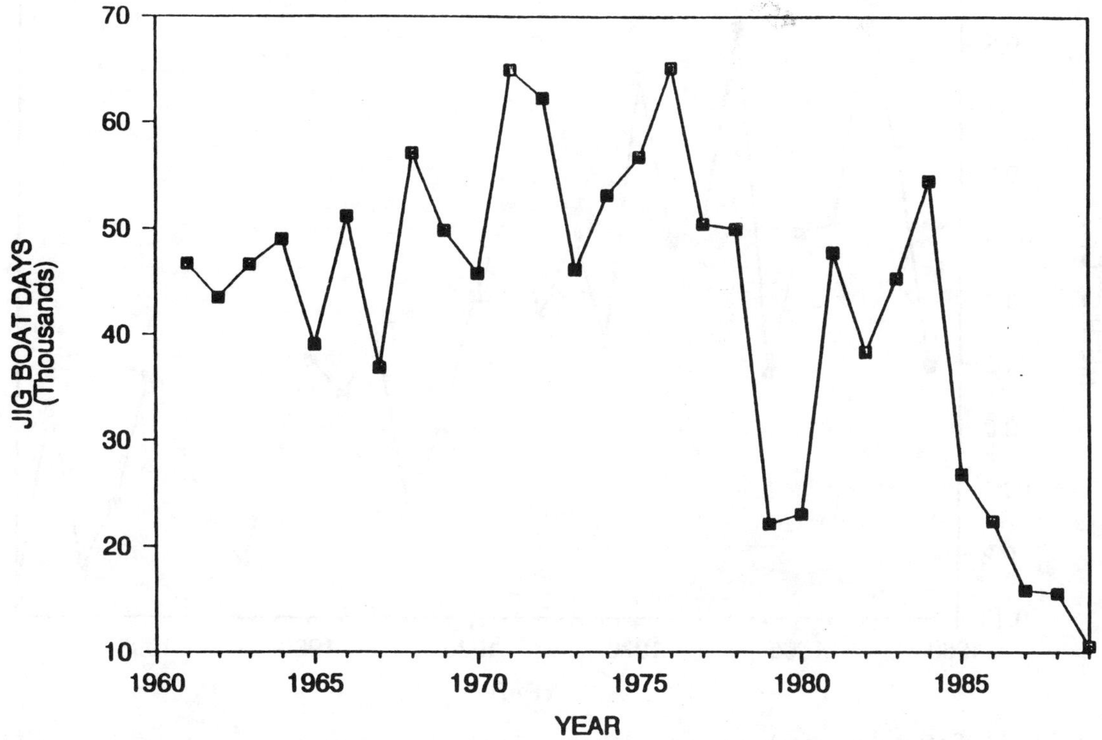

Figure 9. Estimated effort in the USA troll fishery.

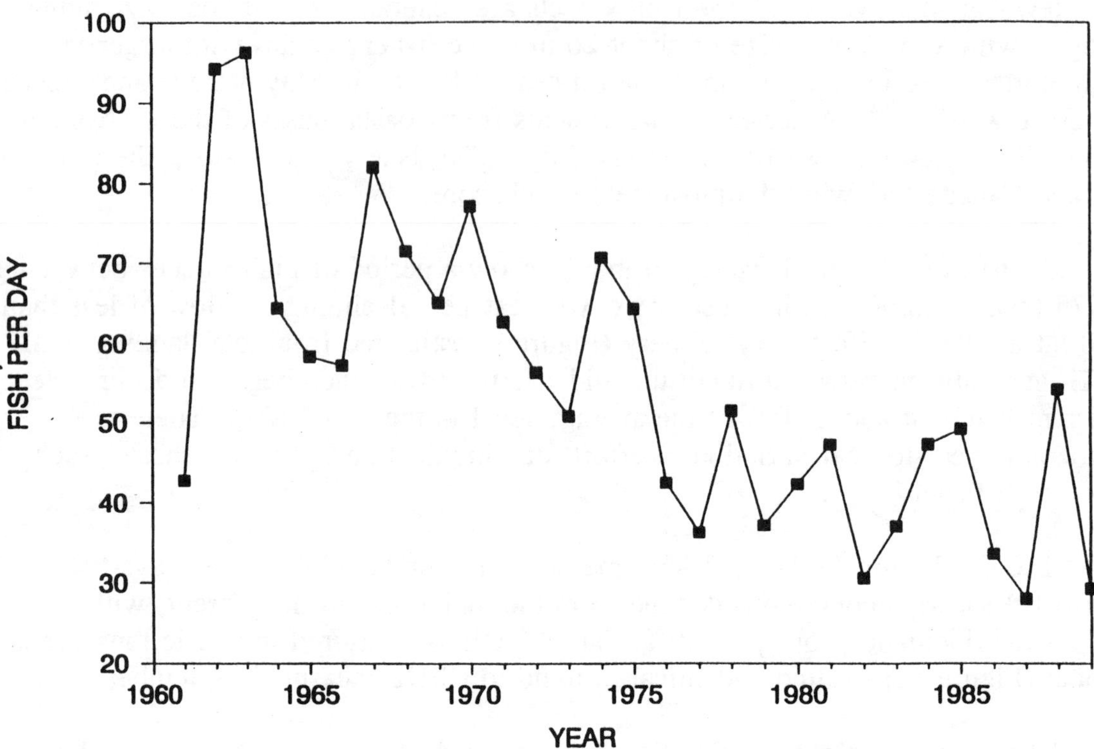

Figure 10. Estimated catch-per-unit-effort in numbers per day for the USA troll fleet.

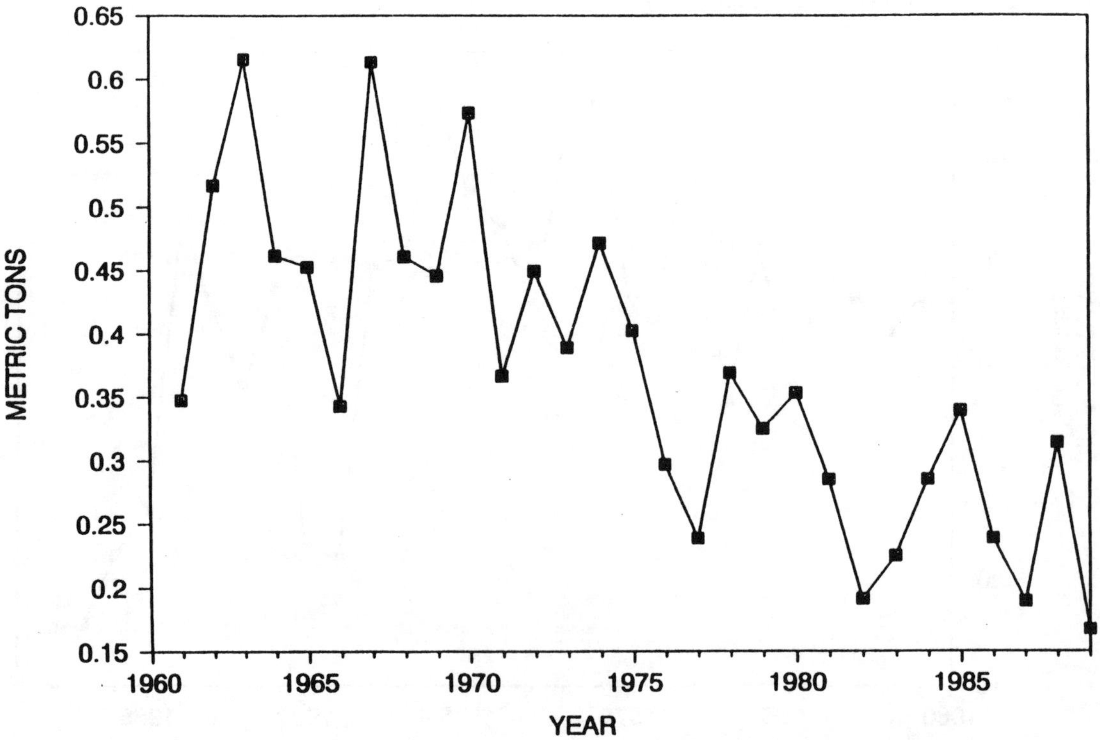

Figure 11.   Estimated catch-per-unit-effort in metric tons per day per boat for the USA troll fleet.

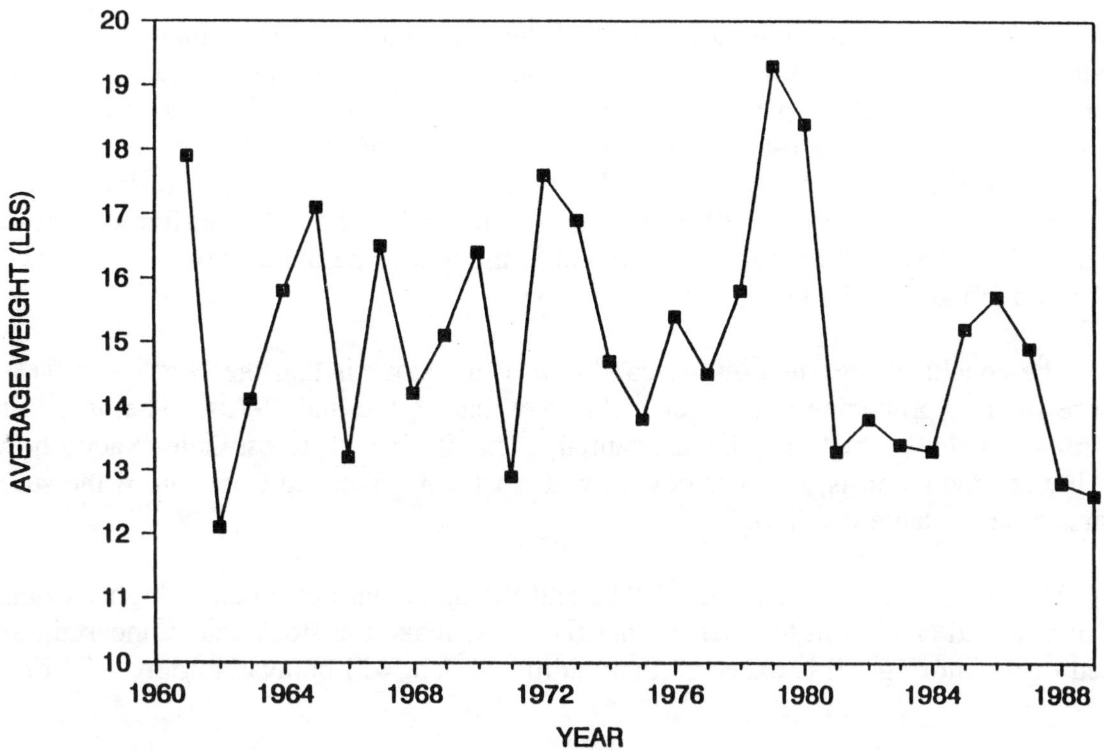

Figure 12. Estimated average weight per fish in the USA troll fishery.

10.4   Drift Gillnet

Albacore catches for the drift gillnet fisheries, both large-mesh and squid, are shown in Table 2 and are incomplete for Taiwan and Korea. Catches by the Japanese drift gillnet fishery began in 1972 and remained relatively small until 1981 when effort expanded. From 1981 to 1989 catches were in the 6,000 to 12,000 mt range. Taiwan's catch is reported only for 1987 - 1989 and ranges from 4,000 mt to 11,000 mt. None of the reported catches include estimates of non-landed catches which may have been lost or discarded at sea.

There are no reliable estimates of total effort or CPUE for the drift gillnet fleets. The total number of large-mesh drift gillnet vessels fishing in 1990 is estimated at 160.

11.   POPULATION DYNAMICS

The condition of the North Pacific albacore stock was reviewed at the 12th North Pacific Albacore Workshop in 1991. The following is based on the findings of the workshop which considered additional information than that presented here.

Firstly, no discussion was held concerning maximum sustainable yield (MSY). The 11th albacore workshop concluded that no reliable estimate of MSY is available; previous MSY estimates in the range of 90,000 mt to 200,000 mt may be suspect because of the different CPUEs used for the USA fleet and the pole-and-line fleet and the possibility of changes in the age structure of the catch over time.

Secondly, the changes in abundance of the population of young albacore are best indicated by data from the troll and pole-and-line fisheries. The trend in adult abundance is best represented by the Japanese longline. It appears that the adult stock has been relatively stable during the 1966 to 1986 period. Since 1986 the trend has been downward, declining as much as 30%. The trend in young fish abundance since 1977 is relatively stable but at a level 1/3 lower than before 1977. The pole-and-line CPUE trend from the mid-1970s to the present (see above) is thought to be biased upward and the actual trend is likely to be level.

The conclusion reached on the condition of the stock is that the North Pacific albacore stock is producing at a lower yield level than in the mid-1970s. The unreported drift gillnet catches, which may be substantial, make it difficult to estimate exactly how much lower production is, but it is down by at least 1/3. The adult portion of the stock appears to be in stable condition.

The unreported catches in the 1980s and the unknown but possibly high mortalities from non-landed catches in the drift gillnet fisheries, make the stock status uncertain and in need of monitoring. It is expected that additional data will be available in 1992 or 1993.

12. INTERACTIONS

There is considerable evidence that fisheries in the North Pacific catching albacore interact with each other. Tag data show that albacore taken off North America are available to the pole-and-line fishery off Japan and vice versa. Further, albacore tagged in surface fisheries have been recovered in the longline fishery.

There are data showing that albacore encountering drift gillnets and escaping are taken in the same and succeeding seasons in the troll fishery (Bartoo *et al.*, 1991). Observations made during the troll-fishing season showed fresh net marks and wounds were present on an average of 7.2% of the fish caught by U.S. trollers. Old scars, presumably from a previous fishing season, were found on 5.2% of the troll-caught fish, bringing the total proportion of the troll catch bearing "tags" from the drift gillnet fisheries to 12.4%. There are no similar data reported for the pole-and-line fishery.

13. REFERENCES CITED

Bartoo, N., D. Holts, and C. Brown. 1991. Report on the 1990 cooperative albacore observer project. *NOAA Admin.Rep.NMFS-SWFC, La Jolla*, LJ-91-09:16 p.

Clemens, H.B. 1961. The migration, age, and growth of Pacific albacore (*Thunnus germo*), 1951-1958. *Fish Bull.Calif.Dep.Fish Game*, (115):128 p.

Foreman, T.J. 1980. Synopsis of biological data on the albacore tuna, *Thunnus alalunga* (Bonnaterre, 1788), in the Pacific ocean. *Spec.Rep.I-ATTC*, (2):17-70.

Laurs, R.M., and R.J. Lynn. 1977. Seasonal migration of north Pacific albacore, *Thunnus alalunga*, into North American coastal waters: distribution, relative abundance, and association with Transition Zone waters. *Fish.Bull.NOAA-NMFS*, 75(4):795-822.

Laurs, R. M., and J.A. Wetherall. 1979. Estimates of growth rates for north Pacific albacore, *Thunnus alalunga* (Bonnaterre), based on an analysis of tag returns. U.S. Nat. Mar. Fish. Serv., unpublished manuscript.

Laurs, R.M., R. Nishimoto, and J.A. Wetherall. 1985. Frequency of increment formation on sagittae of north Pacific albacore (*Thunnus alalunga*). *Can.J.Fish.Aquat.Sci.* 42(9): 1552-5.

Matsumoto, W.M. 1962. Identification of larvae of four species of tuna from the Indo-Pacific region I. Carlsberg Foundation's Oceanographical Expedition Round the World 1928-30 and Previous "Dana" - Expeditions. Dana Report, 55:16 p.

Matsumoto, W.M., E.H. Ahlstrom, S. Jones, W.L. Klawe, W.J. Richards, and S. Ueyanagi. 1972. On the clarification of larval tuna identification particularly the genus *Thunnus*. *Fish.Bull.NOAA-NMFS*, 70(1):1-12.

Nakamura, H. 1969. Tuna distribution and migration. Fishing News (Books) Ltd., London, 76 p.

Otsu, T. and R.N. Uchida. 1959. Sexual maturity and spawning of albacore in the Pacific Ocean. *Fish.Bull.U.S.Fish. Wildl.Serv.*, 59(148):287-305.

Otsu, T. and R.N. Uchida. 1963. Model of the migration of albacore in the north Pacific Ocean. *Fish.Bull.U.S.Fish. Wildl.Serv.*, 63(1):33-44.

Sakagawa, G.T., A.L. Coan, and N.W. Bartoo. 1987. Patterns in longline fishery data and catches of bigeye tuna, *Thunnus obesus*. *Mar.Fish.Rev.*, 49(4):57-66.

Sund, P.N., M. Blackburn, and F. Williams. 1981. Tunas and their environment in the Pacific Ocean: a review. *Rev.Oceanogr.Mar.Biol. Ann.Rev.*, 19:443-512.

Ueyanagi, S. 1957. Spawning of the albacore in the western Pacific. *Rep.Nankai Reg.Fish.Res.Lab.*, 6:113-24.

Yabe, H., S. Ueyanagi, S. Kikawa, and H. Watanabe. 1958. Young tunas found in the stomach contents. *Rep.Nankai Reg.Fish.Res.Lab.*, 8:31-48.

Yoshida, H.O. 1965. New Pacific records of juvenile albacore, *Thunnus alalunga* (Bonnaterre) from the stomach contents. *Pac.Sci.*, 19(4):442-50.

Yoshida, H.O. 1968. Early life history and spawning of the albacore, *Thunnus alalunga*, in Hawaiian waters. *Fish.Bull.U.S.Fish Wildl.Serv.*, 67(2):205-11.

A REVIEW OF THE BIOLOGY AND FISHERIES FOR ALBACORE, *THUNNUS ALALUNGA*, IN THE SOUTH PACIFIC OCEAN

Talbot Murray
Research Centre, MAFF Fisheries
P.O. Box 287
Wellington, New Zealand

1.   INTRODUCTION

This paper relies heavily on the recent work of a number of scientists active in improving the information available for stock assessment. While most of the published information has been included in this review, only a portion of the more recent work presented in the South Pacific Albacore Research (SPAR) workshops could be included.

2.   CLASSIFICATION

The classification scheme used follows Klawe (1980) in modified form:

Phylum Chordata
    Class Osteicthyes
        Order Perciformes
            Family Scombridae
                Subfamily Scombrinae
                    Tribe Thunnini
                        Genus *Thunnus*
                            Species *alalunga*
                                Authority (Bonnaterre, 1788)

3.   EARLY LIFE HISTORY

Albacore larval development and the morphological features distinguishing them from other tunas are described by Ueyanagi (1969). He notes that albacore larvae resemble yellowfin tuna when smaller than 9 mm but can be distinguished on the basis of melanophore patterns; a full complement of fin rays is developed when the larva is about 13 mm total length. Ueyanagi (1969) presents data suggesting albacore larvae are slightly more abundant in night tows and in tows taken at 20-50 m than at the surface. Murray (1986) summarised surface circulation in the South Pacific which indicated that larvae are likely to be transported southwards from the spawning grounds. The distribution of post-larvae and juveniles smaller than 29 cm length to caudal fork (LCF) is unknown but presumed to be in tropical oceanic waters.

4. DISTRIBUTION, MIGRATION, AND STOCK STRUCTURE

4.1 Distribution of Larvae, Juveniles and Adults

Nishikawa *et al.* (1985) summarise albacore larval distribution patterns in the South Pacific Ocean. They indicate larvae occur from northeastern Australia eastward through the French Polynesia EEZ (about 140°W) between 5°S and 25°S. Most larvae have been caught during the austral spring months of October to December (although larvae were also caught in small numbers in all other months except December, January, March, and April).

Catches of juvenile albacore by commercial fishers and research surveys suggest a zonal distribution from about 30° to 45°S across the South Pacific in surface waters where austral summer temperatures are generally 16° to 21°C. Research survey information suggests that juveniles are largely distributed in the upper 100-120 m and that fish tend to stay deeper during daylight hours.

Juvenile albacore usually range in size from about 45 to 85 cm LCF but fish as small as 34 cm are occasionally caught by trolling (Roberts and James, 1974). Fish smaller than 45 cm LCF represent only a small fraction of commercial catches, primarily along the New Zealand continental shelf north of 39°S. Roberts (1980) reports that juveniles are equally abundant in subtropical and Subtropical Convergence Zone (STCZ) waters in the area east of New Zealand but only rarely caught in surface Subantarctic Water. In high seas areas east of the area surveyed by Roberts (1980), juveniles have been caught primarily within the STCZ rather than in subtropical waters (Laurs, 1986, Laurs *et al.*, 1987). There is no information on the distribution of juvenile albacore other than in austral summer months in the South Pacific.

Adult albacore occur throughout the South Pacific from the equator to at least 49°S (Wang, 1988; Bailey and Ross, 1987). Although caught in surface Subantarctic Water, albacore are most common in subtropical and tropical waters. Using commercial longline catch data, Wang (1988) portrays a seasonal pattern where adult albacore are concentrated in the area 15°-25°S in austral summer and 30°-40°S in austral winter months from Australia eastward to 100°W. Adult albacore are reported to be caught most often in depths of 170-220 m off Chile, 100-220 m off Easter Island (Ichikawa and Shirasawa, 1980), and 150-300 m in the area west of Fiji (Saito, 1973). In temperate waters off New Zealand adult albacore are caught shallower than 150 m.

4.2 Migration

The general migration pattern of albacore in the South Pacific is described by Jones (1991). He uses tag returns and seasonal patterns of fishing effort to explain patterns observed in his study of parasite fauna. These data support the hypothesis that juveniles move from the tropics into temperate waters at about 35 cm LCF and then generally eastward along the STCZ. Based on the loss of tropical endemic trematodes (didymozoids of several species) with increasing fish size and their absence in intermediate-sized fish, he infers that albacore do not return to the tropics until they mature at about 85 cm LCF. The presence of didymozoids in sexually mature albacore in temperate waters suggests that these fish return to temperate waters from spawning grounds in tropical waters.

The movements of juveniles comprising the troll catch are unknown other than in austral summer months. Didymozoid trematodes are not known to occur south of about 30°S and no evidence of reinfection of albacore is evident in fish 50-85 cm LCF (Jones, pers. commun.), suggesting that juveniles do not return to tropical waters north of 30°S. Research trolling surveys targeting albacore in austral winter months as far north as 26°S in the South Fiji Basin and in the Tasman Sea south of 30°S in the austral spring have failed to catch albacore. Longline vessels which fish deeper primarily catch juveniles in the austral autumn in temperate and subtropical waters. Based on these observations, albacore appear to move from temperate to subtropical waters during the austral autumn. As they move into subtropical waters they are presumed to move deeper where they are less vulnerable to trolling. Based on commercial trolling catches, juveniles move southwards to temperate waters in late austral spring or early summer months across the South Pacific. Adults occur from the tropics to temperate waters of the South Pacific throughout the year.

4.3   Stock Structure

Lewis (1990) reviews available information relating to likely stock boundaries and exchange between the South Pacific and other albacore stocks. The seemingly discrete spawning areas based on Nishikawa's *et al.* (1985) larval distribution for South Pacific, North Pacific, and Indian Oceans suggest distinct stocks. Further support for considering these to be separate stocks is found in the very low longline catch rates in equatorial waters (Lewis, 1990) and the age-dependent environmental requirements of albacore compiled by Pianet (in Anon., 1991). These indicate that the minimum temperature and dissolved oxygen requirements for albacore larvae and juveniles are not met in surface waters of the equatorial Pacific in any season although adult requirements are met. This suggests that an effective physiological barrier exists for larvae and juveniles which prohibits exchange between the North and South Pacific Oceans of all life history stages other than of adults.

Low longline catch rates of adult albacore in equatorial waters suggest that exchange of adults is limited between oceans. The lack of tag recoveries of adults tagged as juveniles in the North Pacific but recovered in the South Pacific Ocean (and *vice versa*) suggests that exchange between albacore stocks, while possible, is not appreciable. Exchange of adult albacore between the South Pacific and Indian Ocean stocks is also possible south of Australia but is believed to be limited.

These observations and the distribution of longline catch rates support the hypothesis that albacore in the South Pacific constitute a discrete stock. Jones (1991) offers two explanations for observed parasite faunal patterns, one of which suggests separate New Zealand and central STCZ sub-stocks of juvenile albacore. While some degree of separation may exist among juveniles, the recovery of several albacore in the New Zealand area, tagged as juveniles in the central STCZ, suggests a single South Pacific stock.

5. AGE AND GROWTH

Growth rate has been estimated from caudal vertebrae (Murray and Bailey, 1989), otolith increments (Wetherall, *et al.*, 1989), and length frequency data (Hampton *et al.*, 1990). Labelle (1991) compared these approaches by reexamining the data of Murray and Bailey's (1989) and Hampton *et al.*, (1990) in light of additional tag recoveries. He suggests that ages determined from otolith daily increments are likely to be incorrect given the correspondence between growth rates estimated using caudal vertebrae and length frequency and their similarity to growth increments observed in tagged albacore. Relative growth rate based on tag returns, caudal vertebrae, and length frequency analysis suggests rates on the order of 0.5 cm per month.

6. MATURATION AND SPAWNING

Isii and Inoue (1956) report female albacore caught in the Coral Sea in January with ripe ovaries weighing 140-149 gm and containing 812,000 to 870,000 ova. They further report that egg diameters from anterior, central, and posterior sections varied from 0.397 to 0.431 to 0.445 mm, respectively. Egg diameters did not vary from the centre out to the edge of the ovary in any section. Saito (1973) shows that gonad weight increases with fish length in females but not in males in the area west of Fiji. He further indicates that females larger than 87 cm LCF were all mature and had gonads larger than 200 gm, while males with gonads larger than 150 gm were mature (corresponding to fish he sampled larger than 97 cm LCF).

Ratty *et al.* (1989) studied testes morphology and spermatogenesis in troll-caught albacore from the STCZ and found a marked asymmetry in male gonads with the right gonad larger than the left in 72% of fish examined. They further found that 7% of males 50-70 cm LCF were sexually mature and that a substantial portion of 70-80 cm LCF males in the STCZ were mature. However, females 55-95 cm LCF were not in an advanced state of maturity. Bailey (1991) has confirmed the gonad asymmetry for males and found that in longline-caught females from New Caledonia and Tonga the right gonad is also usually larger than the left. He further indicates that high gonad indices (>2.0) were found in males and females 85 cm LCF and larger.

Maturity has been assessed using gonad weight, with female gonads <140 gm reported to be immature (Saito, 1973). Ramon (1991) reports the highest female gonad weights in November and February. Although the gonad weights she reports are considerably lower than Saito's (1973) maturity criterion, throughout the November-February period she found larger egg diameters than those considered to be ripe by Isii and Inoue (1956).

Despite the differences among authors, female albacore appear to mature at about 85 cm LCF while males can mature at much smaller sizes. Females with late-developing ovaries capable of spawning have been found in the South Pacific north of 20°S from May through September (Otsu and Hansen, 1962). However, 90% of females had late-developing ovaries from November through February. Based on these data and the peak in egg diameter and gonad weight reported by Ramon (1991), spawning appears to take place primarily in the November-February period north of 20°S.

7.   NATURAL MORTALITY

Natural mortality rates are unknown for the South Pacific albacore stock. However, Hampton (1990) suggests that a value of *M* of 0.3 to 0.4 per year may be appropriate given the likely longevity of albacore where few age 9+ fish appear in the longline catch. His suggestion assumes that growth rates for South Pacific albacore are similar to those of the Atlantic albacore stock and uses size composition data from the early years of longline fishery in the South Pacific.

8.   OCEANOGRAPHIC FEATURES ASSOCIATED WITH THIS SPECIES

Albacore are with few exceptions reported to be caught in oceanic waters within a relatively narrow temperature range, which differs for larvae, juveniles, and adults. Larvae occur shallower than 60 m where temperatures are higher than 24°C and the thermocline is at about 250 m (Ueyanagi, 1969).

Laurs *et al.* (1977) have shown that juvenile albacore tend to accumulate near coastal upwelling fronts, dispersing when the fronts break up. They also show that albacore movements are related to sea surface temperature, with fish spending little time in waters cooler than 15°C. Roberts (1980) also reports that albacore in the South Pacific are seldom caught in STCZ water cooler than 15°C.

The sea surface temperature range where juveniles can be caught differs between areas, ranging from 18.5° to 21.3°C in New Zealand continental shelf waters (Roberts, 1974) 14° to 19.5°C in oceanic and slope waters of the eastern Tasman Sea off New Zealand (Roberts, 1975), and 15.5° to 20.6°C in the STCZ (Laurs, *et al.* (1987). Highest catch rates occur in a narrower sea surface temperature range within each of these reported ranges, typically 16.1° to 18.3°C (Roberts, 1974; Laurs *et al.*, 1987). Depth of the thermocline also affects juvenile albacore catches, and few fish are caught by trolling when the thermocline is deeper than 95 m. Most albacore catches by trolling occur when the thermocline is shallower than 50 m (Murray and Bailey, 1986).

Adult albacore are primarily caught at depths of 100 to 380 m where *in situ* temperatures range from 9° to 20°C (Saito, 1973; Ichikawa and Shirasawa, 1980). Saito (1973) reports that in the area west of Fiji, albacore are most abundant at depths of 200-300 m where a high salinity, warm water mass intrudes into a low salinity, cold water mass. He suggests that albacore concentrate within a water mass boundary layer where food is particularly abundant.

9.   INTERACTIONS WITH OTHER SPECIES

9.1   Food and Feeding

Albacore in the South Pacific are opportunistic carnivores which feed on a wide variety of small fish, planktonic crustaceans, and squid. Stomach contents of juvenile albacore caught incidental to the New Zealand purse-seine fishery and by trolling in the STCZ east of New Zealand have been analyzed by Bailey and Habib (1982) and Bailey (1983, 1986).

Bailey and Habib (1982) present data suggesting that diet changes with albacore size. In juveniles smaller than 50 cm LCF, diet is almost exclusively planktonic crustaceans. For juveniles 50-75 cm LCF, diet is a mixture of crustacea, squid, and small fish. Albacore 76-95 cm LCF feed primarily on small fish and squid. Bailey (1983) demonstrates that diet can differ substantially in different regions and that diet appears to be more varied in continental slope compared to continental shelf waters. He further demonstrates a relationship between time of day and fullness of albacore stomachs, suggesting that juvenile albacore feed primarily during morning hours (0900-1200 hours) and early evening (1800-2100 hours). In oceanic areas east of New Zealand, Bailey (1986) reports that juvenile albacore caught by trolling feed on planktonic crustacea, squid, and small fish, with small fish predominating throughout the STCZ area. He further reports differences in the fish prey near New Zealand, where myctophids and saury predominate, in contrast to the central STCZ, where Peruvian jack mackerel predominate in the diet.

Examination of stomach contents of adult albacore caught by longline indicates that they also feed primarily on crustacea, squid, and fish (Saito, 1973), with significant differences in stomach fullness occurring in fish caught at different depths. He notes that albacore in the area west of Fiji caught at depths of 80-200 m had less in their stomachs on average than those caught at greater depths (200-380 m). He also noted differences in diet between the shallower and deeper layers, with crustacea predominating in the diet at deeper layers.

9.2   Competitors

Many oceanic pelagic species are opportunistic carnivores with a wide dietary spectrum. Species caught incidentally in surface and longline fisheries targeting albacore (see Murray, 1990 and Sharples *et al.*, 1991) demonstrate an ability to recognise and feed on similar prey items. By virtue of their co-occurrence and similar response to a prey item, in this case a particular bait, they are likely to be competitors of albacore. The species which probably compete with albacore vary with area and season and are primarily skipjack, yellowfin, and bigeye tunas, and pomfrets.

9.3   Predators

Argue *et al.* (1983) report juvenile albacore smaller than 12 cm from the stomachs of skipjack tuna and wahoo; other tunas, tuna-like species, and billfish are also likely to prey on albacore as small juveniles. Several apex predators caught incidentally on longlines set for albacore or observed during trolling are likely predators of albacore. They include mako and blue sharks, billfish, and dolphins. Bailey (1983) has confirmed that of three marlin species (blue, black, and striped marlin) whose stomach contents were analyzed, blue marlin occasionally feed on albacore. Also known to prey on albacore is the small pelagic "cookie cutter" shark (Hampton *et al.* 1991), which inflicts small nonfatal wounds. Hampton *et al.* (1991) also report substantial wounds from blue sharks.

9.4   Parasites

Jones (1991) lists 24 parasites identified from troll-caught and longline-caught albacore in widely separate geographical areas. The parasite fauna includes 12 trematode species (11 of which are didymozoids), 4 cestode species, 1 acanthocephalan, 3 nematodes, 3 copepods, and 1 coccidian. The parasite fauna of South Pacific albacore is generally widespread throughout the area, with little evidence of regional speciation.

10.   DESCRIPTIONS OF FISHERIES

Stable longline and surface troll fisheries for albacore have operated in the South Pacific for several decades. Since 1986, however, high seas surface fisheries including troll and large-mesh pelagic drift gillnet fisheries have expanded. The drift gillnet fishery in the South Pacific ended in June 1991.

10.1   Longline Fisheries

The longline catch of South Pacific albacore is summarised by vessel flag in Table 1. The total catch peaked in 1967 when over 40,000 mt were caught. Since 1967, catches have averaged 29,700 mt (range 20,900 mt to 38,900 mt).

South Pacific albacore fisheries began in 1952 with the expansion of Japan's distant water longline fleet into the South Pacific (Watanabe and Nishikawa, 1990). The Japanese catch rapidly increased to a peak of 34,620 mt by 1962 before declining to less than 7,000 mt in the late 1960s. The Japanese catch since the 1970s has averaged 3,800 mt (range 1,333 to 5,723 mt).

Following the success of the Japanese fleet, Korean longliners began targeting South Pacific albacore in 1958, followed by Taiwan in 1967. Korean catches increased from a few hundred tonnes in the early 1960s to over 10,000 mt in the mid-1960s. The catch by Korean longliners has been variable and except for a few years, when catches have been of the order of 5,000 to 10,000 mt, catches have been 10,000 to 15,000 mt. Catch data since 1987 suggest that catches by this fleet have declined to a few thousand tonnes.

The first reported Taiwanese catch of South Pacific albacore was in 1967 when 11,751 mt were caught. Since 1967, Taiwanese longline catches have consistently been higher (except in 1969 and 1985 when about 9,600 mt were caught). Longline catches, while variable, increased to over 20,000 mt in the late 1970s. Since the peak catch in 1980, catches have averaged 12,600 mt (range 9,600 to 17,100 mt).

Since 1983, South Pacific coastal states have begun longlining for adult albacore on the high seas and within their Exclusive Economic Zones (EEZs). Longline vessels from Tonga and New Caledonia target albacore, while albacore is a significant by-catch in a number of longline and handline fisheries throughout the South Pacific. The combined catch by South Pacific coastal states, while steadily increasing, is only about 2,000 mt.

The combined catch by all fleets targeting adult albacore has been at or below the estimated maximum sustainable yield (MSY) of 31,000 to 33,000 t (Wang *et al.*, 1988) in all years since 1981 except one (1986). This MSY estimate is based on surplus production models using target longline catches and adult size composition, and assumes a small stable surface fishery catch of about 2,000 mt.

Table 1. Longline catches (mt) of South Pacific albacore by vessel flag. South Pacific coastal states includes the catches by Australia, Fiji, French Plynesia, New Caledonia, New Zealand, and Tonga. The symbol "+" denotes small but unknown catches. Data from Anon. (1993).

Year	Japan	Korea	Taiwan	Pacific Coastal States	Total
1952	154				154
1953	803				803
1954	9,578				9,578
1955	8,625				8,625
1956	7,281				7,281
1957	8,757				8,757
1958	18,490	146			18,636
1959	17,385	456			17,841
1960	21,638	610			22,248
1961	23,412	330			23,742
1962	34,620	599			35,219
1963	29,120	1,367			30,487
1964	19,390	2,911			22,301
1965	17,793	6,405			24,198
1966	21,627	10,817			32,444
1967	15,104	13,717	11,751		40,572
1968	6,659	10,138	12,424		29,221
1969	4,894	9,963	9,595	.	24,452
1970	5,297	11,599	14,689	+	31,585
1971	3,472	14,482	15,887	+	33,841
1972	3,027	14,439	16,814	+	34,280
1973	2,550	17,452	17,742	+	37,744
1974	1,868	12,194	17,283	+	31,345
1975	1,333	9,015	17,071	+	27,419
1976	2,045	12,212	13,700	+	27,966
1977	2,328	13,176	21,932	+	37,436
1978	2,845	10,989	20,942	+	34,776
1979	2,274	8,682	15,086	+	26,042
1980	2,216	10,852	18,180	+	38,912
1981	4,203	14,793	14,595	+	33,591
1982	4,899	12,586	12,689	106	30,280
1983	5,723	6.669	12.119	155	24,666
1984	3,804	5,730	11,155	247	20,936
1985	3,868	14,267	9,601	305	28,041
1986	4,426	18,799	11,913	425	35,563
1987	4,490	8,646	15,009	1,015	29,160
1988	7,469	5,600	17,120	1,026	31,215
1989	5,365	3,997	10,867	1,485	21,714
1990	6,428	2,586	9,689	2,212	20,915
1991	4,401	1,225	14,030	2,007	21,663

10.2   Surface Fisheries

Trolling for albacore first developed in the South Pacific in the New Zealand EEZ in the late 1960s. The New Zealand troll fishery operates chiefly in a band 40-80 miles offshore of the west coast of the South Island from January through March. Season length varies from 6 weeks to 6 months depending on weather conditions. The fishery also experiences variable landings due to weather conditions and probably due to climatic factors affecting the southern limit of the distribution of juveniles. In good seasons (warm calm summers) over 200 small vessels enter this fishery and can land 2,000 mt to over 4,000 mt of juvenile albacore. In bad seasons (cold windy summers) fewer vessels fish and landings can be less than 1,000 mt. In the 1990/91 season 142 vessels landed 1,626 mt in New Zealand.

The troll fishery has expanded since 1986 following exploratory fishing along the STCZ east of New Zealand. The STCZ albacore troll fishery has developed from 89 mt and two vessels in 1985/86 to over 4,000 mt and more than 50 vessels since 1988/89. The STCZ troll fishery operates entirely in high seas areas primarily from 39° to 41°S, between 165° and 140°W. Some exploratory fishing in high seas areas of the Tasman Sea was done during the 1989/90 season with limited success. Vessels from the United States, Canada, New Zealand, French Polynesia, and Fiji participate in this fishery. In the 1990/91 season one Japan Marine Fishery Resource Research Center vessel also fished by trolling but reportedly was unsuccessful. Effort in this fishery is not expected to increase much beyond the 1990/91 level.

While the STCZ troll fishery was developing, a large-scale pelagic drift gillnet fishery for juvenile albacore was also developing in high seas areas east and west of New Zealand. This fishery, started by Japan in 1982/83 in high seas areas of the Tasman Sea, expanded to include vessels from Taiwan and Korea fishing most of the high seas temperate waters of the South Pacific. The rapid expansion of the drift gillnet fishery from 10 vessels in 1986/87 to at least 130 in 1988/89 and the magnitude of drift gillnet catches caused considerable concern over the sustainability of continued harvests of juveniles at 1988/89 levels. Concern over the potential for interactions between drift gillnet and troll fleets also arose since both fisheries caught similar sizes of fish in the same areas and in the same months.

The rapid development of the drift gillnet fishery without adequate information on the population dynamics of South Pacific albacore and without collecting catch and effort statistics from the fleets has made impact assessment extremely difficult. However, reductions in drift gillnet fleet size since 1988/89 and the cessation of drift gillnet fishing in the South Pacific since July 1991 has reduced the threat posed by the 1988/89 catch levels (Anon., 1991). Table 2 summarizes surface catches of albacore in the South Pacific by country and fishing method by calendar year.

10.3   Geographical and Seasonal Distribution of Fisheries

As mentioned previously, surface fisheries are restricted to austral summer months, primarily December into April. The geographical extent of areas where surface fishing is effective is also limited to mid-temperate latitudes where austral summer sea surface temperatures tend to be 16° to 21°C. In the South Pacific high seas fishery (east of New Zealand and in the Tasman Sea) most fishing takes place between 39° and 41°S.

Table 2. Surface fishery catches (mt) of South Pacific albacore since 1967 by area (Australia = western Tasman Sea, NZ = eastern Tasman Sea, STCZ = high seas areas east of New Zealand) and gear. The STCZ includes troll catches by vessels from Canada, Fiji, French Polynesia, New Zealand, and the USA. Drift gillnet catches include catches from the Tasman Sea and the STCZ areas combined. Data from Anon. (1993) except for total drift gillnet catches in 1988/89, which include industry sources.

Fishing Year	Japan P/L	Total Driftnet	Australia	NZ Troll	STCZ Troll	Total
1959/60	45					45
1960/61	-					-
1961/62	-					-
1962/63	16					16
1963/64						-
1964/65						-
1965/66						-
1966/67				5		5
1967/68				14		14
1968/69				-		-
1969/70			100	50		150
1970/71			100	-		100
1971/72			100	268		368
1972/73			100	484		584
1973/74			100	898		998
1974/75			100	646		746
1975/76			100	25		125
1076/77			100	621		721
1977/78			100	1,686		1,786
1978/79			100	814		914
1979/80	19		100	1,468		1,587
1980/81	8		50	2,085		2,143
1981/82	1		50	2,434		2,485
1982/83	2	32	50	744		828
1983/84		1,581	50	2,773		4,404
198485		1,928	50	3,253		5,231
1985/86		1,936	50	1,911	89	3,986
1986/87		919	50	1,227	859	3,055
1987/88		5,271	50	330	3,339	8,990
1988/89		25,000- 49,000[1]	50	4,800	4,014	29,000-58,000
1989/90		7,526	50	1,785	4,857	14,218
1990/91		826	50	1,626	6,922	9,419

[1] In the absence of accurate Taiwanese statistics, catch is estimated from number of vessels reported active in the South Pacific (industry sources), season length, and average catch rates per day.

Along the shelf edge of New Zealand, especially in the Tasman Sea, commercial concentrations are more broadly distributed from about 37° to 45°S.

The longline fishery for albacore operates year round, moving from north to south seasonally. These patterns are described by Wang (1988) for the Taiwanese fleet and by Wetherall and Yong (1989) for the Korean fleet. The pattern of movement is assumed to be similar for the Japanese fleet although the fishing area extends further east (Polacheck, 1987). The general pattern is southwards during January to April to subtropical waters and then northwards from July to October. The Taiwanese fleet appears to concentrate fishing from 5° to 45°S while Korean vessels fish from north of the equator to 45°S. Most albacore longlining by Taiwanese and Korean fleets is west of 120°W, with little longline targeting for albacore in the Tasman Sea or in the area adjacent to New Zealand in recent years. Data presented by Wang (1988) and Wetherall and Yong (1989) suggests that the Tasman Sea and New Zealand EEZ were extensively fished prior to the early 1980s, while more recent fishing is predominantly in high seas areas.

Polacheck (1987) indicates that the Japanese longline fishing pattern for albacore differs slightly from those of Korea and Taiwan. Japanese longliners appear to fish a narrower band east of New Zealand from 10° to 35°S and as far south as 50°S in the Tasman Sea and around New Zealand. Japanese vessels also appear to fish further east across the South Pacific than the other fleets.

11. TRENDS IN CATCH, EFFORT, AND CPUE

11.1 Catch

Before the start of the drift gillnet fishery, surface fishery catches were less than 2,500 mt, most caught by the seasonal and weather-limited New Zealand near-shore troll fishery. Between 1983/84, the start of commercial-scale drift gillnet fishing by Japan, and 1985/86, when the STCZ troll fishery began, total surface fishery catches were less than 5300 t. The historically high total catch of 29,000 to 58,000 mt occurred in 1988/89 due to the rapid expansion of drift gillnet fishing, primarily by Taiwanese vessels. Subsequent reductions in drift gillnet fishing (following the consensus adoption of United Nations Resolution 44/225) resulted in the progressive decline in the total surface fishery catch from 29-58,000 mt in 1988/89 to 14,218 mt in 1989/90 and to 9,419 mt in 1990/91. Since no drift gillnet fishing has taken place since 1991, other surface fishing methods tried thus far have had only limited success, and catches by trolling are not expected to exceed about 10,000 mt; it seems unlikely that the surface fishery will return to 1988/89 levels.

Longline catches increased with expanding effort from 1952 to 1967 to reach its historical peak of 40,572 mt. Since 1967, total longline catches have ranged from about 21,000 to nearly 39,000 mt but have usually been less than 35,000 mt. Most longline catches are by distant water vessels from Taiwan, Korea, and Japan, with increasing but still small catches (about 2,000 mt) by vessels from South Pacific states.

11.2 Effort

Depending on weather and other factors, a variable number of New Zealand troll vessels, ranging from 25 to over 200, fish for albacore each summer. In most years over

100 vessels fish from January through March in this near-shore fishery. High-seas troll vessels from the USA, Canada, New Zealand, French Polynesia, and Fiji have been slowly increasing from 44 vessels in 1987/88 to about 70 in 1990/91. Further increases in vessel number are not expected. The other surface fishery, composed of drift gillnet vessels primarily from Taiwan and Japan, increased rapidly from 11 in 1986/87, to 28 in 1987/88, to at least 130 in 1988/89. Since 1989 these fleets have been reduced in equally dramatic fashion to 32 vessels in 1989/90 and 7-9 vessels in 1990/91. Based on agreement, all drift gillnet fishing interests promised not to use drift gillnets in the South Pacific after June 1991.

Longline fishing effort since 1970 has been variable and mostly in the range of 200-250 million hooks set per year. Korean and Taiwanese longline vessels are the major fishing interests in recent years, and declining catches by Korea and Japan probably reflect changing economics for tuna longline fisheries with less effort directed at albacore compared to yellowfin and bigeye tunas.

11.3   CPUE

Variation in weather for the New Zealand troll fishery and changes in species targeting by Japanese and Korean longliners mean that the best data for indices of abundance are the STCZ troll and Taiwanese longline CPUE data. However, work is still ongoing to standardise these data and only nominal CPUE estimates are presently available.

Average CPUE for juvenile fish in the STCZ troll fishery has been high compared with other areas fished (New Zealand and the Tasman Sea), ranging from about 100 fish (1.2 mt) per vessel-day to 250 fish (2.6 mt) per vessel-day between 1986/87 and 1990/91. Coan and Rensink (1991) show that STCZ troll fishery CPUE, expressed as mt per day, has been relatively stable since 1988. Figure 1 shows STCZ troll fishery and Japanese drift gillnet CPUE, expressed as number of fish per vessel-day. The CPUE of the STCZ troll fishery was nearly constant from 1986 to 1990, while drift gillnet CPUE generally increased over the period.

Figure 2 indicates that Taiwanese longline CPUE is highest and the most variable south of 20°S. CPUE in the waters 0°-20°S declined slightly from 1967 to 1975 and has been approximately stable since 1975 at 2-3 fish per 100 hooks. The CPUE in the waters 20°-30°S, also declined to 1975 and has since fluctuated approximately between 2.5-5.5 fish per 100 hooks. Highest and most variable CPUE is found in the 30°-50°S band where most fishing is conducted immediately north of the STCZ surface fishery area. CPUE declined in this area from 1967 to 1976 but generally trends upwards after 1976 from 3 to about 9.5 fish per 100 hooks. Estimates from a different source suggest declines in CPUE in subtropical and temperate latitudes, but within the range observed since the mid-1970s.

12.   POPULATION DYNAMICS

Since 1986, scientists from South Pacific states and distant water fishing interests have worked to increase the information available to model the population dynamics of South Pacific albacore. The absence of large surface fisheries and the inadequate size composition data and ancillary biological information on age and growth, etc. prior to

Figure 1. CPUE trends in South Pacific albacore surface fisheries operating in the STCZ. Reprinted from Anon (1991).

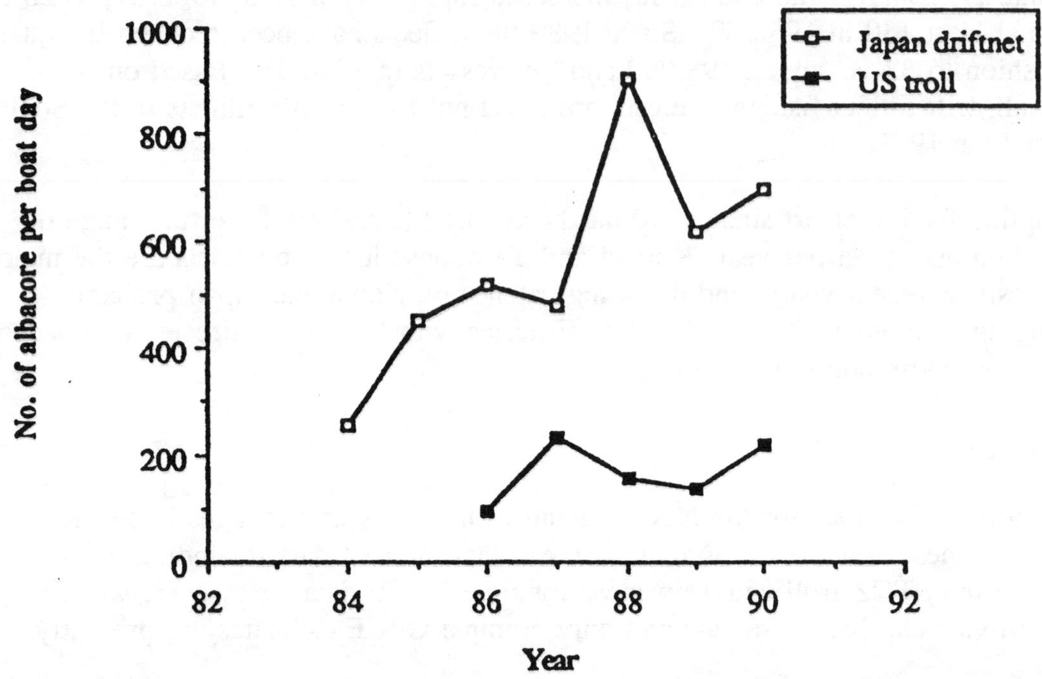

1988 resulted in reliance on relatively simple surplus production models (Skillman, 1975, Wetherall and Yong, 1984, and Wang *et al.*, 1988). These authors present essentially the same conclusion: that the average annual yield to longliners could not be increased by increasing effort and that nearly the same yield would result from lower effort. The estimated MSY of the longline fishery operating in the absence of a surface fishery ranges from 31,000 to 33,000 mt (Wang *et al.*, 1988) up to 37,000 mt (Wetherall and Yong, 1984).

Using available information on total catches and size composition by fishery and making assumptions about M, average recruitment, stock structure, and growth, Hampton (1990) constructed an age-structured simulation model of South Pacific albacore. Using a range of assumed parameter values in combination, he predicted the consequences of continuing the high 1988/89 combined exploitation levels by all fisheries operating at the time (drift gillnet, troll, and longline). He predicted that continued exploitation at the 1988/89 levels would result at best in parental stock declines over the next 5-year period to 61% of 1988/89 levels (equivalent to 32% of preexploitation levels). He concluded that unless South Pacific albacore behaved more like yellowfin tuna rather than albacore in other areas, the 1988/89 catch levels were very unlikely to be biologically sustainable. No other population dynamics modelling has been completed on the South Pacific albacore stock.

13. INTERACTIONS AMONG FISHERIES

Indications that interactions may occur between surface fisheries (drift gillnet-troll and between troll fisheries) and between surface and longline fisheries come from the spatial and temporal distribution of fleets and observations on drift gillnet damaged fish in

Figure 2.   CPUE trends in South Pacific albacore longline fisheries stratified by latitude. Reprinted from Anon (1991).

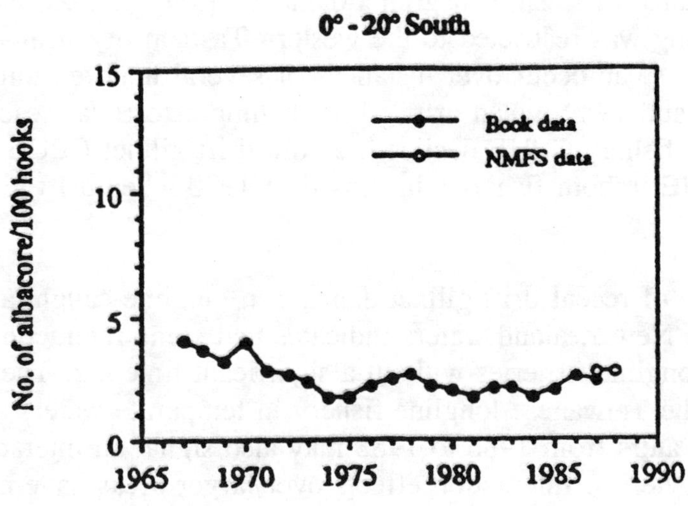

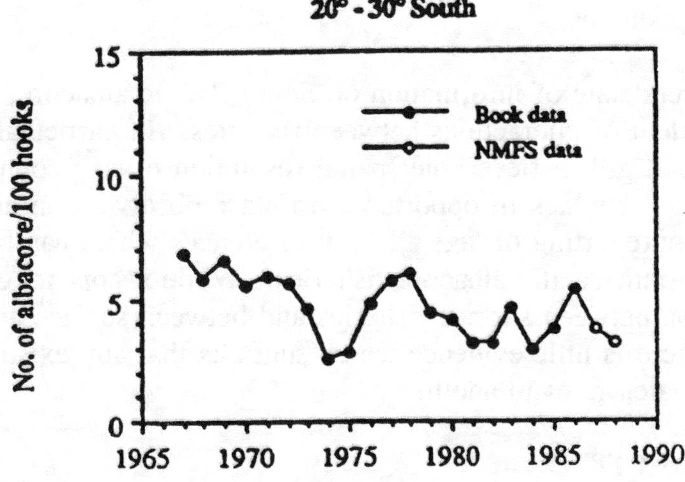

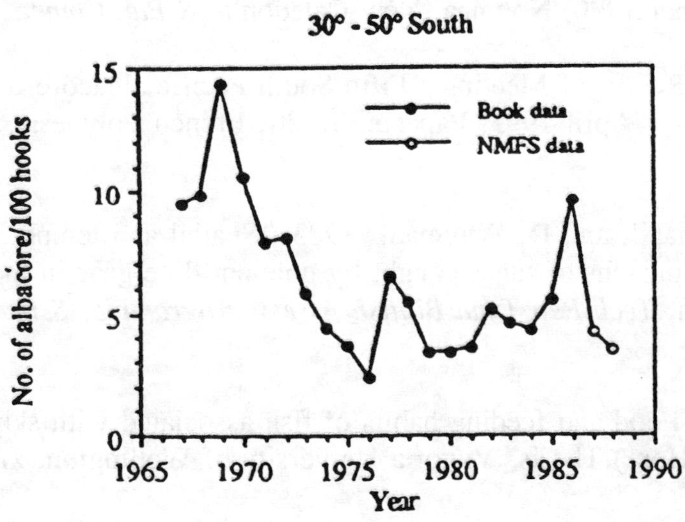

different fisheries. Higher incidence of drift gillnet damaged fish in the STCZ troll fishery when drift gillnet effort was high and also when drift gillnet and troll vessels operated in close proximity suggests interactions between surface fisheries, at least during a season. The appearance of fresh drift gillnet damage in the New Zealand troll fishery when drift gillnet fishing was restricted to the western Tasman Sea near Australia also suggests that interactions can occur over distances of several hundred miles during a season. During 1988 and 1989, when drift gillnet fishing effort was reaching its maximum, STCZ troll fishery CPUE declined, as did drift gillnet CPUE in the year of peak effort. The CPUE in both fisheries increased in 1990 when drift gillnet effort was much reduced.

The appearance of recent drift gillnet damage in longline-caught albacore of intermediate size from New Zealand waters indicates that some interaction also occurs between surface and longline fisheries without a significant time lag. The apparent declines in CPUE in the Taiwanese longline fishery in temperate waters and to a lesser extent in subtropical waters from 1986 to 1988 may also signal an interaction between surface and longline fisheries. Interaction effects over larger areas have not been indicated, and possible interactions between industrial albacore fisheries and small-scale commercial, artisanal, and recreational fisheries have yet to be examined because of a lack of data for these fisheries.

Given the current state of information on South Pacific albacore, it is not possible to better define the extent of interactions between fisheries. In particular the inadequate historical data from drift gillnet fleets, the spatial resolution of size composition data from high seas longline fleets, the lack of opportunity to place observers on high seas longline fleets, and apparent nonreporting of tags all limit inferences which can be drawn about interactions between South Pacific albacore fisheries. While it appears clear that there is potential for interaction between surface fisheries and between surface and longline fisheries, at present there is little evidence which suggests that any existing fishery is appreciably affecting catch rates in another.

14.   REFERENCES CITED

Anonymous. 1991. Report of Meeting. Third South Pacific albacore research (SPAR) workshop, 9-12 October 1990, Noumea, New Caledonia. *S.Pac.Comm.*, 41 p.

Anonymous. 1993. Report of Meeting. Fifth South Pacific albacore research (SPAR) workshop, 29 March - 1 April 1993, Papeete, Tahiti, French Polynesia. *S.Pac.Comm.* (In press)

Argue, A.W., F. Conand, and D. Whyman. 1983. Spatial and temporal distributions of juvenile tunas from stomachs of tunas caught by pole-and-line gear in the central and western Pacific Ocean. *Tech.Rep.Tuna Billfish Assess.Programme, S.Pac.Comm.*, (9):47 p.

Bailey, K.N. 1983. Food and feeding habits of fish associated with skipjack tuna in New Zealand. MSc (Hon.) Thesis, Victoria University of Wellington, xii + 206 p.

Bailey, K.N. 1986. A preliminary analysis of the stomach contents of albacore, *Thunnus alalunga*, from the Subtropical Convergence Zone east of New Zealand. Paper presented at the First South Pacific Albacore Research (SPAR) Workshop, 9-12 June 1986, Auckland, New Zealand. *S.Pac.Comm.*, WP/17.

Bailey, K. 1991. Observations on South Pacific albacore spawning as indicated by gonad indices. Paper presented at the Fourth South Pacific Albacore Reseach (SPAR) Workshop, 4-8 November 1991, Taipei, Taiwan. *S.Pac.Comm.*, WP/4.

Bailey, K.N., and G. Habib. 1982. Food of incidental fish species taken in the purse-seine skipjack fishery, 1976-81. *Occas.Publ.Fish.Res.Div.New Zealand Minist.Agri. Fish., Data Ser.*, (6):24 p.

Bailey, K., and A. Ross. 1987. The foreign longline catch of albacore in the New Zealand EEZ. *Catch* 14(6):20-2.

Coan Jr., A.L., and G.M. Rensink. 1991. U.S. troll fishery for albacore in the South Pacific. Paper presented at the Fourth South Pacific Albacore Research (SPAR) Workshop, 4-8 November 1991. *S.Pac.Comm.*, WP/11.

Hampton, J. 1990. Simulations of the South Pacific albacore population: effects of rapid developments in the surface fishery. Paper presented at the Third South Pacific Albacore Research (SPAR) Workshop, 9-12 October 1990. *S.Pac.Comm.*, IF/1.

Hampton, J., D.A. Fournier, and J.R. Sibert. 1990. MULTIFAN analysis of South Pacific albacore length-frequency data collected by observers, 1989-1990. Paper presented at the Third South Pacific Albacore Research (SPAR) Workshop, 9-12 October 1990. *S.Pac.Comm.*, WP/1.

Hampton, J., T. Murray, and K. Bailey. 1991. South Pacific albacore observer programme on troll vessels, 1989-1990. *Tech.Rep.Tuna Billfish Assess.Programme, S.P.Comm.*, (25):25 p.

Ichikawa, W., and T. Shirasawa. 1980. Report on 1979 tuna longline fishing investigations in higher latitudes of eastern Pacific. *JAMARC* (23).

Isii, K., and M. Inoue. 1956. Some notes on the ovary of albacore, *Germo germo*, taken from the Coral Sea. *Bull.Jap.Soc.Sci.Fish.*, 22:89-93.

Jones, J.B. 1991. Movements of albacore tuna (*Thunnus alalunga*) in the South Pacific: evidence from parasites. *Mar.Biol.*, 111:1-9.

Klawe, W. 1980. Classification of the tunas, mackerels, billfishes, and related species and their geographical distribution. *Spec.Rep.I-ATTC*, (2):5-16.

Labelle, M. 1991. Estimates of age and growth for South Pacific albacore. Paper presented at the Fourth South Pacific Albacore Research (SPAR) Workshop, 4-8 November 1991. *S.Pac.Comm.*, WP/5.

Laurs, R.M. 1986. U.S. albacore trolling exploration conducted in the South Pacific during February-March, 1986. *NOAA Tech.Memo.NMFS-SWFC, La Jolla,* (66).

Laurs, R.M., H.S.H. Yuen, and J.H. Johnson. 1977. Small-scale movements of albacore, *Thunnus alalunga*), in relation to ocean features as indicated by ultrasonic tracking and oceanographic sampling. *Fish.Bull.NOAA-NMFS,* (75):347-55.

Laurs, R.M., K. Bliss, J. Wetherall, and B. Nishimoto. 1987. South Pacific albacore fishery exploration conducted by U.S. jig boats during early 1987. *NOAA Admin.Rep.NMFS-SWFC, La Jolla,* LJ-87-22:31 p.

Lewis, A.D. 1990. South Pacific albacore stock structure: a review of available information. Paper presented at the Third South Pacific Albacore Research (SPAR) Workshop, 9-12 October 1990, Noumeau, New Caledonia. *S.Pac.Comm.,* WP/5.

Murray, T. 1986. South Pacific surface circulation during the austral summer. Paper presented at the First South Pacific Albacore Research (SPAR) Workshop, 9-12 June 1986, Auckland, New Zealand. *S.Pac.Comm.,* BP/2.

Murray, T., and K. Bailey. 1986. Preliminary report of RV Kaharoa cruises K03/86 and K05/86 - oceanography and albacore catch rates. Paper presented at the First South Pacific Albacore Research (SPAR) Workshop, 9-12 June 1986, Auckland, New Zealand. *S.Pac.Comm.,* WP/11.

Murray, T., and K. Bailey. 1989. Preliminary report on age determination of South Pacific albacore using caudal vertebrae. Paper presented at the Second South Pacific Albacore Research (SPAR) Workshop, 14-16 June 1989, Suva, Fiji. *S.Pac.Comm.,* WP/20.

Murray, T. 1990. Review of research and of recent developments in South Pacific albacore fisheries with emphasis on large-scale pelagic driftnet fishing. *FAO Fish.Rep.,* 434:52-77.

Nishikawa, Y., M. Honma, S. Ueyanagi, and S. Kikawa. 1985. Average distribution of larvae of oceanic species of Scombroid fishes, 1956-1981. *S Ser.Far Seas Fish.Res.Lab.,* (12):99 p.

Otsu, T., and R. Hansen. 1962. Sexual maturity and spawning of the albacore in the central South Pacific Ocean. *Fish.Bull.U.S.Fish Wildl.Serv.,* 62(204):151-62.

Polacheck, T. 1987. Atlas of longline catch rates by Japanese longliners in the Pacific Ocean - 1962 to 1985. *Tech.Rep.Tuna Billfish Assess.Programme, S.Pac.Comm.,* (21):109 p.

Ramon, D. 1991. Spawning of albacore, *Thunnus alalunga*, in the South Pacific Ocean. Paper presented at the Fourth South Pacific Albacore Research (SPAR) Workshop, 4-8 November 1991, Taipei, Taiwan. *S.Pac.Comm.,* WP/13.

Ratty, F., R. Kelly, and R.M. Laurs. 1989. Testes morphology, histology and spermatogenesis in South Pacific albacore tuna. Paper presented at the Second South Pacific Albacore Research (SPAR) Workshop, 14-16 June 1989, Suva, Fiji. *S.Pac.Comm.*, WP/19.

Roberts, P.E. 1974. Albacore off the north-west coast of New Zealand, February 1972. *New Zealand J.Mar.Freshwat.Res.*, 8:455-72.

Roberts, P.E. 1975. 1971-72 tuna survey, West Coast South Island. *Occas.Publ.Fish. Res.Div.New Zealand Minist.Agri.Fish.*, 8:16 p.

Roberts, P.E. 1980. Surface distribution of albacore tuna, *Thunnus alalunga* Bonnaterre, in relation to the Subtropical Convergence Zone east of New Zealand. *New Zealand J.Mar.Freshwat.Res.*, 14:373-80.

Roberts, P.E., and G.D. James. 1974. Juvenile albacore from New Zealand. *New Zealand J.Mar.Freshwat.Res.*, 8:437-40.

Saito, S. 1973. Studies on fishing of albacore, *Thunnus alalunga* (Bonnaterre) by experimental deep-sea tuna long-line. *Mem.Fac.Fish. Hokkaido Univ.*, 21:107-84.

Sharples, P., K. Bailey, P. Williams, and A. Allan. 1991. Report of observer activity on board JAMARC driftnet vessel R.V. Shinhoyo maru fishing for albacore in the South Pacific Ocean (22 November-23 December 1989 and 10 February-3 March 1990). *Tech.Rep.Tuna Billfish Assess.Programme, S.Pac.Comm.*, (24).

Skillman, R.A. 1975. An assessment of the South Pacific albacore, *Thunnus alalunga*, fishery, 1953-72. *Mar.Fish.Rev.*, 37(3):9-17.

Ueyanagi, S. 1969. Observations on the distribution of tuna larvae in the Indo-Pacific Ocean with emphasis on the delineation of the spawning areas of albacore, *Thunnus alalunga. Bull.Far Seas Fish.Res.Lab.*, (2):177-256.

Wang, C.-H. 1988. Seasonal changes of the distribution of South Pacific albacore based on Taiwan's tuna longline fisheries, 1971-1985. *Nat.Taiwan Univ.Sci.Rep.Acta Oceanogr.Taiwanica*, 20:13-40.

Wang, C.-H., M.-S. Chang, and M.-C. Lin. 1988. Estimating the maximum sustainable yield of South Pacific albacore, 1971-1985. *Nat.Taiwan Univ.Sci.Rep.Acta Oceanogr.Taiwanica*, 21:67-81.

Watanabe, Y., and Y. Nishikawa. 1990. A review of Japanese albacore fisheries in the South Pacific. Paper presented at the Third South Pacific Albacore Research (SPAR) Workshop, 9-12 October 1990, Noumea, New Caledonia. *S.Pac.Comm.*, WP/7.

Wetherall, J.A., and M.Y.Y. Yong. 1984. Assessment of the South Pacific albacore stock based on changes in catch rates of Taiwanese longliners and estimates of total annual yield from 1964 through 1982. *NOAA Admin.Rep.NMFS-SWFC, Honolulu*, H-84-11:7 p.

Wetherall, J.A., R.N. Nishimoto, and M.Y.Y. Yong. 1989. Age and growth of South Pacific albacore determined from daily otolith increments. Paper presented at the Second South Pacific Albacore Research (SPAR) Workshop, 14-16 June 1989, Suva, Fiji. *S.Pac.Comm.*, WP/18.

Wetherall, J.A., and M.Y.Y. Yong. 1989. Use of longline catch rate statistics to monitor the abundance of South Pacific albacore. Paper presented at the Second South Pacific Albacore Research (SPAR) Workshop, 14-16 June 1989, Suva, Fiji. *S.Pac.Comm.*, WP/11.

A REVIEW OF THE BIOLOGY AND FISHERIES FOR BIGEYE TUNA, *THUNNUS OBESUS*, IN THE PACIFIC OCEAN

Naozumi Miyabe
National Research Institute of Far Seas Fisheries
Shimizu-shi, Japan

1.   INTRODUCTION

This paper is prepared for the Expert Consultation on Interaction of Pacific Ocean Tuna Fisheries.   Thus the contents included here are limited to those related to Pacific bigeye tuna.   In compiling this paper, the "Synopsis of biological data on the bigeye tuna, *Thunnus obesus* (Lowe, 1839), in the Pacific Ocean" by Calkins (1980) and "Fishery biology of the bigeye tuna resource in the Pacific Ocean" by Kume (1979a) are major sources of reference.

2.   CLASSIFICATION

According to Nelson (1976), Genus *Thunnus* is classified as follows:

Phylum Chordata
    Superclass Gnathostomata
        Class Osteichthyes
            Division Euteleostei
                Superorder Acanthopterygii
                    Order Perciformes
                        Superfamily Scombroidae
                          Family Scombridae
                            Subfamily Scombrinae
                              Tribe Thunnini
                                Genus *Thunnus*

Bigeye tuna was first described by Lowe (1839) based on the specimen caught in the area of Madeira, Portugal (cited in Iwai *et al.*, 1965).   The scientific name given to this species varied considerably among the taxonomists until mid-1960s, since then *Thunnus obesus* has been generally accepted by the scientific community.

The classification of bigeye tuna and other tunas in Genus *Thunnus* is discussed by Iwai *et al.* (1965), Gibbs and Collette (1967), Sharp and Pirages (1978) and Collette and Nauen (1983).

3.   EARLY LIFE HISTORY

Currently the egg of this species cannot be differentiated from other tuna eggs. The fertilized egg is known to be pelagic and non-adhesive. Kikawa (1953) reported that the egg had an oil globule and the diameter of running ripe eggs was about 1 mm. In Yuen (1955) and Nikaido *et al.* (1991), the diameter of the most advanced eggs ranges between 0.8 to 1.2 mm. Yasutake *et al.* (1973) described larval development from hatching to 86 hours after hatching of bigeye tuna eggs artificially fertilized on board a research vessel. It is reported that it took 24 to 30 hours before the first hatching at 25.5-29.0°C. The size of larvae when hatched was 2.5 mm and after 26 hours the larvae reached 2.76-3.12 mm in total length. By 86 hours after hatching the early post-larval stage was attained. The development of the larvae in these stages is shown in Figure 1.

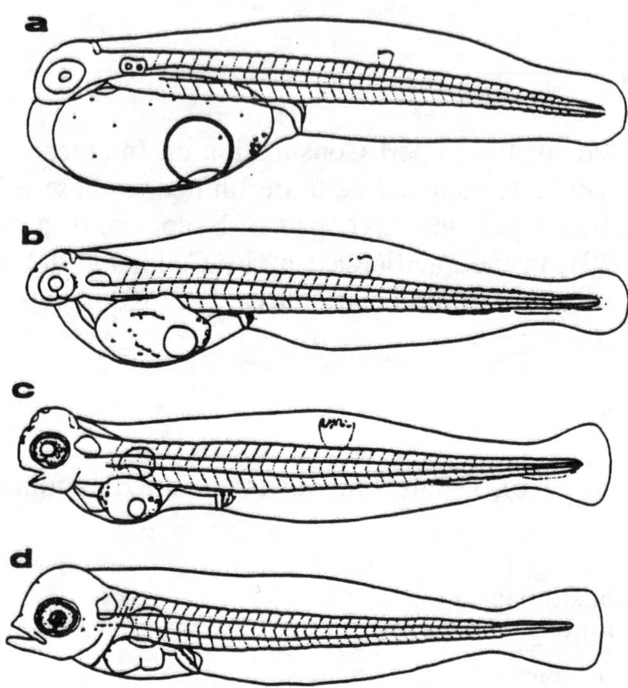

Figure 1.   Larvae of bigeye tuna: a) newly-hatched (2.5 mm in TL); b) 24 hours after hatching (3.0 mm in TL); c) 48 hours after hatching (3.1 mm in TL); d) 86 hours after hatching. (After Yasutake *et. al.*, 1973).

The diagnostic keys to the larvae (3 to 12 mm standard length) of Family Scombridae (including Genus *Thunnus*) based on the morphological characters is given in Nishikawa and Rimmer (1987). Pigmentation is the most important feature for the identification of tuna larvae. The identification of larvae and juveniles from 12 mm to about 60 mm is very difficult because of morphological similarities among the tuna species and the overlapping counts in the meristic characters. Graves *et al.* (1988) conducted an electrophoretic analysis on tuna larvae and early juveniles in an attempt to separate bigeye and yellowfin. Their samples were all identified as yellowfin in spite of the fact that their samples included larvae which have morphological characteristics for bigeye and yellowfin. This result raised a question on the validity of the pigmentation pattern which is currently used for the species separation between bigeye and yellowfin larvae.

Nishikawa *et al.* (1985) reported average tuna larvae distribution based on the total of 63,017 net tows during 1956 to 1981. According to their results, bigeye tuna larvae are distributed very widely in the equatorial area of the western, central and eastern Pacific. The area where they occur is very broad in the western Pacific, extending from off the south coast of Japan to the northern Coral Sea. In the eastern Pacific, there are several 1-degree squares marked with higher larval densities than in the western and central Pacific. It appears, however, that, despite the huge spawning potential of the Pacific bigeye tuna inferred from the studies of sexual activities, the number of their larvae per unit of water strained is generally smaller than expected, especially compared to other tuna larvae such as yellowfin. This difference may suggest a possible different characteristic of bigeye tuna larvae which makes them less vulnerable to the ordinary sampling procedure.

4.   AGE AND GROWTH

Yukinawa and Yabuta (1963) derived the growth curve based on scale samples. A total of 1,622 samples were collected from the central to western Pacific north of 10°S; of the total 463 were readable. In this study up to six rings were counted; the rings were reported to be formed twice a year, one in spring and the other in fall. Yukinawa and Yabuta (1963) reported that the rings were hard to detect for fish over 130 cm in fork length (FL).

Shomura and Keala (1963) estimated the growth curve by sex from the weight-frequency data of the fish unloaded by the Hawaiian longline fishery. They fitted normal distributions to those weight-frequency data with several assumptions, one of which was that fish entering the fishery in September at about 45 pounds were 17 months old. Sexual dimorphism was observed for fish over 130 cm in FL.

For the fish in the eastern Pacific, Kume and Joseph (1966) followed modal progressions of bigeye tuna taken by longline. Their data indicated that males and females grow at approximately the same rate up to 150 cm in length, but that males appear to grow faster at sizes greater than 150 cm. Therefore they did not use modes greater than 150 cm in their growth study.

Suda and Kume (1967) modified slightly the equation estimated by Yukinawa and Yabuta (1963) by taking observations on young bigeye into consideration.

Parameters of von Bertalanffy's growth equation from the above-mentioned papers are listed in Table 1. These growth curves were plotted in Figure 2 for comparison; the growth curve of Suda and Kume (1967) has been excluded since it is quite similar to that of Yukinawa and Yabuta (1963), particularly for 4 year old and older fish. It is noted that although the growth curves do differ from each other, the growth rates appears to be similar.

Length-weight relationship in the form of $W = a \cdot L^b$, where W is round weight in kg and L is fork length (FL) in cm, was estimated by Iversen (1955), Kume and Shiohama (1964), Nakamura and Uchiyama (1966) and Morita (1973). Samples taken in those studies were mostly from the western and central Pacific. Parameter estimates of a and b are listed in Table 2. As the original units of measurement were different among studies, equations are modified so as to use FL in cm and round weight in kg. For the conversion from gilled and gutted weight to round weight, the factor of 1.16 (Morita, 1973) was used.

Table 1.   Von Bertalanffy's growth parameters estimated for Pacific bigeye tuna. Length is fork length in cm and weight is live weight in kg.

Unit	Max. Size	k	t0	Author(s)
Length	215	0.10412	-0.01056	Yukinawa and Yabuta (1963)
Weight	234.7	0.114	1.07	Shomura and Keala (1963), Male
Weight	164.9	0.167	1.06	Shomura and Keala (1963), Female
Length	186.95	0.095	2.11	Kume and Joseph (1966)
Length	214.8	0.2066	-0.12052	Suda and Kume (1967)

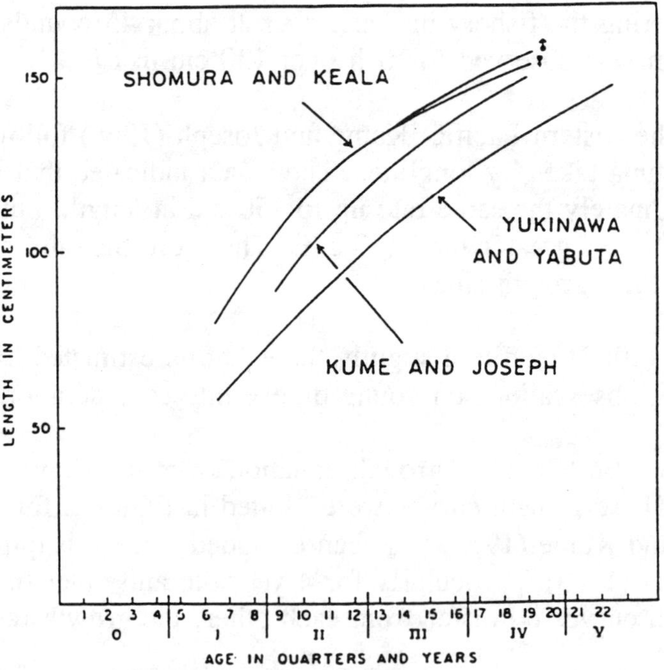

Figure 2.   Growth curves of bigeye tuna estimated by various authors. (After Kume and Joseph, 1966).

These results are very similar in the common size range up to 150 cm, although the difference becomes greater as the fish becomes larger. The maximum difference in length among equations is about 10% when the length is 200 cm; at 200 cm the fish is close to the maximum size reported for bigeye tuna.

Table 2. Length-weight relationship ($W = a\cdot L^b$) estimated for Pacific bigeye tuna. L is fork length in cm and W is live weight in kg.

a	b	Area in Pacific	Sample #	Author(s)
2.9499×10^{-5}	2.9304	Central	?	Iversen (1955)
1.3504×10^{-5}	3.1056	Western-north	4121	Kume and Shiohama (1964)[*]
1.7265×10^{-5}	3.0475	Western equatorial	2538	Kume and Shiohama (1964)[*]
3.3263×10^{-5}	2.9180	Central	1832	Kume and Shiohama (1964)[*]
3.6562×10^{-5}	2.90182	Central	9144	Nakamura and Uchiyama (1966)
1.9729×10^{-5}	3.0247	Western	481	Morita (1973)

[*] Gilled and gutted weight is converted to live weight by a factor of 1.16.

5. MATURATION AND SPAWNING (SEX RATIO)

5.1 Maturation

The minimum size at first maturity of Pacific bigeye tuna is reported as 91-100 cm by Kikawa (1953). Yuen (1955) also reported minimum size at first maturity, but his results were given in weight (14-20 kg). These two values are equivalent and are supported by the later studies (Kikawa 1957; 1961; 1962).

As regards the Gonosomatic Index ($GSI = 10^4 \cdot GW/L^3$; GW = gonad weight in g, L = FL in cm) of longline-caught bigeye tuna by latitude, Kikawa (1957) found the main spawning ground in the western Pacific in the area of the equatorial counter current (2°N-10°N). Kikawa (1961) investigated seasonal and areal change of GSI and size composition of longline-caught fish in the equatorial Pacific. Kikawa (1966) further extended his study with additional data extending as far east as 100°W. He reported that the size of fish, CPUE, and group maturity (rate of mature fish) increased from west to east. Mature fish were seen throughout the year in all areas. The seasonality of group maturity was recognized, but it appeared to be similar among areas. Maturity was highest in June and July in the waters west of 140°E, April to July between 140°E and 180°, April to September between 180° and 140°W and February to July between 140°W and 100°W. Kume and Joseph (1966) confirmed a similar trend of seasonality of group maturity for the longline-caught fish in the eastern Pacific, and reported that in the area between the equator and 5°S the proportion of mature fish was considerably less than in the adjacent areas to the north and south.

5.2 Spawning

Nikaido *et al.* (1991) examined histologically the gonad samples taken by longline southwest offshore of Hawaii during May to July and in the waters off Java in the Indian Ocean during January to March. They limited their study to fish which were alive when they were hauled on the deck in order to investigate the spawning time and the developmental process of gonad with time. Nikaido *et al.* (1991) noted that most of the fish (>100 cm in FL) were mature, *i.e.* in tertiary yolk stage or a more advanced stage of maturity. From the degeneration process of postovulatory follicles (POF), the developmental process of gonad and the change of GSI by time, bigeye was determined to be a multiple spawner. Out of the mature fish more than 90% were identified to have spawned within 24 hours. At the same time it was estimated that spawning took place approximately from 19:00 to midnight. Spawning frequency was calculated 1.00-1.57 (days per spawning) taking the inverse of the number of sample with POF divided by the total number of mature samples.

Batch fecundity (number of eggs spawned per day) was also estimated (Table 3) based on the number of most advanced eggs. The estimated batch fecundity was different between two areas studied. For example, in the area off Hawaii the batch fecundity was estimated at 2.2 million eggs for fish at 150 cm in FL, while the batch fecundity was 2.8 million eggs for the same size fish in the Java area. Currently it is not known whether this reflects a difference in stocks. These spawning figures are smaller than those estimated by Yuen (1955). It is thought that Yuen (1955) might have included eggs in pre-tertiary yolk stage in his calculation.

Table 3.   Batch fecundity (number of eggs spawned per day) by size. (After Nikaido *et al.*, 1991).

$$(\text{Unit}:10^6)$$

Fork length (cm)	Off Java	South-western offshore of Hawaii
100	0.56	0.40
110	0.83	0.63
120	1.17	0.86
130	1.61	1.20
140	2.16	1.64
150	2.85	2.19
160	3.69	2.87
170	4.70	3.69
180	5.90	4.69

Currently information is not available on the length of the spawning season, thus it is not known how long the spawning season lasts for a fish and how many times the same fish spawns during a spawning season.

Kikawa (1966) calculated the spawning potential of bigeye and yellowfin tunas in the equatorial Pacific by two-month intervals utilizing the following equation:

$$K = a \cdot p \cdot s \cdot D$$

where a = mean weight of ovary, p = mean group maturity, s = mean sex ratio, and D = mean hook rate weighted by area surveyed. It was pointed out that the annual spawning potential was much higher in the eastern Pacific (between 140°W and 100°W) than in the western and central Pacific; the eastern Pacific comprising roughly 70% of the total spawning potential of bigeye tuna. This was attributable to not only the high hook rate but also the larger size of fish and the higher rate of group maturity in the eastern Pacific. The seasonal difference in spawning potential, which was considered to reflect the group maturity, was observed with its maximum in April-May in the area north of the equator, and in February-March in the area south of the equator. The minimum spawning potential occurred in October-January in the area north of the equator and August-November in the area south of the equator. It is also indicated that the seasonal difference in spawning potential in the longitudinal direction is similar both north and south of the equator, respectively.

5.3   Sex Ratio

Kikawa (1966) and Kume (1969b) analysed sex ratio of longline-caught bigeye tuna from the broad area of equatorial Pacific. Both studies summarized the data by area and quarter-of-the-year; however, Kume (1969b) provided the data in smaller units. There was a general tendency of predominance of male fish over the entire size range encountered. The dominance of males became more prominent as the size increased. They reported no discernible seasonal change in sex ratio by size. However, Kume (1969b) noted that in the eastern equatorial Pacific the sex ratio was close to 0.5 where sea-surface temperature (SST) was low (<24°C) and fishes were found to be immature, whereas male predominated over female (less than 40 % were female) in the adjacent areas where sea surface temperature (SST) was high (>25°C). For fish of all sizes caught by longline in the Pacific north of 28°N between 140°E and 180° the sex ratio was almost equal to 0.5, although a predominance of males was observed among fish smaller than 130 cm in the area east of 180° (Kume 1969a).

6.   DISTRIBUTION, MIGRATION AND STOCK STRUCTURE

6.1   Distribution

Bigeye tuna inhabit the tropical to temperate waters of the Pacific Ocean. This species is found across the entire Pacific between northern Japan (45°N) and the north island of New Zealand (40°S) on the western side, and from about 40°N to 30°S on the eastern side (Figure 3). Significant bigeye catch has not been reported so far from the area along the coast of Mexico and Central America between about 10°N and 20°N. Isolated specimens have been reported from as far north as 47°10'N on the coast of North America (Radovich, 1961; Meehan, 1965).

6.2   Migration

Most of the information on this subject is obtained from the fisheries, *i.e.* through seasonal and areal change of fishing ground, catch, CPUE, size, *etc.*, assuming that these changes reflect the migration and/or movement of the fish.

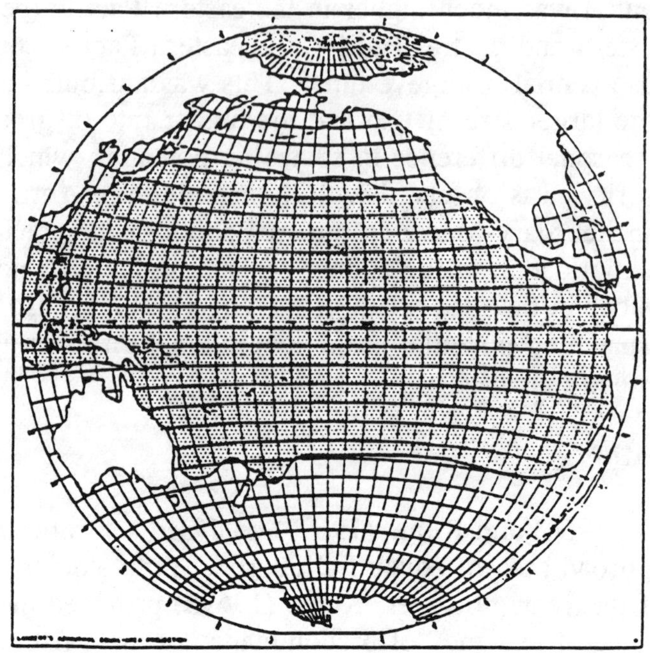

Figure 3.   Distribution of bigeye tuna in the Pacific. (After Calkins, 1980).

Kawasaki (1958; 1960) and Kume and Morita (1967) reported the seasonal shift of baitboat fishing grounds and the size of fish caught around Japan. The Japanese baitboat fishery starts operations in April in Izu-Bonin Islands waters. The fishing grounds move progressively northeastward and reach the Tohoku area (north of 35°N, 140-165°E) in July and August. Subsequently, the fishery turns south to southwest until the end of the fishing season. Similar north-south movements of the fishing grounds are also reported for the longline fishery in higher latitudes of the North Pacific along 30°N from 130°E to 120°W (Kume, 1967), waters around north island of New Zealand (Kume, 1967), and waters off Chile (Miyabe and Bayliff, 1987). Since the main component of the bigeye tuna caught by the Japanese baitboat fishery is small (45-80 cm in FL), it is considered that this represents the movement of young bigeye tuna around Japan (Honma and Kamimura, 1955; Kume and Morita, 1967).

In the North Pacific area along 30°N, the majority of fish caught by longline are between 90-140 cm with more fish prevailing in the smaller size of this range (Kamimura and Honma, 1953; Kume, 1969a). Similar-sized fish are also caught in the area off Chile as shown in Miyabe and Bayliff (1987). In both areas fishing takes place in their respective winter. On the other hand, the fishing grounds in the equatorial Pacific are formed throughout the year and the size of fish caught is larger (110-160 cm). It is a well-known fact that those fish caught in the tropical Pacific are sexually mature but those encountered in the temperate waters are immature. Based on these observations, many investigators hypothesized that the spawning-feeding migration between equatorial waters and temperate waters must be in existence since fish in the both areas have to be intermingled through spawning.

Tagging data are also available but the amount of information is scarce because of fewer releases and resultant fewer recaptures compared to other tunas such as yellowfin. In Table 4, tagged bigeye tuna which were recaptured after substantial time at liberty are listed. Although there is a tendency for tagged bigeye to stay in the vicinity where they were tagged, some fishes did exhibit long-distance movement. It is worth noting that two fishes tagged in the Coral Sea were recaptured in the central Pacific after more than 900 days at liberty. The sizes at recapture of those two fishes are 131 and 160 cm, respectively, and are the largest of those listed in Table 4. The movement of adult bigeye may be more extensive than previously considered.

Table 4. Recapture records of tagged bigeye tuna with long time at liberty. Length and distance are given in fork length (cm) and nautical miles, respectively. LL-longline; HL-handline. (Modified from Calkins, 1980).

Release			Recapture				Dis-tance	Days Free	Direc-tion (Deg.)	Reference
Area	Date	Length	Area	Date	Length	Gear				
30°59'N 171°14'W	Jan. 31, 1955	122.3	32°41'N 155°57'W	Nov. 24, 1955	126.8	LL	785	298	82	Otsu and Uchida (1956)
30°59'N 171°14'W	Jan. 31, 1955	109.0	29°50'N 177°50'W	Feb. 2, 1956	127.5	LL	348	368	259	Kume (1967)
32°59'N 143°19'E	May. 31, 1958	82	29°15'N 133°45'E	Nov. 11, 1959	110-115	LL	540	530	255	Kume (1967)
32°59'N 143°19'W	May. 31, 1958	81	35°27'N 141°10'W	Jan. 17, 1960	119	LL	182	597	324	Kume (1967)
3°18'N 90°50'W	May. 6, 1967	80	2°12'S 81°01'W	Jan. 13, 1968	?	?	664	253	117	IATTC un-published[1]
3°18'N 90°50'W	May. 12, 1967	50	4°32'N 107°50'W	Jun. 18, 1969	128.0	?	1,020	769	274	IATTC (1970)
1°01'S 157°18'E	Dec. 30, 1981	42	2°-4°N 152°-157°E	Feb. 15-Mar. 18, 1986	126	LL	200-700	1,508-1,539	330-360	FSFRL (1988)
15°10'S 146°22'E	Nov. 19, 1986	96	16°58'S 146°93'E	Oct. 29, 1987	112	HL	129	345	148	Peter Ward pers. comm.[2]
15°10'S 146°22'E	Nov. 19, 1986	109	7°17'S 155°67'W	Jun. 2, 1990	160	LL	3,408	1,292	89	Peter Ward pers. comm.
15°10'S 146°22'E	Nov. 19, 1986	108	16°43'S 146°82'E	Nov. 2, 1987	127	HL	110	349	148	Peter Ward pers. comm.
15°27'S 146°13'E	Nov. 21, 1986	78	3°42'S 171°32'W	Jun. 17, 1989	131	LL	2,591	940	79	Peter Ward pers. comm.
15°27'S 146°13'E	Nov. 21, 1986	98	16°42'S 146°72'E	Nov. 1, 1987	124	HL	94	346	143	Peter Ward pers. comm.

[1] Cited in Calkins (1980).
[2] Ward, Peter. Personal Communication. Fisheries Resources Branch. Bureau of Rural Resources. Australia.

6.3   Stock Structure

Except for Fujino and Kang (1968), there have been no other studies on the genetic differences among bigeye tuna population(s) in the Pacific. Fujino and Kang (1968) analyzed serum-esterase groups of bigeye tuna taken in the Hawaiian waters, and found two phenotypes in the sample. This study, however, is not sufficient enough to draw any conclusions regarding the population structure of Pacific bigeye tuna.

The stock structure of Pacific bigeye tuna has been inferred by indirect evidence such as geographical distribution of fish, CPUE and fish size, and maturity and spawning by area and time. Evidence that supports a single stock includes: (a) continuous distribution of bigeye throughout the Pacific, (b) similar size frequencies in neighbouring areas, (c) broad spawning area and time, (d) the concurrent appearance of a dominant year class throughout the North Pacific as observed by the length-frequency studies. On the other hand, the existence of subpopulation(s) is supported by the cline shown in the size of bigeye tuna in the east-west direction and the limited movement of tagged fish; both features suggest a low level of intermingling of fish.

In summary, it might be appropriate to repeat the following interpretation by Suda and Kume (1967) that "It is rather difficult to consider that the mixing of Pacific bigeye is active enough to ensure the unified stock in the whole Pacific. At the same time there is no clear evidence to support the existence of plural subpopulations. Presumably, it seems reasonable to assume a single population in the Pacific where mixing of the fish takes place gradually through the whole life history."

7.   NATURAL MORTALITY

Applying the Paloheimo method (Paloheimo, 1961), Suda and Kume (1967) estimated the instantaneous coefficient of natural mortality (M). Looking at the age composition of the fish taken by the longline fishery, they concluded that bigeye tuna are fully recruited to the fishery after age 4. Therefore the effective effort (f) and CPUE for age 5 and older fish during 1957-1964 were used, and fitted to the following equation assuming catchability (q) is constant among age and year for fully-recruited fish:

$$log_e S = q \cdot f + M$$

where S is survival rate calculated from the ratio of CPUE of the fish older than 4 years old in one year and that of older than 5 years old fish in next year. They obtained the annual coefficient of 0.361.

In general it is likely that natural mortality is high at the younger ages, lower at the intermediate ages and again higher at the older ages.

8.   OCEANOGRAPHIC FEATURES ASSOCIATED WITH SPECIES

In the developing stage of the Japanese longline fishery, Federation of Japan Tuna Fisheries Cooperative Association (1959) hypothesized that (a) tuna species distribute in different current systems, (b) even in the same species, tuna might have a habitat in the different current systems according to their developmental stages, and (c) the change of developmental stage and resultant change in tuna longline fishing grounds take place around March and September. Nakamura and Yamanaka (1959) also discussed the tuna distribution in relation to the ocean current and noted that bigeye tuna had a characteristic to aggregate around the boundary of currents. Yamanaka and Anraku (1962) discussed the relation between the distribution of tunas and water types in the central and western Pacific based on the T-S diagram.

Among oceanographic features associated with bigeye, water temperature is the most common feature that has been addressed by many workers to explain the spacial/temporal distribution of fish or fishing grounds. Uda (1957) cited the optimum water temperature (17.5°-22°C) for bigeye. This range of optimum water temperature for bigeye was supported by Suda *et al.* (1969) as coinciding with the temperature of the permanent thermocline, in which bigeye were considered to inhabit. In later studies (Saito and Sasaki, 1974; Saito, 1975; Hanamoto, 1976), it was found that bigeye were caught more efficiently in much deeper waters (133-245 m in waters southwest of Hawaii and 290-380 m in the western waters off Fiji) by the experimental vertical longline. In fact, the Japanese small-size class longliners introduced the so-called "deep longlining" to the western equatorial Pacific during the mid-1970s (Suzuki *et al.*, 1977) and it immediately prevailed among other boats and areas thereafter (Suzuki and Kume, 1981; Miyabe, 1989).

Hanamoto (1987) analyzed the oceanographic data (water temperature, dissolved oxygen, salinity) in relation to longline catch data. The latter included information on branch number, from which the depth at capture was estimated on the assumption that the main line formed a catenary curve. The results (Figure 4) showed the optimum water temperature to lie between 10° and 15°C which was much lower than that previously considered. Hanamoto (1987) stated that the same finding on water temperature could be applied to salinity, since, in the above-mentioned optimum temperature range, there was positive correlation between temperature and salinity. The optimum salinity was between 34.5 and 35.5 per mill in the South Pacific, between 34.0 and 34.7 per mill in the North Pacific, and between 34.7 and 35.2 per mill in the equatorial Pacific.

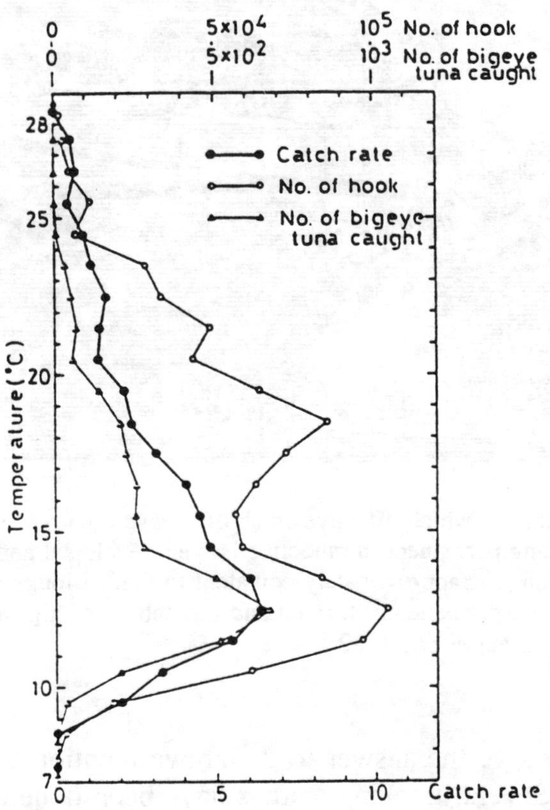

Figure 4.   The catch, catch rate of bigeye tuna and the number of hooks used by water temperature at the depth of capture. (After Hanamoto, 1987).

As shown by Sund *et al.* (1980) and Sharp (1978), the minimum dissolved-oxygen requirements for bigeye are considered to be 0.5-1.0 ml/l. Hanamoto (1987) constructed the diagram showing the depth of the 1 ml/l surface of dissolved oxygen in the Pacific, and found that the poor catch of bigeye based on the long-term average of longline catch coincided with the shallow depth area (about 100 m) of the 1 ml/l surface. These poor catches were located in the eastern Pacific between 10°N and 20°N and in the area off Chile to 120°W in the south of equator.

It has been mentioned from fishing experience that the swimming depth of bigeye is deeper than that of other tuna species. Suda *et al.* (1969) compared the areas of high hook rate by longline with the depth of thermocline and found a significant correspondence between areas of high hook rate and the location of thermocline at the depth of 100-150 m, which was identical to the depth of hook set at the longline operation (Figure 5). Taking these into consideration, Suda *et al.* (1969) postulated that bigeye inhabit waters in or just below the thermocline. It is thought this assumption could explain to some extent the observed west-east cline in both size of fish and CPUE. If the assumption holds, ordinary longline gear is less effective and tends to catch smaller fish since hooks do not reach the thermocline in the western Pacific where larger fish are supposed to inhabit. Hanamoto (1987) also discussed the effect of oceanographic condition on the longline catch along this line. He suggested the possible distribution of this species in the area where the optimum water temperature was located. Those suggested areas were located in the middle latitudes along 20° in both hemispheres and high latitudes between 30°S and 40°S in the South Pacific.

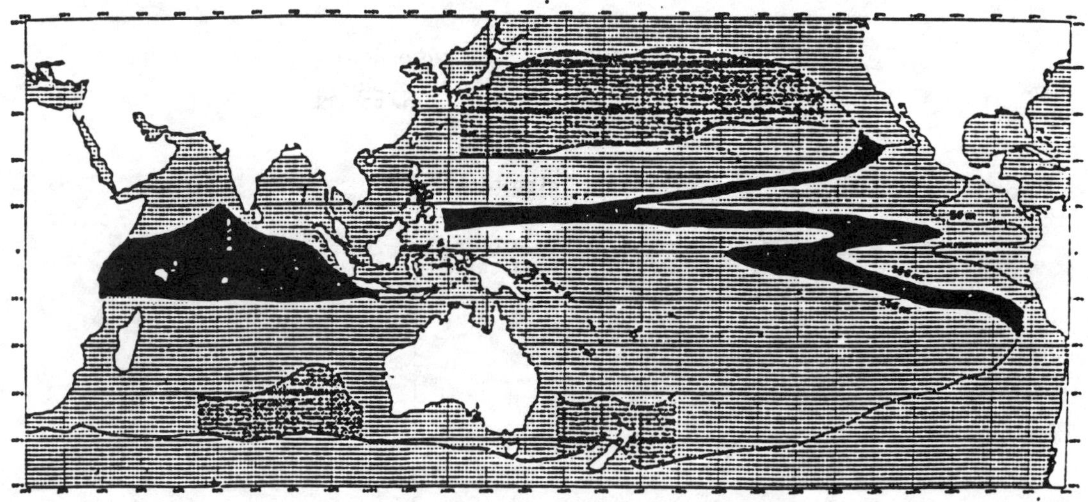

Figure 5.   Illustration of the area in which effective catch of bigeye tuna by longline is supposedly expected. Shaded area indicating permanent thermocline is well developed and located at depth between 100 and 150 m. This range is approximately eqivalent to that of longline hooks. Dotted area shows that thermocline is not permanently formed and the habitat of bigeye tuna ranges from surface to some depth. (After Suda *et al.*, 1969).

Sonic tracking may give the answer to the above-mentioned hypotheses on the behavior of bigeye tuna. Several tracking studies have been done on this species, however, most tracks have been on small fish less than 70 cm. One exception was a 70 kg bigeye which was tracked by Carey and Lawson (1973) off the Atlantic coast of

Canada. Generally speaking, although the results were different among fishes tracked, the swimming depth ranged from the surface to 300 m and fish tended to stay in deeper waters during the day than during the night (Holland *et al.*, 1990; Koido and Miyabe, 1990).

The effect of large-scale oceanographic change on bigeye and its fishery in the eastern Pacific, known as El-Niño, was briefly analyzed after the very strong 1982-1983 event. Kume and Miyabe (1987) compared the location of the Japanese longline fishing grounds in the eastern tropical Pacific during El-Niño with that during normal conditions. It was reported that, during El-Niño periods, good fishing grounds were formed around the equator between 95°W and 130°W; fishing seldom took place in these areas during normal conditions. The possible influence of the El-Niño event on bigeye tuna stock was studied by Kiyota *et al.* (1988) based on CPUE and cohort analysis. It was suggested that cohorts born in the El-Niño year or one year before El-Niño were larger and that CPUEs were higher during El-Niño than during normal conditions. These effects might be caused by the changes of oceanographic condition, such as water temperature and the depth of thermocline, although the mechanism involved in the process is not well known.

9. INTERACTION WITH OTHER SPECIES

Based mainly on observations during fishing, many authors report that bigeye tuna are mixed with yellowfin, skipjack, kawakawa, and frigate tuna, particularly when they are young (<100 cm). Around Japan, bigeye are caught by baitboat and purse-seine fisheries in pure or mixed school with yellowfin and/or skipjack. It is known bigeye tuna is seldom caught with albacore in the same operation. In the Coral Sea, small to medium bigeye (50-120 cm) as well as yellowfin are caught by handlining by the Japanese longliners during October to December (Hisada, 1973). It is reported that the types of school sought after are shark-associated, and, to a lesser extent, free-swimming schools with birds.

Recently the bigeye catches by the industrialized purse-seine fisheries in the western and eastern Pacific (Table 10) have accounted for about 5-10% and 0.2-5.0% of the total tuna catch by weight for the two areas, respectively. The former value is a rough estimate because purse-seine catch data of bigeye tuna for Korea, Philippines, Taiwan, USA, *etc.*, are lacking.

10. GENERAL DESCRIPTION OF FISHERIES

The longline fishery has landed the largest share of bigeye tuna catch since the initiation of the fishery in the early 1950s. The baitboat fishery has also accounted for a significant share of the bigeye tuna catch and has a long history as well. The purse-seine fishery, which was developed in later years (during the early 1960s in the eastern Pacific and during the mid-1970s in the western Pacific) and has replaced the baitboat fishery thereafter, has landed a substantial amount of the bigeye catch. Other local fisheries which catch small amounts of bigeye tuna exist in the coastal countries throughout the Pacific; these fisheries include trolling, hand-lining and small-scale purse seine.

10.1 <u>Longline Fishery</u>

Japan, Korea and Taiwan are the major countries catching bigeye tuna by longline gear. Some very minor catches are recorded by Australia, New Caledonia and Tonga. The size of distant-water longliners now operating ranges from 50 to 1000 gross tonnages (GT), and is mostly between 200 and 500 GT. In Japan and Taiwan there have been coastal to offshore tuna-longline fisheries carried on by the small-size vessels up to 50 GT.

10.1.1 Japan

The Japanese longliners operate in almost the entire Pacific; the principal exception is the central part of the South Pacific Ocean, south of 20°S. Fishing grounds are located in both tropical (10°N-15°S) and temperate (around 30°N and 30°S) waters extending in an east-west direction (Figure 6). In tropical waters the fishing takes place almost year around while in temperate waters it takes place in winter. Bigeye tuna is the most important tuna species for the Japanese tuna fishery in terms of the catch in weight (except skipjack) and value (Ministry of Agriculture, Forestry and Fishery of Japan, 1990). Bigeye tuna is the species most frequently aimed at and there are by-catches of yellowfin, albacore and billfishes. Catches by distant-water longliners are deeply frozen to below -60°C, and consumed as sashimi in the domestic market. In this fishery a limited-entry system has been adopted and vessels larger than 20 GT are required to obtain a permit from the Government. All longliners which have the permit are not allowed to unload catches in foreign countries except as transshipment to Japan; the total amount of transshipment by longliners is limited by the Government. The numbers of longliners fishing since 1977 are shown in Table 5. Smaller-sized boats (20-100 GT) are decreasing, while numbers of 100 to 200 GT class have doubled during that period. The number of longliners in the largest size class has been stable.

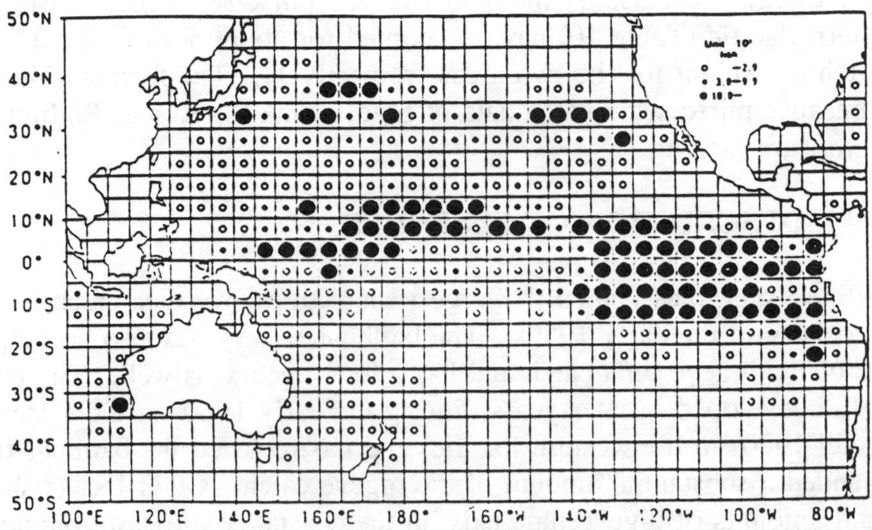

Figure 6.   Distribution of bigeye catch (in number) for Japanese longline fishery by 5-degree square in 1987. (After Miyabe, 1991).

Table 5.   Annual change in the number of tuna longliners by country, 1977-1989.

	Japan					Korea (100-500 ton)			Taiwan (50-1000 ton)			
	20-50 ton	50-100 ton	100-200 ton	200< ton	Total	Home based	Foreign based	Total	Pacific based	Takao based	Other	Total
1977	86	658	72	612	1,428	214	287	501	319		305	624
1978	87	707	69	617	1,480	216	266	482	302		313	615
1979	69	720	82	624	1,495	217	234	451	157	185	297	639
1980	57	715	103	645	1,520	219	220	439				
1981	55	706	100	661	1,522	208	199	407	146	204	289	639
1982	43	634	90	589	1,356	185	141	326	115	39	340	494
1983	38	593	89	550	1,270	169	101	270	65	88	298	451
1984	32	546	100	610	1,288	157	68	225	61	111	266	438
1985	28	534	109	628	1,299	156	64	220	44	72	307	423
1986	25	471	132	632	1,260	167	61	228	51	28	343	422
1987	23	398	147	649	1,217	189	61	250	60	39	308	407
1988	21	368	154	649	1,192	199	69	268	62	35	307	404
1989	20	334	152	653	1,159							

Data source
   Japan : MAFFJ (1978-1990).
   Korea and Taiwan : FAJ (1989).

10.1.2   Korea

The geographical distribution of fishing effort is shown in Figure 7. The main fishing ground is located in the South Pacific (0°-30°S, 160°E-120°W) and bigeye tuna is the target species. Less extensive operations are conducted in the North Pacific around Hawaii in winter. The number of longliners has decreased from 501 in 1977 to 268 in 1988 (Table 5).

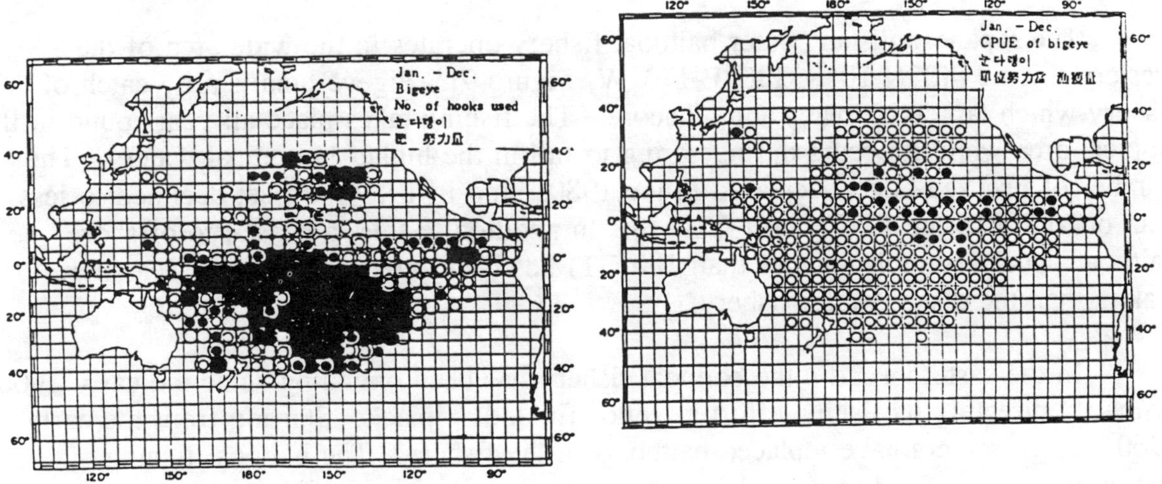

Figure 7.   Distribution of fishing effort and bigeye CPUE for 1981 by Korean longline fishery. (After NFRDA, 1986).

10.1.3   Taiwan

Fishing grounds for Taiwanese longliners are similar to those of Korean longliners, but the centre is located from the central to western South Pacific (Figure 8). Seasonal change of fishing ground is clear for the fleet fishing in the southern part (20°S-40°S) during winter in the southern hemisphere (TRC, 1983). This is because albacore is the target rather than bigeye tuna. The number of longliners has also declined; in particular the number of Pacific-based boats declined from 319 in 1977 to 62 in 1988 (Table 5).

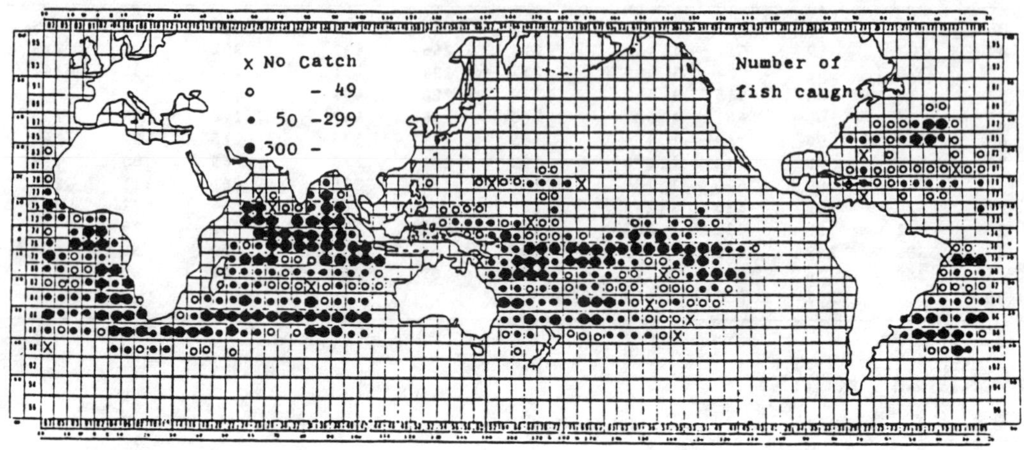

Figure 8. Distribution of bigeye catch (in number) for 1981 by 5-degree square for the Taiwanese longline fishery. (After TRC, 1982).

10.2 Baitboat Fishery

The Japanese distant-water baitboat fishery operates in the wide area of the western Pacific (40°N-10°S, 120°E-170°W, Figure 9). Bigeye tuna is a by-catch of this fishery which targets skipjack and albacore. The fishing takes place all year round in the tropical area while it occurs during spring to fall in the higher latitudinal waters. The number of licensed boats decreased in the 1980s, and in 1989 the fleet declined to less than one-third of the 1977 total (Table 6). In addition to this, a coastal to offshore baitboat fishery (vessels smaller than 100 GT) has been operating in more nearshore waters than the distant-water fishery.

In the eastern Pacific the baitboat fishery has been operating since the early 1900s primarily directing its fishing effort at yellowfin and skipjack. Starting from the early 1960s, purse seiners have replaced baitboats in the fishery. The baitboat fishing operations are confined to coastal waters and to the vicinity of offshore islands.

10.3 Purse-seine Fishery

There are two major fishing areas for this fishery, one in the western Pacific and the other in the eastern Pacific. In this fishery bigeye is a by-catch as in the baitboat fishery.

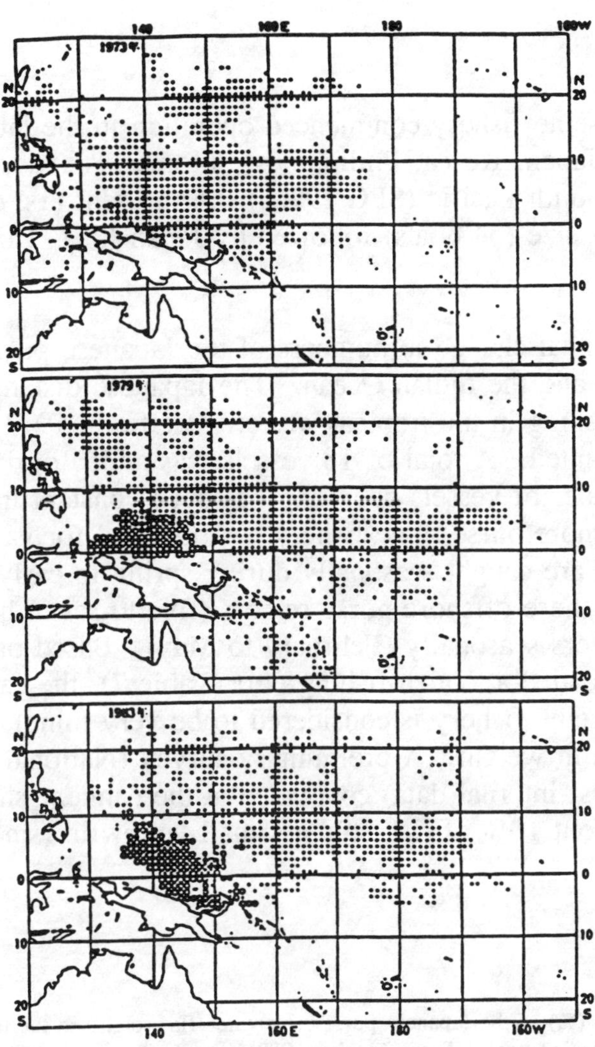

Figure 9.   The fishing ground of the Japanese baitboat and purse-seine fisheries in the tropical Pacific. Large and small circles show purse-seine and baitboat fishing areas, respectively. (After Tanaka, 1989).

Table 6.   Annual change in the number of fishing vessels for the Japanese distant-water baitboat and purse-seine fisheries, 1977-1989.

	Baitboat Distant-water fishery				Purse-seine Tropical fishery		
	<100ton	<200ton	>=200ton	Total	<200ton	>200ton	Total
1977	19	14	260	293	50	14	64
1978	24	10	251	285	47	14	61
1979	29	13	228	270	46	17	63
1980	32	10	198	240	50	16	66
1981	31	6	179	216	50	23	73
1982	35	6	138	179	52	33	85
1983	32	9	116	157	59	36	95
1984	27	10	105	142	54	33	87
1985	25	9	95	129	47	35	82
1986	20	9	91	120	53	38	91
1987	17	9	89	115	47	34	81
1988	16	11	70	97	48	39	87
1989	15	12	67	94	43	37	80

Data source : MAFFJ (1978-1990).

10.3.1   Western Pacific

The industrial purse-seine fishery commenced operation in the late 1970s, and by the mid-1980s vessels from Japan, Korea, Philippines, Solomon islands, Taiwan and the USA were operating in the South Pacific (SPC 1990a). At present (1st quarter of 1990), the USA has the largest fleet size (34 boats) followed by Japan (31), Taiwan (25), and Korea (8) (SPC 1990b).

Table 6 shows the annual change in numbers of the Japanese purse seiners operating both in the Pacific and the Indian Ocean. The Japanese distant-water purse-seine fishery started fishing in the tropical western Pacific in 1973 (Tanaka, 1989). In 1980, 14 vessels were counted. A total of 18 vessels was added during the following 2 years and thereafter the number of vessels operating has been constant in the Pacific. In the vicinity of Japan the offshore purse-seine fishery has operated for various pelagic fish species. Tunas and skipjack are caught seasonally during spring to early fall in the Pacific off Japan. A few of these offshore purse seiners (five to seven boats) are allowed to operate in the tropical waters seasonally (February to May). Based on the species composition of the catch unloaded at Yaizu fishing port (Table 7), the catch of bigeye tuna by the industrial purse-seine fishery is considered to be very minor (0.8-4.2% of total catch) in terms of catch in weight. A preliminary survey (National Research Institute of Far Seas Fisheries, internal data) conducted at the cannery showed that, in terms of numbers of fish, about 10% of fish sold as small yellowfin (smaller than 2 kg) were identified as bigeye.

Table 7.   Species composition (%) of the Japanese purse-seine catch (in weight) in the tropical Pacific unloaded at Yaizu fishing port. (From Tanaka, 1989).

Year	Skipjack	Yellowfin[1] large	Yellowfin[1] small	Bigeye[2]	Juvenile[3]
1976	71.49	24.35		4.16	0.00
1977	68.48	28.19		3.33	0.00
1978	73.24	13.87	9.28	3.61	0.00
1979	65.62	12.71	18.54	3.06	0.07
1980	71.32	12.88	13.15	2.55	0.10
1981	57.71	9.35	28.82	3.38	0.74
1982	65.93	16.31	14.72	2.36	0.68
1983	77.79	6.04	14.11	1.75	0.31
1984	72.55	12.23	14.44	0.78	0.00
1985	70.82	13.81	14.06	1.31	0.00

[1] Large and small yellowfin are larger than 10 kg or less than 10 kg, respectively.
[2] Over 2 kg in round weight.
[3] This category includes juvenile yellowfin and bigeye (less than 2 kg) as well as small amount of damaged fish in other categories.

10.3.2   Eastern Pacific

Table 8 shows the numbers of vessels and capacities in the eastern Pacific purse-seine fleet during 1950-1988. In 1987 the fleet was composed of vessels from 10 countries. Major countries were the USA (54 vessels), Mexico (54), Ecuador (28), and Venezuela (25) (IATTC,1989). In the early 1960s the purse-seine fishery replaced the baitboat fishery and to date has been the dominant fishery. The fishing grounds were confined to the coastal area in the early stage of the fishery, but it began to expand into the offshore areas in the mid-1960s. By the mid-1970s the fishery covered the coastal waters between about 30°N to 15°S and extended offshore to about 145°W between 5°N and 15°N (Calkins *et al.*, 1988). The geographical distribution of catch by the surface fishery during 1982-1986 is shown in Figure 10. The most concentrated area of catch was adjacent to the coast of South America from the equator to about 7°S and from the coast to about 87°W. The geographical catch distribution during 1967-81 was similar to that during 1982-86. The catch was small (less than 3,000 short ton (ST)) until 1974 (Table 10) when it jumped to over 4,000 ST). During 1976-81 it was the highest (8,000-17,000 ST) and then decreased to 2,000-6,000 ST during 1982-86.

The length distributions for 1975-1986 show that the length for the fish in nearshore areas ranged from 30 to 180 cm. In general, most of them occurred between 60 to 120 cm with major modes between 80-110 cm (Figure 11). The size range in the offshore area is about the same as in the nearshore area, but more of the catch appears to fall in the 30 to 70 cm range.

11.   TRENDS IN FISHING EFFORT, CATCH, AND CATCH-PER-UNIT OF FISHING EFFORT

11.1 Trends in Fishing Effort

Not all the data from the various fisheries are available. Information which is available is the number of fishing vessels operated or registered (Tables 5, 6 and 8). The annual change in fishing effort by the Japanese distant-water longline fishery in terms of the number of nominal hooks exerted is shown in Figure 12. The Japanese distant-water longline fishery is the largest component among fisheries for bigeye tuna. The nominal effort increased dramatically from 1952 to 1963 when it reached a high of 3.4×10^8 hooks. Thereafter it fluctuated between 2.5 and 3.0×10^8 hooks up until the mid-1970s, and then went up to a record high in 1981. Since then it has stayed at the highest level at around 3.5×10^8 hooks. Because of the limited-entry system adopted for this fishery, the recent change in the fishing effort is attributed to the increase in the number of hooks set per operation (day), and to the movement of the fleet into the Pacific Ocean or to other Oceans.

11.2 Trends in Catch

The annual catch of Pacific bigeye tuna is shown in Table 9 by country (FAO, 1965-1988), and in Table 10 by major gear, and in Figure 13. There is some difference between these two statistics (Tables 9 and 10) reflecting the difference in the source of data. Generally the catch shown in Table 10 is slightly larger except for 1987. As the Japanese longline catch accounted for over 70% and over 80% of the total catch and of

Table 8.   Numbers and carrying capacities, in short ton, of vessels of the eastern Pacific tuna fleet. Bolicheras are small purse seiners with limited ranges. There is no information available on bolicheras or trollers for 1950-1954. The 1988 data are preliminary. (After IATTC, 1989).

Year	Seiners		Baitboats		Bolicheras		Trollers		Total	
	No.	Tons	No.	Tons	No.	Tons	No.	Tons	No.	Tons
1950	67	7,890	204	39,967	-	-	-	-	271	47857
1951	78	8,731	255	44,160	-	-	-	-	303	52891
1952	64	7,371	202	40,631	-	-	-	-	266	48002
1953	64	7,508	191	42,895	-	-	-	-	255	50,403
1954	69	7,960	182	40,647	-	-	-	-	251	48,607
1955	65	7,880	183	41,729	15	375	0	0	263	49,984
1956	66	7,999	182	41,425	55	1,375	0	0	303	50,799
1957	55	7,019	193	40,785	40	1,000	0	0	288	48,804
1958	49	6,614	180	39,220	29	725	0	0	258	46,559
1959	87	12,224	185	36,066	18	450	5	98	295	48,838
1960	112	22,806	117	16,820	15	375	0	0	244	40,001
1961	124	30,011	93	10,510	1	25	0	0	218	40,546
1962	130	33,945	89	6,725	27	675	0	0	246	41,345
1963	141	39,834	108	5,964	18	450	3	55	270	46,303
1964	134	40,307	88	4,712	3	65	0	0	225	45,084
1965	146	42,283	109	5,777	17	395	7	182	279	48,637
1966	126	39,869	113	6,217	7	140	2	29	248	46,255
1967	122	40,221	108	5,862	8	170	0	0	238	46,253
1968	139	50,613	89	5,743	4	100	2	24	234	56,480
1969	149	57,008	69	4,957	4	95	3	71	225	62,131
1970	162	67,508	49	4,302	0	0	9	177	220	71,987
1971	185	88,770	102	5,569	6	150	66	1,514	359	96,003
1972	206	112,361	108	6,707	4	100	74	1,946	392	121,114
1973	216	131,910	106	6,856	3	75	28	729	253	139,570
1974	230	146,990	111	7,766	4	110	7	150	352	155,016
1975	249	163,766	102	7,403	4	105	9	181	364	171,455
1976	250	176,469	99	7,071	4	115	38	909	391	184,564
1977	250	178,813	79	5,436	3	87	37	953	371	185,289
1978	262	180,781	68	5,044	9	272	50	1,303	389	187,400
1979	268	183,673	45	3,979	14	405	5	112	332	188,169
1980	258	184,647	46	3,838	12	355	4	106	320	188,840
1981	247	183,729	39	3,063	3	72	2	55	291	186,919
1982	221	167,780	36	2,713	2	60	4	131	263	170,684
1983	199	137,842	52	3,470	12	300	8	244	271	141,856
1984	165	113,168	40	3,055	0	0	9	232	214	116,455
1985	175	127,272	25	2,424	0	0	1	14	201	129,710
1986	165	122,564	17	1,939	0	0	0	0	183	124,503
1987	177	143,765	29	2,237	0	0	0	0	206	146,002
1988	182	147,931	36	3,081	0	0	3	70	221	151,082

the total longline catch, respectively, the trend of total catch is similar to that of the Japanese longline fishery (Table 10). Before 1957 the annual total catch was less than 50,000 mt. It increased to between 82,000 and 92,000 mt during 1958-1960, and then showed a sudden increase in 1961 to 136,000 mt. After a high of about 150,000 mt in 1963, it remained at a lower level between 73,000 mt and 100,000 mt up until 1974 except 1969 and 1973. Since then it has reached a higher level of over 100,000 mt. The catch was higher during 1976-1980 and 1985-1987.

It should be noted that in the western Pacific there has been no catch record of this species in the FAO statistics from countries such as the Philippines and Indonesia where substantial catch was reportedly made. Several possible reasons for this may be considered. One is lack of a system whereby catch may be estimated. The other is that species separation between tunas, especially for the small fish, is often difficult because of the similarity of their morphological appearances and because, due to the nature of the catch, bigeye tuna sometimes get caught with other tunas at the same time. In the near future this should be clarified and catches from these countries included in the statistics.

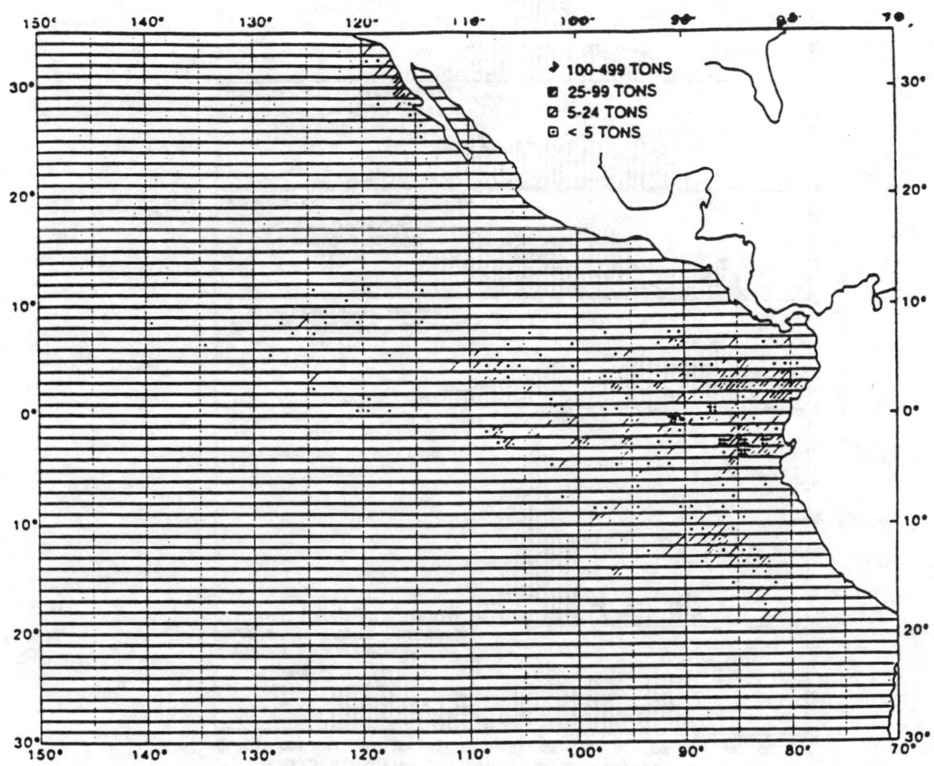

Figure 10.   Annual average of distribution of bigeye catch by the eastern Pacific surface fishery, 1982-1986. (After Calkins *et al.*, 1988).

11.3   Trends in Catch-per-Unit of Fishing Effort (CPUE)

The series of studies (Suda and Schaefer, 1965; Kume and Schaefer, 1966; Kume and Joseph, 1969; Shingu *et al.*, 1974; Miyabe and Bayliff, 1987) on the Japanese longline fishery in the eastern Pacific analyzed the relationship between nominal fishing effort and catch (CPUE). It is shown in Figure 14 for the equatorial area east of 150°W. From this figure three periods can be separated, *i.e.* 1957-1961, 1962-1964 and 1965-1980. During 1957-1961 CPUE was high at about 3 fish per 100 hooks. In the following period in 1962-1964 it quickly decreased to less than half of this. Thereafter, it has been very constant at slightly less than one fish per 100 hooks.

Miyabe (1991) estimated the standardized CPUE for the total Pacific stock, applying the Honma method (Honma, 1974) based on the Japanese longline catch and effort data. Deep longlining, a newly-introduced longline gear setting, was also adjusted to the conventional one employing the ratio of CPUEs from both gear settings and the proportion of fishing effort deployed by deep longline. The results showed that CPUE decreased from the late-1950s to the mid-1960s, but after that it fluctuated moderately without any appreciable trend until 1988 (Figure 15). The current level of CPUE is about 45% of the initial level in the 1950s.

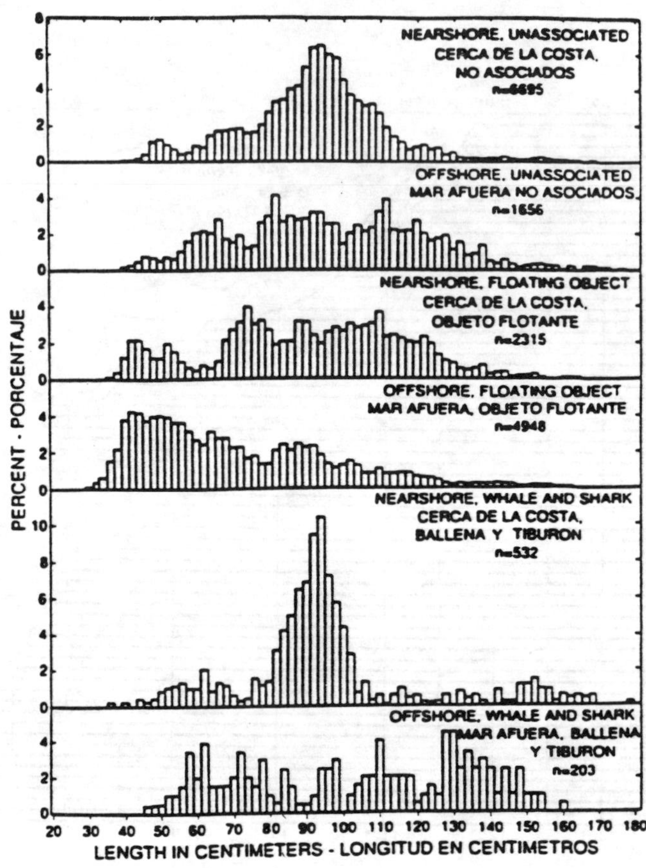

Figure 11.   Combined annual length-frequency distributions for 1975-1986, by school type and area, of bigeye tuna caught by the surface fishery in the eastern Pacific. (After Calkins, *et al.*, 1988).

Kume (1979a) presented CPUE by age and area for 1955-1976 (Figure 16). Nominal CPUEs by area and 5-year intervals averaged for 2 succeeding years were constructed into age-specific CPUEs applying the length frequency data. It is observed that CPUE decreased greatly between 1960-61 and 1965-66 for fish older than 3 years in all areas. After that, CPUE for older age group is generally decreasing although the trend is not so clear as before.

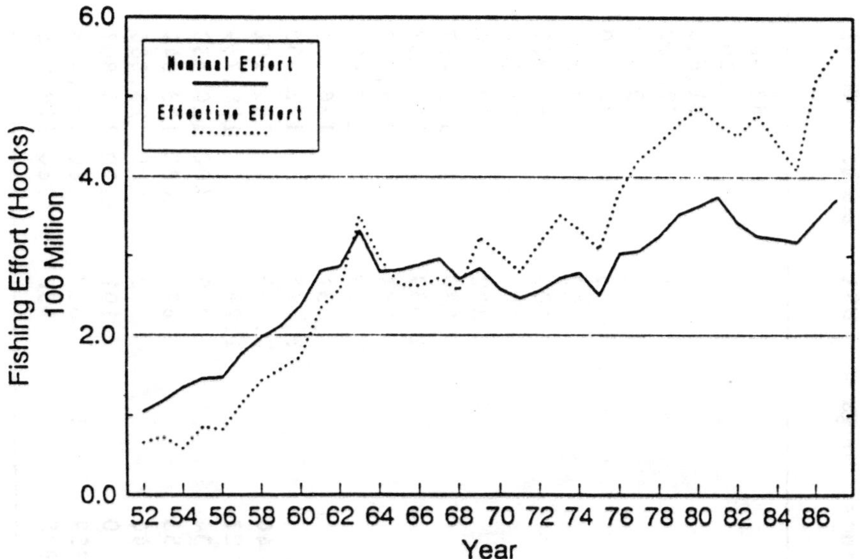

Figure 12.   Annual trend of fishing effort by Japanese longline fishery. (After Miyabe, 1991).

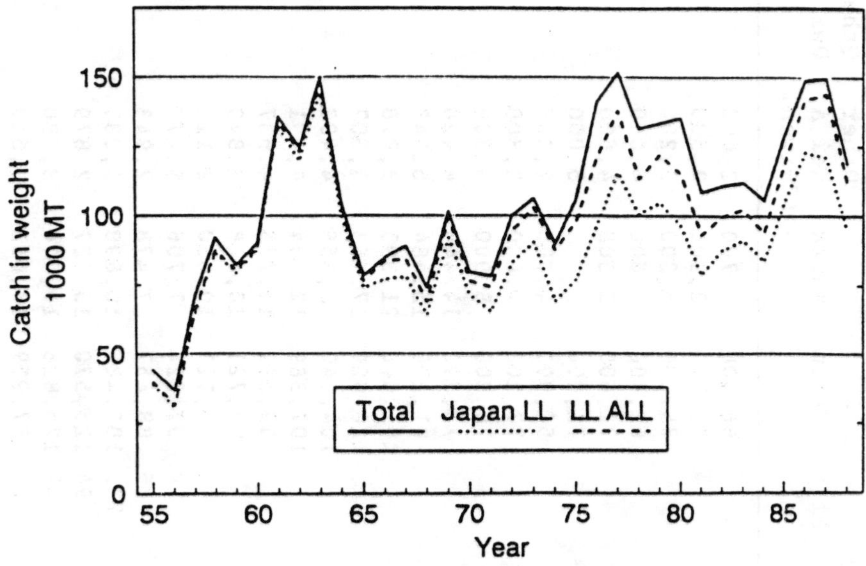

Figure 13.   Annual catch trend of Pacific bigeye tuna.

12.   POPULATION DYNAMICS

Because of the shortage of data essential for the comprehensive analysis of Pacific bigeye tuna, studies done on this subject so far are almost exclusively based on the Japanese longline data.

Table 9. Annual catch (MT) of Pacific bigeye tuna by country.

Year	Bermuda	Cuba	Ecuador	Fiji	Japan	Korea	Other Nei A (Taiwan)	Other Nei B	Panama	Solomon Islands	Tonga	USA	Venezuela	Pacific Total
1965					66,200	700	2,000							68,900
1966					70,700	2,900	3,500							77,100
1967					75,200	3,200	3,200							81,600
1968					62,400	600	4,000							67,000
1969					72,600	2,500	4,600							79,700
1970					71,000		5,000					0		76,000
1971		100			57,900	4,700	4,300					800		67,800
1972					77,200	7,800	3,800					100		88,900
1973					76,300	8,900	3,700					400		89,300
1974					70,392	14,444	4,420					132		89,388
1975	85				81,170	15,484	5,348					330		102,417
1976	0				101,040	21,395	3,078					1,039		126,552
1977	307				120,929	17,663	4,507					581		143,987
1978	0				104,640	8,456	4,402					423		117,921
1979	0				107,389	12,804	4,491					1,331		126,015
1980	0				99,692	13,975	4,637		1,465			3,196		122,965
1981	0				83,721	10,608	3,849		770	40		2,113		101,101
1982	0		1,100	8	94,113	10,050	2,111		177	23		1,207		108,789
1983	0		1,249	14	97,224	7,706	3,477			34		726	400	110,830
1984	0		1,814	16	88,867	7,478	2,943			55		696	1680	103,549
1985	0		2,410	133	106,486	10,898	3,031			46		62	820	123,886
1986	0		1,116	94	125,570	15,927	2,879			0		101	1120	146,807
1987	0		240	49	125,816	19,544	3,280	130		259	14	867	260	150,459
1988	0		240	27	87,959	13,681	3,610	130		1,085	7	1,956	260	108,955

Data source : FAO (1965-1988).

Table 10. Annual catch (MT) of Pacific bigeye tuna by fishing gear, 1955-1988. LL=longline, BB=baitboat, PS=purse seine.

| | Longline | | | | | Surface fishery | | | | | | Grand | % of |
| | | | | Sub- | % of | Japan | | | Solomon | IATTC | Sub- | | |
Year	Japan	Korea	Taiwan	Total	Japan LL	BB	Trop. PS	Others	PS+BB	PS+BB	Total	Total	Japan LL
1955	39,200		800	40,000	98	4,009		342		117	4,468	44,468	88
1956	30,700		900	31,600	97	4,373		957		40	5,370	36,970	83
1957	64,400		900	65,300	99	5,198		435		68	5,701	71,001	91
1958	86,500		1,000	87,500	99	4,196		114		232	4,542	92,042	94
1959	79,300		800	80,100	99	1,729		74		150	1,953	82,053	97
1960	87,600		700	88,300	99	1,524		152		183	1,859	90,159	97
1961	132,200		1,500	133,700	99	1,837		111		213	2,161	135,861	97
1962	119,800		3,400	123,200	97	824		213		328	1,365	124,565	96
1963	144,400		3,600	148,000	98	1,822		39		75	1,936	149,936	96
1964	99,500		3,500	103,000	97	1,142		260		68	1,470	104,470	95
1965	73,500	700	2,000	76,200	96	1,254		231		118	1,603	77,803	94
1966	76,900	2,900	3,500	83,300	92	1,108		96		267	1,471	84,771	91
1967	77,700	3,200	3,200	84,100	92	2,803		314		1,663	4,780	88,880	87
1968	63,939	600	4,000	68,539	93	2,272		250		2,559	5,081	73,620	87
1969	91,885	2,500	4,600	98,985	93	1,679		158		576	2,413	101,398	91
1970	71,165		5,000	76,165	93	1,579		247		1,332	3,158	79,323	90
1971	65,059	4,700	4,300	74,059	88	931		218		2,567	3,716	77,775	84
1972	82,632	7,800	3,800	94,232	88	2,364		781		2,238	5,383	99,615	83
1973	90,313	8,900	3,700	102,913	88	852		251		1,978	3,081	105,994	85
1974	68,730	14,444	4,420	87,594	78	729		456		889	2,074	89,668	77
1975	76,913	15,484	5,348	97,745	79	3,522		743		3,722	7,987	105,732	73
1976	96,816	21,395	3,078	121,289	80	7,982		889		10,185	19,056	140,345	69
1977	115,833	17,663	4,507	138,003	84	5,096		970		7,054	13,120	151,123	77
1978	100,557	8,456	4,402	113,415	89	3,330		1,987		11,710	17,027	130,442	77
1979	104,776	12,804	4,491	122,071	86	1,967		1,239		7,530	10,736	132,807	79
1980	96,637	13,975	4,637	115,249	84	2,205		1,021		15,417	18,643	133,892	72
1981	78,630	10,608	3,849	93,087	84	2,357		1,733	40	10,089	14,219	107,306	73
1982	87,571	10,050	2,111	99,732	88	4,057		2,516	23	4,103	10,699	110,431	79
1983	91,200	7,706	3,477	102,383	89	3,847	1,559	645	34	3,260	9,345	111,728	82
1984	83,504	7,478	2,943	93,925	89	3,447	884	745	55	5,853	10,984	104,909	80
1985	104,208	10,898	3,031	118,137	88	2,895	1,163	1,382	46	4,531	10,017	128,154	81
1986	123,103	15,927	2,879	141,909	87	2,227	1,672	1,009	0	1,979	6,887	148,796	83
1987	121,386	19,544	3,280	144,210	84	1,834	1,765	1,002	259	771	5,631	149,841	81
1988	94,666	13,681	3,610	111,957	85	2,900	1,044	953	1,085	1,053	7,035	118,992	80
1989	103,326					2,472	1,741	1,317		1,807			

Data source :
 Japan LL >20 GT for 1955-1973 : Kume (1979b) and FAO (1974-1988).
 Japan LL <20 GT for all years : MAFFJ (1955-1990).
 Japan surface catch : MAFFJ (1955-1990).
 Korea LL : FAO (1965-1988). All are assumed by LL.
 Taiwan LL : FAO (1965-1988). other nei A. All are assumed by LL. Before 1965 data are taken from Kume (1979b).
 Solomon : FAO (1981-1988).
 IATTC : IATTC (1989).

12.1 Production Model Analysis

It may not be appropriate to apply the method of production model analysis if the assumptions that (a) the rate of natural increase of the stock responds immediately to changes in population density, and (b) the rate of natural increase of the stock at a given level of biomass is independent of the age (or size) composition of the stock, are not satisfied. In the case of Pacific bigeye tuna it is considered that it takes 4 years for the fish to recruit into the longline fishery (Suda, 1970b) so that the first assumption is not matched satisfactorily. Furthermore, the second assumption also appears seldom satisfied for long-lived species such as tunas. On the other hand, the fact that the fishing condition and the fishery have been stable, and that the changes in the age (or size) composition of the catch seem to be smaller than for other tunas, may provide a basis for the application of this method. Irrespective of the above-mentioned drawbacks, production model analysis might give a general picture of stock status, MSY, *etc.*, for the species under consideration, especially since there is a lack of complete and detailed data for more sophisticated analysis such as cohort analysis.

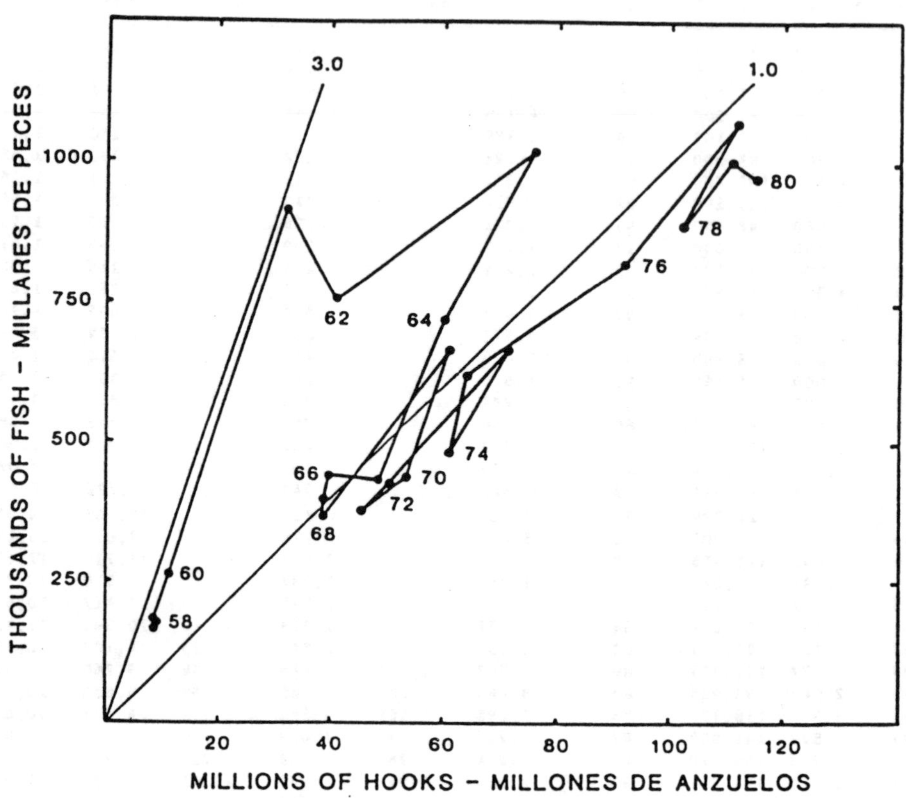

Fig. 14. Relationship between estimated Japanese longline catch and effort for bigeye tuna in the eastern tropical Pacific for 1957-1980. The fine lines and figures in two digits denote hook rates in numbers of fish per 100 hooks and year, respectively. After Miyabe and Bayliff (1987).

Production model analysis requires catch and fishing effort or CPUE. Since complete fishing effort data are not available, Kume (1979b) and Miyabe (1989; 1991) used the Japanese longline effort as basic data. The effective fishing effort, standardized by the Honma method (Honma 1974) of the Japanese longline fishery, was raised to the total effective effort using the proportion of the Japanese longline catch relative to the total catch. As stated above, the proportion of the Japanese longline catch has been very high (70-90%). The programme "PRODFIT" of Fox (1975) is used, applying the number of year classes that contributes significantly in the catch set at four. The results are shown in Figure 17 and Table 11. Kume (1979b) estimated the MSY between 100,000-106,000 mt with best fit at shape parameter $m = 0.0$. Miyabe (1991) estimated the MSY between 130,000-167,000 mt with best fit at $m=0.0$. Although the shape of the production curve is not known, and the current level of fishing effort is the highest recorded to date, the fishing effort does not appear to exceed considerably the level that gives the MSY.

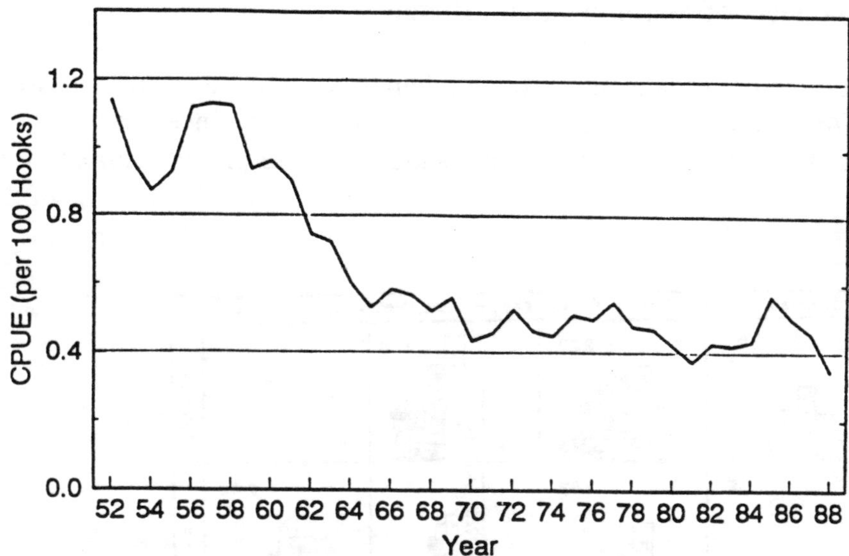

Fig. 15. Annual trend in the Japanese longline CPUE standardized by Honma method. After Miyabe (1991).

12.2 <u>Virtual Population Analysis (VPA)</u>

Kume (1979b) estimated the catch-at-age for fish caught by the Japanese longline fishery for 1957-1975 using the length-frequency samples and the growth equation by Suda and Kume (1967). Then, assuming that the Taiwanese and Korean longline fleets caught the same size of fish in a given area, catch-at-age was prorated to include the catch by Taiwan and Korea. Minimum stock size analysis (Honma 1978), which is one of the variety of VPA analyses, was applied in order to estimate the recruitment. Natural mortality rate (0.361) was employed from Suda and Kume (1967). The estimated recruitment (age 1 fish) was about 9 millions for 1956-57 cohorts and 6 to 6.5 millions for 1964-66 cohorts. The recruitment at age 1 was also estimated assuming the constant recruitment number using the relationship between fishing effort and reciprocal of CPUE (Suda 1970a). The estimate was 7.4 million and very close to Kume's (1979b) results.

Similarly, Miyabe (1989) constructed the catch-at-age for 1965-1987 but solely for the fish caught by the Japanese longline fishery. Miyabe (1989) tuned the VPA with standardized CPUE from the Japanese longline fishery in a way described by Parrack (1986). A value of 0.4 was used for M. The objective function to be minimized is:

$$SSQ = \Sigma \; (CPUE_{cal} - CPUE_{obs})^2$$

where SSQ = sum of squares, $CPUE_{cal}$ = calculated CPUE by VPA, and $CPUE_{obs}$ = observed CPUE. $CPUE_{cal}$ can be calculated by regressing population number (N) from the VPA to observed CPUE applying the equation $CPUE_{cal} = q \cdot N$. Here q is catchability coefficient.

The estimated population number at age 1 ranges between 11 to 13 million with smaller fluctuations (10-20%) among years. This is similar to the findings of Kume (1979b) although the level of recruitment is different.

It should be noted that data, in particular length samples, are often less than the desired level, and the assumptions used may not be appropriate since they cannot be proved practically. Because of this, the results should be interpreted with caution.

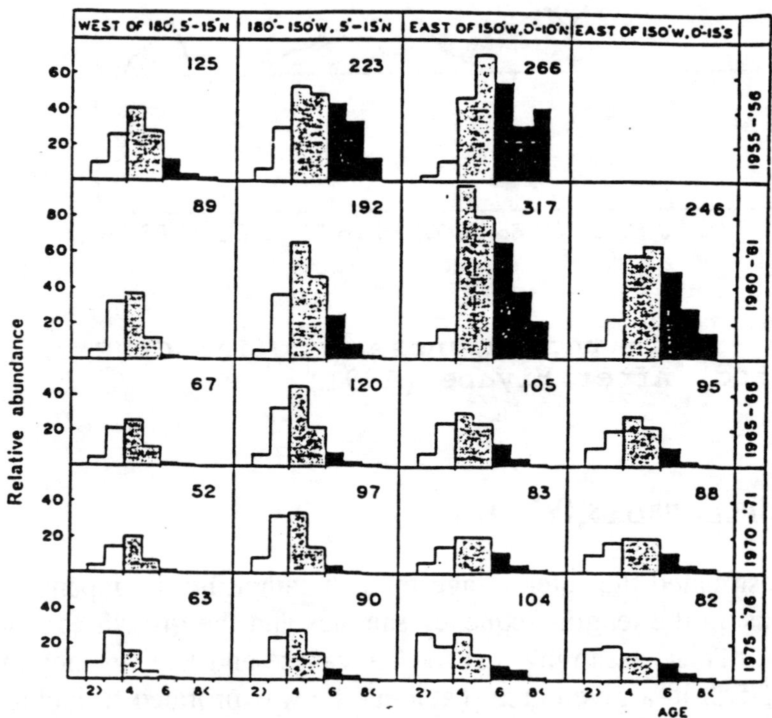

Fig. 16. Changes in CPUE by age of bigeye tuna in the equatorial Pacific, shown by 5-year intervals. Numbers in the panels are total CPUE. After Kume (1979a).

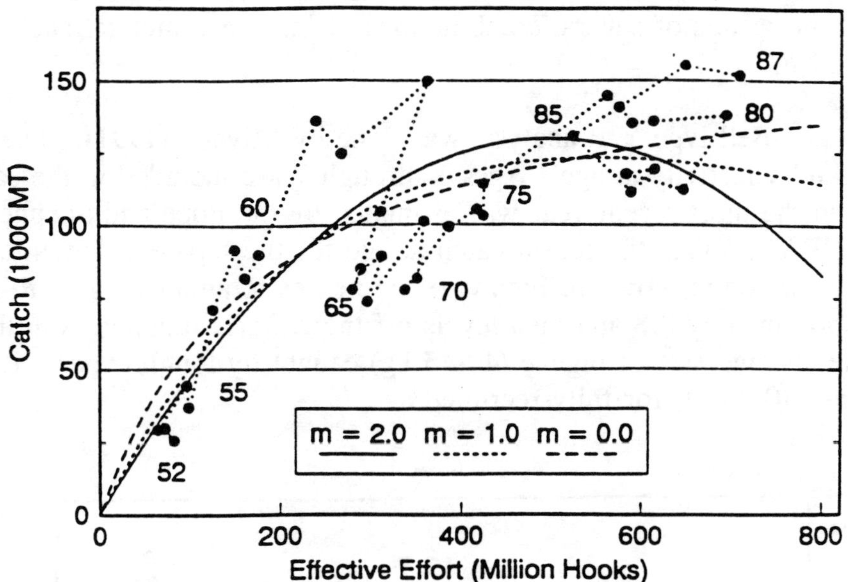

Fig. 17. Annual catch against annual fishing effort and estimated production curves for Pacific bigeye tuna. After Miyabe (1991).

Table 11. Results of production model analysis on Pacific bigeye tuna. After Miyabe (1991).

m (shape parameter)	MSY (1,000 MT)	F_{opt} (million hooks)
0.0	167	∞
1.001	130	500
2.0	130	550

12.3   Yield-per-Recruit (Y/R) Analysis

Suda (1970b) presented the results of Y/R analysis incorporating the Ricker-type stock-recruitment relationship, which was estimated from the spawning potential. Effective fishing effort and M (0.361) were taken from Suda and Kume (1967) for 1957-1964. The estimated equilibrium yield curve is shown in Figure 18. The MSY is about 90,000 mt when *F* is 0.5-0.6 assuming the knife-edge-type recruitment at age 4. He also presented the results of several combinations of input parameters, such as recruitment age and *F*.

Thompson-and-Bell-type Y/R analysis was done by Miyabe (1991). The inputs are *M*, weight-at-age and selectivity-at-age. Ages 1 through 7 are included in the calculation. Selectivity-at-age in the most recent year was estimated by the Pope and Shepherd's (1982) separable VPA applying the recent catch-at-age for the Japanese longline catch. The estimated Y/R, which is shown in Figure 19, increases to about 8 kg as *F* becomes larger up until approximately 0.8 and then levels off thereafter. Judging from the current information on the average size of bigeye (40-45 kg) caught by longline gear, it appears *F* is in moderate range (0.2-0.4) for fully-recruited ages.

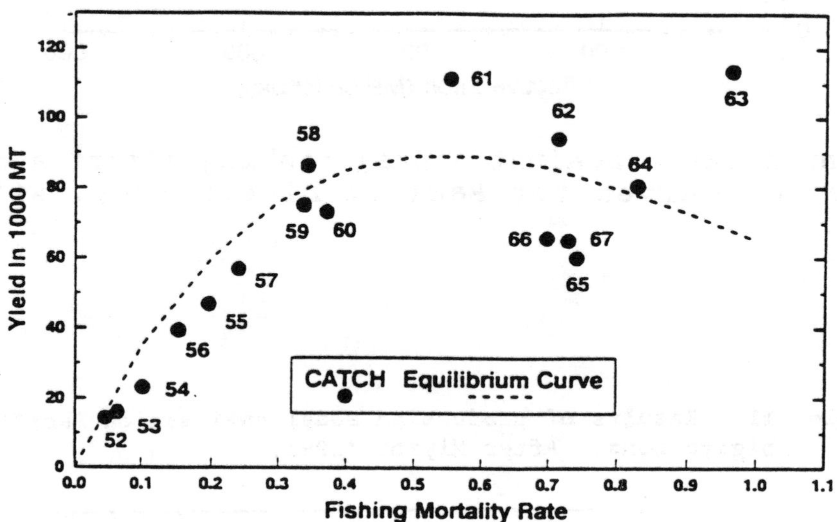

Fig. 18. The estimated equilibrium curve for Pacific bigeye tuna. After Suda and Kume 1967.

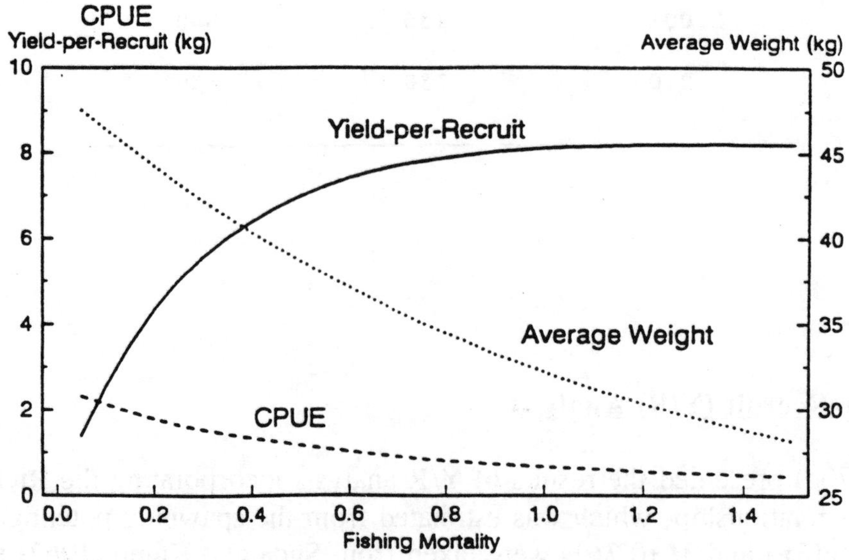

Fig. 19. Thompson and Bell type Y/R curve estimated for the Pacific bigeye tuna. After Miyabe (1991).

13. INTERACTIONS

Among tunas, bigeye seems to be one of the species with the lowest level of gear interaction. There are several fishing gears which harvest bigeye tuna, such as longline, purse seine, baitboat and other miscellaneous gears (trolling, hand-lining, ring net, gill net, *etc.*). In Table 10, the catch was shown divided into surface and longline catch, which consists of smaller to medium and medium to large fish, respectively. It indicated that the longline catch accounted for more than 85% of the total catch. This means there is less within-generation interaction between fisheries. In addition to this, the major distributional pattern of catch by two gear categories differs geographically. The greater catch occurs in coastal or island areas for the surface fishery but in high seas for the longline fishery.

14. REFERENCES CITED

Calkins, T., 1980. Synopsis of biological data on the bigeye tuna, *Thunnus obesus* (Lowe, 1839), in the Pacific Ocean. *Spec.Rep.I-ATTC*, (2):213-59.

Calkins, T., M. Yamaguchi, and N. Miyabe. 1988. Some observations of bigeye tuna (*Thunnus obesus*) caught by the surface and longline fisheries for tunas in the eastern Pacific Ocean. *I-ATTC,*, 48 p. (Unpubl.).

Carey, F.G., and K.D. Lawson. 1973. Temperature regulation in free-swimming bluefin tuna. *Comp.Biochem.Physiol.*, 44(2A):375-92.

Collette, B.B., and C.E. Nauen. 1983. FAO species catalog. Vol. 2. Scombrids of the world. *FAO Fish.Synop.*, (125) Vol.2:137 p.

Fishery Agency of Japan. 1989. Statistics on tuna fisheries. Internal data, Fishery Agency of Japan.

Food and Agriculture Organization of the United Nations. 1965-1988. Catches and landings, 1965-1988, *FAO Yearb.Fish.Statist.*, Vol.26-66:(varying pagination).

Federation of Japan Tuna Fisheries Cooperative Association. 1959. Average year's fishing condition of tuna longline fisheries, 1958 edition, edited by Nankai Regional Fisheries Research Laboratory. *Fed.Jap.Tuna Fish.Coop.Assoc.*, 414 p.

Fox, W.W. Jr. 1975. Fitting the generalized stock production model by least squares and equilibrium approximation. *Fish.Bull.NOAA-NMFS*, 73(1):23-36.

Far Seas Fishery Research Laboratory. 1988. Report of tagging activity on tuna and skipjack for the fiscal year 1984-1986, released by Japan Marine Resources Research Center. [In Japanese] *Far Seas Fish.Res.Lab.*

Fujino, K., and T. Kang. 1968. Serum esterase groups of Pacific and Atlantic tunas. *Copeia*, (1):56-63.

Gibbs, R.H. Jr., and B.B. Collette. 1967. Comparative anatomy and systematics of the tunas, genus *Thunnus*. *Fish.Bull.U.S.Fish Wild.Serv.*, 66(1):65-130.

Graves, J.E., M.A. Simovich, and K.M. Schaefer. 1988. Electrophoretic identification of early juvenile yellowfin tuna, *Thunnus albacares*. *Fish.Bull.NOAA-NMFS*, 86(4):835-38.

Hanamoto, E. 1976. The swimming layer of bigeye tuna. *Bull.Jap.Soc.Fish.Oceanogr.*, (29):41-4.

Hanamoto, E. 1987. Effect of oceanographic environment on bigeye tuna distribution. *Bull.Jap.Soc.Fish.Oceanogr.*, 51(3):203-16.

Hisada, K. 1973. Investigation on tuna hand-line fishing ground and some biological observations on yellowfin and bigeye tunas in the northwestern Coral Sea. *Bull.Far Seas Fish.Res.Lab.*, (8):35-69.

Holland, K.N., R.W. Brill, and R.K.C. Chang. 1990. Horizontal and vertical movements of yellowfin and bigeye tuna associated with fish aggregating devices. *Fish.Bull.NOAA-NMFS*, 88(3):493-507.

Honma, M. 1974. Estimation of overall effective fishing intensity of tuna longline fishery - Yellowfin tuna in the Atlantic Ocean as an example of seasonally fluctuating stocks. *Bull.Far Seas Fish.Res.Lab.*, (10):63-86.

Honma, M. 1978. Calculation of minimum stock size. In Collective volume of computer programs for fisheries stock analysis. *Fish.Agency Japan*, 19:193-97.

Honma, M., and T. Kamimura. 1955. Biology of the big-eyed tuna, *Parthunnus mebachi* (Kishinouye) - II. A consideration on the size composition of the big-eyed tuna caught by pole and line. *Bull.Jap.Soc.Sci.Fish.*, 20(10):863-69.

Inter-American Tropical Tuna Commission. 1970. Annual report of the Inter-American Tropical Tuna Commission, 1969. *Annu.Rep.IATTC*, (1969):117 p.

Inter-American Tropical Tuna Commission. 1989. Annual report of the Inter-American Tropical Tuna Commission, 1988. A*nnu.Rep.IATTC*, (1988):288 p.

Iversen, E.S. 1955. Size frequencies and growth of central and western Pacific bigeye tuna. *Spec.Sci.Rep.U.S.Fish Wildl.Serv. (Fish.)*, (162):1-40.

Iwai, T., I. Nakamura, and K. Matsubara. 1965. Taxonomic study of the tunas. *Spec.Rep.Misaki Mar.Biol.Inst.Kyoto Univ.*, (2):51 p.

Kamimura, T., and M. Honma. 1953. Biology of the big-eyed tuna, *Parathunnus mebachi* (Kishinouye) - I. Length frequency of the big-eyed tuna caught in the North Pacific with special reference to biennial frequency. *Contr.Nankai Reg.Fish.Res.Lab.*, (1):18 p.

Kawasaki, T. 1958. Biological comparison between the Pacific tunas. Part I.
Bull.Tohoku Reg.Fish.Res.Lab., (12):46-79.

Kawasaki, T. 1960. Biological comparison between the Pacific tunas. Part II.
Bull.Tohoku Reg.Fish.Res.Lab., (16):1-40.

Kikawa, S. 1953. Observation on the spawning of the big-eyed tuna (*Parathunnus mebachi*, Kishinouye) near the southern Marshall Islands. *Contr.Nankai Reg.Fish.Res.Lab.*, 1(42):10 p.

Kikawa, S. 1957. The concentrated spawning area of bigeye tuna in the western Pacific.
Rep.Nankai Reg.Fish.Res.Lab., (5):145-57.

Kikawa, S. 1961. The group maturity of bigeye tuna *Parathunnus mebachi* (Kishinouye) in the spawning areas of the Pacific. *Rep.Nankai Reg.Fish.Res.Lab.*, (13):35-46.

Kikawa, S. 1962. Studies on the spawning activity of the Pacific tunas, *Parathunnus mebachi* and *Neothunnus macropterus*, by the gonad index examination.
Occas.Rep.Nankai Reg.Fish.Res.Lab., (1):43-56.

Kikawa, S. 1966. The distribution of maturing bigeye and yellowfin and an evaluation of their spawning potential in different areas in the tuna longline grounds in the Pacific.
Rep.Nankai Reg.Fish.Res.Lab., (23):131-208.

Kiyota, M., T. Koido, N. Miyabe, K. Mizuno, Y. Nishikawa, Z. Suzuki, Y. Warashina, and M. Yukinawa (in alphabetical order). 1988. Report of Juten-Kiso-Kenkyuu.
Jap.Sci.Tech.Agency, (59):13 p.

Koido, T., and N. Miyabe. 1990. II. Field observation, 5. Tunas. Application of Telemetry to Aquatic Animal Behavior, edited by H. Soeda. *Suisangaku Ser.Kouseisha-kouseikaku*, Tokyo, (80):55-66.

Kume, S. 1967. Distribution and migration of bigeye tuna in the Pacific Ocean.
Rep.Nankai Reg.Fish.Res.Lab., (25):75-80.

Kume, S. 1969a. Ecological studies on bigeye tuna - V. A critical review on distribution, size composition and stock structure of bigeye tuna in the North Pacific Ocean (north of 16°N). *Bull.Far Seas Fish.Res.Lab.*, (1):57-75.

Kume, S. 1969b. Ecological studies on bigeye tuna - VI. A review on distribution and size composition of bigeye tuna in the equatorial and South Pacific Ocean. *Bull.Far Seas Fish.Res.Lab.*, (1):77-98.

Kume, S. 1979a. Fishery biology of the bigeye tuna resources in the Pacific Ocean.
Jap.Fish.Res.Cons.Assoc. Suisan Kendyuu Sousho, (32):54 p.

Kume, S. 1979b. Bigeye tuna resource in the Pacific Ocean, its fishery biology and a status of the stock. Paper presented at the Tuna and Billfish Stock Assessment Workshop, Shimizu, Japan, June 1979, Working Paper:28 p.

Kume, S., and J. Joseph. 1966. Size composition, growth and sexual maturity of bigeye tuna, *Thunnus obesus* (Lowe), from the Japanese longline fishery in the eastern Pacific Ocean. Bull.I-ATTC, 11(2):45-99.

Kume, S., and J. Joseph. 1969. The Japanese longline fishery for tunas and billfish in the eastern Pacific Ocean east of 130°W, 1964-1966. Bull.I-ATTC, 13(2):275-418.

Kume, S., and N. Miyabe. 1987. On the relation of El Niño with the formation of bigeye tuna fishing grounds in the eastern equatorial Pacific. The 22nd Symposium on the Tuna Fisheries. *Bull.Jap.Soc.Fish.Oceanogr.*, 51(1):62-8.

Kume, S., and Y. Morita. 1967. Ecological studies on bigeye tuna - IV. Size composition of bigeye tuna *Thunnus obesus* (Lowe), caught by pole-and-line fishery in the northwestern Pacific Ocean. *Rep.Nankai Reg.Fish.Res.Lab.*, (25):81-90.

Kume, S., and M.B. Schaefer. 1966. Studies on the Japanese long-line fishery for tuna and marlin in the eastern tropical Pacific Ocean during 1963. *Bull.I-ATTC*, 11(3):103-170.

Kume, S., and T. Shiohama. 1964. On the conversion between length and weight of bigeye tuna landings in the Pacific Ocean (Preliminary report). *Rep.Nankai Reg. Fish.Res.Lab.*, 20:59-67.

Lowe, R.T. 1839. A supplement to a synopsis of the fishes of *Madeira.Proc.Zool. Soc.Lond.*, (7:)76-92.

Ministry of Agriculture, Forestry, and Fishery of Japan. 1955-1990. Annual report of statistics on fishery and aquaculture. Statistics and Information Division. Ministry of Agriculture, Forestry and Fishery of Japan, (1955-1990):(varying pagination).

Meehan, J.M. 1965. First occurrence of bigeye tuna on the Oregon coast. *Res.Briefs Oregon Fish Comm.*, 11(1):53-4.

Miyabe, N. 1989. Preliminary stock assessment of Pacific bigeye tuna. Rept. The 3rd Southeast Asian tuna conference. Indo-Pacific Tuna Development and Management Programme, FAO, 122-130.

Miyabe, N. 1991. Stock status of Pacific bigeye tuna. The 24th Symposium on Skipjack and Tuna. *Bull.Jap.Soc.Fish.Oceanogr.* 55(2):141-44.

Miyabe, N., and W.H. Bayliff. 1987. A review of the Japanese longline fishery for tunas and billfishes in the eastern Pacific Ocean, 1971-1980. *Bull.I-ATTC*, 19(1):1-163.

Morita, Y. 1973. Conversion factors for estimating live weight from gilled-and-gutted weight of bigeye and yellowfin tunas. *Bull.Far Seas Fish.Res.Lab.*, (9):109-21.

Nakamura, E.L. and J.H. Uchiyama. 1966. Length-weight relations of Pacific tunas. In Proceedings of Governor's Conference on Central Pacific Fishery Resources, edited by T.A. Manar. Hawaii, pp. 197-201.

Nakamura, H., and H. Yamanaka. 1959. Relation between the distribution of tunas and the ocean structure. *J. Oceanogr.Soc.Jap.*, 15(3):1-7.

Nelson, J.S. 1976. Fishes of the world. New York, Wiley-Interscience, John Wiley and Sons, 416 p.

National Fisheries Research and Development Agency. 1986. Annual report of catch and effort statistics and fishing grounds for the Korean tuna longline fishery, 1981-1982. *Nat.Fish.Res.Dev.Agency*, 523 p.

Nikaido, H., N. Miyabe, and S. Ueyanagi. 1991. Spawning time and frequency of bigeye tuna, *Thunnus obesus*. *Bull.Nat.Res.Inst.Far Seas Fish.*, 28:47-73.

Nishikawa, Y., and D.R. Rimmer. 1987. Identification of larval tunas, billfishes, and other scombroid fishes (Suborder Scombroidei): an illustrated guide. *Rep.CSIRO Mar.Lab.*, 186:20 p.

Nishikawa, Y., M. Honma, S. Ueyanagi, and S. Kikawa. 1985. Average distribution of Larvae of oceanic species of scombroid fishes, 1956-1981. *S Ser.Far Seas Fish.Res.Lab.*, (12):99 p.

Otsu, T., and R.N. Uchida. 1956. Tagged bigeye tuna recovered. *Pac.Sci.*, 10(2):236.

Paloheimo, J.E. 1961. Studies on estimation of mortalities. I. Comparison of a method described by Beverton and Holt and a new linear formula. *J.Fish.Res.Board Can.*, 18(5):645-62.

Parrack, M.L. 1986. A method of analyzing catches and abundance indices from a fishery. *Collect.Vol.Sci.Pap.ICCAT*, 24:209-21.

Pope, J.G., and J.G. Shepherd. 1982. A simple method for the consistent interpretation of catch-at-age data. *J.Cons.CIEM,* 40:176-184.

Radovich, J. 1961. Relationships of some marine organisms of the northeast Pacific to water temperatures particularly during 1957 through 1959. *Fish.Bull.Calif.Dep.Fish Game*, (112):62 p.

Saito, S. 1975. On the depth of capture of bigeye tuna by further improved vertical long-line in the tropical Pacific. *Bull.Jap.Soc.Sci.Fish.*, 41(8):831-41.

Saito, S., and S. Sasaki. 1974. Swimming depth of large sized albacore in the south Pacific Ocean - II. Vertical distribution of albacore catch by an improved vertical long-line. *Bull.Jap.Soc.Sci.Fish.*, 40(7):643-49.

Sharp, G.D. 1978. Behavioral and physiological properties of tuna and their effects on vulnerability to fishing gear. In The physiological ecology of tunas, edited by G.D. Sharp and A.E. Dizon. New York, Academic Press, pp. 397-449.

Sharp, G.D., and S. Pirages. 1978. The distribution of red and white swimming muscles, their biochemistry, and the biochemical phylogeny of selected scombrid fishes, edited by G.D. Sharp and A.E. Dizon. New York, Academic Press, pp. 41-78.

Shingu, C., P.K. Tomlinson, and C.L. Peterson. 1974. A review of the Japanese longline fishery for tunas and billfishes in the eastern Pacific Ocean, 1967-1970. *Bull. I-ATTC*, 16(2):65-230.

Shomura, R.S., and B.A. Keala. 1963. Growth and sexual dimorphism in growth of bigeye tuna (*Thunnus obesus*), a preliminary report. Proc. of the World Scientific Meeting on Biology of Tunas and Related Species. *FAO Fish.Rep.*, 6(3):1409-17.

South Pacific Commission. 1990a. Catches of tuna in the western tropical Pacific, 1965-1988. Paper presented at Third Standing Committee on Tuna and Billfish, S.Pac.Comm., Noumea, New Caledonia, WP/7:29 p.

South Pacific Commission. 1990b. Regional Tuna Bulletin, First quarter, 1990. *Reg.Tuna Bull.Tuna Billfish Assess.Programme, S.Pac.Comm.*, 49 p.

Suda, A. 1970a. Approximate estimation of parameters in dynamics of fish population utilizing effort and catch statistics with little informations on biological features. *Bull.Far Seas Fish.Res.Lab.*, (3):1-14.

Suda, A. 1970b. Methods of dealing with sustainable yields for fish species with marked differentiations of living pattern on a course of life history-I. Basic considerations on the conditions to realize a sustainable yield and applications of techniques to calculate amounts of sustainable yield from fish populations which are exploited in simple ways. *Bull.Far Seas Fish.Res.Lab.*, (3):115-46.

Suda, A., and S. Kume. 1967. Survival and recruit of bigeye tuna in the Pacific Ocean, estimated by the data of tuna longline catch. *Rep.Nankai Reg.Fish.Res.Lab.*, (25):91-104.

Suda, A., and M.B. Schaefer. 1965. General review of the Japanese tuna long-line fishery in the eastern Pacific Ocean 1956-1962. *Bull.I-ATTC*, 9(6):307-462.

Suda, A., S. Kume. and T. Shiohama. 1969. An indicative note on a role of permanent thermocline as a factor controlling the longline fishery for bigeye tuna. *Bull.Far Seas Fish.Res.Lab.*, (1):99-114.

Sund, P.N., M. Blackburn, and F. Williams. 1980. Tunas and their environment in the Pacific Ocean: a review. *Oceanogr.Mar.Biol.Ann.Rev.*, (18):443-512.

Suzuki, Z., and S. Kume. 1981. Fishing efficiency of deep longline for bigeye tuna in the Atlantic as inferred from the operations in the Pacific and Indian Oceans. *Collect.Vol.Sci.Pap.ICCAT*, 17(2):471-86.

Suzuki, Z., Y. Warashina, and M. Kishida. 1977. The comparison of catches by regular and deep tuna longline gears in the western and central equatorial Pacific. *Bull.Far Seas Fish.Res.Lab.*, 15:51-73.

Tanaka, T. 1989. Shift of the fishing ground and features of shoals caught by purse seine fishery in the tropical seas of the western Pacific Ocean. *Bull.Tohoku Reg.Fish.Res.Lab.*, 51:75-88.

Tuna Research Center. 1983. Annual catch statistics of Taiwan's tuna longline fishery, 1982. *Tuna Res.Cent.*

Uda, M. 1957. A consideration on the long years trend of the fisheries fluctuation in relation to sea conditions. *Bull.Jap.Soc.Sci.Fish.*, 23(7-8):368-72.

Yamanaka, H., and N. Anraku. 1962. Relation between the distribution of tunas and water masses of the North and South Pacific Oceans west of 160°W. *Occas.Rep.Nankai Reg.Fish.Res.Lab.*, (1):23-34.

Yasutake, H., G. Nishi, and K. Mori. 1973. Artificial fertilization and rearing of bigeye tuna (*Thunnus obesus*) on board, with morphological observations on embryonic through to early post-larval stage. *Bull.Far Seas Fish.Res.Lab.*, (8):71-8.

Yuen, H.S.H. 1955. Maturity and fecundity of bigeye tuna in the Pacific. *Spec.Sci.Rep.U.S.Fish Wild.Serv. (Fish.)*, (150):30 p.

Yukinawa, M., and Y. Yabuta. 1963. Age and growth of bigeye tuna, *Parathunnus mebachi* (Kishinouye). *Rep.Nankai Reg.Fish.Res.Lab.*, (19):103-18.

A REVIEW OF THE BIOLOGY AND FISHERIES FOR NORTHERN BLUEFIN TUNA, *THUNNUS THYNNUS*, IN THE PACIFIC OCEAN

William H. Bayliff
Inter-American Tropical Tuna Commission
La Jolla, California, USA

1. INTRODUCTION

Parts of this report are similar to parts of a species synopsis prepared by Bayliff (1980). The material in Sections 6.2 and 7 is similar to that in sections entitled **LIFE HISTORY AND STOCK STRUCTURE** and **ATTRITION** in Bayliff *et al.* (1991). The material in Section 12.2 is similar to part of the section on stock assessment of northern bluefin in Bayliff (1993b), and that in Section 12.3 is similar to the section entitled *Cohort analysis* in Bayliff (1993a).

2. CLASSIFICATION

Phylum Chordata
  Class Osteichthyes
    Order Perciformes
      Family Scombridae
        Subfamily Scombrinae
        Tribe Thunnini
          Genus *Thunnus*
            Species *thynnus*
              Subspecies *orientalis*

The classification of northern bluefin and other tunas is discussed by Gibbs and Collette (1967), Sharp and Pirages (1978), Collette and Nauen (1983), and Collette *et al.* (1984).

3. EARLY LIFE HISTORY

The larvae of northern bluefin have been described by Matsumoto (1962), Yabe and Ueyanagi (1962), Matsumoto *et al.* (1972), and Richards and Pottoff (1973). The Kuroshio Current probably plays an important role in transporting the larvae, postlarvae, and juveniles northward from the spawning grounds between Japan and the Philippines and southeast of Japan to waters off Japan. Age-0 fish about 15 to 60 cm in length are caught in the vicinity of Japan during the second half of the year (Yabe *et al.*, 1966; Yukinawa and Yabuta, 1967). (In this report fish in their first year of life are referred to as 0-year-olds, age-0 fish, or fish 0 years of age, and so on.)

4. AGE AND GROWTH

The maximum lengths and weights recorded for northern bluefin in the Pacific Ocean, from Foreman and Ishizuka (1990), are as follows:

Location	Date	Length (cm)	Weight (kg)
between 27°N and 29°N and 130°E and 133°E	April 1986	ca. 300	ca. 555.5
San Nicolas Island (*ca.* 33°30'N-119°30'W)	December 18, 1988	271	458

The parameters of the von Bertalanffy growth equation for northern bluefin in the Pacific Ocean have been estimated by various workers from analyses of hard part, length-frequency, and tagging data. These estimates, and estimates of the average lengths at various ages, are shown in Table 1. The estimates of the average lengths at ages 1-6 are fairly close to one another, but the estimates of L_∞ and K of Bayliff (1986) are considerably greater and less, respectively, than those of the other workers.

Bayliff *et al.* (1991) studied the growth of northern bluefin from tagging experiments. They had 721 fish with data for the dates and lengths at both release and recapture. Because the data included many fish less than 30 cm in length at release and many fish at liberty less than 1 year, and relatively few larger fish or fish at liberty more than 1 year, the data were combined into 84 groups. The times at liberty and lengths at release selected were 1-30, 31-60, 61-90, ... days and 151-200, 201-250, 251-300, ... mm, respectively, and the means of each group were calculated and substituted for the individual values. Bayliff *et al.* (1991) found that the growth of bluefin in the length range of about 16 to 153 cm is best described by a two-stage model, the growth of the fish less than or equal to 564 mm following the Gompertz model and that of the fish equal to or greater than 564 mm being linear. The relationship between length in millimeters and time in days is expressed by the following equations:

fish less than or equal to 564 mm

$$x_{t2} = x_\infty + (x_{t2} - x_\infty)e^{-K(t_2 - t_1)}$$

where

$$x_t = (L_t{}^\lambda - 1)/\lambda$$

in which, with their estimate of λ, x was essentially a logarithmic transformation,

L_t = length at time t_1 (t_2 is a time prior to t),
$\lambda = 0.001$,
$L_\infty = 581$ mm, and
K = coefficient of growth = 0.01184 per day.

fish equal to or greater than 564 mm

$$L_t = 564 + 0.709\Delta t$$

TABLE 1. Age and growth data for bluefin in the Pacific Ocean. The lengths are given in centimeters. Schultze and Collins' (1977) estimates are attributed to an unpublished thesis (their Table 1). The sample of Bell (1963b) consisted of 247 fish, of which approximately half had legible scales.

Method	Sample size	Age										von Bertalanffy parameters			Reference
		1	2	3	4	5	6	7	8	9	10	L_∞	K (annual)	t_0 (annual)	
vertebrae	21	43	69	94	118	145	168	190	210	230	250				Aikawa and Katô, 1938
length frequencies	4,163	50.0	90.0	125.0	154.0	178.0	198.0	215.0	229.0	241.0		300.0			Yokota et al., 1961: 217
scales	about 124	57.10	72.08	90.65	106.95	128.50	142.00								Bell, 1963b
scales	97	51.8	78.2	102.0	123.5	142.9	160.3	176.1	190.3	203.1	214.6	320.5	0.1035	-0.7034	Yukinawa and Yabuta, 1967
scales	2,743	53.0	80.4	104.8	126.3	145.5	162.4					295.4			Schultz and Collins, 1977
tagging												219	0.211		Peterson, 1985: 258
vertebrae	232	49.1	71.1	92.4	113.1	133.0	152.2					703.6	0.0343	1.107	Bayliff, 1986: 34
length frequencies	93,141	76.4	101.2	124.8											Bayliff, 1993a

where

$$\Delta t = \text{time, in days, elapsed since } L_t = 564 \text{ mm.}$$

Bayliff (1993a) used length-frequency data to obtain an estimate of 0.675 mm per day for the growth of bluefin in the eastern Pacific Ocean. He found the growth to be more rapid in the summer than in the winter. Such has also been found to be the case for northern bluefin in the Atlantic Ocean (Cort, 1990: Figure 68) and for southern bluefin (Burgess *et al.*, 1991). Monthly estimates of the lengths are shown in Table 2.

TABLE 2. Estimated growth of northern bluefin in the eastern Pacific Ocean (from Bayliff, 1993a).

Year	Month	Length (cm)	Year	Month	Length (cm)	Year	Month	Length (cm)
1	July	56.8	2	September	88.5	3	November	115.8
1	August	60.2	2	October	90.0	3	December	117.4
1	September	63.7	2	November	91.7	4	January	118.8
1	October	65.2	2	December	93.3	4	February	120.2
1	November	66.9	3	January	94.8	4	March	121.8
1	December	68.5	3	February	96.4	4	April	123.2
2	January	70.0	3	March	98.0	4	May	124.8
2	February	71.6	3	April	99.6	4	June	127.2
2	March	73.2	3	May	101.2	4	July	130.8
2	April	74.8	3	June	102.8	4	August	134.2
2	May	76.4	3	July	106.1	4	September	137.8
2	June	78.2	3	August	109.4	4	October	139.2
2	July	81.6	3	September	112.8	4	November	140.8
2	August	85.0	3	October	114.3	4	December	142.2

5. MATURATION AND SPAWNING

Fragmentary data on the sex ratios of northern bluefin are given in Table 3.

Hirota *et al.* (1976) reported that a male northern bluefin reared in captivity matured at 3 years of age and that a female of the same age appeared to be approaching maturity. The lengths of these fish (male, 1190 mm; female, 1353 mm) were considerably in excess of the average lengths of 3-year-old fish given in Table 1. Nakamura (1943; cited by Hirota *et al.*, 1976) stated that bluefin mature at 3 years of age in nature. Harada (1980), however, stated that spawning of bluefin occurs at about 5 years of age, and that the lengths and weights of these fish are about 150 cm and 60 kg, respectively. The estimate of 5 years is in agreement with estimates for northern bluefin in the Atlantic Ocean (Clay, 1991) and for southern bluefin (Davis, 1991).

Fish of 270 to 300 kg have about 10 million eggs (Yamanaka and staff, 1963). No information is available on the frequency of spawning.

TABLE 3. Data on the sex ratios of bluefin tuna in the Pacific Ocean. The length range for 1988 is the length range for the 987 fish in the catch, rather than for the 45 fish in the sample.

Area	Year	Length range (mm)	Gear	Males	Females	Reference
eastern Pacific	1960	770-892	purse seine	40	35	Bell, 1963a
eastern Pacific	1988	1300-2710	purse seine	45	0	Foreman and Ishizuka, 1990
Lingaen Bay (west. Pacif.)	1938		long-line	46	36	Yamanaka and staff, 1963
Formosa (west. Pacif.)	1952-1953		long-line	28	47	Yamanaka and staff, 1963
Sanriku (west. Pacif.)	1952-1953		long-line	29	22	Yamanaka and staff, 1963

Spawning occurs between Japan and the Philippines in April, May, and June and off southern Honshu in July (Yamanaka and staff, 1963; Yabe *et al.*, 1966; Nishikawa *et al.*, 1985). Okiyama (1974 and 1979) found larvae in the Sea of Japan in August, and considered them to have resulted from spawning in that area. In addition, Miller (1979) reported larvae of this species from the Hawaiian Islands. Bruce C. Mundy, U.S. National Marine Fisheries Service, Honolulu, Hawaii (pers. commun.), however, reported that no bluefin larvae were found in 460 plankton samples collected 1 to 15 nautical miles off Oahu during 1985 and 1986.

6. DISTRIBUTION, MIGRATION, AND STOCK STRUCTURE

6.1 Distribution

6.1.1 Eastern Pacific Ocean

In the eastern Pacific, northern bluefin are caught mostly between Cabo San Lucas, Baja California, Mexico, and Point Conception, California, USA (Figure 1). Within the area between Cabo San Lucas and Point Conception the fish tend to be distributed further to the north in years when the sea-surface temperatures are above normal and further to the south in years when those temperatures are below normal (Hester, 1961). Squire (1983) presents some maps showing distributions of schools of bluefin sighted from aircraft off North America north of 30°N during 1962-1978. Statistics for small amounts of bluefin landed at California ports north of Los Angeles during 1977-1986 are given by Oliphant *et al.* (1990) and during years previous to 1977 in earlier versions of the same series. On September 24, 1960, a purse seiner caught 50 mt of bluefin about 50 miles west of Cape Mendocino, more than 300 miles north of Point Conception. This species has also been caught off Oregon and Washington (Brock, 1938; Oregon Fish Commission, 1948; Radovich, 1961), off British Columbia (Neave, 1959), and in Shelikoff Strait, Alaska (Radovich, 1961). These

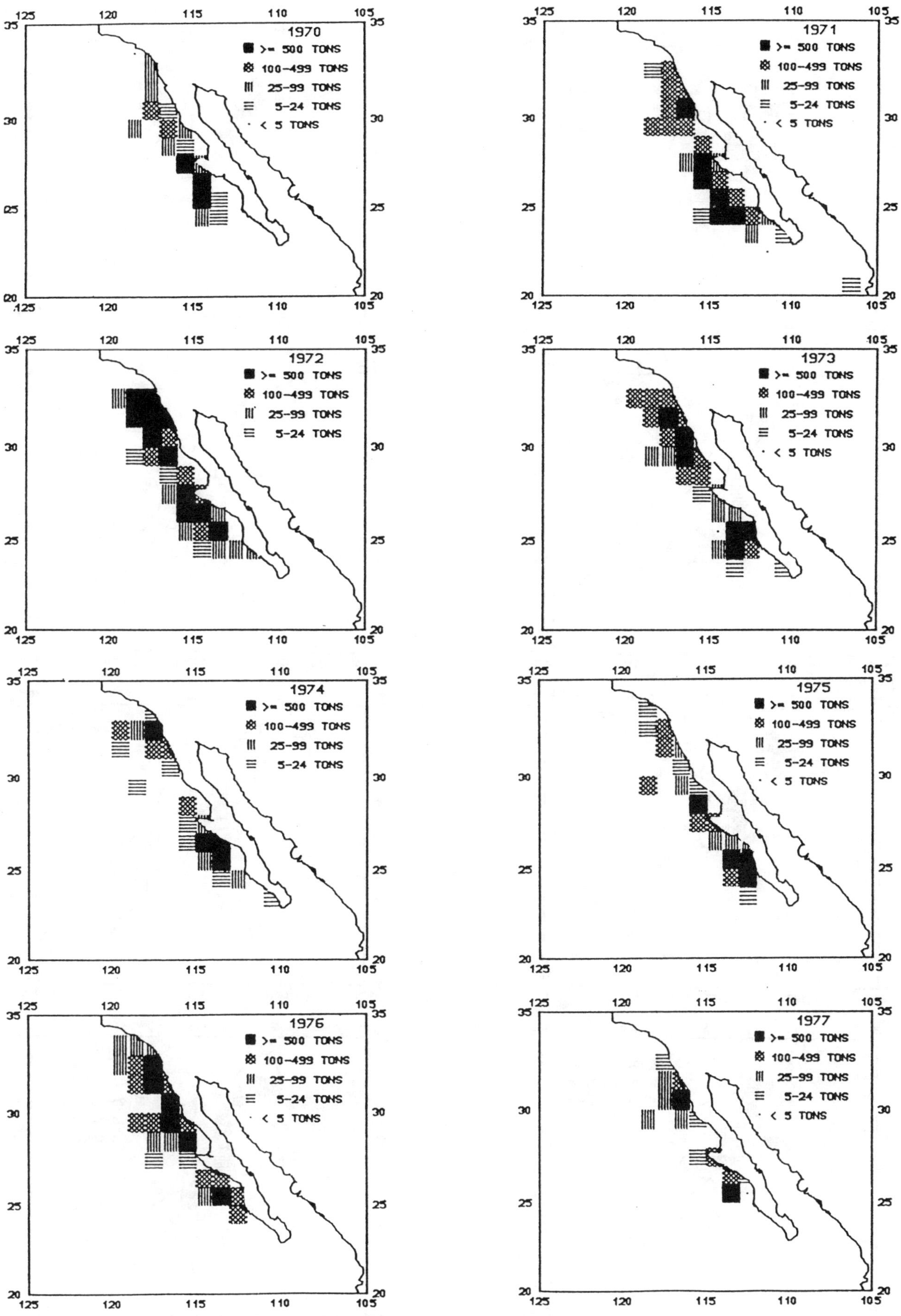

FIGURE 1. Annual distributions of northern bluefin catches in the eastern Pacific Ocean, 1970-1989.

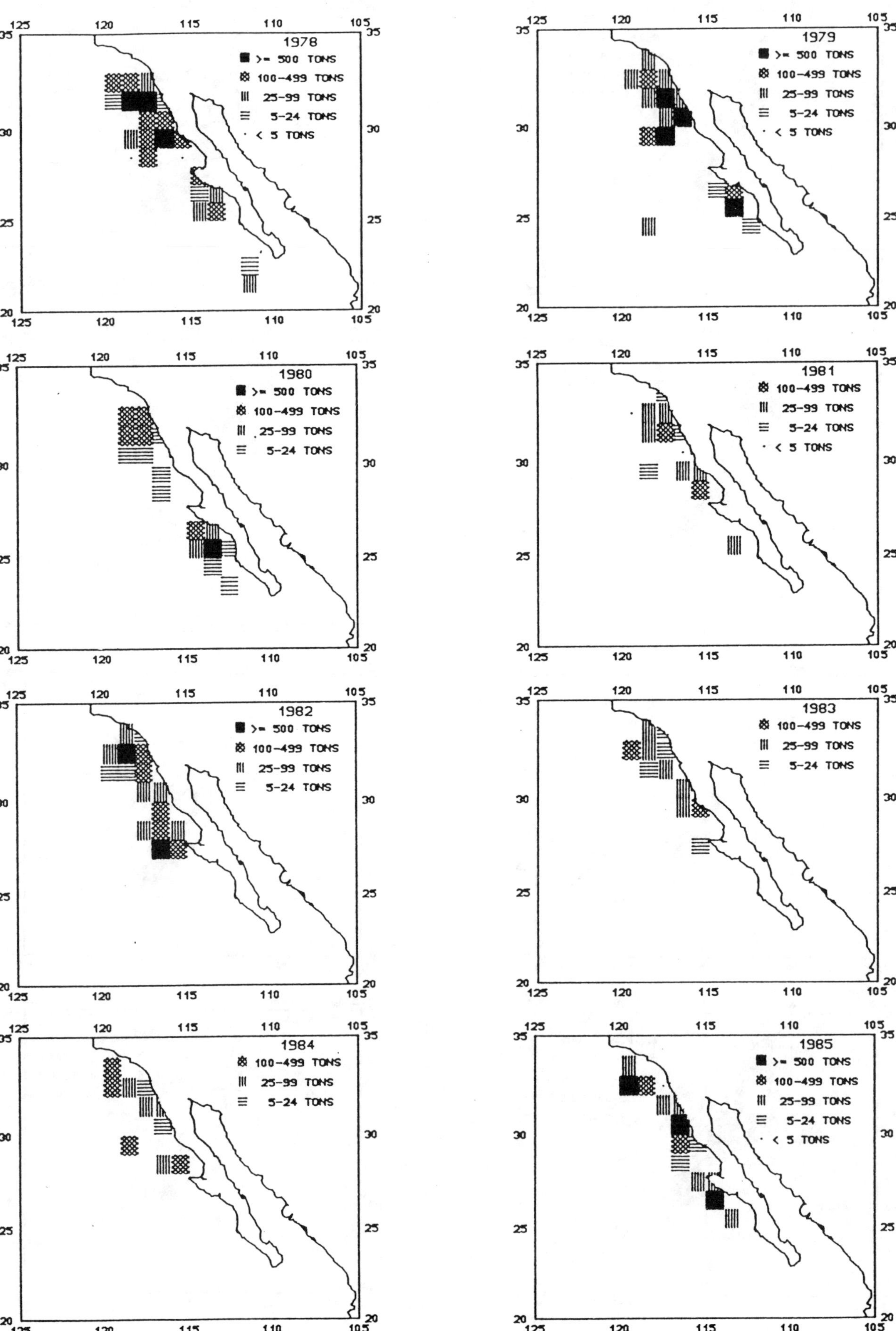

FIGURE 1. (continued)

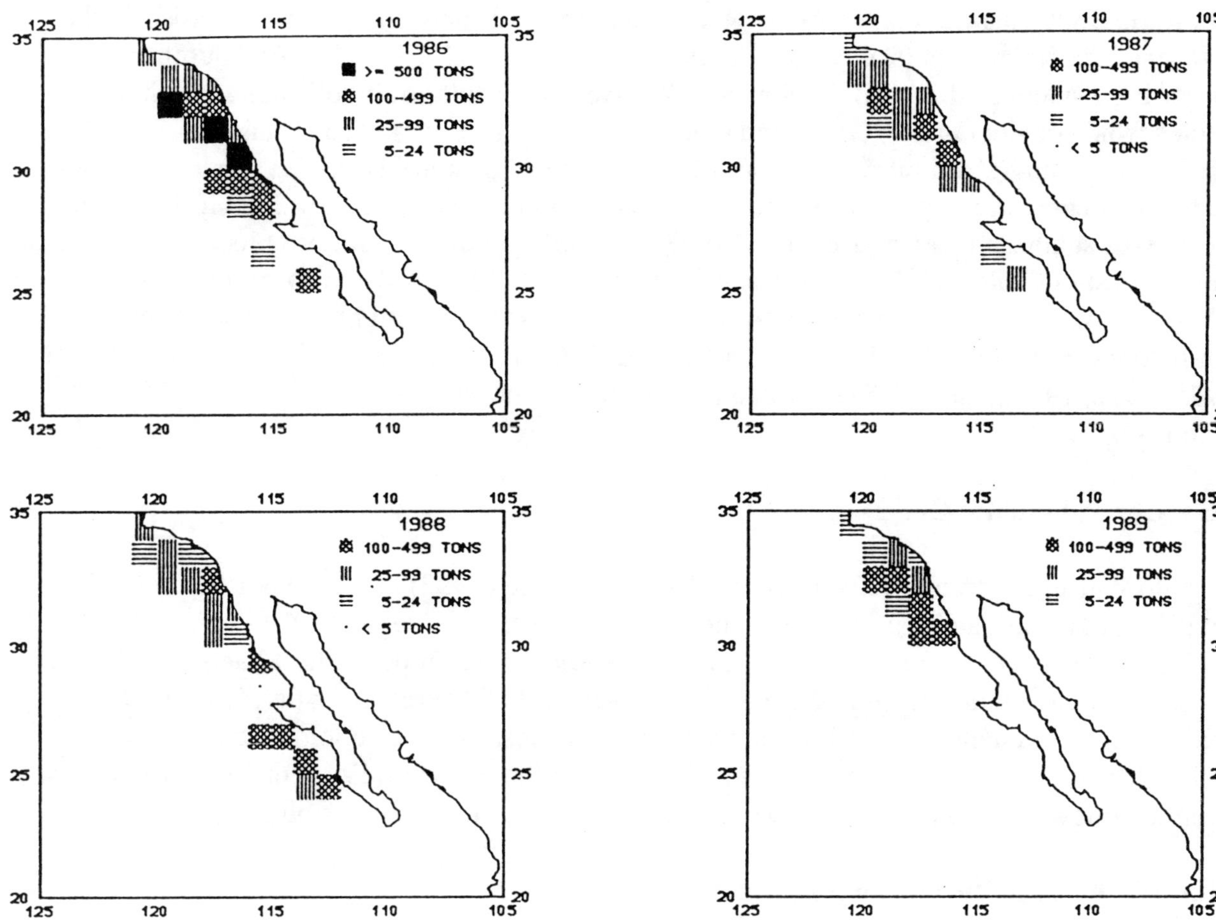

FIGURE 1. (continued)

occurrences far to the north of the usual range of this species have been attributed to greater-than-normal sea-surface temperatures (Radovich, 1961). Bluefin have been recorded in the vicinity of the Galapagos Islands (Snodgrass and Heller, 1905; Herre, 1936), "but there is no supporting evidence which would eliminate *T. maccoyii* [southern bluefin] or any other species from consideration" (Gibbs and Collette, 1967). Two large (1553- and 1640-mm) bluefin were caught by a purse seiner at 3°23'S-81°09'W on June 14, 1969. Their identity as bluefin was confirmed by the striations on the livers, but it was not determined if they were northern or southern bluefin. De Buen (1953 and 1958) recorded northern bluefin off the central coast of Chile, and Nakamura and Warashina (1965) recorded a specimen caught at 37°11'S-114°41'W. Also Kume (1974) published data for fish caught off the west coast of South America between 80°W and 115°W during 1968-1970, and Fisheries Agency of Japan (1975, 1976, 1977a, and 1978) recorded this species in the same area during the 1972-1976 period (Figure 2).

6.1.2 Central Pacific Ocean

Jordan and Jordan (1922), Fowler (1928), and June (1952) recorded northern bluefin caught near Hawaii, and Kume (1974) and Fisheries Agency of Japan (1974, 1975, 1976, 1977a, and 1978) recorded other catches of this species made in the central Pacific both north and south of the equator (Figure 2). Fowler's (1928 and 1932) records of a 355-mm bluefin from the Society Islands and of bluefin from the Marquesas Islands could be based upon misidentifications, however, as Klawe (1959) showed that Fowler (1944) had identified juvenile yellowfin, *Thunnus albacares*, caught in the eastern Pacific, as bluefin.

6.1.3 Western Pacific Ocean and Indian Ocean

In the western Pacific, northern bluefin occur from Sakhalin Island to southwestern Australia and New Zealand (Shimada, 1951; Okachi, 1963; Yamanaka and staff, 1963; Kume, 1974; Shingu *et al.*, 1974; Fisheries Agency of Japan, 1974, 1975, 1976, 1977a, and 1978; Collette and Smith, 1981), but are most abundant in the vicinity of Japan (Figure 2). They are much more plentiful off Japan in years when the sea-surface temperatures are above normal than when they are below normal (Uda, 1962 and 1973). According to Yamanaka and staff (1963), "Fishes having body weight over 8.3 kg live far out at sea in summertime."

In the Indian Ocean, northern bluefin have been caught west of Australia (Nakamura and Warashina, 1965).

6.2 <u>Migration and Stock Structure</u>

A model of the life history and stock structure of northern bluefin in the Pacific Ocean has been formulated. This model was first described by Bayliff (1980), but it has been modified considerably since then due to the acquisition of additional data for returns of tagged fish.

Figure 3 is a diagram of the model. The migrations shown by dashed lines are more speculative than those shown by solid lines. The diagram is intended to show the general areas where the migrations are believed to begin and end, but not the precise routes, as space limitations would make this impractical even if the routes were known. For example, it appears that the route of migration of juveniles bound for the eastern Pacific is south of the

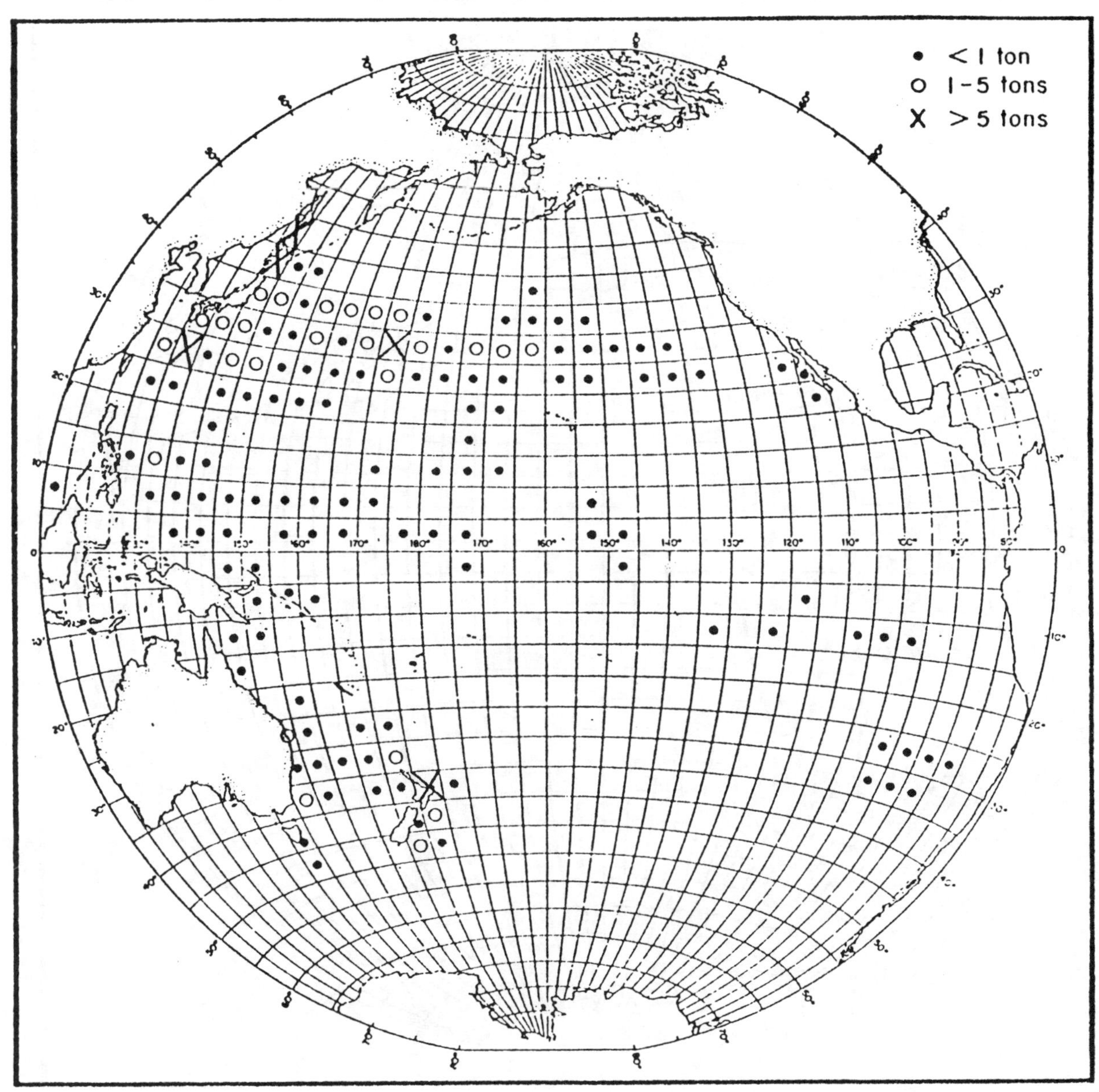

FIGURE 2. Average annual distribution of catches of northern bluefin in the Pacific Ocean by Japanese longliners during 1972-1976 (from Bayliff, 1980).

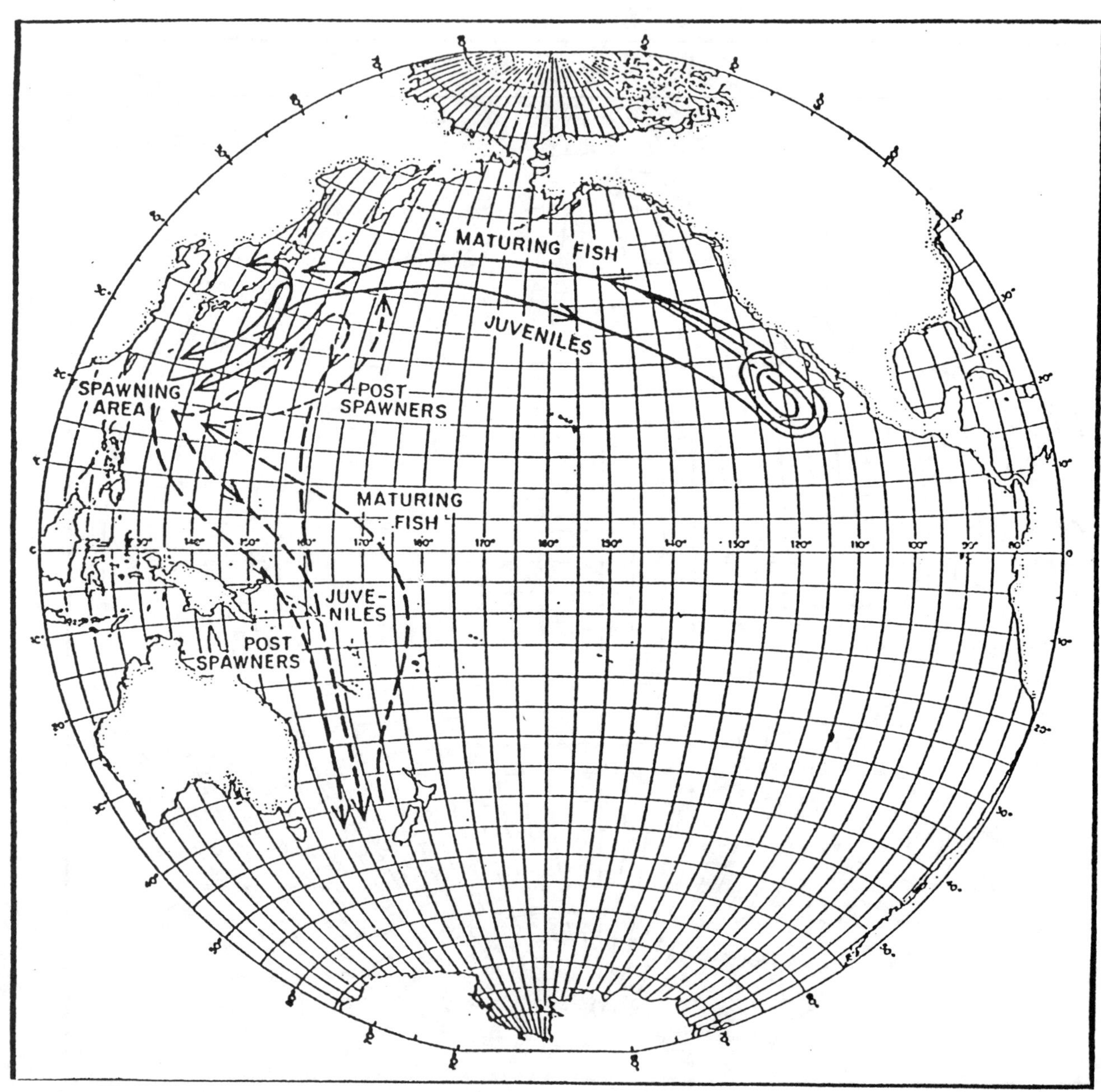

FIGURE 3. A model for northern bluefin migration in the Pacific Ocean (from Bayliff, 1980).

route of migration of maturing fish bound in the opposite direction, but such is not necessarily the case.

The spawning of northern bluefin occurs between Japan and the Philippines in April, May, and June, off southern Honshu in July, and in the Sea of Japan in August (Yamanaka and staff, 1963; Yabe *et al.*, 1966; Okiyama, 1974 and 1979; Nishikawa *et al.*, 1985). The larvae, postlarvae, and juveniles produced south of Japan are transported northward by the Kuroshio Current toward Japan. Fish of age 0 about 15 to 60 cm in length are caught in the vicinity of Japan during the summer, fall, and winter of their first year of life (Yabe *et al.*, 1966; Yukinawa and Yabuta, 1967). The results of tagging experiments indicate that some of these remain in the western Pacific Ocean and others depart for the eastern Pacific during the fall or winter of their first year of life or the summer, fall, or winter of their second year of life. If the fish are restricted to the temperature range given in Section 8.1 when crossing the ocean, they probably occur mostly in the region shown in Figure 4, which more or less coincides with the North Pacific Subarctic-Subtropical Transition Zone (Roden, 1991). The journey from the western to the eastern Pacific takes as little as 7 months, or perhaps even less. Of 121 such migrants for which dates of recapture were obtained, 23 had been at liberty 215 to 358 days and 98 had been free for 368 to 999 days. It is possible that other fish migrate from the western to the eastern Pacific later in life, but there is no information concerning this because few tagged fish greater than about 1 year of age have been released in the western Pacific. It is also possible that some fish travel part of the way from the western to the eastern Pacific, and then turn back.

The fish which migrate from the western to the eastern Pacific form the basis for the fishery in the eastern Pacific, which takes place principally during May through October. In Figure 3 it appears that the fish in the eastern Pacific occur further and further offshore and have an increasingly restricted north-south distribution as they grow older. It was necessary for the sake of clarity to make the diagram this way, but actually such is not necessarily the case. Fish less than about 100 cm in length, which make up the bulk of the eastern Pacific catch, may or may not leave the eastern Pacific Ocean each fall or winter. Northern bluefin of that size are seldom caught in the eastern Pacific during November-April, which might indicate that they have left that region. If so, they probably do not go all the way to the western Pacific, however, as the minimum time at liberty observed for a tagged fish released in the eastern Pacific and recaptured in the western Pacific was 674 days (Table 4); if most of them migrated to the western Pacific each fall or winter and back to the eastern Pacific each spring some would probably have been recaptured in the western Pacific in the winter or early spring after less than about 100 to 150 days at liberty. Also, the energy costs of making such a long migration are so great that it would probably not be feasible for a fish to make two such migrations each year for several years.

The length of the sojourn in the eastern Pacific appears to be variable. Tagged fish believed to have been 1 year of age at the time of release in the eastern Pacific have been recaptured 2 years later in the western Pacific, but other tagged fish released in the eastern Pacific have been recaptured in the eastern Pacific after as long as 2 years at liberty. Fish of at least 6 or 7 age groups are caught in the eastern Pacific (Bell, 1963b; Schultze and Collins, 1977), so it is possible that some fish stay in that region for at least 5 or 6 years.

After a sojourn in the eastern Pacific, which may or may not be interrupted by visits to the central or western Pacific, the survivors return to the western Pacific, where they

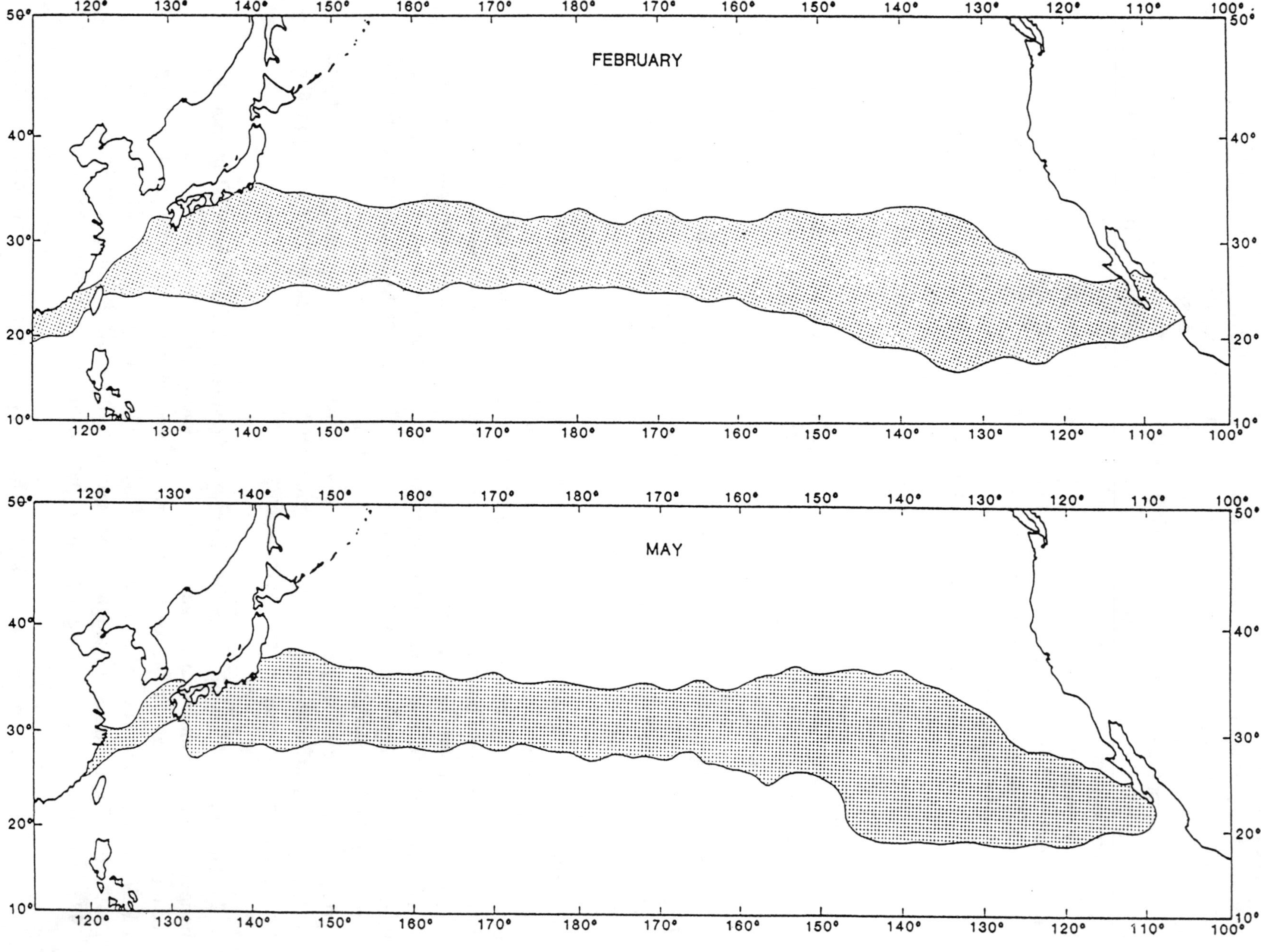

FIGURE 4. Areas of the north Pacific Ocean bounded by the 17° and 23°C surface isotherms (after Robinson, 1976).

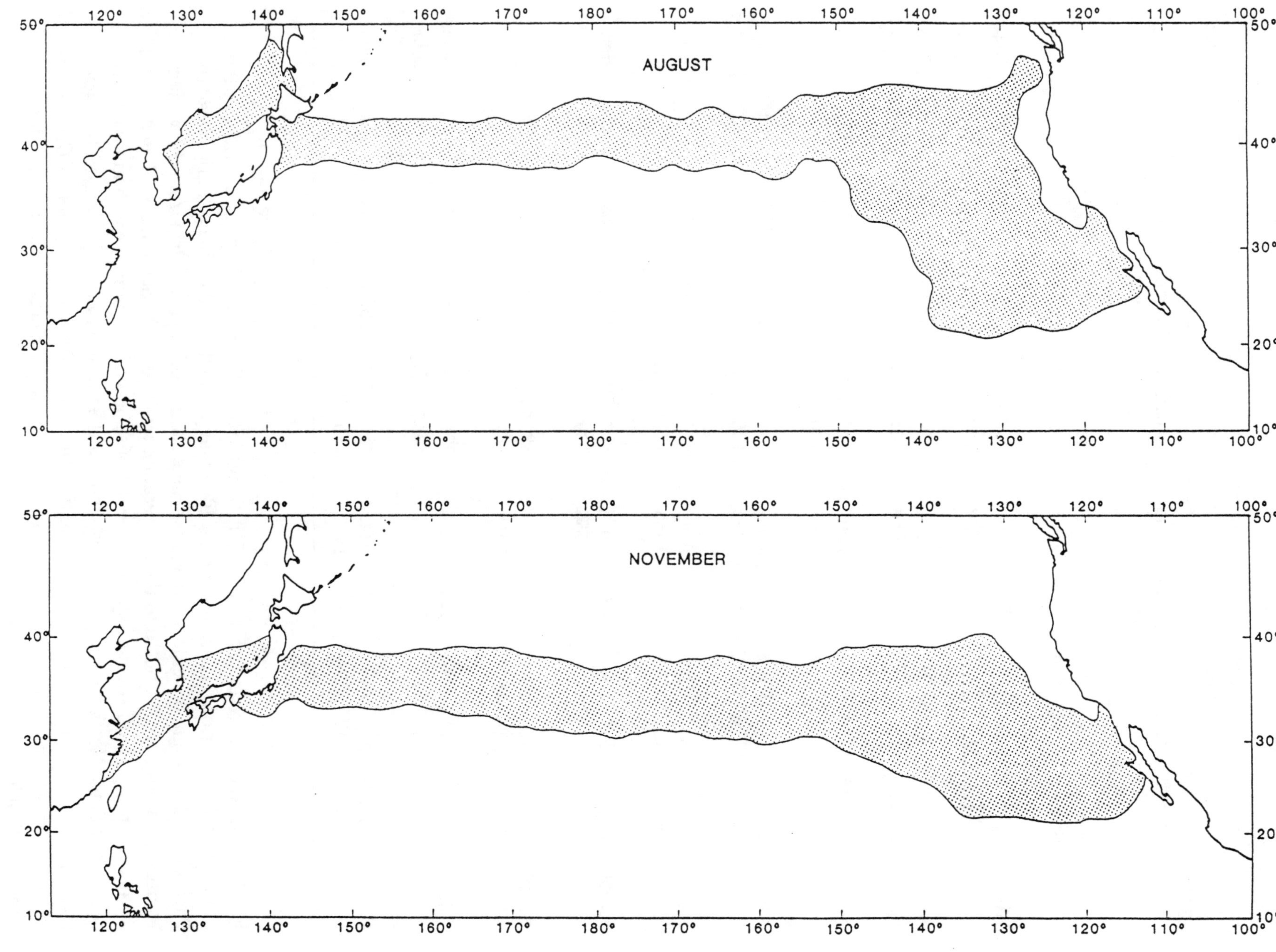

FIGURE 4. (continued)

TABLE 4. Data on releases and recaptures of tagged bluefin which crossed the Pacific Ocean from
east to west (from Bayliff *et al.*, 1991).

Tag number	Release			Recapture			Days free	Distance (nm)	Gear
	Date	Location	Length (mm)	Date	Location	Length (mm)			
N1033	Feb. 2, 1958	28°45'N- 118°15'W	?	Apr. 23, 1963	29°03'N- 139°42'E	1825	1907	5147	long-line
A0374	Aug. 15, 1962	29°43'N- 117°20'W	?	Jun. 18, 1964	40°44'N- 140°00'E	?	674	4781	trap
A0405	Aug. 15, 1962	29°43'N- 117°20'W	?	Aug. 17, 1964	41°39'N- 141°09'E	1150	734	4708	long-line
A0575	Aug. 16, 1962	29°50'N- 117°13'W	?	Aug. 29, 1964	41°15'N- 140°43'	?	745	4737	sport
A0189	Aug. 14, 1962	29°48'N- 116°57'W	?	Jun. 23, 1965	34°16'N- 136°54'E	1380	1045	5124	trap
A0603	Aug. 21, 1962	33°21'N 119°01'W	?	Aug. 10, 1965	41°43'N- 141°03'E	1200	1086	4514	troll
A1950	Aug. 20, 1964	32°41'N 117°55'W	600	Jul. 1, 1966	38°25'N- 139°15'E	1100	681	4746	trap
A1608	Aug. 13, 1964	30°30'N- 116°45'W	850	Jul. 4, 1968	38°49'N- 142°28'E	?	1422	4734	purse seine
A2123	Aug. 20, 1964	32°41'N- 117°55'W	650	Jul. 14, 1968	39°37'N- 143°16'E	1580	1425	4549	purse seine
P3134	Aug. 16, 1979	31°19'N- 117°50'W	850	Apr. 29, 1982	33°00'N- 136°30'E	1526	988	5092	purse seine

eventually spawn. Nakano and Bayliff (1992: Figure 16) show catches of bluefin by
longlines between 25°N and 35°N and 120°W and 150°W during the first and fourth quarters
of the 1981-1987 period. These were most likely migrating from the eastern to the western
Pacific, but they might have been arriving in the eastern Pacific after a trip from the western
Pacific. If the fish are restricted to the temperature range given in Section 8.1 when crossing
the ocean, they probably occur mostly in the regions shown in Figure 4, which more or less
coincide with the North Pacific Subarctic-Subtropical Transition Zone (Roden, 1991). The
return journey from the eastern to the western Pacific may take nearly 2 years, as 674 days is
the minimum time recorded between release and recapture of a tagged fish making this
migration (Table 4).

Large fish are occasionally caught in the eastern Pacific Ocean, especially in the vicinity of Guadalupe Island, Mexico, and the Channel Islands, off Southern California (Calkins, 1982: Figures 5 and 6; Foreman and Ishizuka, 1990; Bayliff, 1993b: Figure 19). The largest of these are probably over 10 years old. These have not necessarily resided in the eastern Pacific Ocean since they were about 1 to 2 years old. They may have arrived for the first time shortly before they were caught, or they may have made more than one round trip across the Pacific. It seems unlikely that all the large fish could spawn in the western Pacific each year, as some have been caught during or shortly before or after the spawning season, *e.g.* July 1978 (Calkins, 1982: Figure 6).

Many of the fish caught by longliners in the mid-Pacific (Figure 2) are presumably *en route* from the eastern to the western Pacific. Upon arriving in the western Pacific they presumably proceed to one or more of the spawning areas to spawn, either immediately or eventually.

Northern bluefin are also caught by longline vessels east of the Philippines, northeast of Papua New Guinea, and southeast of Australia, especially in the vicinity of New Zealand (Figure 2), and by sport gear in the Gulf of Papua (Collette and Smith, 1981). The question arises as to whether these fish come from juveniles which went south from one of the spawning areas, from immature fish which migrated south from Japan after a brief sojourn there, or from older fish which migrated south from the spawning area after spawning. If either the first or second possibility is the case there should be small bluefin south of 20°N. Small numbers of baitboat-caught northern bluefin have been recorded north of Papua New Guinea and in the vicinity of the Solomon and Marshall Islands (Fisheries Agency of Japan, 1977b, 1977c, 1977d, 1977e, and 1977f; this report: Figure 5), so it appears that at least some of the northern bluefin caught south of the spawning grounds by longline vessels are the result of movement of juveniles from the spawning area or the result of migration of immature fish south from Japan. This does not mean, however, that none of them are the result of movement of adult fish south from the spawning areas after spawning.

Larvae or postlarvae of northern bluefin have not been found in the Australia-New Zealand area, so it does not appear that the fish which were caught there by longlines would have spawned there. Rather, they would have to migrate back to the spawning area or not spawn at all. It seems unlikely that all the large fish found in the southwestern Pacific spawn in the northwestern Pacific each year, as the large bluefin reported by Collette and Smith (1981) off Papua New Guinea was caught in April. Also, the bluefin reported by de Buen (1958) in the southeastern Pacific were caught in April and May, during the spawning season.

After spawning, the fish probably disperse from the spawning areas to other areas of the western Pacific. Some may even migrate to the eastern Pacific, as large fish are found there (see above). The following year, if they have not travelled too far, they presumably return to the spawning areas to spawn again.

7. NATURAL MORTALITY

There are no data available for estimating directly the coefficient of natural mortality of northern bluefin in the Pacific Ocean. Pauly (1980) used data on the growth, mortality, and mean environmental temperatures of 175 stocks of fish to calculate the following equation:

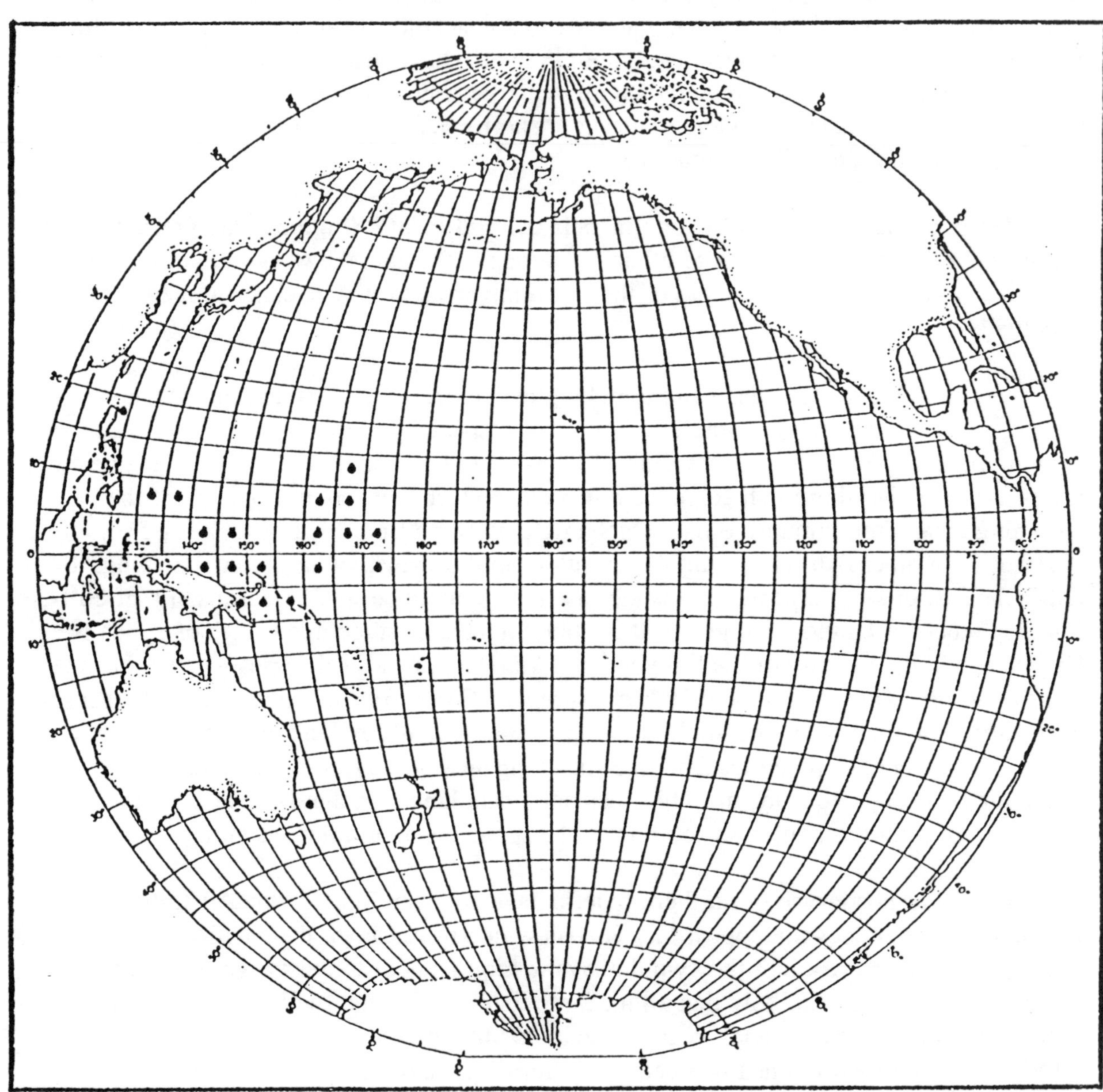

FIGURE 5. Five-degree areas south of 20° N in the Pacific Ocean in which northern bluefin were caught by surface gear during 1972-1976 (from Bayliff and Calkins, 1979).

$$\ln M = -0.0066 - 0.279\ln L_{\infty} + 0.6543\ln K + 0.4634\ln T$$

where

M = coefficient of natural mortality,
L_{∞} = asymptotic length in von Bertalanffy equation,
K = coefficient of growth in von Bertalanffy equation, and
T = mean environmental temperature.

Values of 223 cm and 0.18 for L_{∞} and K, respectively, are given by Bayliff *et al.* (1991: Table 5, line 1). (It should be noted, however, that the best fit to the growth data was not a von Bertalanffy curve, but a Gompertz curve for fish between 160 and 564 mm and a straight line for fish from 564 to 1530 mm.) Pauly (1980) states that, "the values of L_{∞} used [to derive the equation] pertain mostly to total length," so the value of 223 cm for the fork length was converted to 257 cm for the total length by the formula

$$TL = -17.42 + 1.158FL$$

where TL = total length and FL = fork length, both in millimeters. Bell (1963a) states that bluefin are found in water with surface temperatures of 17° to 23°C, so 20 was substituted for T. This gives an estimate of 0.276 for M for fish in the range of about 16 to 153 cm. The standard deviation for $\ln M$ is 0.245 (Pauly, 1980), so the 90-percent confidence limits of the estimate are about 0.161 and 0.471.

Northern bluefin of the Pacific Ocean have lived as long as 16 years in captivity (INFOFISH, 1991), but the life span of wild fish may be more or less than that.

8. OCEANOGRAPHIC FEATURES ASSOCIATED WITH THE SPECIES

8.1 Eastern Pacific Ocean

During their sojourn in the eastern Pacific Ocean, northern bluefin are residents of the California Current, a region of upwelling off California and Baja California. The California Current is described by Reid, Roden, and Wyllie (1958) and Reid (1988). According to Bell (1963a), bluefin are found most often in the eastern Pacific Ocean in waters with surface temperatures between 17° and 23°C (Figure 4). Flittner (1966) stated that 95 percent of the temperatures recorded for 777 purse-seine sets in which bluefin were caught during 1959-1963 were between 17° and 22°C. Hester (1961) noted that bluefin are distributed further to the north in years when the sea-surface temperatures are above normal and further to the south in years when those temperatures are below normal.

8.2 Central Pacific Ocean

During their trans-Pacific migrations northern bluefin probably inhabit the North Pacific Subarctic-Subtropical Transition Zone (Roden, 1991), which more or less coincides with the zones with sea-surface temperatures of 17° to 23°C shown in Figure 4.

8.3 Western Pacific Ocean

In the western Pacific Ocean most of the northern bluefin inhabit the Kuroshio Current (Marr, 1970; Stommel and Yoshida, 1972; Sugawara, 1972; Takenouti, 1980). This current probably plays an important role in transporting the larvae and postlarvae northward from the spawning grounds between Japan and the Philippines and southeast of Japan to waters off Japan. According to Uda (1957), bluefin are found most often in the western Pacific Ocean in waters with surface temperatures between 14° and 19°C. Kida (1936) gives this range as 15° to 17°C for waters off Hokkaido. He said that the temperature range increases with increasing size of the fish. Bluefin are much more plentiful off Japan in years when the sea-surface temperatures are above normal than when they are below normal (Uda, 1962 and 1973). Data on the sea-surface temperatures and lengths of fish tagged during the Inter-American Tropical Tuna Commission's tagging experiments of 1980-1982 in the western Pacific Ocean are given in Table 5. Fish 15 to 31 cm in length were caught in the Kochi summer fishery at temperatures of about 24° to 29°C, fish 31 to 60 cm in length were caught by the Kagoshima winter and spring troll and trap fishery at temperatures of about 15° to 21°C, and fish 44 to 52 cm in length were caught by the Toyama Bay spring and summer trap fishery at temperatures of about 20° to 22°C.

Ogawa and Ishida (1989a and 1989b) reported that peak catches of adult northern bluefin occur off northeastern Honshu during May-July and peak catches of juvenile bluefin occur in the same area during June-August and October-December. They stated that the declines in the catches of both adults and juveniles since the mid-1960s have coincided with decreasing latitudes of the southern limit of the cold waters of the First Oyashio Intrusion.

9. INTERACTIONS WITH OTHER SPECIES

9.1 Feeding and Food

Uotani *et al.* (1990) found that bluefin larvae collected near the Nansei Islands, Japan, had fed on small zooplankton, mainly copepods. Those less than 5 mm long consumed copepod nauplii, while those more than 5 mm long ate larger copepods, especially *Corycaeus*.

Blunt (1958), referring to surface-caught fish, stated: "Of 168 bluefin taken off California [during 1957], 70 percent had been feeding on anchovies [*Engraulis mordax*] ... Bluefin sometimes feed close inshore near the surf. Tuna captured in shallow water had eaten starfish, kelp, sandabs [*Citharichthys*], surf perches [Embiotocidae] and white croakers [*Genyonemus lineatus*]." Bell (1963a) listed 14 species of fish found in northern bluefin stomachs. Pinkas (1971) reported that surface-caught bluefin sampled in 1968 and 1969 consumed, in order of importance, anchovies, red crabs (*Pleurocodes planipes*), sauries (*Cololabis saira*), squid (*Loligo opalescens*), and hake (*Merluccius productus*). Anchovies made up 80 percent of the stomach contents by volume. In addition, other species of fish, cephalopods, and crustaceans, *Atlanta* (a gastropod), and pieces of kelp were found in the stomachs. He remarked that bluefin occurred further offshore in 1968 and 1969 than in 1957, which probably accounted at least partly for the differences in his results and those of Blunt (1958). Scott and Flittner (1972) found that red crabs are a significant part of the diet only south of 29°N, and that boiling and jumping schools of fish are much more common north of that latitude, where fish are the principal item of the diet. They stated that the differences in

TABLE 5. Lengths (cm) and sea-surface temperatures (°C) of fish tagged during the IATTC tagging experiments of 1980-1982 in the western Pacific Ocean.

| | 1980 | | | 1981 | | | 1982 | | |
	Area	Range of lengths (cm)	Range of temperatures	Area	Range of lengths (cm)	Range of temperatures	Area	Range of lengths (cm)	Range of temperatures
July				Kochi	15-24	26.2-28.2	Kochi	19-26	24.2-25.2
August	Kochi	22-31	27.0-28.0	Kochi	16-28	26.8-28.9	Kochi	17-29	25.2-27.8
September	Kochi	27	28.0	Kochi	19-27	27.3-27.8			
January	Kagoshima	31-55	16.5-19.0						
February	Kagoshima	32-50	15.0-17.5						
March	Kagoshima	33-58	15.0-19.0						
April	Kagoshima	37-57	17.0-21.0						
May	Kagoshima	45-60	18.0-20.0						
June	Toyama Bay	50	20.0						
July	Toyama Bay	44-52	20.0-22.0						

behavior in the two areas could be due to differences in the food, *i.e.* "modified filter feeding" might be employed while feeding on red crabs, while "vigorous pursuit" would be required for feeding on fish.

Yamanaka and staff (1963) summarized the available information on the feeding and food of northern bluefin in the western Pacific. Fish 20 to 65 cm in length consume anchovies and other fish, plus crustaceans and squid, while longline-caught (larger) fish eat fish and squid. Fish of unspecified size were reported to consume both pelagic and demersal fish, *Calamarius* (probably a squid), pteropods, *Pyrosoma* (a tunicate), and crustaceans. In addition to the above, Doi (1960) reported that bluefin eat round herring (*Etrumeus teres*), pampanos (Carangidae), and mackerel (*Scomber japonicus*). Yokota *et al.* (1961) stated that bluefin consume frigate mackerel (*Auxis* spp.), spotted mackerel (*Scomber tapeinocephalus*), jack mackerel (*Trachurus japonicus*), scad (*Decapterus muroadsi*), sauries, skipjack tuna (*Katsuwonus pelamis*), anchovies (*Engraulis japonica* and *Stolephorus zollingeri*), beaked salmon (*Gonorhynchus abbreviatus*), *Acinacea notha*, argentines (*Argentina semifasciata*), sprat (*Spratelloides japonicus*), lanternfishes (Myctophidae), and common squid (*Todarodes pacificus*). Mori (1972) also found skipjack in the stomachs of bluefin. Kida (1936) said that a "certain connexion seems to exist between the incoming schools of sardine [*Sardinops melanosticta*] ... on one side and the [bluefin] fishery on the other," and, according to Yamanaka and staff (1963), "Nakamura (1949) reported that ... the route of the northward migration of this species coincides with that of the migration of the sardine; ... when sardine suddenly decreased, the fishing of [bluefin] simultaneously declined. Uda (1960) reported that (i) the resources of both this species and of sardine were large from 1933 to 1940, but on and after 1941 decreased simultaneously; (ii) with the decrease in resources, the northern limitation of the area of occurrence for both kinds of fishes shrank southward." The catches of sardines in Japan increased greatly during the 1976-1985 period (Yamanaka *et al.*, 1988: Table 4); the catches of bluefin in the western Pacific increased from 1976 to 1981, but then decreased from 1981 to 1985 (Table 6).

9.2 Competitors

Bell (1963a) stated that albacore (*Thunnus alalunga*), yellowtail (*Seriola dorsalis*), barracuda (*Sphyraena argentea*), and mackerel (*Scomber japonicus*) compete with northern bluefin for food. According to Yamanaka and staff (1963), "it can be supposed that other kinds of tunas, spearfish [*Tetrapturus angustirostris*], Spanish mackerel [*Scomberomorus* spp.], and some others are competitors."

9.3 Predators

Yokota *et al.* (1961) mentioned the presence of juvenile northern bluefin in the stomachs of skipjack. Yamanaka and staff (1963) stated that "immature forms ... have many enemies--seals, dolphins, spear-fishes, sword-fishes [*Xiphias gladius*], sharks, and large forms of their own or allied species. Tunas are devoured by killer-whales which are said to catch them at the nape and kill them immediately. They fear killers so greatly that they are frightened several miles away from the spots where these ferocious enemies are found. Thus catches made by pound-nets [traps] vary greatly according to the favorable or unfavorable proximity of killers. Sometimes tunas leap recklessly on to beaches to escape their enemies."

TABLE 6. Catches, in metric tons, of northern bluefin in the western Pacific Ocean. ROC and ROK stand for Republic of China and Republic of Korea, respectively.

Year	Japan	ROC	ROK	Year	Japan	ROC	ROK
1951	15,400	*	*	1972	7,831	14	0
1952	11,556	*	300	1973	10,562	33	0
1953	16,573	*	300	1974	18,951	14	0
1954	15,095	*	0	1975	11,174	61	2
1955	19,138	*	200	1976	8,208	17	0
1956	27,742	*	0	1977	12,427	130	0
1957	20,269	*	0	1978	20,341	66	21
1958	8,277	*	0	1979	23,506	58	0
1959	7,864	*	0	1980	20,149	100	0
1960	10,924	*	0	1981	30,057	179	0
1961	10,089	*	0	1982	24,726	204	0
1962	10,327	*	0	1983	17,278	168	0
1963	9,948	*	0	1984	9,758	476	0
1964	9,309	*	0	1985	11,819	808	0
1965	11,330	*	0	1986	13,183	147	0
1966	13,486	0	0	1987	19,321	412	43
1967	13,892	52	0	1988	6,832	189	0
1968	16,591	32	0	1989	11,951	323	0
1969	9,022	14	0	1990	5,406	358	0
1970	7,715	0	0	1991	16,921	361	0
1971	10,046	1	0	1992	10,838	471	*

* data not available

9.4 Parasites

Information on parasites of northern bluefin is given by Bell (1963a), Silas (1967), Silas and Ummerkutty (1967), Love and Moser (1977), and Cressey and Cressey (1980).

10. GENERAL DESCRIPTION OF THE FISHERIES

10.1 Eastern Pacific Ocean

Sport fishing for northern bluefin began in California in 1898 (Bell, 1970). Prior to World War I, before the advent of the commercial fishery, many large fish were taken, particularly by vessels based at Catalina Island. The largest of these fish weighed 251 pounds (114 kg). More recently, the average size of the sport-caught fish has been roughly 25 pounds (11 kg), although large fish are still taken (Dotson and Graves, 1984). Many of the sport-caught fish are taken by fishermen who are directing their efforts primarily toward albacore.

The purse-seine fishery for northern bluefin began in 1918 (Whitehead, 1931), and since then the catches by purse seiners have far exceeded those by any other type of gear. From 1918 until about 1959 or 1960 most of the vessels were relatively small, with capacities less than about 200 short tons (181 mt). None of them fished exclusively for bluefin. The smaller ones fished chiefly for sardines, mackerel, and other pelagic fish other than tunas, and the larger ones fished mostly for yellowfin and skipjack. During 1959 and 1960 most of the larger tuna baitboats were converted to purse seiners, and during the ensuing years many new purse seiners were built. During the 1960s, 1970s, and 1980s many of the smaller, older vessels sank or dropped out of the fishery, and the new vessels which replaced them tended to be larger. As a result, there are now more larger purse seiners and fewer smaller ones than had been the case during the early 1960s. Bluefin are now taken by vessels of all sizes, but the smaller ones account for a proportionally greater share of the catch. The proportion of the catch made by vessels which fish primarily for pelagic fish other than tunas is less now than it was during the early years of the fishery, as the centre of abundance of the fish appears to have shifted southward beyond the area where those vessels normally fish. Most of the fish caught by purse seiners weigh less than about 50 pounds (23 kg), but larger ones are sometimes caught, including one weighing 1,009 pounds (458 kg) (Section 4).

Most of the information regarding the distribution of the catches by tuna purse seiners after 1947 has been obtained from the logbook records of these vessels. The vessel captains or navigators usually record the catch of each set in short tons, the location of each set to the nearest degree and minute of latitude and longitude, the date, and the times each set was initiated and completed. An abstract of the vessel logbook is prepared at the end of each trip. These abstracts are incorporated into the Inter-American Tropical Tuna Commission's computer records, from which summaries of the northern bluefin catch, stratified by any combination of area, time, vessel size class, *etc.*, can be prepared. During the 1961-1991 period the annual logged catches ranged from 69 to 100 percent of the corresponding weighed-out catches.

Northern bluefin are rarely encountered south of Cabo San Lucas, Baja California, or north of Point Conception, California. Within this area a considerable change has taken place during the 20th century (Bell, 1970). Fishing was conducted only off California prior to 1930. During that year bluefin were discovered off Isla Guadalupe, Baja California, located at about 29°N-118°W, and about 40 percent of the catch in 1930 was made in that area (Whitehead, 1931). From 1930 through 1947, fishing was conducted off both California and Baja California, but in most years the majority of the catch came from off California. From 1948 to the present, however, most of the catch has been made off Baja California (Calkins, 1982: Figure 2; this report: Figure 1). In 1989, for the first time in many years, no catches were made south of 30°N (Figure 1). The average annual catches made off California during the 1960s, 1970s, and 1980s have been considerably less than the average annual catches made in the same area during the 1918-1929 period. Part of the reason for this may be that during more recent years the larger vessels fishing off Baja California have tended to intercept the fish before they reach the waters off California.

The distributions of the catches by months are shown in Figure 3 of Calkins (1982), Figure 4 of Hanan (1983), and Figure 6 of this report. During January through April there are typically only light and sporadic catches. Most of these are made off the coast of Baja California between 24°N and 26°N and in the vicinity of Isla Guadalupe. In May and June the catches increase, and most of them are made between 24°N and 27°N. During July the fishing area spreads to the north and is at its widest extent of the year; most of the catch is made between 25°N and 33°N. In August there are usually only light catches at the southern end of the fishing area, most of the catch being made between 28°N and 33°N. During

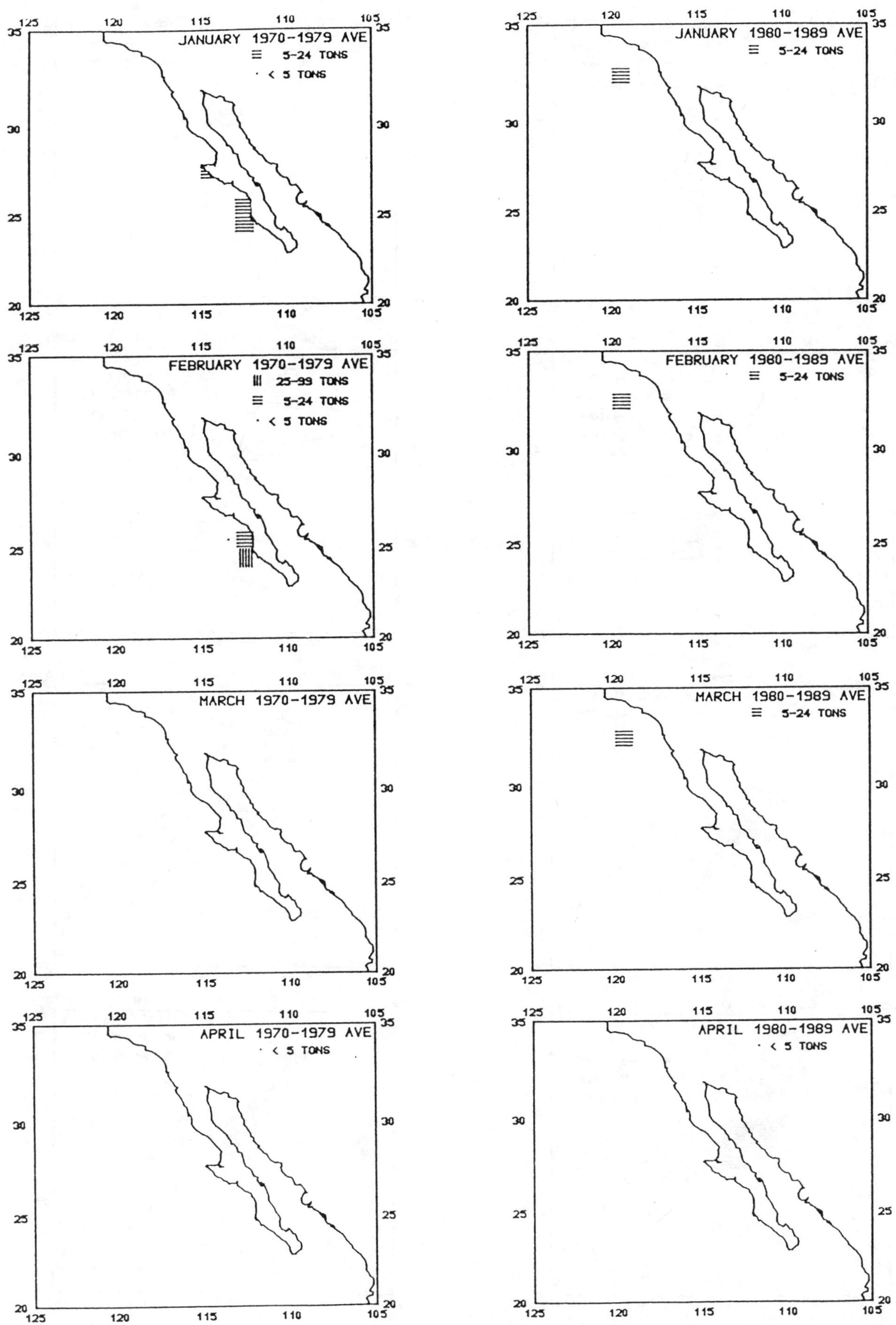

FIGURE 6. Average monthly distributions of northern bluefin catches in the eastern Pacific Ocean, 1970-1979 and 1980-1989.

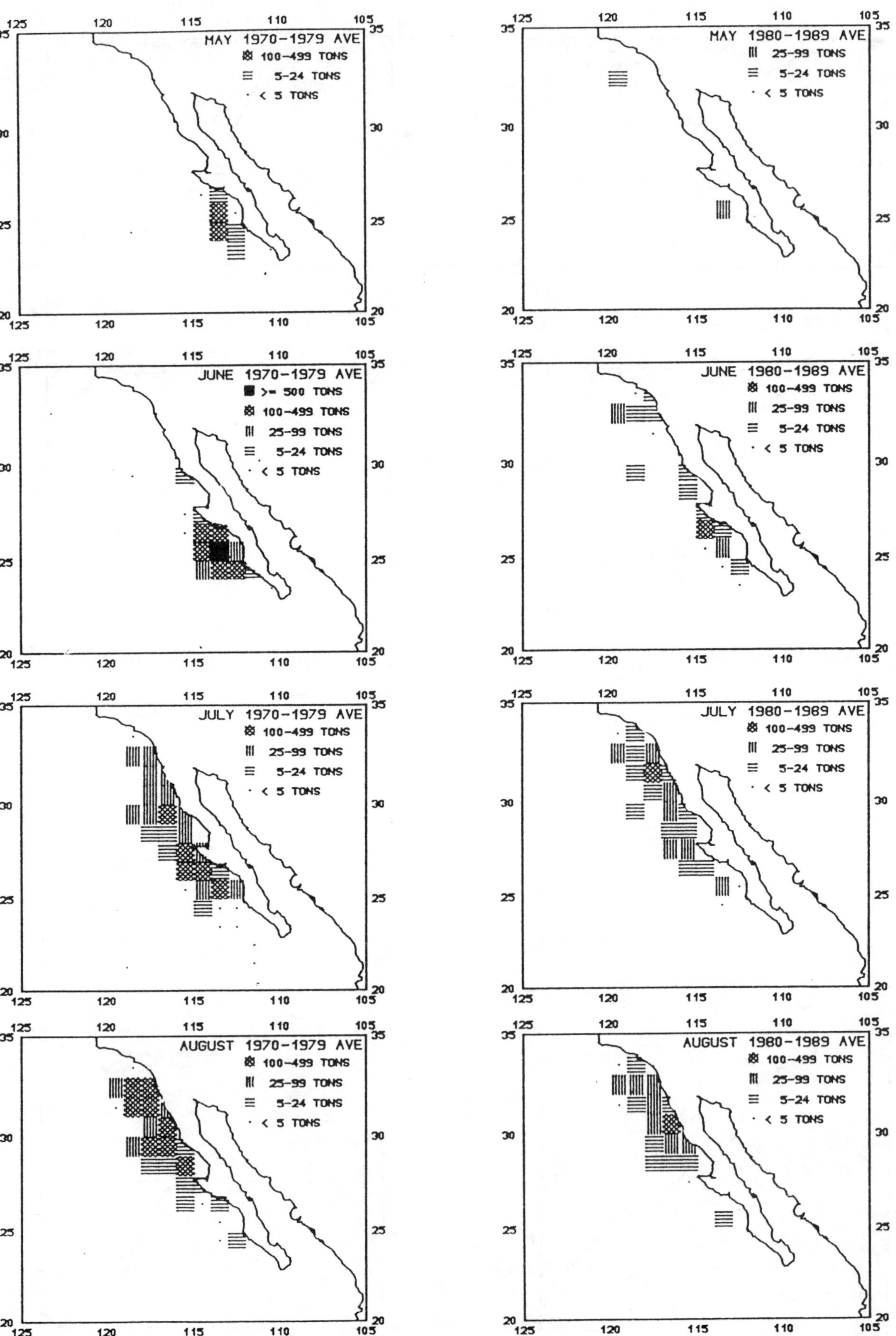

FIGURE 6. (continued)

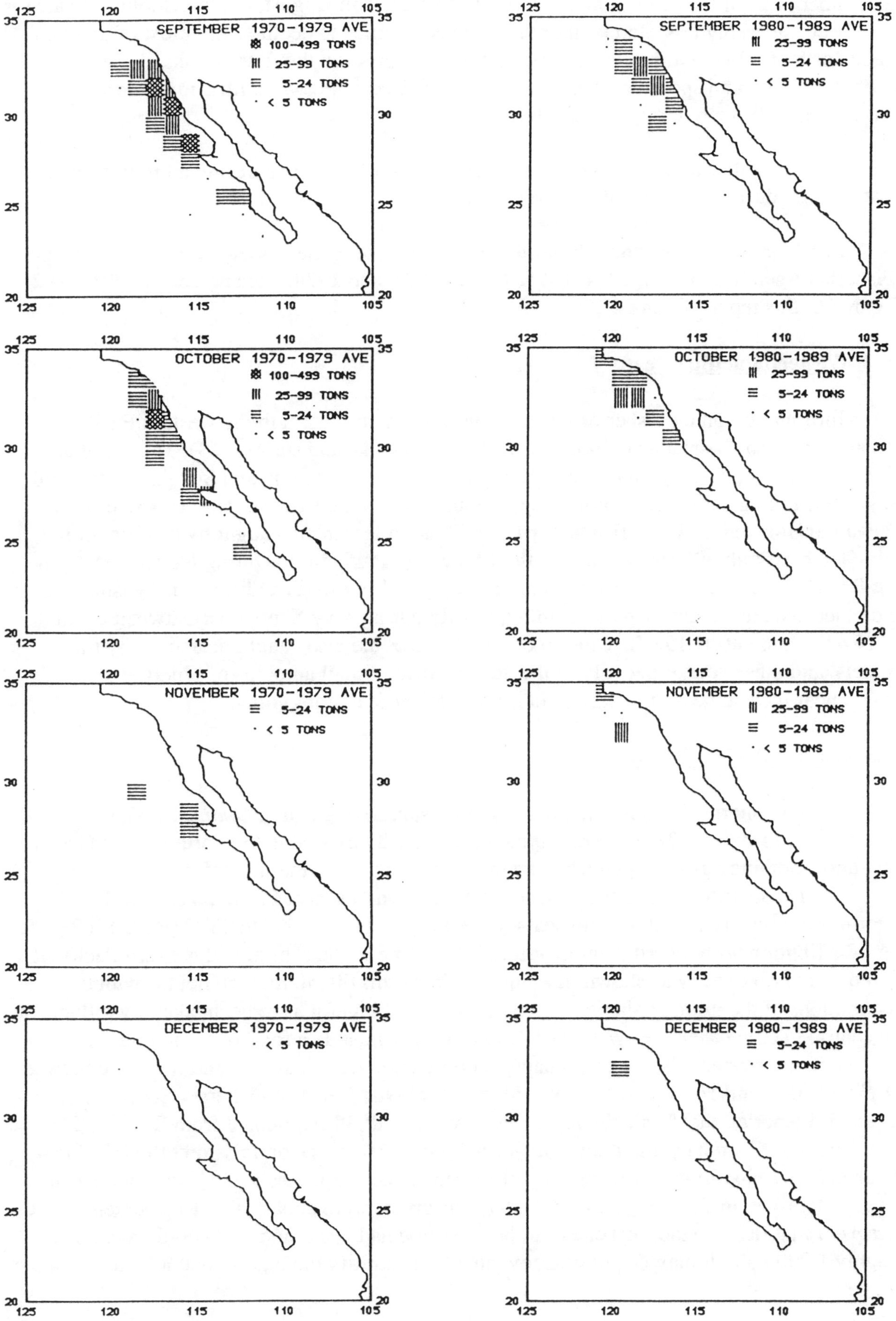

FIGURE 6. (continued)

September most of the catch is made in the same area as in August, but the amount of catch is usually considerably less. In October the catches continue to decline, and most of them are made north of 30°N. During November and December, as in the first months of the year, the catches are light and sporadic. The distribution of the catches was a little more northerly during the 1980s than during the 1970s.

Northern bluefin are caught in the vicinity of Isla Guadalupe during all months of the year, but not during every month of every year.

Small amounts of northern bluefin are caught by longline vessels in the eastern Pacific (Fisheries Agency of Japan, 1974, 1975, 1976, 1977a, and 1978; Nakano and Bayliff, 1992: Figure 16; this report: Figure 2).

10.2 Western Pacific Ocean

Information on the fisheries which exploit northern bluefin in the western Pacific Ocean is given by Yamanaka (1958 and 1982), Yamanaka and staff (1963), Tatsuki *et al.* (1963), Yukinawa and Yabuta (1967), Shingu *et al.* (1974), Honma and Suzuki (1978), and Bayliff (1980). Bluefin are exploited by various gears in the western Pacific Ocean from Taiwan to Hokkaido. Age-0 fish about 15 to 50 cm in length are caught by trolling during July-October south of Shikoku and south of Shizuoka Prefecture. During November-April age-0 fish about 35 to 60 cm in length are taken by trolling south and west of Kyushu. Age-1 and older fish are caught by purse-seining, mostly during May-September between about 30°N-42°N and 140°E-152°E. Bluefin of various sizes are also caught nearshore by traps, gillnets, and other gear, especially in the Sea of Japan. Small amounts of bluefin are also caught near the southeastern coast of Japan and offshore by longlining.

10.3 Central Pacific Ocean

Gillnet fisheries for salmon, *Oncorhynchus* spp.; flying squid, *Ommastrephes bartrami*; and albacore, *Thunnus alalunga*, have recently existed in the north Pacific Ocean, but almost nothing has been published about the catches of northern bluefin by these fisheries. The distributions of these fisheries are shown in Figure 7. In the eastern Pacific, bluefin occur mostly in waters with sea-surface temperatures of 17° to 23°C (Bell, 1963a). If their distribution with regard to temperature is the same in the central and western Pacific it would be approximately as shown in Figure 4. Since this distribution coincides with the distributions of the gillnet fisheries, it seems likely that bluefin have been taken by gillnets in the central Pacific Ocean. Bayliff *et al.* (1991: Table 6) list data for five tagged bluefin released in the western Pacific and recaptured by gillnet vessels in the central Pacific between 34°N and 40°N and 161°E and 161°W. International North Pacific Fisheries Commission (1986: 55) reported that "a single day's catch [of a squid gillnet] ranged from 871 to 8,256 squid and 5 to 530 tuna." International North Pacific Fisheries Commission (1991: Table 4) stated that 107 bluefin were recorded by observers aboard Japanese squid gillnet vessels in 1990. According to San Diego Union (1991), observers were aboard about 10 percent of the vessels of that fleet, so the total catch, if the sampling had been random, would have been roughly 1,000 fish. It may be possible, eventually, to stratify the data so that a better estimate can be calculated.

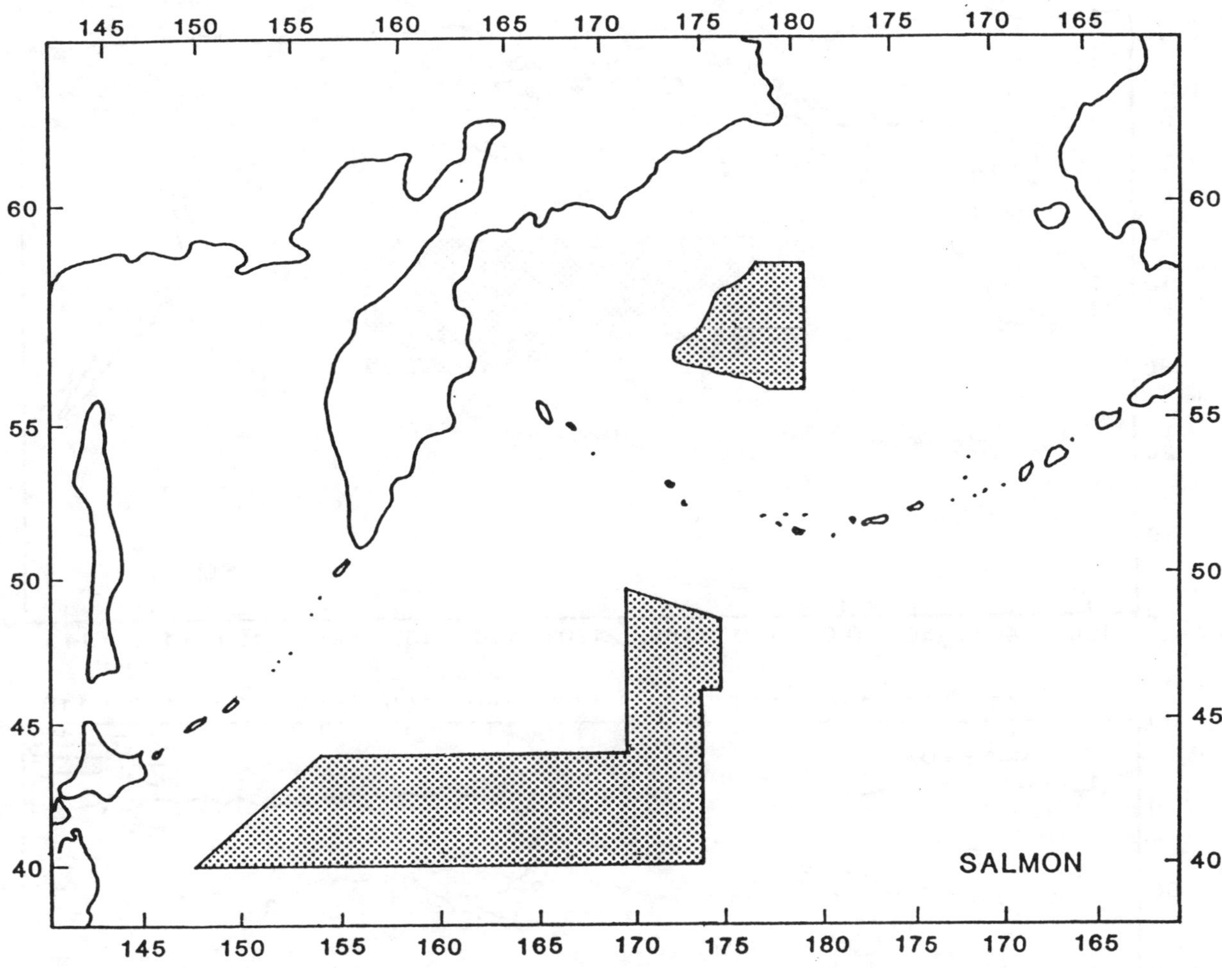

FIGURE 7. Areas in which offshore gillnet fishing by Japanese vessels takes place (after Suzuki, 1990).

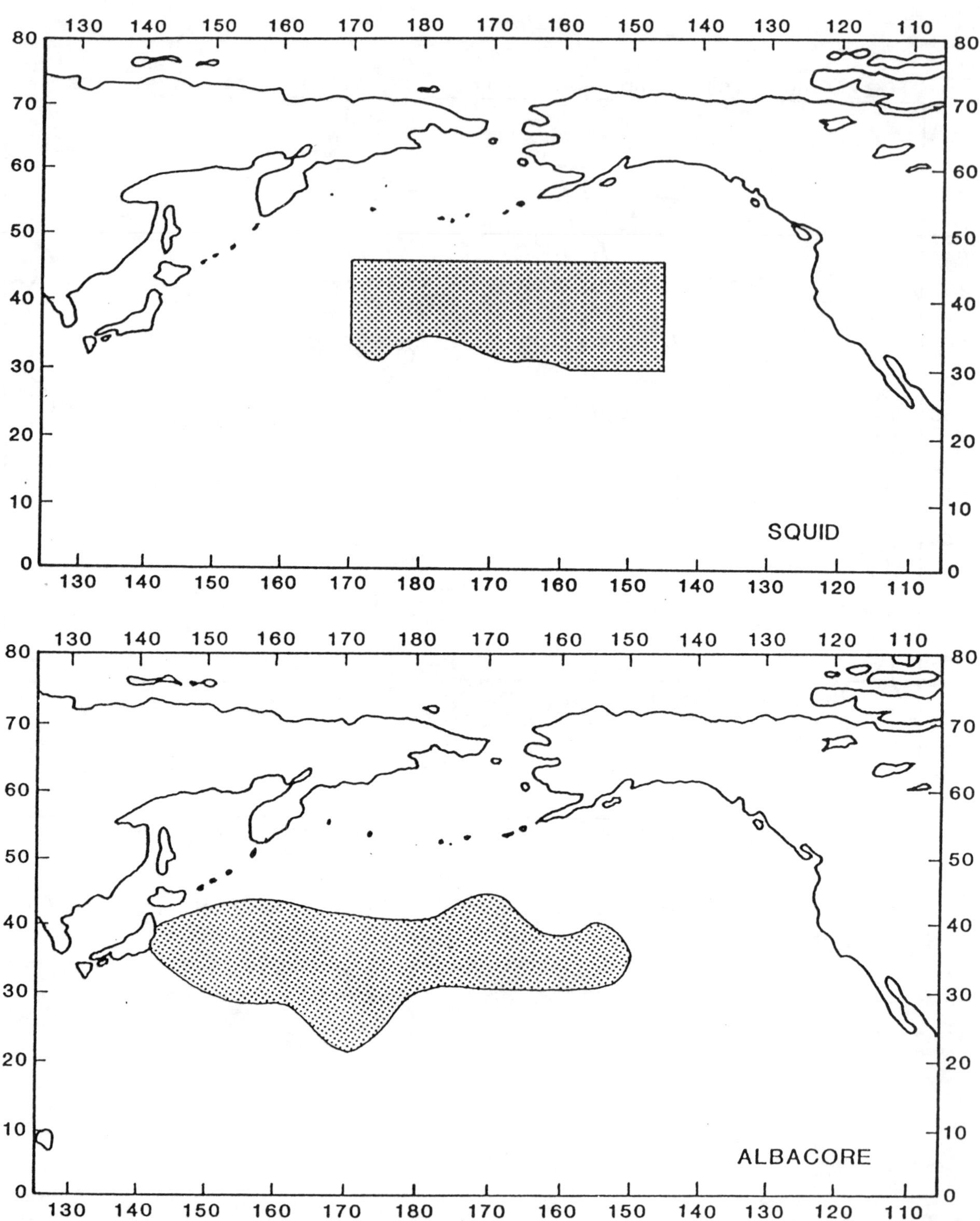

FIGURE 7. (continued)

The Japanese baitboat fishery for albacore extends as far to the east as 175°W (Majors *et al.*, 1984: Figure 2). Bayliff *et al.* (1991: Table 6) reported that one tagged northern bluefin released in the western Pacific was recaptured by a baitboat at approximately 38°N-167°E.

11. TRENDS IN CATCH, EFFORT, AND CATCH PER UNIT OF FISHING EFFORT

11.1 Eastern Pacific Ocean

11.1.1 Annual catches

The total annual catches of northern bluefin by commercial and sport vessels in the eastern Pacific Ocean are shown in Table 7. The catches prior to 1918 were negligible. The

TABLE 7. Catches, in metric tons, of bluefin in the eastern Pacific Ocean. The sources of the data are as follows: commercial - 1918-1960, Bell, 1963a; 1961-1992, Bayliff, 1993b; sport - 1936-1946, Bell, 1963a; 1947-1990, Leet *et al.*, 1992.

Year	Commercial	Sport	Year	Commercial	Sport	Year	Commercial	Sport
1918	2,722		1943	4,617		1968	5,976	1
1919	6,800		1944	9,228		1969	6,926	17
1920	4,776		1945	9,341		1970	3,966	21
1921	894		1946	9,993		1971	8,360	8
1922	1,275		1947	9,452	25	1972	13,348	17
1923	1,460		1948	2,961	1	1973	10,746	61
1924	1,470		1949	1,991	22	1974	5,617	65
1925	1,725		1950	1,242	0	1975	9,583	38
1926	2,960		1951	1,752	81	1976	10,646	23
1927	2,222		1952	2,076	2	1977	5,473	21
1928	6,215		1953	4,433	48	1978	5,396	5
1929	3,414		1954	9,537	11	1979	6,118	12
1930	9,943		1955	6,173	93	1980	2,938	8
1931	1,603		1956	5,727	388	1981	1,090	6
1932	486		1957	9,215	73	1982	3,150	8
1933	254		1958	13,934	10	1983	853	22
1934	8,327		1959	6,914	15	1984	882	32
1935	11,418		1960	5,422	1	1985	4,054	56
1936	8,584	33	1961	8,136	26	1986	5,084	8
1937	5,758	46	1962	11,268	28	1987	995	21
1938	8,041	135	1963	12,271	8	1988	1,423	4
1939	5,369	112	1964	9,218	8	1989	1,169	74
1940	9,058	78	1965	6,887	1	1990	1,541	43
1941	4,318		1966	15,897	23	1991	462	*
1942	5,826		1967	5,889	36	1992	1,895	*

* data not available

data for 1918 through 1960 include only the catches landed in California, but it is believed that the catches landed elsewhere prior to 1961 were inconsequential. The eastern Pacific subsurface catches for all years are omitted, but these are also believed to be inconsequential.

The catches tended to be greater during the 1960s and 1970s than during the previous period. This is probably due to the fact that during 1959 and 1960 most of the tuna baitboats were converted to purse seiners, and during the 1960s and 1970s many new purse seiners were built. During the earlier period the catch was made by smaller purse seiners, most of which fished part-time for tunas and part-time for other pelagic fish, whereas during the later period most of the catch has been made by larger purse seiners which fish full-time for tunas (but not exclusively for northern bluefin, of course).

11.1.2 Monthly catches

Data on the estimated monthly catches of northern bluefin in the eastern Pacific surface fishery, estimated from logged monthly catch and total annual landings data, are given in Table 8. The proportions of the annual catches taken during the various months vary considerably from year to year.

11.1.3 Effort and catch per unit of effort

Calkins (1982) stated that, since the area in which northern bluefin are caught is subjected to intensive searching for tunas and other pelagic fish throughout the year, the "total catch may be the best indicator of bluefin abundance." Nevertheless, he attempted to ascertain the fishing effort directed toward bluefin during 1961-1980. As stated previously, the vessels which catch bluefin also catch considerable quantities of other species of pelagic fish, so it cannot be considered that all their effort is directed toward bluefin. Accordingly, he adopted the following rules: (1) no effort made south of 23°N was considered to be bluefin effort; (2) no effort during the November-April period was considered to be bluefin effort; (3) no effort in 1-degree area-month strata in which no sets were made on bluefin was considered to be bluefin effort. In the area-time strata in which bluefin effort was assumed to occur during the 1961-1980 period, the catches of bluefin ranged from 19 to 83 percent of the total catch, with an average of 51 percent (Calkins, 1982: Table 5). These effort data are listed in his Table 6. No attempt was made to standardize these data by vessel size class. Hanan's (1983) rules were similar, but not identical, to those of Calkins (1982). He stated that "one boat-day or part of a boat-day of effort is assigned to a seiner for each day or partial day of purse seining or searching for tuna in the bluefin fishing range (north of lat. 22°N) during months in which bluefin were caught."

Calkins (1982) calculated the catches per unit of effort (CPUEs) by dividing the catches of northern bluefin in the area-time strata in which bluefin effort was assumed to occur by the effort in those strata. These results are also shown in his Table 6. The total annual catches of bluefin during 1961-1989, plus the logged bluefin effort and the CPUEs for those years, are shown by Bayliff (1991: Figure 7). The catches and CPUEs for 1961 through 1984 were highly correlated, but after that the relationship broke down, high CPUEs occurring in years with low to medium catches. This is due to the fact that in the years with high CPUEs the fish tended to occur in only a few area-time strata, with high CPUEs in them. In 1989, for example, no bluefin were caught south of 30°N (Figure 1), and 70 percent of the catch was made during August.

TABLE 8. Estimated catches, in metric tons, of northern bluefin in the eastern Pacific Ocean (after Bayliff, 1993a).

Year	Jan.	Feb.	Mar.	Apr.	May	Jun.	Jul.	Aug.	Sep.	Oct.	Nov.	Dec.	Total
1959	0	203	18	449	15	0	2751	3478	0	0	0	0	6914
1960	115	22	0	0	0	1896	2103	1032	222	0	32	0	5422
1961	0	0	0	0	31	1145	1100	4464	1187	141	70	0	8136
1962	24	0	97	34	342	2099	3496	4773	403	0	0	0	11268
1963	48	27	0	109	89	2156	1764	4691	3338	7	15	26	12271
1964	9	0	0	7	21	1704	2454	4116	895	5	6	0	9218
1965	0	7	21	11	4	616	955	2663	1577	1033	0	0	6887
1966	0	0	0	0	230	5012	7007	3485	162	0	0	0	15897
1967	12	0	4	0	52	3076	2392	91	244	18	0	0	5889
1968	0	0	0	0	0	880	2438	1608	869	181	0	0	5976
1969	12	0	1	260	568	775	2564	1980	714	0	0	52	6926
1970	12	0	0	30	0	1415	2164	291	54	0	0	0	3966
1971	0	35	0	0	2151	1941	1435	788	1109	886	0	13	8360
1972	0	0	0	0	380	2530	2652	5368	775	1644	0	0	13348
1973	0	11	0	0	19	2067	5677	2717	255	0	0	0	10746
1974	0	0	0	0	0	1652	1963	866	1045	84	0	6	5617
1975	0	3	0	0	74	3576	2162	516	2245	809	165	32	9583
1976	192	378	0	2	267	1825	309	3750	3399	344	153	25	10646
1977	0	0	0	0	1566	443	70	2213	734	376	31	41	5473
1978	103	0	0	0	0	738	1787	2634	122	14	0	0	5396
1979	0	0	0	0	1938	462	907	2390	277	144	0	0	6118
1980	0	0	0	0	370	1340	290	610	328	0	0	0	2938
1981	0	0	0	0	6	381	160	390	152	0	0	0	1090
1982	0	0	0	0	0	0	1106	856	755	434	0	0	3150
1983	115	0	0	0	0	0	337	220	132	50	0	0	853
1984	8	0	0	54	25	91	357	142	138	66	0	0	882
1985	0	82	0	0	28	1582	332	806	132	582	432	79	4054
1986	0	0	104	15	15	331	1681	1684	392	520	234	108	5084
1987	102	3	0	0	0	38	298	280	213	52	10	0	995
1988	0	0	0	0	0	227	882	106	17	63	89	40	1423
1989	8	0	0	0	0	0	7	814	236	39	20	45	1169
1990	12	6	0	0	0	0	488	305	547	84	100	0	1541
1991	0	0	0	0	0	112	19	331	0	0	0	0	462

Another index of effort and catch per unit of effort, the "habitat index," is described by Bayliff (1993b). Bluefin are most often caught by purse-seine vessels in the eastern Pacific Ocean in waters with surface temperatures of 17° to 23°C (Bell, 1963a), so it was assumed that fishing effort exerted by tuna purse seiners off California and Baja California in waters in that temperature range during May through October is bluefin effort, regardless of whether bluefin were caught. This avoids the problem caused by concentration of bluefin in a small part their range, as discussed above for the index of Calkins (1982). The sums of the logged catches of bluefin in the eastern Pacific Ocean during May-October were divided by the sums of the logged purse-seine effort in the 1-degree areas which were suitable bluefin habitat during the same months to get the annual CPUEs. These data are shown in Figure 8. They appear to indicate that both reduced abundance and reduced fishing effort have contributed to the reduced catches of bluefin in the eastern Pacific Ocean during the 1980s and early 1990s. Two years of this period, 1985 and 1986, were apparently years of high abundance of bluefin, but the catches were only average because the fishing effort was so low. Unless the fishing effort increases, it appears unlikely that catches in the eastern Pacific Ocean greater than those of 1985 and 1986 will be taken in the future.

The relative abundance of many pelagic ocean fishes can also be monitored by aerial surveys. Squire (1983 and 1993) gives information on northern bluefin recorded during aerial surveys of pelagic fishes carried out over the eastern Pacific Ocean, but the surveys covered only waters off Southern California and Baja California north of 30°N.

11.2 Western Pacific Ocean

11.2.1 Annual catches

The total annual catches of northern bluefin in the western Pacific Ocean by vessels of Japan, the Republic of China, and the Republic of Korea are shown in Table 6. The data for Japan are from the National Research Institute of Far Seas Fisheries of Japan. They include small amounts of bluefin caught in the central and eastern Pacific Ocean. The data for 1951-1965 include only large fish (greater than about 15 kg or 33 pounds), while those for 1966-1992 include both small and large fish. The data for the small fish were estimated from the catches of *meji* (small bluefin, yellowfin, and bigeye) by a proration process based upon the catches of large bluefin, yellowfin, and bigeye. The data for the Republic of China are from the yearbooks of fisheries statistics of the Taiwan Fisheries Bureau and those for the Republic of Korea are from the yearbooks of fisheries statistics of the Food and Agriculture Organization of the United Nations. Data on the Japanese catches by gear are shown in Table 9.

11.2.2 Monthly catches

No information on this subject, other than that given in Section 10.2, is available.

11.2.3 Effort and catch per unit of effort

Except for data on the numbers of Japanese purse-seine vessels and traps (Skillman and Shingu, 1980), no information on this subject is available.

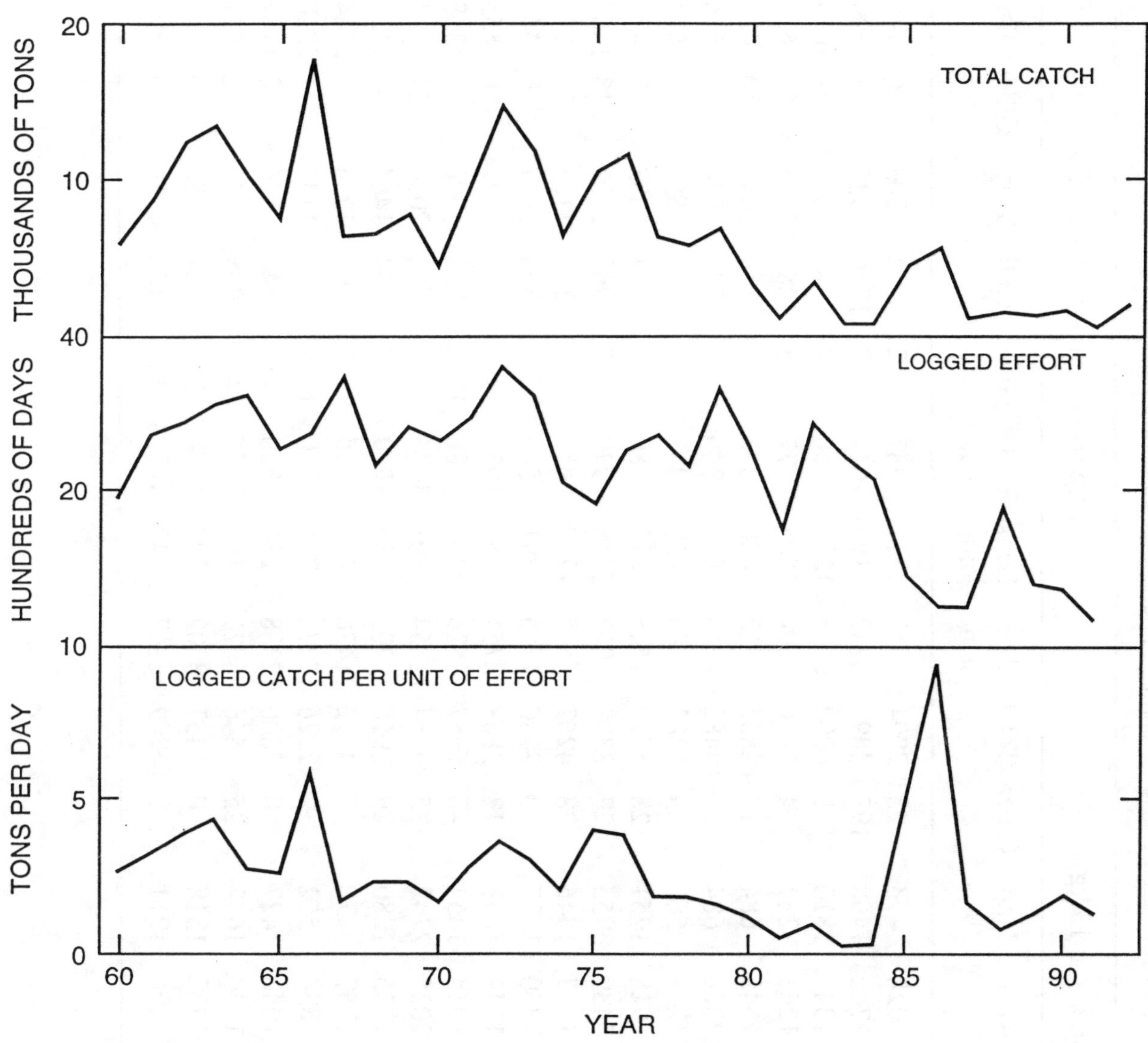

FIGURE 8. Total catch, logged fishing effort, and logged catch per unit of fishing effort for northern bluefin in the eastern Pacific Ocean (after Bayliff, 1993b).

TABLE 9. Catches, in metric tons, of bluefin in the Pacific Ocean by Japanese vessels (after Ishizuka, 1989).

	Small fish (less than about 15 kg)								Large fish (more than about 15 kg)								Grand total
	Purse seine	Long-line	Bait-boat	Gill-net	Troll	Trap	Other	Total	Purse seine	Long-long	Bait-boat	Gill-net	Troll	Trap	Other	Total	
1966	315	107	0	37	1028	493	23	2004	8517	1311	437	2	392	803	20	11482	13486
1967	1380	104	0	10	1742	1468	193	4897	4742	1193	1011	29	1016	937	67	8995	13892
1968	2705	134	0	3	1341	1810	7	6000	6178	1283	850	2	947	1300	31	10591	16591
1969	202	103	0	13	1261	1237	4	2820	2504	1005	377	5	1285	1015	11	6202	9022
1970	556	62	0	0	1545	684	10	2856	1812	800	378	0	813	1046	10	4859	7715
1971	897	48	0	15	2826	684	12	4482	2665	643	740	1	579	934	2	5564	10046
1972	366	96	0	131	1217	565	9	2383	3445	637	472	3	331	559	1	5448	7831
1973	1032	179	0	160	2928	1251	28	5579	1618	1077	180	115	903	1083	7	4983	10562
1974	927	695	0	165	2439	3955	127	8309	3012	3448	929	70	963	2207	13	10642	18951
1975	1321	241	0	195	1137	1346	28	4267	3505	1219	659	68	307	1138	11	6907	11174
1976	139	115	0	481	860	1185	0	2780	1785	391	565	352	308	2027	0	5428	8208
1977	746	111	0	399	1741	858	12	3868	3751	483	1971	402	469	1475	8	8559	12427
1978	559	139	0	551	3915	1453	40	6657	8526	753	986	1344	752	1308	15	13684	20341
1979	483	313	0	564	2028	2742	35	6164	11854	808	1144	756	672	2084	24	17342	23506
1980	1886	318	0	668	1735	1280	29	5918	9205	838	1279	909	494	1464	42	14231	20149
1981	1089	118	0	970	1757	970	8	4912	21476	619	675	772	363	1233	7	25145	30057
1982	555	143	0	240	834	552	2	2326	17272	756	1710	1337	177	1147	1	22400	24726
1983	1227	84	0	28	1040	427	0	2806	12388	286	320	786	146	546	0	14472	17278
1984	384	77	0	78	1340	1642	25	3546	4021	162	456	430	295	846	2	6212	9758
1985	1522	57	0	228	1332	1318	0	4457	3302	324	1700	332	324	1380	0	7362	11819
1986	1341	21	0	128	808	1911	0	4209	6367	142	1022	141	182	1120	0	8974	13183

11.3 Central Pacific Ocean

The catch data in Tables 6 and 9 include catches made by Japanese vessels in the central Pacific. No further information on this subject, other than that given in Section 10.3, is available.

12. POPULATION DYNAMICS

12.1 Predator-prey Relationships

Doi (1960) used data on the catches of northern bluefin and of squid and five species of smaller fishes as indices of their abundance, and then applied Volterra's predator-prey equations to the data to compare the observed and theoretical abundances. The actual and expected results were fairly close for the bluefin-squid model for 1951-1955 and for the bluefin-squid plus smaller fishes model for 1951-1956. Additional information on predator-prey relationships is given in Section 9.1.

12.2 Yield-per-recruit Relationship

A yield-per-recruit analysis has been performed for northern bluefin, using the data in Table 10 and Figure 9. The following assumptions were made: (1) the fish are hatched on May 16; (2) they are recruited to the fishery on July 1 of the same year at a length of 15 cm; (3) they attain a length of 223 cm on their 15th birthday, at which time they disappear from the fishery. (The 223-cm value comes from an estimate of the asymptotic length given by Bayliff *et al.* (1991).) The annual coefficient of natural mortality was set at 0.2 and 0.3 because those values bracket the estimate of 0.276 calculated in Section 7. Actually, M probably varies among age groups. The annual coefficients of fishing mortality were set at 1.0 for the periods of substantial catches, 0.0 for the periods of zero or very low catches, and 0.5 for the periods of transition. The data in Figure 9 are based upon what are believed to be three of the most common scenarios for individual fish: remaining throughout life in the western Pacific Ocean (Option 0); beginning a west-east migration during the first year of life and then beginning an east-west migration during the third year of life (Option 1); and beginning a west-east migration during the second year of life and then beginning an east-west migration during the third year of life (Option 2). Graphs showing the yields per recruit possible with different ages of entry into the fishery, and with three multipliers of the vector of fishing effort, are shown in Figure 10. Because the values of the natural and fishing mortality, especially the latter, are little more than guesses, the results of these analyses should not be taken literally. Nevertheless, it appears that increasing the age at entry into the fishery to about 2 1/2 years (about 90-100 cm) would maximize the overall yield per recruit of fish which migrate to the eastern Pacific Ocean, and increasing the age at entry to about 4 years (about 130-140 cm) would maximize the yield per recruit of fish which remain in the western Pacific.

12.3 Cohort Analysis

Cohort analyses were run, using the data in Tables 11 and 12 and the methods described by Tomlinson (1970). In these tables all fish caught during the calendar year in which they were hatched are referred to as age-0 fish, all fish caught during the calendar year after that in which they were hatched are referred to as age-1 fish, and so on. The cohort

TABLE 10. Growth and weight-length data used for estimation of yields per recruit of northern bluefin.

Age in years	Equation	Units of measurement	Source
Growth			
0.125-0.917	$L_{t2}^{0.001} = 581^{0.001} + (L_{t1}^{0.001} - 581^{0.001})e^{0.01184(t_2 - t_1)}$	millimeters, days	this report, Section 5.1
0.917-4.500	$L_t = 564 + 0.709\Delta t$	millimeters, days	this report, Section 5.1
5.500-13.500	$L_t = 1482 + 105.9\Delta t$	millimeters, years	linear interpolation
5.00-15.00	$L_t = 320.5(1 - e^{-0.1035(t + 0.7034)})$	centimeters, years	this report, Table 1
Weight-length			
0.125-1.042	$w = (1.82335 \times 10^{-9})l^{3.39759}$	millimeters, kilograms	Bayliff, 1991, Table 7
1.042-14.500	$w = (2.01711 \times 10^{-7})l^{2.79130}$	millimeters, pounds	Bayliff, 1991, Table 7

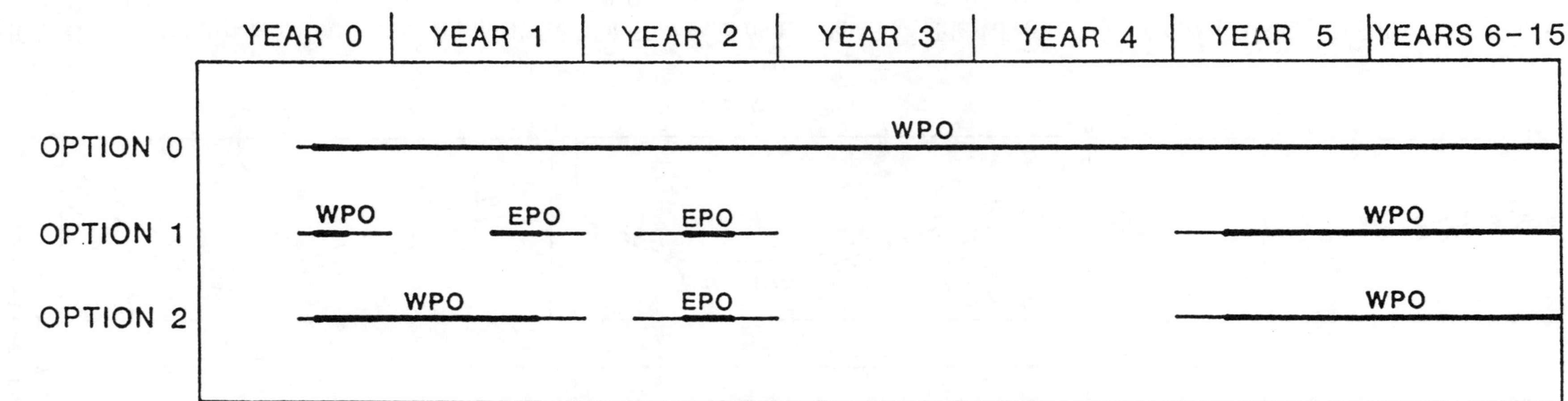

FIGURE 9. Fishing mortality rates used for estimation of yield-per-recruit curves for northern bluefin. WPO and EPO stand for western Pacific Ocean and eastern Pacific Ocean, respectively. The heavy and light segments of the lines indicate annual instantaneous fishing mortality rates of 1.0 and 0.5, respectively, and the blank areas between segments of the lines indicate no fishing. The options are described in the text.

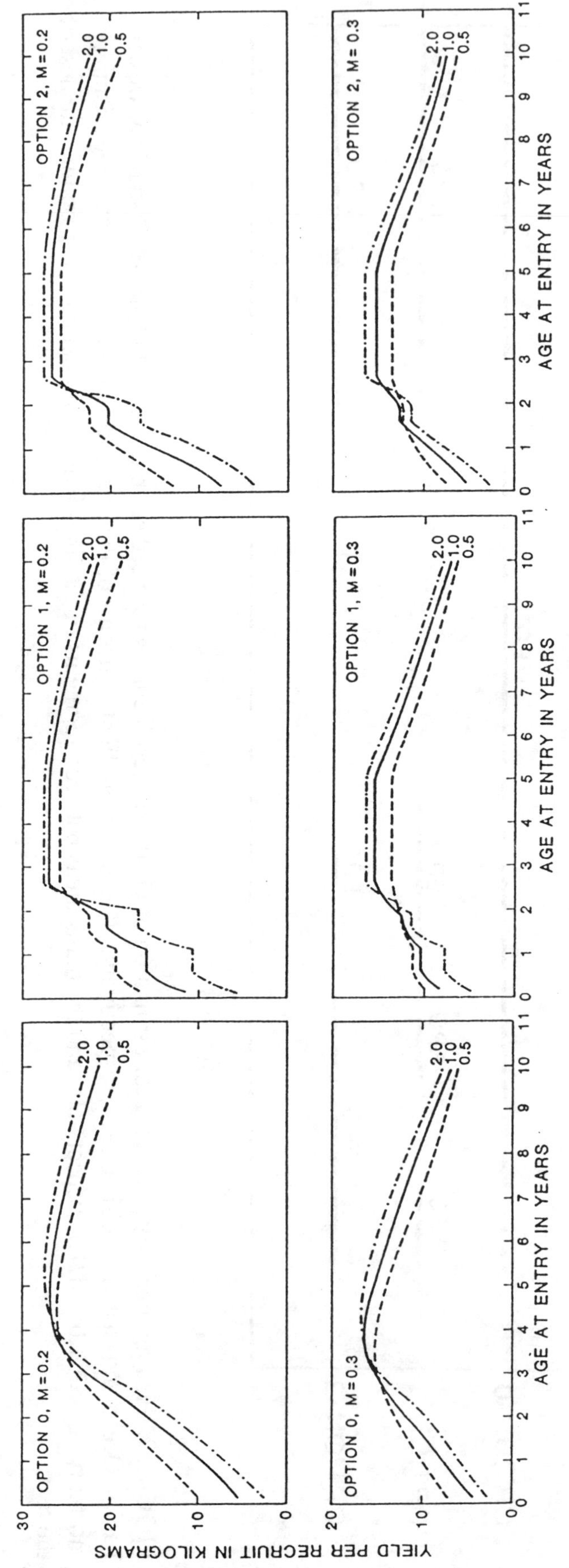

FIGURE 10. Yield-per-recruit curves for northern bluefin, based on the data in Table 10 and Figure 9 (after Bayliff, 1993b). The three values, 0.5, 1.0, and 2.0, represent multipliers of the vectors of fishing mortality shown in Figure 9. The horizontal portions of the curves are caused by lack of fishing effort during these periods.

TABLE 11. Estimated numbers of bluefin, in thousands, caught in the eastern Pacific Ocean (after Bayliff, 1993a: Appendix 1).

						Year caught						
Age	1959	1960	1961	1962	1963	1964	1965	1966	1967	1968	1969	1970
1	12	58	524	829	526	972	327	622	97	416	290	14
2	269	66	347	608	521	256	305	715	308	155	311	258
3	109	33	6	9	80	29	16	35	34	5	32	35
4	4	69	<1	1	5	4	1	0	1	0	1	1
>4	<1	4	0	1	3	0	1	0	<1	0	1	<1

						Year caught						
Age	1971	1972	1973	1974	1975	1976	1977	1978	1979	1980	1981	1982
1	467	-	609	51	260	488	55	508	78	48	4	249
2	711	-	589	440	722	150	250	81	539	284	70	120
3	7	-	2	1	2	172	21	7	<1	1	11	2
4	0	-	0	0	<1	24	21	8	<1	0	<1	1
>4	0	-	0	0	0	3	11	12	<1	0	0	<1

					Year caught				
Age	1983	1984	1985	1986	1987	1988	1989	1990	1991
1	12	33	76	22	7	34	45	5	52
2	62	44	315	388	74	45	40	78	8
3	6	6	<1	5	6	21	6	8	1
4	<1	1	0	<1	<1	<1	<1	1	<1
>4	<1	<1	0	<1	<1	1	<1	<1	0

analyses were begun at the end of the 12th year in the fishery (December 31). Initiating rates of F of 0.5 and 0.4 were used with M values of 0.2 and 0.3, respectively (see Section 7). The results are shown in Table 13. It can be seen that the catches in the western Pacific usually exceed those in the eastern Pacific for all ages except the 2-year olds. The 1969 year class, which was about average (Table 13, second column), produced very poor catches of age-2 fish in the western Pacific and excellent catches of age-2 fish in the eastern Pacific. The 1972 year class, which was a little below average, produced very poor catches of age-0 and -1 fish in the western Pacific and excellent catches of age-1 and -2 fish in the eastern Pacific. The 1981 year class, which was about average, produced above-average catches of age-2 fish in the western Pacific and very poor catches of age-2 fish in the eastern Pacific.

The cohort analysis is highly speculative. The errors in assigning the fish to ages 0, 1, 2, and 3 are probably minimal, but those in assigning them to greater ages are almost certainly greater. In addition, the fact that many fish migrate back and forth across the Pacific Ocean probably causes the results to be less accurate than they would be otherwise.

TABLE 12. Estimated numbers of bluefin, in thousands, caught in the western Pacific Ocean (after Ishizuka, 1989).

	Year caught											
Age	1966	1967	1968	1969	1970	1971	1972	1973	1974	1975	1976	1977
0	1270	3607	2300	2970	1938	3316	498	4875	3953	1277	1784	2542
1	266	461	964	371	378	443	682	124	1403	676	222	698
2	91	3	1	78	48	2	15	20	28	46	96	61
3	62	5	52	13	23	14	18	4	12	115	61	44
4	30	8	17	1	17	24	18	16	3	29	26	95
5	25	22	41	5	5	3	34	15	11	10	8	10
6	63	7	28	19	16	7	7	19	27	10	6	14
7	1	22	22	9	13	15	6	8	30	9	2	2
8	2	21	7	2	1	11	4	5	13	6	4	1
9	4	4	4	1	<1	2	1	2	7	2	4	1
10	3	2	3	1	<1	1	<1	1	5	1	1	1
11	1	1	2	1	<1	<1	<1	1	2	<1	<1	<1
12	<1	<1	1	1	<1	<1	<1	<1	<1	<1	<1	<1
13	<1	<1	<1	<1	<1	<1	0	<1	<1	<1	<1	<1
14	<1	<1	0	0	0	0	<1	<1	<1	<1	0	<1
15	0	<1	0	0	0	<1	0	<1	0	0	0	<1
16	0	<1	0	0	0	0	0	<1	0	0	<1	0

	Year caught								
Age	1978	1979	1980	1981	1982	1983	1984	1985	1986
0	5091	2088	2810	1975	665	1362	2417	2046	1470
1	478	1452	611	605	785	213	421	757	760
2	151	98	119	180	200	139	44	49	61
3	14	38	76	584	64	54	21	86	123
4	10	28	7	84	99	56	15	75	30
5	87	15	11	32	73	71	15	4	24
6	20	72	3	26	22	43	4	3	3
7	31	61	27	5	19	13	4	2	2
8	5	4	51	10	11	4	5	2	3
9	2	1	7	15	10	3	3	2	4
10	2	1	3	2	13	2	3	2	4
11	1	1	2	1	3	2	4	2	4
12	1	<1	1	<1	1	1	1	1	2
13	<1	<1	<1	<1	<1	<1	<1	<1	1
14	<1	<1	<1	<1	<1	<1	<1	<1	<1
15	<1	<1	<1	0	<1	<1	<1	<1	<1
16	<1	<1	<1	0	<1	0	<1	0	<1

TABLE 13a. Results of the cohort analysis described in the text, with an annual coefficient of natural mortality of 0.2 (from Bayliff, 1993a). IPS, WPO, EPO, F, and Mort. stand for initial population size, catch in western Pacific Ocean, catch in eastern Pacific Ocean, annual coefficient of fishing mortality, and natural mortality, respectively. All the values except those for F are in thousands of fish.

Year class	Age 0					Age 1				
	IPS	WPO	EPO	F	Mort.	IPS	WPO	EPO	F	Mort.
1966	2692	1270	0	0.72	351	1071	461	97	0.84	133
1967	6959	3607	0	0.83	869	2483	964	416	0.93	298
1968	4221	2300	0	0.90	512	1409	371	290	0.71	184
1969	5156	2970	0	0.98	604	1582	378	14	0.32	247
1970	4012	1938	0	0.75	519	1555	443	467	1.01	181
1971	6217	3316	0	0.87	764	2137	682	292	0.59	190
1972	3532	498	0	0.17	591	2443	124	609	0.40	369
1973	10139	4875	0	0.74	1314	3950	1403	51	0.52	565
1974	6331	3953	0	1.13	700	1678	676	260	0.93	201
1975	3331	1277	0	0.54	469	1585	222	488	0.67	212
1976	3750	1784	0	0.73	489	1477	698	55	0.81	186
1977	5950	2542	0	0.63	809	2599	478	508	0.54	368
1978	9958	5091	0	0.81	1252	3615	1452	78	0.62	495
1979	3927	2088	0	0.86	483	1356	611	48	0.75	175
1980	4601	2810	0	1.09	519	1272	605	4	0.74	165
1981	4510	1975	0	0.65	607	1928	785	249	0.88	236
1982	1685	665	0	0.56	235	785	213	12	0.38	120
Average	5116	2527	0	0.76	652	1937	622	232	0.68	254

Year class	Age 3					Age 4					Age 5				
	IPS	WPO	EPO	F	Mort.	IPS	WPO	EPO	F	Mort.	IPS	WPO	EPO	F	Mort.
1966	380	1	155	0.59	52	172	13	32	0.34	27	100	17	1	0.21	15
1967	805	78	311	0.75	104	312	23	35	0.23	50	204	24	0	0.14	35
1968	564	48	258	0.89	68	190	14	7	0.13	33	136	18	6	0.19	19
1969	943	2	711	1.67	84	146	18	24	0.31	16	88	16	0	0.22	15
1970	464	15	269	1.43	89	91	4	2	0.08	16	69	3	0	0.05	12
1971	973	20	589	1.13	108	256	12	1	0.06	45	198	29	0	0.18	33
1972	1341	28	440	0.48	194	679	115	2	0.21	112	450	26	24	0.13	76
1973	1931	46	722	0.57	269	894	61	172	0.34	139	522	95	21	0.28	83
1974	541	96	150	0.68	71	224	44	21	0.38	34	125	10	8	0.17	21
1975	663	61	250	0.72	87	265	14	7	0.09	46	198	28	0	0.18	34
1976	538	151	81	0.64	73	233	38	0	0.20	40	155	7	0	0.05	27
1977	1245	98	539	0.81	155	453	76	1	0.20	74	302	84	0	0.36	46
1978	1590	119	284	0.33	247	940	584	11	1.16	104	241	99	1	0.60	33
1979	522	180	70	0.74	68	204	64	2	0.44	30	108	56	0	0.83	13
1980	498	200	120	1.18	53	125	54	6	0.74	16	49	15	1	0.41	6
1981	658	139	62	0.41	99	358	21	6	0.09	63	268	75	0	0.37	41
1982	440	44	44	0.24	70	282	86	0	0.41	42	154	30	0	0.24	25
Average	829	78	297	0.78	111	343	73	19	0.32	52	198	37	4	0.27	31

TABLE 13b. Results of the cohort analysis described in the text, with an annual coefficient of natural mortality of 0.3.

Year class	Age 0					Age 1				
	IPS	WPO	EPO	F	Mort.	IPS	WPO	EPO	F	Mort.
1966	3271	1270	0	0.59	651	1350	461	97	0.64	263
1967	8132	3607	0	0.70	1541	2984	964	416	0.75	555
1968	4965	2300	0	0.75	923	1742	371	290	0.57	348
1969	6004	2970	0	0.82	1082	1952	378	14	0.26	448
1970	4656	1938	0	0.64	904	1814	443	467	0.84	325
1971	7428	3316	0	0.71	1404	2708	682	292	0.46	464
1972	5336	498	0	0.11	1311	3527	124	609	0.27	806
1973	12453	4875	0	0.59	2472	5106	1403	51	0.40	1105
1974	7285	3953	0	0.95	1246	2086	676	260	0.71	393
1975	4316	1277	0	0.41	925	2114	222	488	0.48	440
1976	4539	1784	0	0.59	900	1855	698	55	0.62	364
1977	7224	2542	0	0.51	1482	3200	478	508	0.44	680
1978	11622	5091	0	0.69	2212	4319	1452	78	0.52	885
1979	4532	2088	0	0.74	845	1599	611	48	0.63	312
1980	5168	2810	0	0.95	884	1474	605	4	0.64	287
1981	5610	1975	0	0.52	1152	2483	785	249	0.64	482
1982	2285	665	0	0.40	491	1129	213	12	0.26	260
Average	6166	2527	0	0.63	1201	2438	622	232	0.54	495

Year class	Age 3					Age 4					Age 5				
	IPS	WPO	EPO	F	Mort.	IPS	WPO	EPO	F	Mort.	IPS	WPO	EPO	F	Mort.
1966	529	1	155	0.41	113	260	13	32	0.22	61	154	17	1	0.14	36
1967	1049	78	311	0.55	212	448	23	35	0.16	108	282	24	0	0.10	69
1968	733	48	258	0.65	142	285	14	7	0.09	71	193	18	6	0.13	44
1969	1112	2	711	1.27	167	232	18	24	0.19	48	142	16	0	0.14	35
1970	579	15	269	1.02	140	155	4	2	0.05	39	110	3	0	0.03	28
1971	1270	20	589	0.79	232	429	12	1	0.04	110	306	29	0	0.12	75
1972	1988	28	440	0.32	445	1075	115	2	0.13	262	696	26	24	0.09	173
1973	2547	46	722	0.42	544	1235	61	172	0.25	286	716	95	21	0.21	169
1974	757	96	150	0.47	159	352	44	21	0.24	82	205	10	8	0.11	50
1975	964	61	250	0.46	202	451	14	7	0.06	114	316	28	0	0.11	79
1976	738	151	81	0.45	156	350	38	0	0.14	86	226	7	0	0.04	58
1977	1534	98	539	0.64	297	600	76	1	0.16	144	379	84	0	0.29	85
1978	1904	119	284	0.28	434	1067	584	11	1.00	180	292	99	1	0.50	61
1979	628	180	70	0.60	124	254	64	2	0.35	56	132	56	0	0.66	25
1980	578	200	120	0.98	197	161	54	6	0.55	32	69	15	1	0.29	15
1981	967	139	62	0.27	222	544	21	6	0.06	137	380	75	0	0.26	87
1982	644	44	44	0.17	153	403	86	0	0.28	92	225	30	0	0.17	54
Average	1089	78	297	0.57	232	488	73	19	0.23	112	284	37	4	0.20	67

13. INTERACTIONS AMONG FISHERIES

There are two basic types of interactions among the fisheries which take northern bluefin, interactions between ocean areas (eastern, central, or western Pacific) and interactions within ocean areas.

Northern bluefin are recruited to the troll fishery for small fish in the western Pacific. Some of them remain in the western Pacific and others, after exposure to the troll fishery and other fisheries in the western Pacific, migrate to the eastern Pacific. In the eastern Pacific they are exploited by the purse-seine fishery. After a sojourn in the eastern Pacific, those which are not caught or succumb to natural mortality return to the western Pacific, where they are exposed to further exploitation. Accordingly, there is important interaction between the fisheries of the western and eastern Pacific. It is known that bluefin are caught by the gillnet and baitboat fisheries of the central Pacific (Bayliff *et al.*, 1991: Table 6) and by longlining, but reasonably complete estimates of the amounts of bluefin caught are not available, so the importance of the interactions between the western and central Pacific or the eastern and central Pacific are not known.

Northern bluefin are caught almost exclusively by purse seining in the eastern Pacific, so the principal interaction is that between Mexican- and U.S.-flag vessels. The fish usually appear first off Mexico in about May or June, and later are taken in waters off the United States in July, August, September, and October (Figure 6).

Northern bluefin are taken by trolling vessels, traps, purse seines, baitboats, gillnets, handlines, and longlines in the western Pacific (Table 9), and the activities of each type of vessel affect the catches of those which take fish of the same size or larger fish.

14. ACKNOWLEDGEMENTS

The cohort analyses were performed by Mr. Patrick K. Tomlinson of the Inter-American Tropical Tuna Commission. Suggestions for improvement of the manuscript were received from Drs. Richard B. Deriso, Ziro Suzuki, and Alexander Wild, and Messrs. Doyle A. Hanan and Richard S. Shomura.

15. REFERENCES CITED

Aikawa, H., and M. Kato. 1938. Age determinations of fish. I. *Bull.Japan.Soc.Sci.Fish.*, 1 (7): 79-88 [translation by W. G. Van Campen, 1950, *Spec.Sci.Rep.U.S.Fish Wild.Serv. (Fish.)*, (21):22 p.].

Bayliff, W.H. 1980. Synopsis of biological data on the northern bluefin tuna, *Thunnus thynnus* (Linnaeus, 1758), in the Pacific Ocean. *Spec.Rep.I-ATTC*, (2):261-93.

Bayliff, W.H. (ed.) 1986. Annual report of the Inter-American Tropical Tuna Commission 1985. *Annu.Rep.I-ATTC*, (1985):248 p.

Bayliff, W.H. 1991. Status of northern bluefin tuna in the Pacific Ocean. *Spec.Rep.I-ATTC*, (7):29-88.

Bayliff, W.H. 1993a. Growth and age composition of northern bluefin tuna, *Thunnus thynnus*, caught in the eastern Pacific Ocean, as estimated from length-frequency data, with comments on trans-Pacific migrations. *Bull.I-ATTC*, 20(9):501-40.

Bayliff, W.H. (ed.) 1993b. Annual report of the Inter-American Tropical Tuna Commission 1992. *Annu.Rep.I-ATTC*, (1992), (in preparation).

Bayliff, W.H., and T.P. Calkins. 1979. Information pertinent to stock assessment of northern bluefin tuna, *Thunnus thynnus*, in the Pacific Ocean. *Internal Rep.I-ATTC*, (12):78 p.

Bayliff, W.H., Y. Ishizuka, and R.B. Deriso. 1991. Growth, movements, and mortality of northern bluefin, *Thunnus thynnus*, in the Pacific Ocean, as determined from tagging experiments. *Bull.I-ATTC*, 20(1):1-94.

Bell, R.R. 1963a. Synopsis of biological data on California bluefin tuna *Thunnus saliens* Jordan and Evermann 1926. *FAO Fish.Rep.*, 6(2):380-421.

Bell, R.R. 1963b. Preliminary age determination of bluefin tuna, *Thunnus thynnus*. *Calif.Fish Game*, 49(4):307.

Bell, R. R. 1970. Bluefin tuna *Thunnus thynnus orientalis* in the northeastern Pacific Ocean. Calif.Dep.Fish Game (unpublished).

Blunt, C.E., Jr. 1958. California bluefin tuna--wary wanderer of the Pacific. *Outdoor Calif.*, 19(9):14.

Brock, V.E. 1938. A new tuna record from Washington. *Copeia*, (2):98.

de Buen, F. 1953. Contribuciones a la ictiología-VII. Sobre la presencia del atún (*Thunnus thynnus*) en la costa de Chile central. *Inves.Zool.Chilenas*, 1(10):3-4.

de Buen, F.L. 1958. Peces del suborden Scombroidei en aguas de Chile. *Rev.Biol.Mar.*, 7(1-3):3-38.

Burgess, D., A. Caton, J. Gunn, W. Hearn, T. Murray, and C. Proctor. 1991. Aging and growth of juveniles and adults. *In* Review of aspects of southern bluefin tuna biology, population and fisheries, edited by A.E. Caton. *Spec.Rep.I-ATTC*, (7):210-24.

Calkins, T.P. 1982. Observations on the purse-seine fishery for northern bluefin tuna (*Thunnus thynnus*) in the eastern Pacific Ocean. *Bull.I-ATTC*, 18(2):121-225.

Clay, D. 1991. Atlantic bluefin tuna (*Thunnus thynnus* thynnus (L.)): a review, edited by D. Clay. *Spec.Rep.I-ATTC*, (7):89-179.

Collette, B.B., and C.E. Nauen. 1983. FAO species catalogue. Vol. 2. Scombrids of the world. *FAO Fish.Synop.*, 125(2):137 p.

Collette, B.B., and B.R. Smith. 1981. Bluefin tuna, *Thunnus thynnus orientalis*, from the Gulf of Papua. *Japan.J.Ichthyol.*, 28(2):166-8.

Collette, B.B., T. Pothoff, W.J. Richards, S. Ueyanagi, J.L. Russo, and Y. Nishikawa. 1984. Scombroidei: development and relationships. *In* Ontogeny and systematics of fishes, edited by H.G. Moser, W.J. Richards, D.M. Cohen, M.P. Fahey, A.W. Kendall, Jr., and S.L. Richardson, *Spec.Publ.Am.Soc.Ichthyol.Herpetol.*, 1:591-620.

Cort, J.L. 1990. Biología y pesca del atún rojo, *Thunnus thynnus* (L.), del Mar Cantábrico. *Publ.Esp.Inst.Español Ocean.*, (4):272 p.

Cressey, R., and H.B. Cressey. 1980. Parasitic copepods of mackerel- and tuna-like fishes (Scombridae). *Smithson.Contrib.Zool.*, (311):186 p.

Davis, T. 1991. Maturation and spawning. *In* Review of aspects of southern bluefin tuna biology, population and fisheries, edited by A.E. Caton. *Spec.Rep.I-ATTC*, (7):225-9.

Doi, T. 1960. On the predatory relationships among bluefin tuna and coastal fishes in the southern waters of Japan. *Bull.Japan.Soc.Sci.Fish.*, 26(2):99-102.

Dotson, R.C., and J.E. Graves. 1984. Biochemical identification of a bluefin establishes a new California size record. *Calif.Fish Game*, 70(1):62-4.

Fisheries Agency of Japan, Research and Development Department. 1974. Annual report of effort and catch statistics by area on Japanese longline fishery 1972, 279 p.

Fisheries Agency of Japan, Research and Development Department. 1975. Annual report of effort and catch statistics by area on Japanese longline fishery 1973, 265 p.

Fisheries Agency of Japan, Research and Development Department. 1976. Annual report of effort and catch statistics by area on Japanese longline fishery 1974, 267 p.

Fisheries Agency of Japan, Research and Development Department. 1977a. Annual report of effort and catch statistics by area on Japanese longline fishery 1975, 269 p.

Fisheries Agency of Japan, Research and Development Department. 1977b. Annual report of effort and catch statistics by area Japanese skipjack bait fishery 1972, 266 p.

Fisheries Agency of Japan, Research and Development Department. 1977c. Annual report of effort and catch statistics by area Japanese skipjack bait fishery 1973, 272 p.

Fisheries Agency of Japan, Research and Development Department. 1977d. Annual report of effort and catch statistics by area Japanese skipjack bait fishery 1974, 293 p.

Fisheries Agency of Japan, Research and Development Department. 1977e. Annual report of effort and catch statistics by area Japanese skipjack bait fishery 1975, 310 p.

Fisheries Agency of Japan, Research and Development Department. 1977f. Annual report of effort and catch statistics by area Japanese skipjack bait fishery 1976, 291 p.

Fisheries Agency of Japan, Research and Development Department. 1978. Annual report of effort and catch statistics by area on Japanese longline fishery 1976, 264 p.

Flittner, G.A. 1966. Bluefin tuna in the north Pacific Ocean. *Pac.Sci.Congr.*, 11(7):3.

Foreman, T.J., and Y. Ishizuka. 1990. Giant bluefin off Southern California, with a new California size record. *Calif.Fish Game*, 76(3):181-6.

Fowler, H.W. 1928. The fishes of Oceania. *Mem.Bernice Bishop Mus.*, (10):540 p.

Fowler, H.W. 1932. The fishes obtained by the Pinchot South Seas expedition of 1929, with description of one new genus and three new species. *Proc.U.S.Nat.Mus.*, 80(6):16 p.

Fowler, H.W. 1944. The fishes. *In* Results of the fifth George Vanderbilt expedition (1941). *Monogr.Philad.Acad.Nat.Sci.*, (6):57-529.

Gibbs, R.H., Jr., and B.B. Collette. 1967. Comparative anatomy and systematics of the tunas, genus *Thunnus*. *Fish.Bull.U.S.Fish.Wild.Serv.*, 66(1):65-130.

Hanan, D.A. 1983. Review and analysis of the bluefin tuna, *Thunnus thynnus*, fishery in the eastern north Pacific Ocean. *Fish.Bull.NOAA-NMFS*, 81(1):107-19.

Harada, T. 1980. Development and future outlook of studies on the aquaculture of tunas. *Maguro Gyogyo Kyogikai Gijiroku, Suisancho-Enyo Suisan Kenkyusho* (Proceedings of the Tuna Fishery Research Conference, Japan Fisheries Agency-Far Seas Fisheries Research Laboratory):50-8.

Herre, A.W. 1936. Fishes of the Crane Pacific Expedition. *Zool.Ser.Field Mus.Nat.Hist.*, (21):472 p.

Hester, F.J. 1961. A method of predicting tuna catch by using coastal sea-surface temperatures. *Calif.Fish Game*, 47(4):313-26.

Hirota, H., M. Morita, and N. Taniguchi. 1976. An instance of the maturation of 3 full years old bluefin tuna cultured in the floating net. *Bull.Jap.Soc.Sci.Fish.*, 42(8):939.

Honma, M., and Z. Suzuki. 1978. Japanese tuna purse seine fishery in the western Pacific. *S Ser.Far Seas Fish.Res.Lab.*, (10):66 p.

INFOFISH. 1991. Progress in tuna farming. *INFOFISH International*, 2/91:39.

International North Pacific Fisheries Commission. 1986. *Annu.Rep.INPFC*, (1985):67 p.

International North Pacific Fisheries Commission. 1991. Final report of 1990 observations of the Japanese high seas squid driftnet fishery in the north Pacific Ocean, 198 p.

Ishizuka, Y. 1989. Estimates of catch and age compositions for northern bluefin tuna (*Thunnus thynnus*) caught by Japanese fisheries in the Pacific Ocean during 1966 to 1986. Nat.Res.Inst.Far Seas Fish., 16 p. (unpublished).

Jordan, D. S., and E. K. Jordan. 1922. A list of the fishes of Hawaii, with notes and descriptions of new species. *Mem. Carneg.Mus.*, 10(1):1-92.

June, F. C. 1952. Observations on a specimen of bluefin tuna (*Thunnus thynnus*) taken in Hawaiian waters. *Pac.Sci.*, 6(1):75-6.

Kida, T. 1936. On the surface temperature of water in the tunny fishing grounds off Kusiro and Urakawa in summer. *Bull.Jap.Soc.Sci.Fish.*, 5(2):87-90.

Klawe, W. L. 1959. Nuevo examen de atunes jovenes capturados frente a la costa occidental de México durante la quinta expedición George Vanderbilt (1941). *Mexico City Ciencia*, 18(11-12):245-7.

Kume, S. 1974. Tuna fisheries and their resources in the Pacific Ocean. *Proc.Indo-Pac.Fish.Coun.*, 15(3):390-423.

Leet, W.S., C.M. Dewees, and C.W. Haugen (eds.). 1992. California's Living Marine Resources and their Utilization. California Department of Fish and Game, 257 p.

Love, M.S., and M. Moser. 1977. Parasites of California, Oregon, and Washington marine and estuarine fishes. Mar.Sci.Inst.Univ.Calif.Santa Barbara, 530 p.

Majors, A.P., A.L. Coan, Jr., N. Bartoo, and F. Miller. 1984. Summary of the 1983 north Pacific albacore fishery data. *Admin.Rep.NMFS-SWFC La Jolla*, (LJ-84-13):44 p.

Marr, J.C. (ed.) 1970. The Kuroshio: a Symposium on the Japan Current. Honolulu, East-West Center Press, 611 p.

Matsumoto, W.M. 1962. Identification of larvae of four species of tuna from the Indo-Pacific region I. *Dana Rep.*, (55):16 p.

Matsumoto, W.M., E.H. Ahlstrom, S. Jones, W.L. Klawe, W.J. Richards, and S. Ueyanagi. 1972. On the clarification of larval tuna identification particularly the genus *Thunnus*. *Fish.Bull.NOAA-NMFS*, 70(1):1-12.

Miller, J.M. 1979. Nearshore abundance of tuna (Pisces: Scombridae) larvae in the Hawaiian Islands. *Bull.Mar.Sci.*, 29(1):19-26.

Mori, K. 1972. Geographical distribution and relative apparent abundance of some scombroid fishes based on the occurrences in the stomachs of apex predators caught on tuna longline-I. Juvenile and young of skipjack tuna (*Katsuwonus pelamis*). *Bull.Far Seas Fish.Res.Lab.*, (6):111-68.

Nakamura, I., and Y. Warashina. 1965. Occurrence of bluefin tuna, *Thunnus thynnus* in the eastern Indian Ocean and the eastern south Pacific Ocean. *Rep.Nankai Reg.Fish.Res.Lab.*, (22):9-12.

Nakano, H., and W.H. Bayliff. 1992. A review of the Japanese longline fishery for tunas and billfishes in the eastern Pacific Ocean, 1981-1987. *Bull.I-ATTC*, 20(5):185-355.

Neave, F. 1959. Records of fish from waters off the British Columbia coast. *J.Fish.Res. Board Can.*, 16(3):383-4.

Nishikawa, Y., M. Honma, S. Ueyanagi, and S. Kikawa. 1985. Average distribution of larvae of oceanic species of scombroid fishes, 1956-1981. *S Ser.Far Seas Fish.Res.Lab.*, (12):99 p.

Ogawa, Y., and T. Ishida. 1989a. Distinctive features of fluctuations in the catch of *Thunnus thynnus* by set-nets along the Sanriku coast. *Bull.Tohoku Reg.Fish.Res.Lab.*, (51):11-21.

Ogawa, Y., and T. Ishida. 1989b. Hydrographic conditions governing fluctuations in the catch of *Thunnus thynnus* by set-nets along the Sanriku coast. *Bull.Tohoku Reg.Fish.Res. Lab.*, (51):23-39.

Okachi, I. 1963. Studies on the distribution and structure of the fish fauna in the Japan Sea by catch statistics-II. Supplement of seasonal distribution and fishing condition of bluefin tuna. *Bull.Japan Sea Reg.Fish.Res.Lab.*, (11):9-21.

Okiyama, M. 1974. Occurrence of the postlarvae of bluefin tuna, *Thunnus thynnus*, in the Japan Sea. *Bull.Japan Sea Reg.Fish.Res.Lab.*, (25):89-97.

Okiyama, M. 1979. Successful spawning of some holoepipelagic fishes in the Sea of Japan and zoogeographical implications. *In* Japan-Soviet Joint Symp.Aquaculture, (7):223-33.

Oliphant, M.S., P.A. Gregory, B.J. Ingle, and R. Madrid. 1990. California marine fish landings for 1977-1986. *Fish Bull.Calif.Dep.Fish Game*, (173):52 p.

Oregon Fish Commission. 1948. Blue-fin tuna taken in Oregon waters. *Res.Briefs Oregon Fish Comm.*, 1(2):32.

Pauly, D. 1980. On the interrelationships between natural mortality, growth parameters, and mean environmental temperatures in 175 fish. *J.Cons.CIEM*, 39(2):175-92.

Peterson, C.L. (ed.) 1985. Annual Report of the Inter-American Tropical Tuna Commission 1984, *Annu.Rep.I-ATTC*, (1984):270 p.

Pinkas, L. 1971. Bluefin tuna food habits. *Fish Bull.Calif.Dep.Fish Game*, (152):47-63.

Radovich, J. 1961. Relationships of some marine organisms of the northeast Pacific to water temperatures particularly during 1957 through 1959. *Fish Bull.Calif.Dep.Fish Game*, (112):62 p.

Reid, J.L. 1988. Physical oceanography, 1947-1987. *Rep.CCOFI*, 29:42-65.

Reid, J.L., Jr., G.I. Roden, and J.G. Wyllie. 1958. Studies of the California Current system. *Prog.Rep.CCOFI*, (1 July 1956-1 January 1958):27-57.

Richards, W.J., and T. Pothoff. 1973. Analysis of the taxonomic characters of young scombrid fishes, genus *Thunnus*. *In* The Early Life History of Fish, edited by J.H.S.Blaxter, Berlin, Springer-Verlag:623-48.

Robinson, M.K. 1976. Atlas of north Pacific Ocean monthly mean temperatures and mean salinities of the surface layer. *Ref.Publ.U.S.Naval Ocean.Office*, 2.

Roden, G.I. 1991. Subarctic-Subtropical Transition Zone of the North Pacific: large-scale aspects and mesoscale structure. *NOAA Tech.Rep.NMFS*, (105):1-38.

San Diego Union. 1991. Strip-mining the sea. *San Diego Union*, July 7, 1991, page C2.

Schultze, D.L., and R.A. Collins. 1977. Age composition of California landings of bluefin tuna, *Thunnus thynnus*, 1963 through 1969. *Tech.Rep.Mar.Res.Calif.Dep.Fish Game*, (38):44 p. (manuscript).

Scott, J.M., and G.A. Flittner. 1972. Behavior of bluefin tuna schools in the eastern north Pacific Ocean as inferred from fishermen's logbooks. *Fish.Bull.NOAA-NMFS*, 70(3):915-27.

Sharp, G.D., and S. Pirages. 1978. The distribution of red and white swimming muscles, their biochemistry, and the biochemical phylogeny of selected scombrid fishes. *In* The Physiological Ecology of Tunas, edited by G.D. Sharp and A.E. Dizon. New York, San Francisco, and London, Academic Press:41-78.

Shimada, B.M. 1951. Contributions to the biology of tunas from the western equatorial Pacific. *Fish.Bull.U.S.Fish Wild.Serv.*, 52(62):111-19.

Shingu, C., Y. Warashina, and N. Matsuzaki. 1974. Distribution of bluefin tuna exploited by longline fishery in the western Pacific Ocean. *Bull.Far Seas Fish.Res.Lab.*, (10):109-40.

Silas, E.G. 1967. Parasites of scombroid fishes. Part I. Monogenetic trematodes, digenetic trematodes, and cestodes. *Proc. Symp.Scombroid Fishes, Mar.Biol.Assoc.India*, (3):799-875.

Silas, E.G., and A.N.P. Ummerkutty. 1967. Parasites of scombroid fishes. Part II. Parasitic Copepoda. *Proc. Symp.Scombroid Fishes, Mar.Biol.Assoc.India*, (3):876-993.

Skillman, R.A., and C. Shingu (rapporteurs). 1980. Pacific northern bluefin tuna. *FAO Fish.Tech.Pap.*, (200):24-30.

Snodgrass, R.E., and E. Heller. 1905. Papers from the Hopkins-Stanford Galapagos Expedition, 1898-1899. XVII. Shore fishes of the Revillagigedo, Clipperton, Cocos, and Galapagos Islands. *Proc.Wash.Acad.Sci.*, (6):333-427.

Squire, J.L., Jr. 1983. Abundance of pelagic resources off California, 1963-78, as measured by an airborne fish monitoring program. *NOAA Tech.Rep.NMFS SSRF*, (762):75 p.

Squire, J.L., Jr. 1993. Relative abundance of pelagic resources utilized by the California purse-seine fishery: results of an airborne monitoring program, 1962-90. *Fish.Bull.NOAA-NMFS*, 91(2):348-61.

Stommel, H., and K. Ishida (eds.). 1972. Kuroshio: Physical Aspects of the Japan Current. Seattle and London, University Washington Press, 517 p.

Sugawara, K. (ed.). 1972. The Kuroshio II: Proceedings of the Second Symposium on the Results of the Cooperative Study of the Kuroshio and Adjacent Regions. Tokyo, Saikon Publishing Company Limited, 562 p.

Suzuki, Z. 1990. Description of Japanese pelagic driftnet fisheries and related information. *FAO Fish.Rep.*, (434):37-51.

Takenouti, A.Y. (ed.). 1980. The Kuroshio IV: Proceedings of the Fourth Symposium for the Co-operative Study of the Kuroshio and Adjacent Regions. Tokyo, Saikon Publishing Company Limited, 1029 p.

Tatsuki, K., M. Uto, and K. Kitamura. 1963. On the landings of the yokowa (young bluefin) and its size composition off the Cape Shionomisaki. *Rep.Nankai Reg.Fish Res.Lab.*, (17):55-61.

Tomlinson, P.K. 1970. A generalization of the Murphy catch equation. *J.Fish.Res.Board Can.*, 27(4):821-5.

Uda, M. 1957. A consideration on the long years trend of the fisheries fluctuation in relation to sea conditions. *Bull.Jap.Soc.Sci.Fish.*, 23(7-8):368-72.

Uda, M. 1962. Cyclical fluctuation of the Pacific tuna fisheries in response to cold and warm water intrusions. *Spec.Sci.Rep.U.S.Fish Wild.Serv. (Fish.)*, (415):39.

Uda, M. 1973. Pulsative fluctuations of oceanic fronts in association with the tuna fishing grounds and fisheries. *J.Fac.Mar.Sci.Technol.Tokai Univ.*, (7):245-66.

Uotani, I., T. Saito, K. Hiranuma, and Y. Nishikawa. 1990. Feeding habit of bluefin tuna *Thunnus thynnus* larvae in the western north Pacific Ocean. *Nippon Suisan Gakkaishi*, 56(5):713-7.

Whitehead, S.S. 1931. Fishing methods for bluefin tuna (*Thunnus thynnus*) and an analysis of the catches. *Fish Bull.Calif.Dep.Fish Game*, (33):32 p.

Yabe, H., and S. Ueyanagi. 1962. Contributions to the study of the early life history of the tunas. *Occas.Rep.Nankai Reg.Fish.Res.Lab.*, (1):57-72.

Yabe, H., S. Ueyanagi, and H. Watanabe. 1966. Studies on the early life history of bluefin tuna *Thunnus thynnus* and on the larvae of the southern bluefin tuna *T. maccoyii*. *Rep.Nankai Reg.Fish.Res.Lab.*, (23):95-129.

Yamanaka, H. 1958. Studies on the bluefin tuna. I. The modes of the weight compositions of the catch. *Rep.Nankai Reg.Fish.Res.Lab.*, (9):125-35.

Yamanaka, H. 1982. Fishery Biology of the Bluefin Tuna Resource in the Pacific Ocean. Tokyo, Japan Fisheries Resources Conservation Association, 140 p.

Yamanaka, H., and staff. 1963. Synopsis of biological data on kuromaguro *Thunnus orientalis* (Temminck and Schlegel) 1942 (Pacific Ocean). *FAO Fish.Rep.*, 6(2):180-217.

Yamanaka, I., S. Ito, K. Niwa, R. Tanabe, Y. Yabuta, and S. Chikuni. 1988. The fisheries forecasting system in Japan for coastal pelagic fish. *FAO Fish.Tech.Pap.*, (301):72 p.

Yokota, T., M. Toriyama, F. Kanai, and S. Nomura. 1961. Studies on the feeding habit of fishes. *Rep.Nankai Reg.Fish.Res.Lab.*, (14):1-234.

Yukinawa, M., and Y. Yabuta. 1967. Age and growth of bluefin tuna, *Thunnus thynnus* (Linnaeus), in the north Pacific Ocean. *Rep.Nankai Reg.Fish.Res.Lab.*, (25):1-18.

REVIEW OF ASPECTS OF SOUTHERN BLUEFIN TUNA BIOLOGY, POPULATION, AND FISHERIES

A.E. Caton (editor)
Fisheries Resources Branch
Bureau of Resource Sciences
P.O. Box E11, Queen Victoria Terrace
PARKES A.C.T. 2600, Australia

1.   INTRODUCTION

Scientists of the southern bluefin tuna (SBT) working group commenced a review for the Food and Agriculture Organization Expert Consultation on Interactions in Pacific Tuna Fisheries. When the need arose for a background review on SBT for the May 1990 World Meeting on Bluefin Tunas at the Inter-American Tuna Commission in La Jolla, California, drafts of completed chapters of the Expert Consultation SBT review were distributed to participants. A more complete and edited version was included in the world meeting report[1]. The review presented contained a considerable amount of new data on SBT and its fisheries, updating material in the earlier comprehensive reviews of Shingu (1978)[2] and Olson (1980)[3]. It included a comprehensive reference list, but other references may also be located in the bibliography of Eckert *et al.* (1987)[4].

The SBT review from the World Bluefin Meeting report was not distributed in its entirety for the purposes of the Expert Consultation; instead, the 'Overview' chapter from the review was provided as a background document, with a listing of the review's contents (Appendix 1). That overview follows below, updated slightly and including figures and tables.

2.   OVERVIEW

Southern bluefin tuna (SBT), *Thunnus maccoyii* (Figure 1), are large (up to 200 cm and 200 kg), long-lived (more than 20 years) and highly migratory pelagic fish. The

---

[1] [Caton, A.E. (editor). 1991. Review of aspects of southern bluefin tuna biology, population and fisheries. *In* World meeting on stock assessments of bluefin tunas: strengths and weaknesses, edited by R.B. Deriso, and W.H. Bayliff. 1991. *Spec.Rep.I-ATTC*, (7):181-350.]

[2] [Shingu, C. 1978. Ecology and stock of southern bluefin tuna. Japan Association of Fisheries Resources Protection. Fisheries Study 31:81 p. (In Japanese. English translation in *Rep.CSIRO Div.Fish.Oceanogr.*, 131:79 p.)]

[3] [Olson, R.J. 1980. Synopsis of biological data on the southern bluefin tuna, *Thunnus maccoyii* (Castelnau 1872). *Spec.Rep.I-ATTC*, (2):151-212.]

[4] [Eckert, G.J., J. Kalish, J. Majkowski, and R. Pethebridge. 1987. An indexed bibliography of the southern bluefin tuna (*Thunnus maccoyii* [Castelnau, 1872]). *Rep.CSIRO Mar.Lab.*, 185:49 p.]

species has a general circum-global distribution between 30°S and 50°S, with a spawning ground between 7°S and 20°S in the northeastern Indian Ocean south of Java (Figure 2).

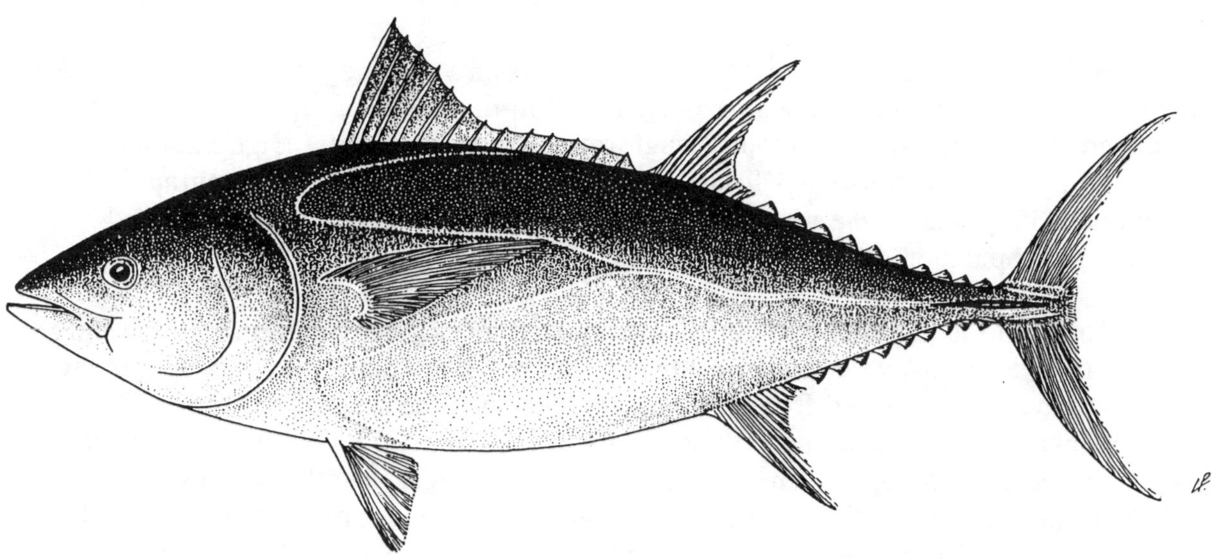

Figure 1.   Southern bluefin tuna, *Thunnus maccoyii* (Castelnau, 1872). (Line drawing courtesy FAO; from: FAO species catalogue, Vol.2.Scombrids ofthe world; an annotated and illustrated catalogue of tunas, mackerels, bonitos and related species known to date; by B.B. Collette and C.E. Nauen, *FAO Fisheries Synopsis* No. 125 (1983), 137 p.)

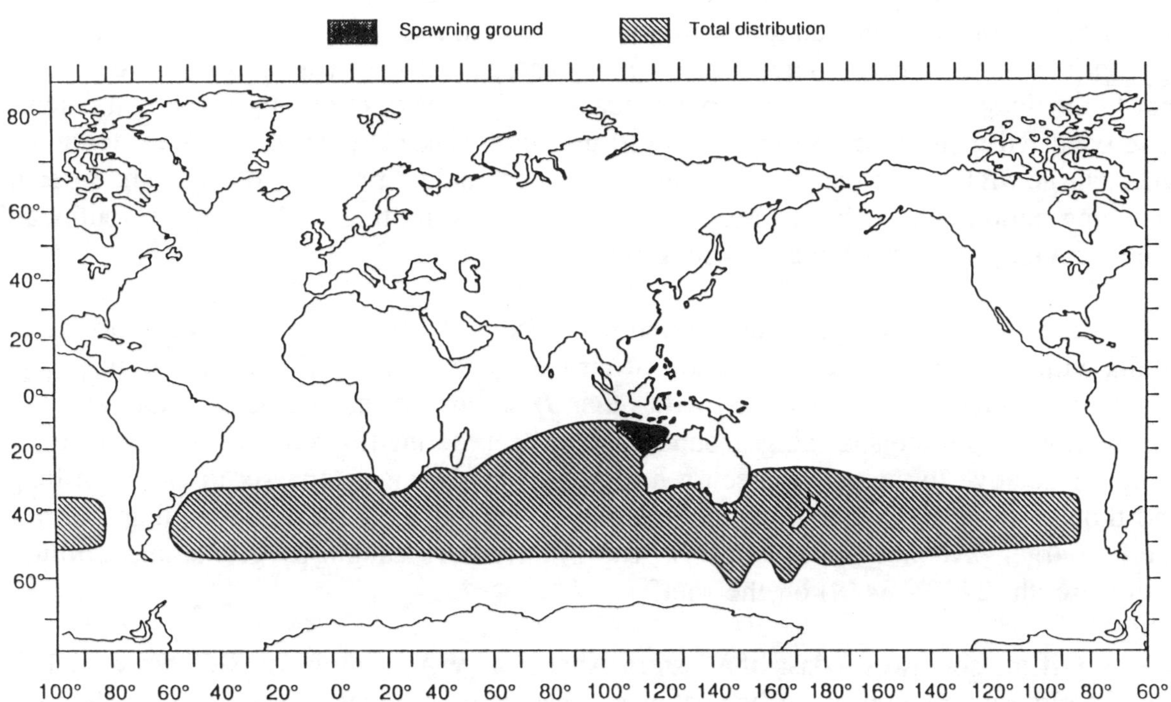

Figure 2.   Known worldwide distribution of southern bluefin tuna (hatched) and the spawning ground (shaded). (After Collette and Nauen, 1983, and Nishikawa *et al.*, 1985.)

Southern bluefin tuna can be distinguished in the field from the bluefin tuna *Thunnus thynnus*, mainly found in the northern hemisphere, by its yellow median caudal keel (dark in *T. thynnus* and other thunnids) and the relatively longer adult pectoral fin (20-23% of length to caudal fork in *T. maccoyii*, compared with 17-22% in *T. thynnus*).

A single spawning area, morphological uniformity, and tag-return data suggest a single population. Because spawning is restricted in time and space, there is little chance for different populations to develop through reproductive isolation. High vagility (as evidenced by the numerous long-distance tag returns) in relation to the geographical distribution (indicated by the range of longline catches) would also suggest limited potential for population differentiation.

Southern bluefin tuna with an elevated gonad index, are found only in the Indian Ocean. In spite of many years of worldwide tuna longline fishing, no data suggesting spawning have been reported from any waters other than the area of the Indian Ocean south of Java. On the spawning grounds, only large, mature SBT with well-developed gonads are caught. Maturity can occur at about 120 cm fork length, but more commonly at 130 cm (about 8 years old). Fecundity of one fish 158 cm long with a gonad index of 4.31 was estimated to be 14-15 million. Ripe eggs remaining in a 143 cm, spent fish were 0.66-1.05 mm in diameter. Examinations of sex ratio in catches in various fishing areas show that, as juveniles, females outnumber males; the situation is reversed in adults.

Little detailed research has been carried out on the reproductive biology of SBT. To remedy this, biological sampling on the Oka (spawning area) and Oki (adjacent area to the south) fishing grounds is required. It is not known whether spawning has a particular periodicity (such as a lunar cycle), or how spawning intensity is spread throughout the spawning season; ichthyoplankton sampling in the area has not been carried out at the appropriate spatial and temporal scales. However, recent ichthyoplankton surveys suggest that the main spawning period is from January to February, as the highest larval catches have occurred then. The fishing ground to the south (Oki) appears to be the staging area where large SBT with ripening or spent gonads accumulate prior to, during, and after the spawning period (September to March). It is believed that these SBT migrate southward from the Oki ground after the spawning season.

The limits of the spawning ground have been established by the distribution of adults with ripe gonads and the capture of larvae. The identification of larvae is, however, unreliable, and confusion with other *Thunnus* species can occur. Larval distribution within the spawning ground has not been defined because of insufficient sampling relative to the patchy occurrence of larvae. Juvenile tuna six to eight cm long, and thought to be SBT, have been collected by Bouke-ami net on the Australian northwest shelf. Larger juveniles (20-60 cm fork length) have been caught by gillnet and trolling further south (22°S - 34°S) on the shelf.

Off the southwest coast of Western Australia (WA; Figure 3), SBT between 1 and 3 years old are common in surface catches. Eastward, originally as far as New South Wales (NSW; Figure 3), SBT in surface catches commonly range from 2 to 6 years old, with individuals to at least 9 years old also taken at times. The younger-aged individuals (1 to 3 years old) seem more closely associated with waters of the coast and continental shelf than are older individuals. Some immature fish (from 3 years old, and progressively for older ages) leave the warmer waters of coastal areas for the wide region of the West Wind Drift (40°-45°S) where feeding adults are mainly found.

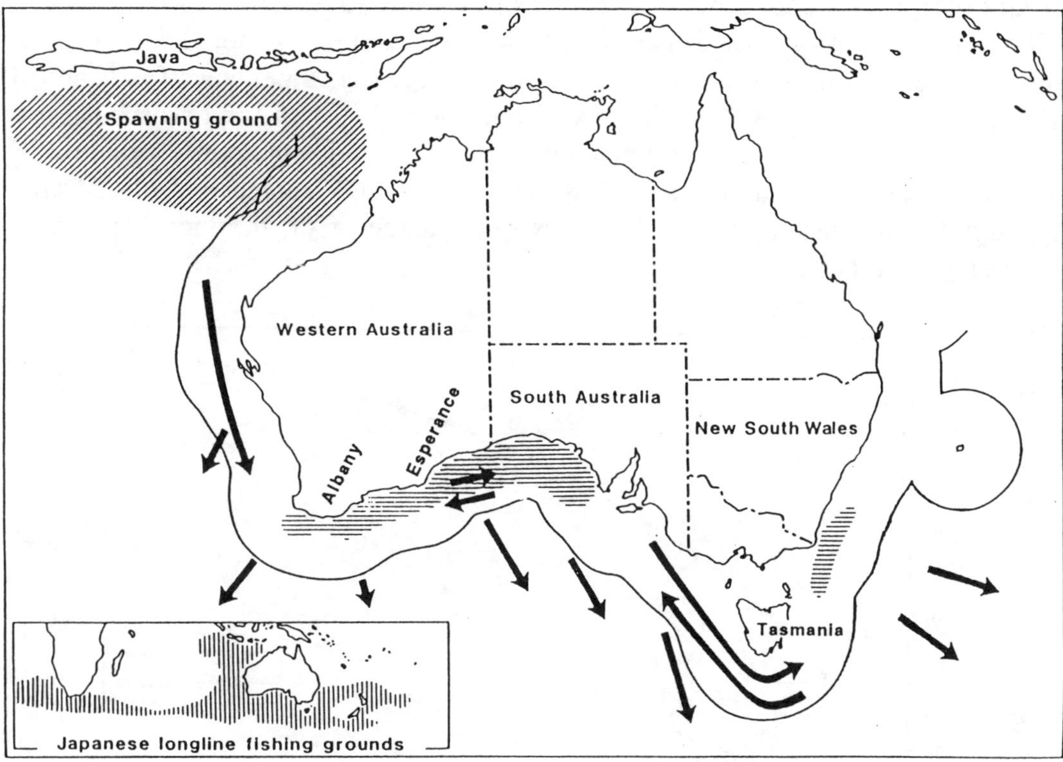

Figure 3. Southern bluefin tuna spawning ground and migration pattern within and out of Australian waters. The 200-mile Australian fishing zone is indicated by the solid line and the horizontal hatching indicates the composite distribution of the Australian surface fishery. The general distribution of Japanese longline fishing is inset. (Modified from Majkowski *et al.*, 1988.)

A promising technique for the study of migration patterns, involving the correlation of physical environmental features (particularly temperature) with otolith microchemistry, is under investigation. Also, the adoption of new, high technology tagging techniques - in particular the use of archival tags - could provide new data on continuous three-dimensional movements.

Information about the nutrition of SBT is largely anecdotal. The larval diet, which consists of mainly microcrustacea and macrozooplanktonic crustacea, is similar to that of related tuna species. It includes a degree of cannibalism. Studies show that survival and growth are dependent on the densities of larvae and other prey. Juveniles and adults are opportunistic feeders, chiefly on cephalopods, crustacea, fish, and salps. In general, smaller SBT feed mainly on crustacea. Adults feed mainly on fish. Sharks, other tunas, other teleosts, seabirds, whales, and seals are possible competitors, whilst various species of sharks and the killer whale are possible predators.

Tag-recapture data and length-frequency modal analysis have traditionally provided information on SBT age structure and growth rates, and length-at-age estimates for the fished population. There are different age-at-length relationships because of the low numbers of recoveries of fish at liberty for long periods, and the questionable reliability of data provided by some recoveries. The von Bertalanffy growth parameters derived have ranged from 171.5 to 207.6 cm for L_∞, 0.128 to 0.187 per year for K and -0.554 to 0 years for t_0. They produce lengths at 2, 8 and 14 years old ranging between 46 and 56 cm (2 years old), 124.7 and 136.7 cm (8 years old), and 157.4 and 174.7 cm (14 years old). The von Bertalanffy growth parameters which have been adopted for stock assessments in recent years have been 207.6 cm for L_∞, 0.128 per year for K and -0.394 years for t_0 (Figure 4). Growth variations among individual fish, and seasonal growth (faster during summer and early autumn) have been reported, apparently synchronised with ambient water-surface temperature. Recoveries of SBT tagged in 1983-84 all indicated significantly faster growth than previously tagged fish, the increase persisting with increasing time at liberty.

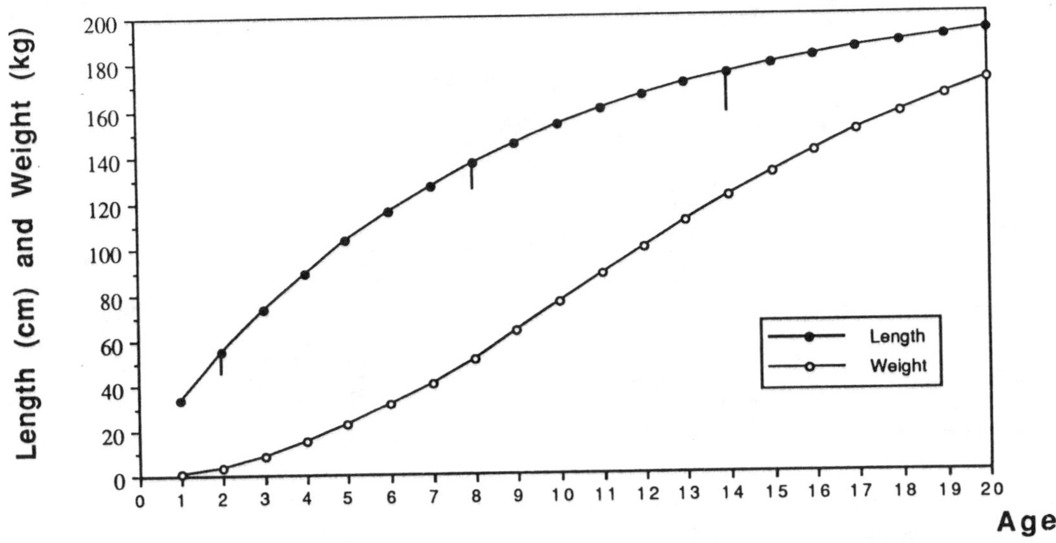

Figure 4.   Growth in length* and weight of southern bluefin tuna. Values of von Bertalanffy growth parameters for the length curve area K=0.128 per year; t-0.394 years; and L =207.6 cm. Growth in weight is derived from the length-weight relationships in Figure5. (* The bars on the length curve represent the range of length at ages 2, 8 , and 14 years for the different von Bertalanffy parameters derived for southern bluefin tuna.)

Annual banding has been reported on scales from fish up to 130 cm fork length (which had seven bands) but clarity decreased with increasing fish size. Bands on scales from larger fish (> 130 cm fork length) were unreadable. Daily growth of larvae has been determined from otolith-increment microstructure, validated using marginal-increment analysis. A burning technique, (validated for 2-4 years old, inclusive) showed annual growth-bands in otoliths from SBT up to 167 cm fork length. For fish younger than about 8 years old, the relationship of age and mean-length was similar to those derived previously by the methods mentioned above. Recently, X-ray examination of SBT otolith microchemistry has shown conspicuous episodic variations in strontium-calcium ratios for both small and very large SBT, apparently consistent with the expected age of a fish, based on its size and previous estimates of growth rates. This technique may enable reliable age determination of large (> 170 cm fork length) SBT, thus filling the gap left by previous aging techniques. Finally, trial use of computer-based, simultaneous analyses of a time series of length-frequency distributions to separate modes into age groups (MULTIFAN) has given results consistent with age-at-length relationships from other methods.

The age composition of the catch has been determined from length composition, using an age-at-length relationship. Where length composition is unavailable, catch-weight composition and a length-weight relationship have been used to derive length composition and, thereby, age composition. For fish less than 130 cm in length to caudal fork, the relationship between length and weight used during the later 1980s for stock assessments has been $W=3.13088 \times 10^{-5} \times L^{2.9058}$, where W is whole weight and L is length to caudal fork (Figure 5). For fish greater than 130 cm in length, the relationship used has been $W=1.4036 \times 10^{-6} \times L^{3.5399}$. It is more difficult to classify larger, older fish by this method however, raising doubt about the accuracy of estimates of age-distribution in the catch for such fish.

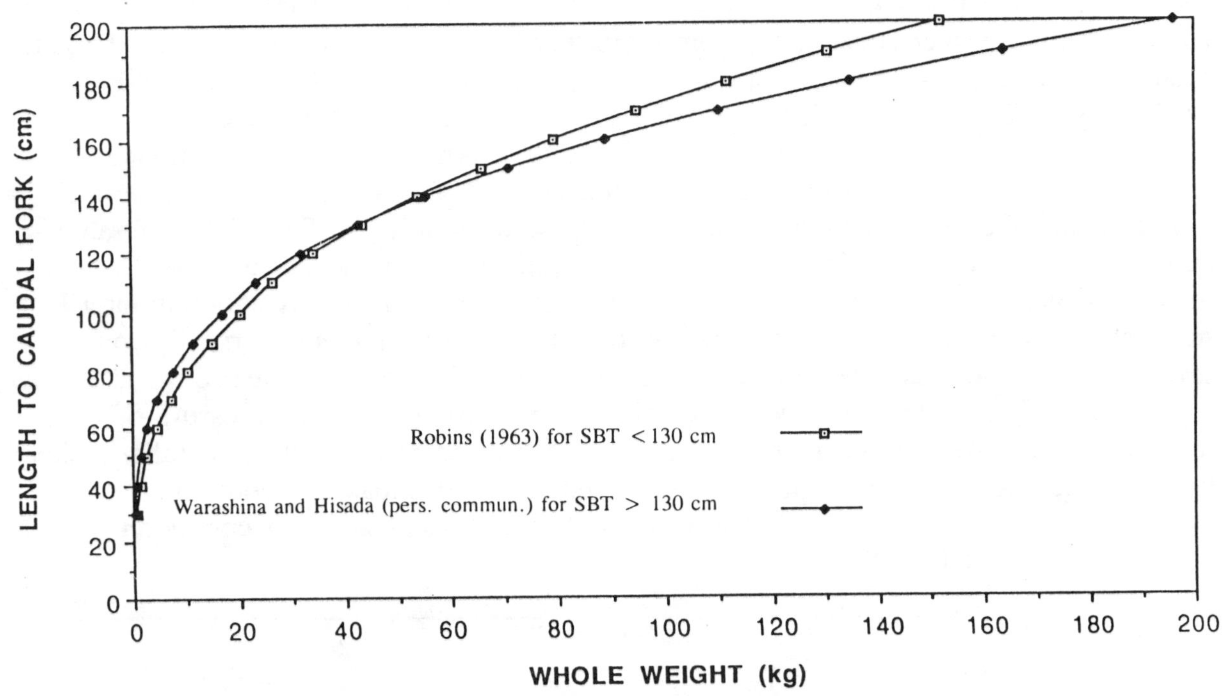

Figure 5.   Relationships currently used to convert southern bluefin tuna weight to length; the relationship for southern bluefin tuna > 130 cm has been adjusted from a processed-weight; length relationship to a whole-weight:length relationship using a 1:1.15 ratio of processed to whole weight.

The natural mortality rate (M) of SBT is an input parameter to Virtual Population Analysis (VPA) and is required for yield-per-recruit analyses. Again, there is considerable uncertainty about the appropriate value for SBT and whether M varies by area or age. The most commonly used value has been 0.2. However, assessments have generally incorporated tests to examine their sensitivity to the uncertainty in the parameter. All reliable analyses indicate that the rate of natural mortality is within the range 0.2 ± 0.1 per year. Further analyses may yield closer bounds of M.

Australia, Japan and New Zealand have been the main nations fishing for or controlling access to fishing zones for SBT. Incidental troll catches of SBT had been taken off south-eastern Australia for many years before pole-and-live-bait fishing for surface aggregations of juveniles commenced in the early 1950s. Catches steadily increased, and peaked at 21,500 mt (about 2.37 million fish) in 1982 (Figures 6 and 7; Table 1). Quotas have progressively forced down catches to just over 5,000 mt at the end of the 1980s (Table 2). The fishery commenced off NSW, where SBT 2 to 5 years old were common at the edge of the continental shelf. South Australian (SA; Figure 3) activity, on SBT 2 to 4 years old, developed in the mid-1950s, and a WA fishery, mainly for SBT 2 years old, commenced at the end of the 1960s. In the NSW and SA fisheries, aerial spotting developed, activity spread widely and further offshore, purse-seining commenced, and combined poling and purse-seine sets on schools evolved. The age of fish taken increased to about 9 years old. Off WA, in contrast, vessels remained comparatively small (7-15 m) and the fishery was predominantly confined to the continental shelf and to SBT 2 years old. In the early 1980s the NSW fishery failed and until 1991 no surface schools were reported there. However, Japanese longline catches continue, albeit at lower hook rates than at the end of the 1970s. The WA fishery has virtually stopped, catches declining from more than one million fish in the early 1980s to about 5,000 in 1992 because of the transfer of quota holdings to SA. Areas of SBT availability off SA also dwindled during the 1980s but increased to some extent after 1987. Scientists had expressed concern in the early 1980s about declining parental biomass, and as a result a quota was set for the 1983-84 season. In order to generate finance within the industry by the sale and purchase of quotas, in 1984-85 transferable quotas, allocated to vessels individually, were introduced. Another measure at the time required fishermen to re-direct activities towards older fish. This was only temporarily successful because the declining availability of surface aggregations of large fish over several years forced resumption of SA operations on predominantly 3-year-old and 4-year-old SBT, despite their lower value and the increasing economic pressure of reduced quotas. After 1990, there was a major shift in emphasis in the Australian fishery from surface operations to longlining in joint ventures with Japan. In addition, experimental cage rearing of SBT was successful to the extent that four commercial ventures commenced in 1992.

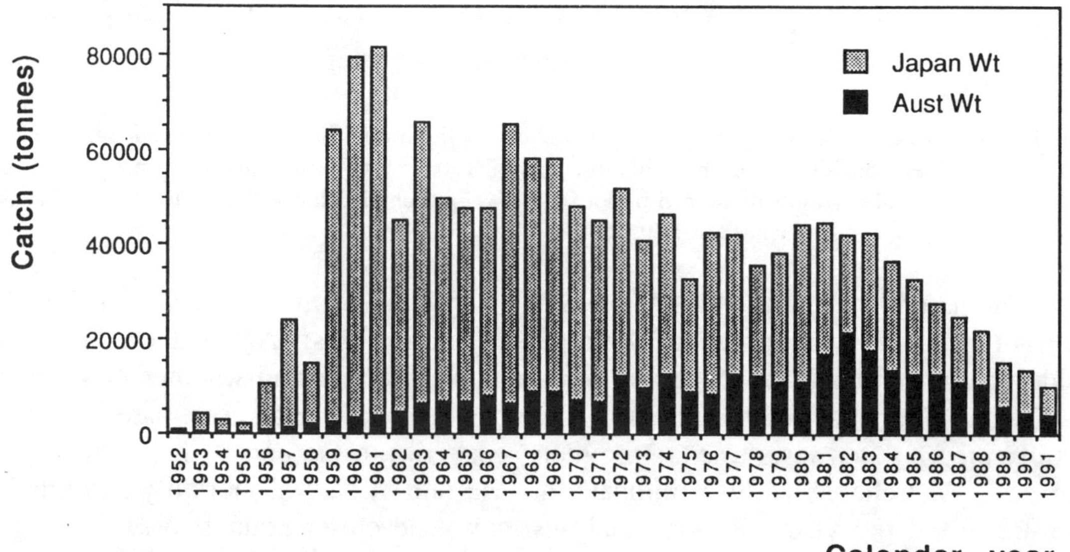

Figure 6. Japanese and Australian southern bluefin tuna catch (tonnes), 1952 to 1991 caledar years [Japanese catch for 1991 is not known, so quota for that year (see Table 2) is used].

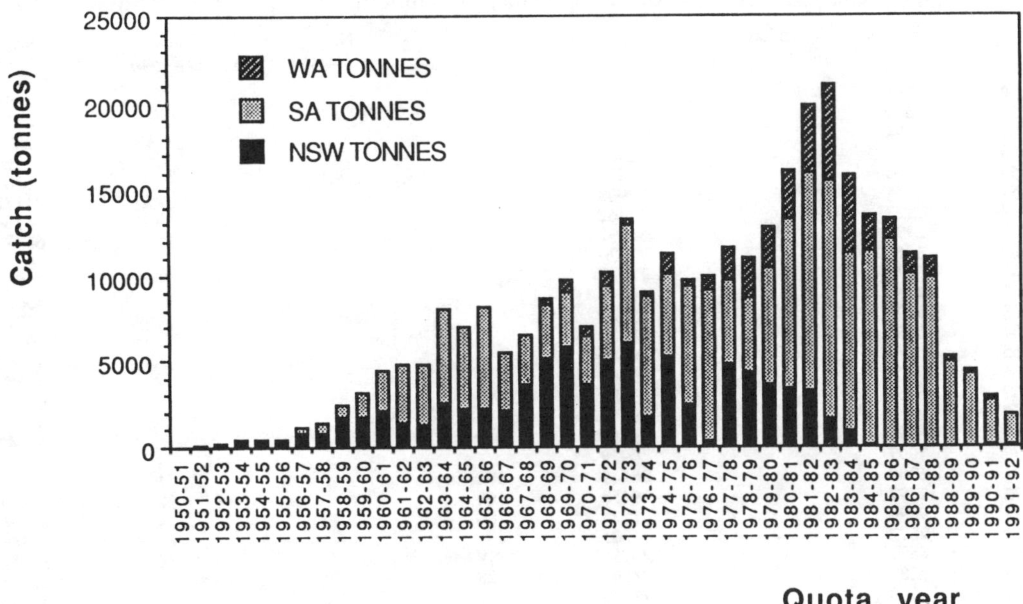

Figure 7.   Western Australian, South Australian, and New South Wales southern bluefin tuna catches, 1951-52 to 1991-92 quota years (1991-92 catch is an estimate using data from Australian Fisheries Management Authority).

The Japanese SBT longline fishery commenced in 1952, the catch peaking sharply by 1960-61 at more than 75,000 mt (about 1.2 million fish), then declining steadily to 11,000 mt by 1988 (Figure 6 and Table 1). Quotas reduced the catch to less than 7,000 mt in 1990 (Table 2). Japanese fishing operations developed on the spawning ground and later off eastern New Zealand, spreading southward in the Indian Ocean to the West Wind Drift region. They eventually ranged from the east of New Zealand, westward throughout the West Wind Drift region, to the southern Atlantic (Figure 8). Fishing effort increased steadily from commencement of the fishery, slowed between 1970 and 1980, then declined because of a 20% reduction in the number of vessels. It subsequently increased slightly as a result of an increasing trend in number of hooks per set, then declined again (Figure 9). The longline grounds east of New Zealand and off NSW are not as important for SBT as they used to be. The former area is a geographically marginal area for SBT, and the latter has been subject to closures and has seen an increase in fishing activity directed at yellowfin tuna (*T. albacares*), bigeye tuna (*T. obesus*), broadbill swordfish (*Xiphias gladius*), and marlin (*Makaira* spp.). Activity on the spawning ground and the Oki ground to its south is now directed mainly at bigeye tuna. The most important area for the SBT fishery since the 1960s has been the fishing ground south of Africa, where effort reached 50 million hooks in 1979. The hook numbers there remained around 30-40 million annually until 1990 when less than 25 million hooks were set. Globally, effort levels were in the order of 100-125 million hooks annually between 1980 and 1990, when the level declined to 63 million hooks. The hooking rate peaked soon after commencement of the fishery, then declined during and after the expansion of fishing grounds in the 1960s. In 1987 it was about 25% of the early 1970s rate. In contrast to the Australian fishing operations on surface aggregations of juveniles, the Japanese target medium-sized and large-sized SBT (70-180 cm fork length), with the proportion of medium-sized fish increasing when the fishing grounds expanded from the spawning ground. There has been a partial overlap in size

Table 1. Calendar year catch (whole weight, mt) of southern bluefin tuna by country, 1952 to 1991.

	Australia	Indonesia	Japan	Korea	New Zealand	China (Taiwan)	Total
1952	264		565				829
1953	509		3890				4399
1954	424		2447				2871
1955	322		1964				2286
1956	964		9603				10567
1957	1264		22908				24172
1958	2322		12462				14784
1959	2486		61892				64378
1960	3545		75826				79371
1961	3678		77927				81605
1962	4636		40397				45033
1963	6199		59724				65923
1964	6832		42838				49670
1965	6876		40689				47565
1966	8008		39644				47652
1967	6357		59281				65638
1968	8737		49657				58394
1969	8679		49769				58448
1970	7097		40929			8	48126
1971	6969		38149	500		51	45635
1972	12397		39458	100		38	51967
1973	9890		31225	100		4	41216
1974	12672		34005	182		47	46868
1975	8833		24134	99		14	33068
1976	8383		34099	28		10	42511
1977	12569		29600	7		24	42196
1978	12190	4	23632	94		50	35973
1979	10783	10	27828	0		57	38685
1980	11195	8	33353	0	130	42 (179)[E]	44746[F]
1981	16843	3	27981	0	173	23 (259)[E]	45048[F]
1982	21501	2	20789	6	305	42 (366)[E]	42640[F]
1983	17695	3	24881	0	132	21 (323)[E]	42731[F]
1984	13411	1	23328	1	93	97 (607)[E]	36930[F]
1985	12589		20396	0	94	13 (2389)[E]	33091[F]
1986	12531	4	15182	0	82	266 (531)[E]	28065[F]
1987	10821	3 (14)[B]	13964	0	59	109 (1001)[E]	24909[F]
1988	10591	3 (14)[B]	11422	0	93	146 (1046)[B]	22219[F]
1989	6118[A]	n.a. (104)[B]	9222[C]	0	431[D]	145 (1486)[B]	15671[F]
1990	4719[G]	n.a. (63)[B]	7056[C]	0	480[H]	157 (1231)[J]	n.a.
1991	4162[I]	n.a. (165)[B]	n.a.	n.a.	160[K]	n.a. (960)[J]	n.a.

Sources: Australia, Japan, and New Zealand: data provided to trilateral scientific discussions on southern bluefin tuna; data for Japan are values which have been amended from the historical catch series, taking account of a 1990 revision of Japanese statistical areas for the southern bluefin tuna fishery; Koreak, China (Taiwan): Indian Ocean and Southeast Asian tuna fisheries data summary for 1990; IPTP Data Summary No. 12, May 1992, Colombo, Sri Lanka (note that it modifies the Data Summary 11 series); Indonesia: Fishing ground and distribution of southern bluefin tuna (*Thunnus maccoyii*) in southJava and Nusatenggara waters. (Doc. No. 16) Third Southeast Asian Tuna Conference, Bali, Indonesia, 22-24 August 1989; n.a.=not available.

[A] Total of 5,435 mt from domestic surface fishery; 684 mt from Australia/Japan joint-venture longline vessels.

[B] Indonesia catch is uncertain; tonnages in parentheses are Japanese imports of bluefin tuna from Indonesia.

[C] This tonnage had been calculated using a different length:weight relationship to that for previous years; use of the previous relationship would generage a tonnage close to 10,500 mt for 1989, but this is inconsistent with recent length:weight observations. Currently there is no basis for revising the time series to take account of changes in the relationship over time.

[D] Total of 134 mt from domestic handline/troll fishery; 290 mt from NZ/Japan joint-venture longline vessels.

[E] Values in parentheses were provided to the Australian Fisheries Service by the Taiwan Fisheries Bureau as southern bluefin tuna catch, but appear to include all bluefins. For 1989, statistics on catch by ocean and gear suggest that 879 mt of the 1,486 mt were southern bluefin tuna.

[F] Provisional total; ingnores values in parentheses for China (Taiwan) and Indonesia.

[G] Total of 4,319 mt from domestic fishery; 400 mt attributed to Australia/Japan joint-venture longline vessels.

[H] Total 247 from domestic fishery; 233 mt from New Zealand/Japan joint-venture longline vessels.

[I] Total 2,871 mt from domestic fishery; 1,290 mt from Australian Japanese-style charter, and Asutralia/Japan joint-venture longline vessels.

[J] Values in parentheses are taken from Fisheries Yearbook Taiwan Area, Taiwan Fisheries Bureau; they are a continuation of the series in parentheses for 1981 to 1989.

[K] Total of 35 mt from domestic fishery; 125 mt from New Zealand/Japan joint-venture longline vessels.

Table 2. Southern bluefin tuna catch limits, periods of catch-limit operation, and catches realised for Australia, Japan, and New Zealand since 1983.

Negotiation Year	Australia			Japan			New Zealand		
	Operating Period	Catch Limit	Catch Realised	Operating Period	Catch Limit	Catch Realised	Operating Period	Catch Limit	Catch Realised
1983	1.10.83-30.9.84	21,000 (A)	15,843		no limit	(23,323) (1984)	1.1.84-31.12.84	10,000 (B) fish	93
1984	1.1084-30.9.85	14,500 (C)	13,486		no limit	(20,393) (1985)	1.1.85-31.12.85	10,000 fish	94
1985	1.10.85-30.9.86	14,500	13,237	1.3.86-28.2.87	23,150	15,522 (1986)	1.1.86-31.12.86	1,000	82
1986	1.10.86-30.9.87	11,500 (D)	11,308	1.3.87-29.2.88	19,500 (D)	13,955 (1987)	1.1.87-31.12.87	1,000 (D)	59
1987	1.10.87-30.9.88	11,500 (D)	10,976	1.3.88-28.2.89	19,500 (D)	11,422 (1988)	1.1.88-31.12.88	1,000 (D)	93
1988	1.10.88-30.9.89	6,250 (E)	5,984 (F)	1.3.89-29.2.90	8,800 (E)	9,150 (G)	1.1.89-31.12.89	450 (E)	424 (H)
1989	1.10.89-30.9.90	5,265 (I)	4,848 (J)	1.3.90-28.2.91	6,065 (I)	7,056 (K)	1.1.90-31.12.90	420 (I)	519(L)
1990	1.10.90-30.9.91	5,265 (M)	4,315 (N)	1.3.91-28.2.92	6,065 (O)	n.a.	1.1.91-31.12.91	420 (P)	160 (Q)
1991	1.10.91-31.10.92(T)	5,265 (R)	s.n.c.	1.3.92-28.2.93	6,065 (S)	s.n.c.	1.1.92-31.12.92	420 (P)	s.n.c.

Definitions:

Negotiation Year - the year when the catch limit indicated was determined.
Operating Period - the period for which the catch limit applied. The Australian limit commences on 1 October of the negotiation year, the New Zealand limit on 1 January of the subsequent year, and the Japanese limit on 1 March of that subsequent year.
Catch Limit and *Catch Realised* are in tonnes unless indicated otherwise.

Abbreviations: *s.n.c.* = season not completed; *n.a.* = not available.

Keys to Table 2:

(A)   The 1983-84 Australian catch limit provisions consisted of a total quota of 21,000 tonnes, with 4,000 mt allocated to the Western Australian fishery (where no purse seining was permitted), and 15,000 mt to the southeastern (*i.e.*, New South Wales and South Australian) fishery; the purse seine component in the latter was limited to 5,000 mt. A further 2,000 mt was set aside as reserve/developmental quota for activities specially diverted towards larger fish; 500 mt of fish larger than 70 cm from the Western Australian fishery were debited against that component.

(B)   New Zealand had previously maintained a quota of 5,000 fish for its domestic southern bluefin tuna catch.

(C)   The 1984-85 Australian catch limit was distributed among vessels as individual transferable quotas (Franklin, 1989); this arrangement has continued for subsequent Australian seasons.

(D)   Australia and Japan maintained national quotas of 14,500 mt and 23,150 mt respectively, but agreed to a global catch limit of 31,000 mt. Additionally, they agreed that their national catches would not exceed 11,500 mt and 19,500 mt respectively. A "development" quota of 1,000 mt was reserved for New Zealand.

(E)   Joint-venture longlining arrangements were established between Japan and Australia, and Japan and New Zealand. The subsequent distribution of catch limits was: Australia 5,480 mt domestic; 770 mt joint-venture (6,250 mt collectively); New Zealand 150 mt domestic; 300 mt joint-venture (450 mt collectively); and Japan 8,800 mt domestic; 1,070 mt joint-venture (9,870 mt collectively).

(F)   Value comprises 5,300 mt from domestic surface/small longliner fishery and 684 mt from Australia/Japan joint-venture longline vessels.

(G)   This tonnage has been calculated using a different length-weight relationship to that for previous years. Use of the previous relationship would generate a tonnage close to 10,500 mt, but this is inconsistent with recent length-weight observations. Currently there is no basis for revising the time series to take account of changes in the relationship over time.

(H)   Value comprises 134 mt from domestic handline/troll fishery and 290 mt from New Zealand/Japan joint-venture longline vessels.

(I)   Joint-venture longlining arrangements were established between Japan and Australia and Japan and New Zealand. The subsequent distribution of catch limits was: Australia 4,865 mt domestic;400 mt joint-venture (5,265 mt collectively); New Zealand approximately 220 mt domestic, 200 mt joint-venture (420 mt collectively); and Japan 6,065 mt domestic; 600 mt joint-venture (6,665 mt collectively).

(J)   Value comprises 4,448 mt from domestic surface/small longliner fishery and 400 mt from Australia/Japan joint-venture longline vessels.

(K)   This tonnage has been calculated using the length-weight relationship used for 1989/90.

(L)   Value comprises 286 mt from domestic handline/troll fishery and 233 mt from New Zealand/Japan joint-venture longline vessels.

(M)   Total of 1,000 mt frozen, of which 300 mt was assigned to Japanese longliners participating in a real-time monitoring programme (RTMP) of catch length composition; 966 mt assigned to longline joint venture with Japan, 434 mt assigned to Australian-chartered Japanese-style longliners, leaving 2,865 mt as surface/small longliner component.

(N)   Value comprises 2,879 mt from domestic surface/small longliner fishery and 1,136 mt from Australian chartered and Australia/Japan joint-venture longline vessels. The 300 mt RTMP allocation is included.

(O)   Japanese longliners also had access to an additional 300 mt of quota for the RTMP, and to joint-venture catches (966 mt in an Australia/Japan joint-venture, and approximately 200 mt in a Japan/New Zealand joint-venture).

(P)   Joint-venture longlining arrangements were established between Japan and New Zealand. The subsequent distribution of catch limits was approximately 220 mt to the domestic fishery and 200 mt to New Zealand/Japan joint-venture longliners.

(Q)   Value comprises 125 mt from domestic handline/troll fishery and 35 mt from New Zealand/Japan joint-venture longline vessels.

(R)   Total of 800 mt of Australian quota was assigned to the expanded RTMP; 2,070 mt of the remainder was made available to Australia/Japan joint-venture longline vessels, leaving a balance of 2,395 mt for the domestic and Australian chartered longline vessel fisheries.

(S)   Japanese longliners also had access to the 800 mt RTMP quota, and to joint-venture catches (2,070 mt in an Australian/Japan joint-venture, and an unspecified component of the New Zealand 420 mt.

(T)   The 1991/92 Australian quota year has been extended to 13 months; subsequent quota years will commence on 1 November and finish on 30 October.

composition of the surface and longline fishery catches (Figure 10). In 1971, because of increasing catches of small SBT, voluntary area-closures were adopted in the longline fishery. An area south of the spawning ground was also closed seasonally for the second half of the spawning season to protect migrating adults (Figure 11). Following recommendations by scientists for catch restraints, and several subsequent years of trilateral management discussions among Australia, Japan and New Zealand, Japan introduced a catch limit of 23,150 mt for its 1986-87 season. The Japanese catch limit was reduced, progressively, to 6,250 mt for the 1990-91 season (Table 2).

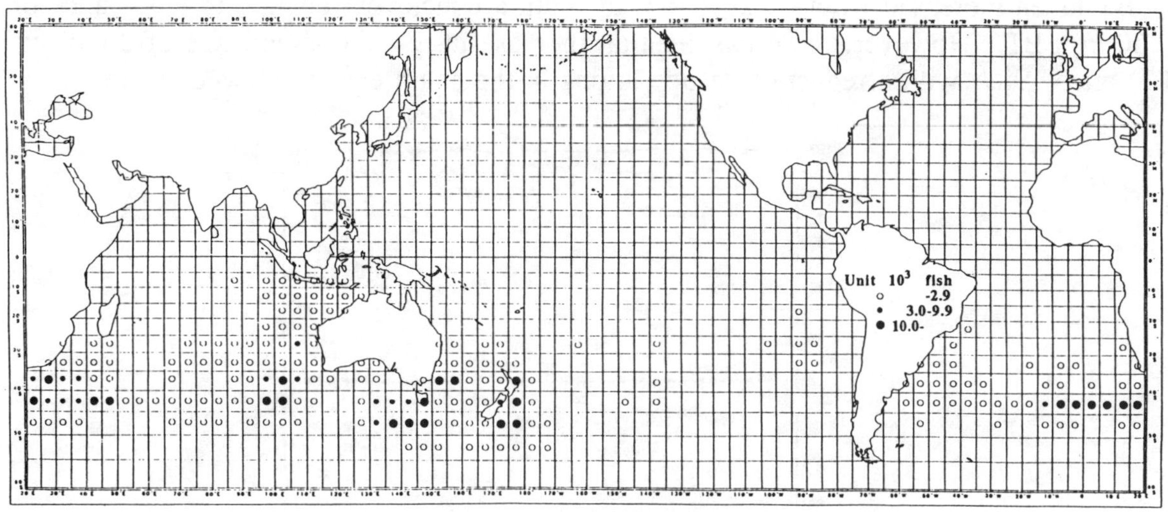

Figure 8.   Distribution of southern bluefin tuna catch in numbers by Japanese longline tuna fishery, 1970. (From: Fisheries Agency of Japan, 1972. *Annual report of effort and catch statistics by area on Japanese tuna longline fishery 1970.*)

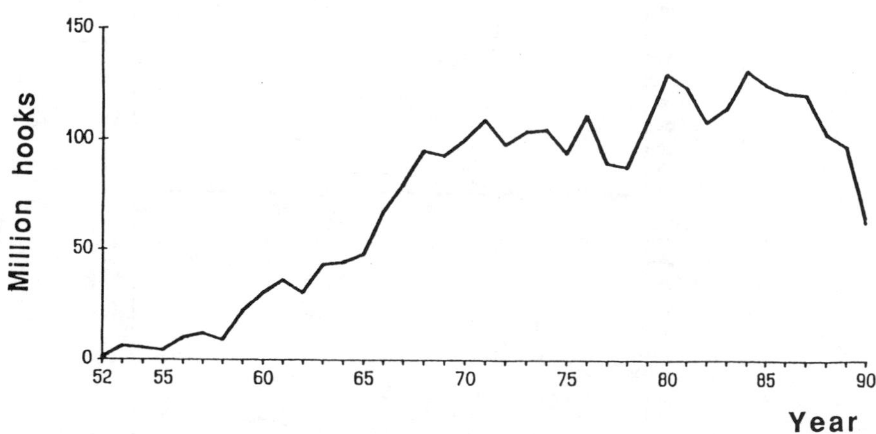

Figure 9.   Trends of annual fishing effort (hooks) of Japanese longline fishery in the southern bluefin tuna fishing grounds. (Source: Nishida, T. and Y. Ishizuka, 1992. *Japanese southern bluefin tuna (Thunnus maccoyii) fishery in recent years (1985-90).* Working document for the eleventh meeting of Australian, Japanese, and New Zealand scientists on southern bluefin tuna, National Research Institute of Far Seas Fisheries, Shimizu, Japan.)

The small, New Zealand domestic troll and handline fisheries for adult SBT, operating alongside trawlers taking demersal species off the west of the South Island, commenced in 1980 (Figure 12). Catches reached 305 mt in 1982, declined until 1987, increased following the diversification of some vessels to small-vessel longlining, then fluctuated (Table 1). From 1989, domestic activities included the operation of four to five chartered Japanese longliners. Since the 1950s, Japanese longliners had fished in areas adjacent to New Zealand (Figure 12), and their operations there came under New Zealand control when the New Zealand 200-mile exclusive economic zone (NZEEZ) was established in 1978. Japanese catches, number of hooks set, and hooking rate in the NZEEZ declined steadily from 1980 to 1988. Catches declined from 7,600 mt to 762 mt. Although the number of hooks set halved, the hooking rate declined to one fifth over the same period. Vessels have increased the average number of hooks per set from 2,400 to 2,900 in an attempt to counteract this decline. Between 1980 and 1989 the average size of SBT taken increased from 63 kg to 95 kg, with a concurrent decline in abundance of 40-50 kg SBT. However, there has been an increase in relative abundance of 20-40 kg SBT since 1989, with a reduction in operations in the south east of New Zealand.

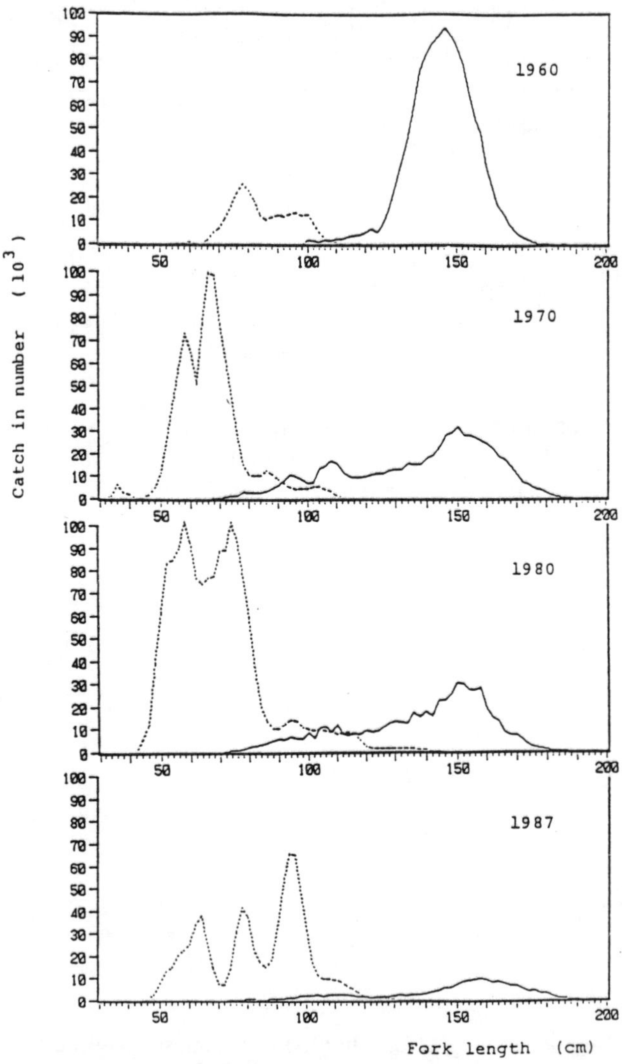

Figure 10. Length-frequency distribution of southern bluefin tuna caught by Australian surface fishery (dotted line) and Japanese longline fishery (solid line) in 1960, 1970, 1980, and 1987.

The access areas for Japanese longliners in the 200-mile Australian fishing zone (AFZ) have decreased progressively since the zone came into effect in 1979. The region around Tasmania (Figure 3) has become the main area of longliner operation. With the continued decline of hooking rates in the area to the south west of Tasmania, which supported a summer fishery for adults, effort has been re-directed to a winter fishery off

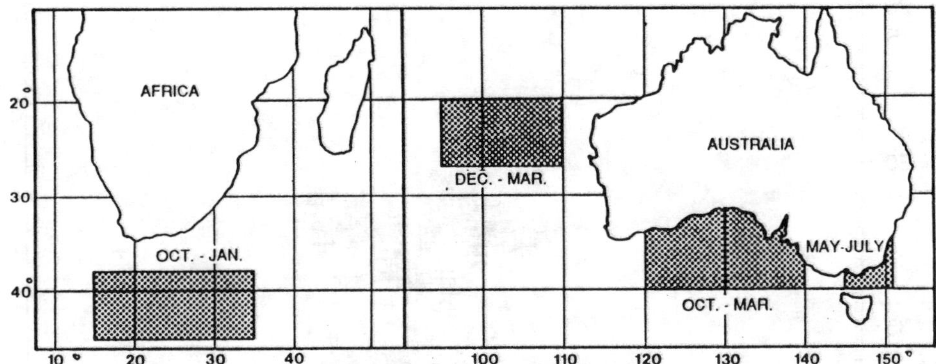

Figure 11.   Regions and periods of the southern bluefin tuna fishing grounds closed by voluntary restrictions. (Source: Singu, 1978.)

eastern Tasmania. In this area there has been an increase in the proportion of pre-adult SBT in the longline fishery (Figure 13), probably reflecting increased escapement from the Australian surface fishery because of the progressive reduction in Australian quotas. However, the increase has been later than anticipated. In keeping with Australia's objective of diverting SBT fishing effort to larger fish, an Australia-Japan longline joint venture involving Japanese longliners and Australian industry and quotas commenced in 1989.

Southern bluefin tuna catches by countries other than Australia, Japan and New Zealand have, in the past, been considered relatively trivial (Table 1). However, the SBT by-catch of driftnetters from Taiwan, and of longliners from Taiwan, Indonesia, and previously, Korea may be more significant than previously thought, especially in comparison with the considerably reduced catches of Australia, Japan, and New Zealand. Imports of bluefin to Japan (Figure 14) and anecdotal information suggest that the 1991 SBT catch by Indonesia may have been in the order of 250 mt, and that the Taiwan catch was possibly in the order of 1,300 mt.

Landed catch by boat, date of landing, port of landing, and destination (purchaser) have been recorded for the Australian surface fishery from its inception, and is currently recorded in conjunction with the monitoring of cumulative catch, during a season, in relation to quota. A logbook (Figure 15) to record details of fishing location and effort was not firmly established until the early 1980s. Southern bluefin tuna length composition has been sampled routinely in the south-eastern fishery from 1963-64, the sampling taking account of the sorting of fish at sea. Length frequency of the catch, by one cm length class and for each half-month, is available for NSW, SA, Albany and Esperance (Figure 3) (or WA, when these could not be discriminated). Spotter aircraft have been used to some extent in the NSW and SA sectors of the Australian SBT fishery since the very early days. Attempts to introduce an aircraft logbook met with little success until 1989 but some spotters' private records are available for recent years. By means of radio reports and log books (Figure 16), catch and effort data for SBT have been collected from Japanese longline vessels operating in the AFZ since 1979. From 1989 longliners have been required to measure length to caudal fork of each SBT taken in the AFZ. Observers occasionally join vessels to check that catch-reporting procedures are understood and properly maintained.

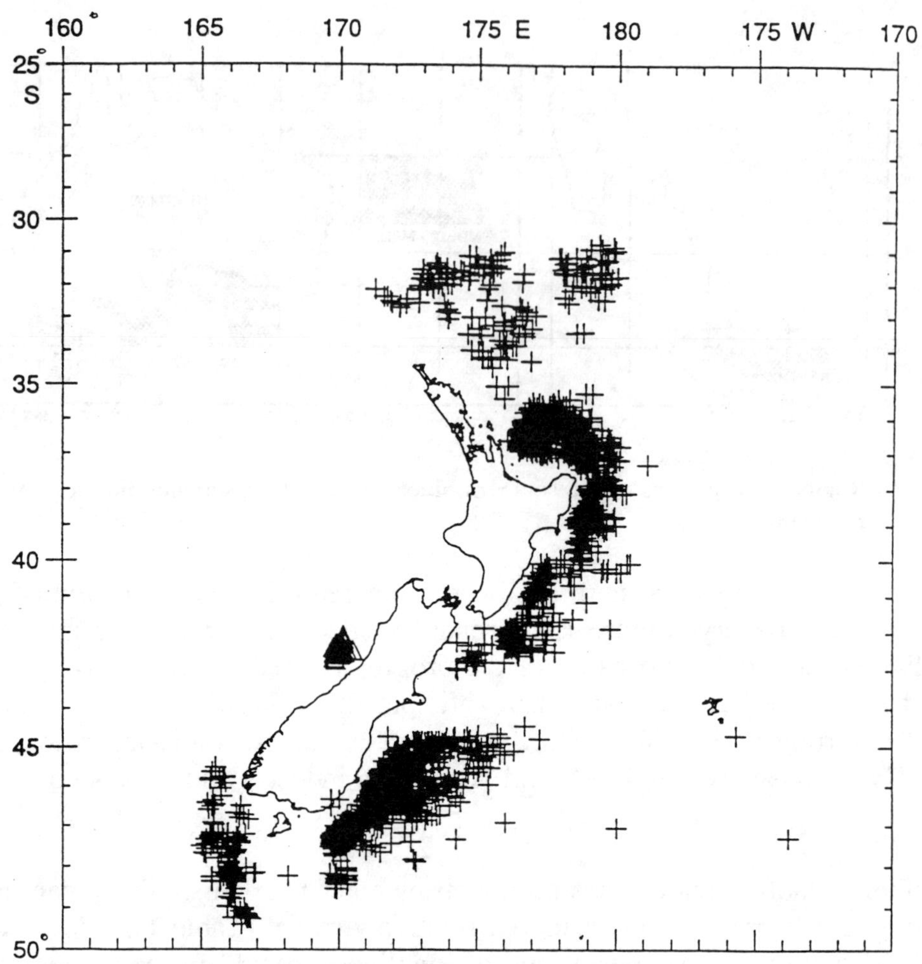

Figure 12.   Locations of catch of southern bluefin tuna by New Zealand domestic (triangles) and foreign (crosses) vessels; positions respresent locations of daily position reports by domestic vessels and start-of-set locations for longliners, for the period 1980 to1988.

Japanese tuna longliners maintain logbooks of daily activity, detailing fishing location, catches (in number of fish by species) and hooks set (Figure 17). Data tabulated by five degree grid squares, were published from 1962 to 1980. Since 1981, data have been provided directly from the Far Seas Fisheries Research Laboratory to CSIRO Australia and Fisheries Research Centre, New Zealand. At first, catch length composition had to be sampled at Yaizu and Tokyo markets but subsequently lengths were measured at sea by cooperative fishermen or derived from individual-weight data from fishing masters' private logbooks. The data have been used to generate catch length frequency for each statistical area and for each yearly quarter in two cm length classes. For up-to-date stock assessment and fishery management, timely collection and processing of catch, effort and length data are needed. A lag of 1.5 to 2.5 years had occurred routinely before Japanese longline fishery catch and length distribution data became available. However, for 1992 assessment work about 70% of the 1991 data was available, representing a major improvement in timeliness. Additionally, a real time catch, effort and length/weight monitoring programme was introduced, involving 12 longliners in 1991 and 17 in 1992. It is still desirable to improve the recovery rate of length data from longliners.

New Zealand vessels are required to complete a catch, effort and landings logbook (Figure 18). Catch data are verified independently against records maintained by licensed

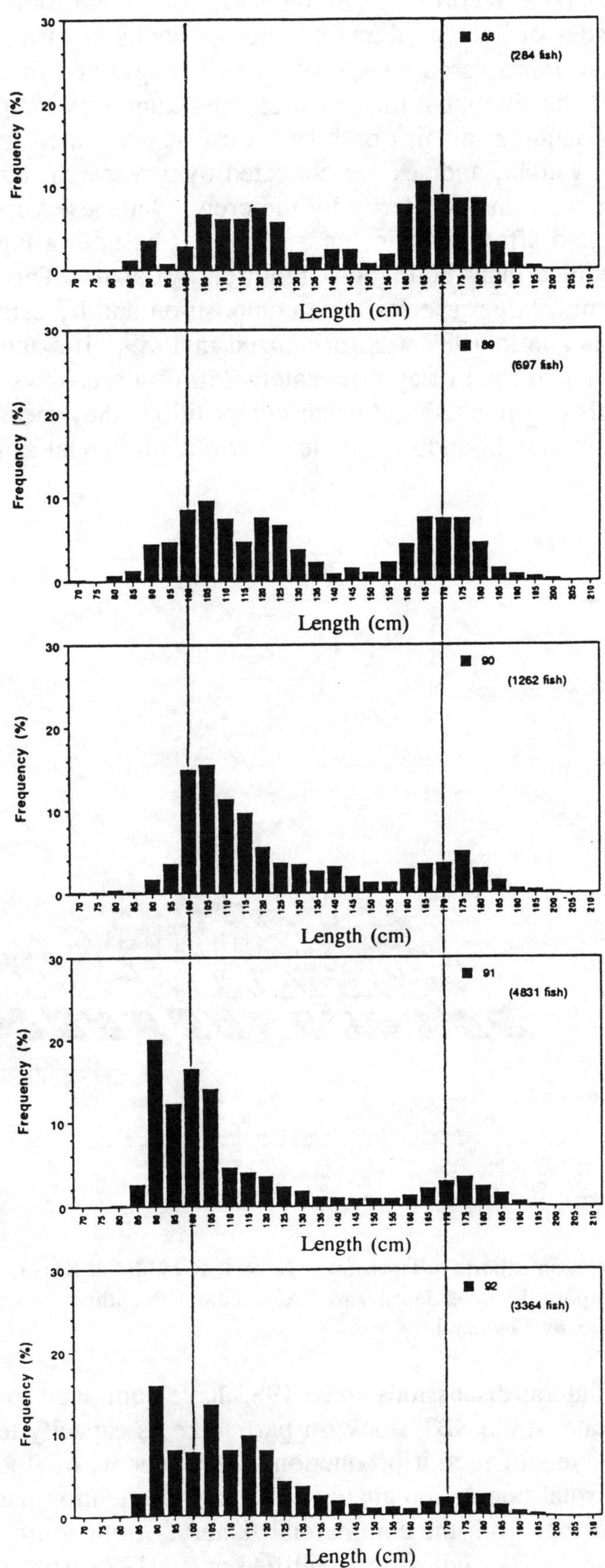

Figure 13.   Southern bluefin tuna length-frequency distribution in the southeastern Australian winter (Southern Hemisphere) Japanese longline catch, 1988 to 1992.

fish receivers. The logbook records fishing method, area or position of fishing, number of hours fishing, number of lines used and number of hooks or lures on each line, target species, estimated total unprocessed weight of all fish caught and the estimated unprocessed weight of the five most important species caught by weight in decreasing order. At the time of landing, all fish on board must be accounted for. Data on individual fish length, weight, and sex are collected by a research logbook distributed to the freezer vessel and filled in voluntarily by the crew. Japanese vessels fishing in the NZEEZ record catch and effort statistics on a set-by-set basis in a logbook (Figure 19) considerably more detailed than the logbook required by Japan. This includes information on environmental factors, target species, size composition and by-catch. A limited observer programme begun in 1987 was formalized in 1989. It extends from April to September for an estimated total of approximately 240 observer-days per year. Observers record details of the fishing method and catch composition, they measure and weigh all tuna and billfish species caught and they collect various biological samples for ongoing research programmes.

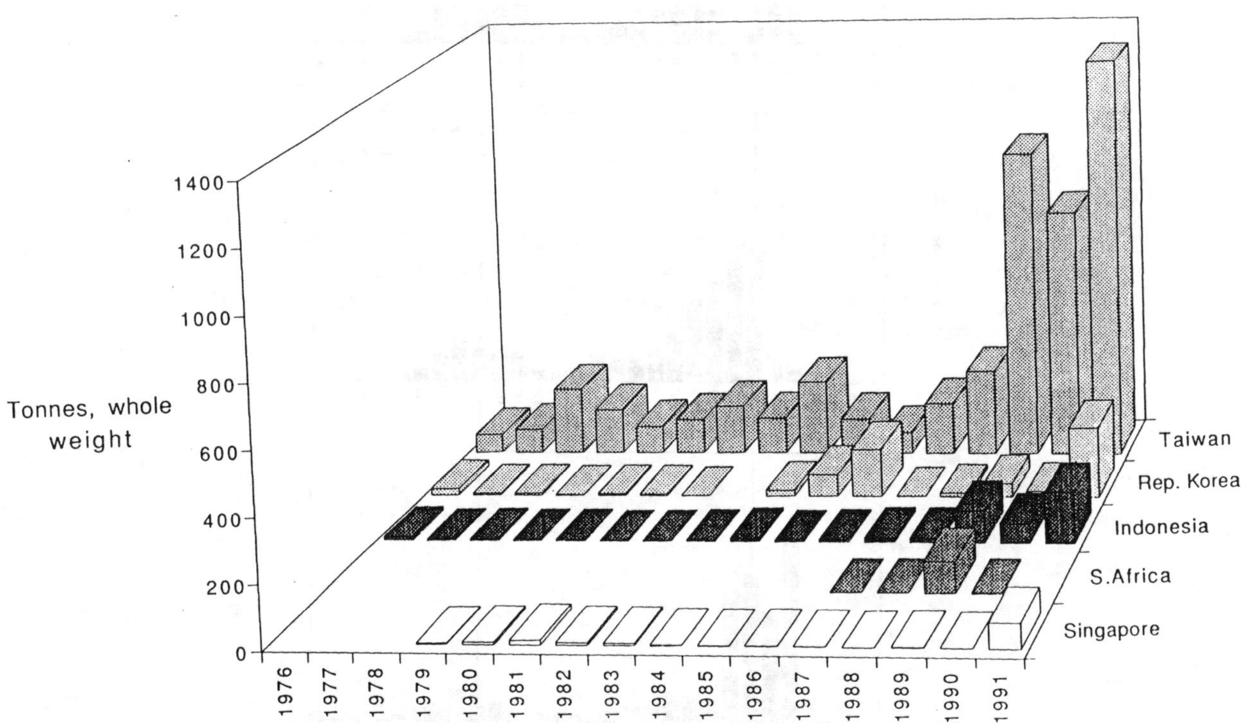

Figure 14.   Annual bluefin exports to Japan from Taiwan, Republic of Korea, Indonesia, South Africa, and Singapore. [Source: Japan Tariff Association. (monthly). *Japan Exports and Imports Commodity by Country.*]

Scientists at trilateral discussions since 1982 have continued to express concern about the biological state of the SBT stock, in particular its capacity to continue to maintain satisfactory levels of recruit production. Assessments of stock condition have been carried out by virtual population analysis (VPA) but attention has also been drawn to a range of direct indicators from the commercial fishery. Both sources point to a severe decline in the population. The indicators identified at the 1988 trilateral scientific discussions were: the extent and persistence of the decline in catches and hooking rate in conjunction with high fishing effort; contraction in the area of the surface fishery and the extent of productive areas of the longline fishery; and major declines in the abundance of pre-adult SBT. Catch and hooking rates have declined further, but there has been a slight increase in the extent of productive areas in the longline fishery, apparently linked to an increase in the pre-adult SBT population, mainly off the coast of Tasmania.

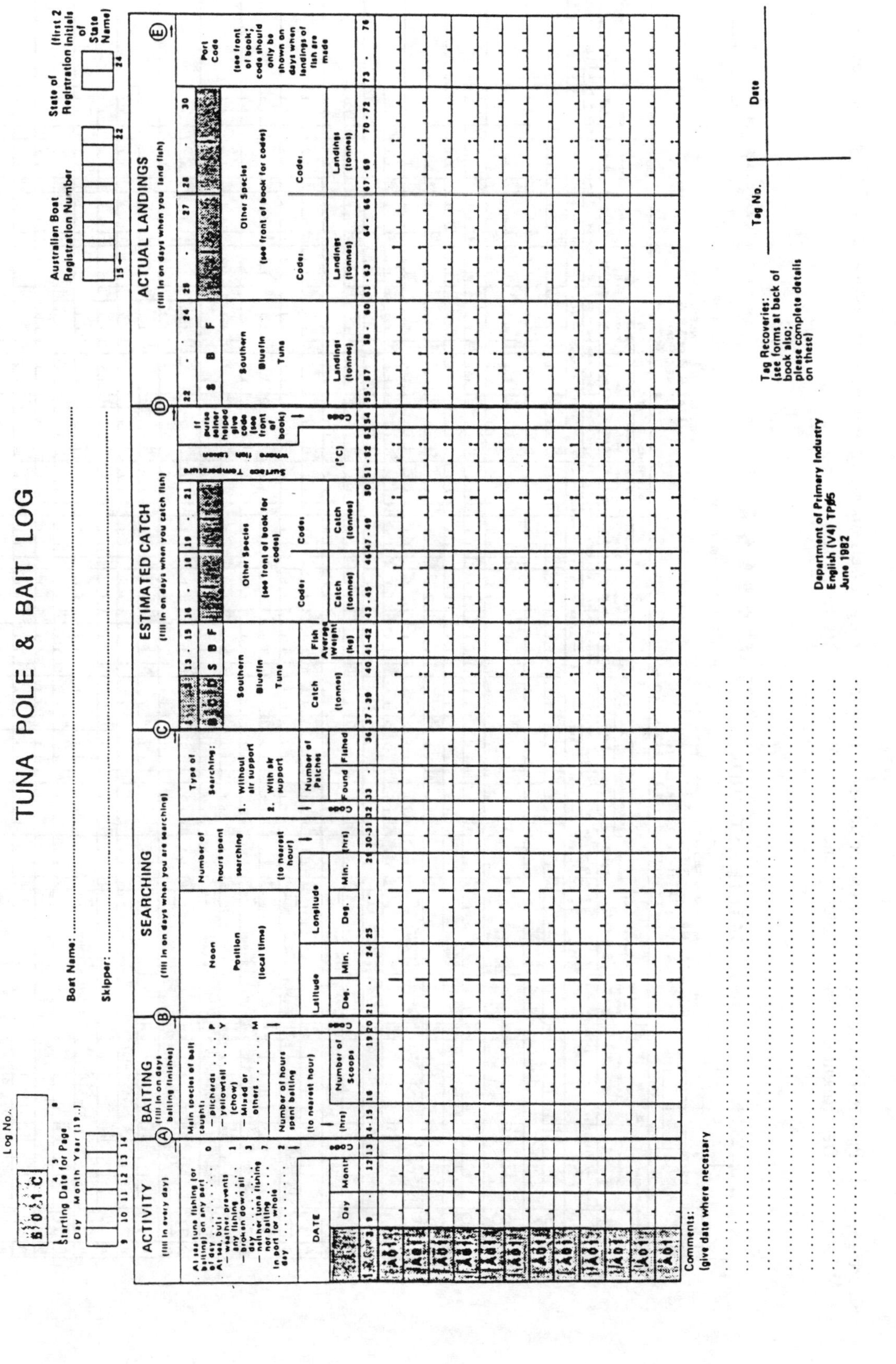

Figure 15. Prescribed logbook form for use by southern bluefin tuna pole-and-line fishermen operating off southeastern Australia. (Source: Australian Department of Primary Industries and Energy.)

Figure 16.   Prescribed logbook form for use on Japanese-longline fishing vessels operating in the 200-mile Australian fishing zone. (Source: Australian Department of Primary Industry and Energy.)

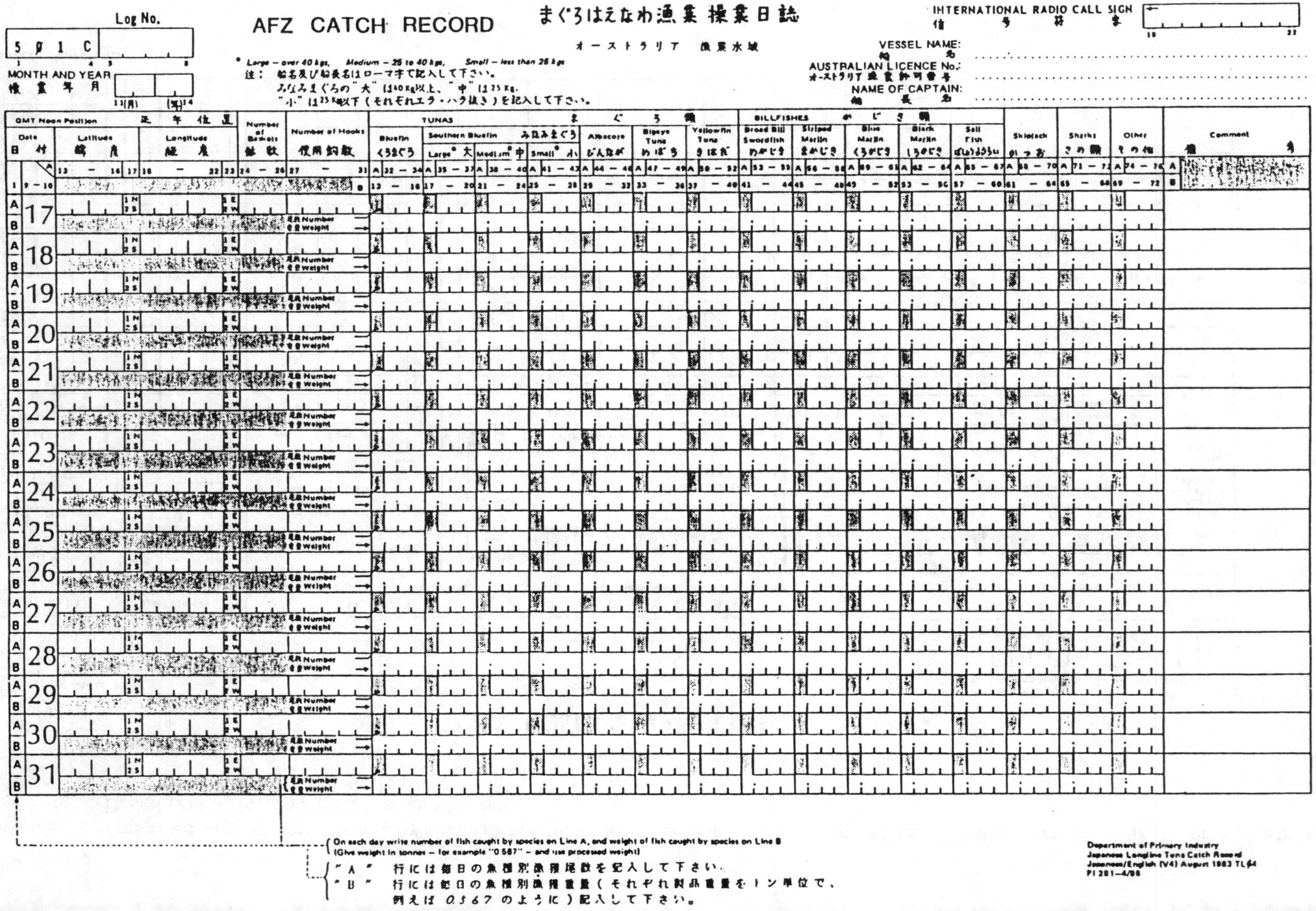

Figure 17. Logbook form used globally on Japanese longline fishing vessels. (Source: Japan Fisheries Agency.)

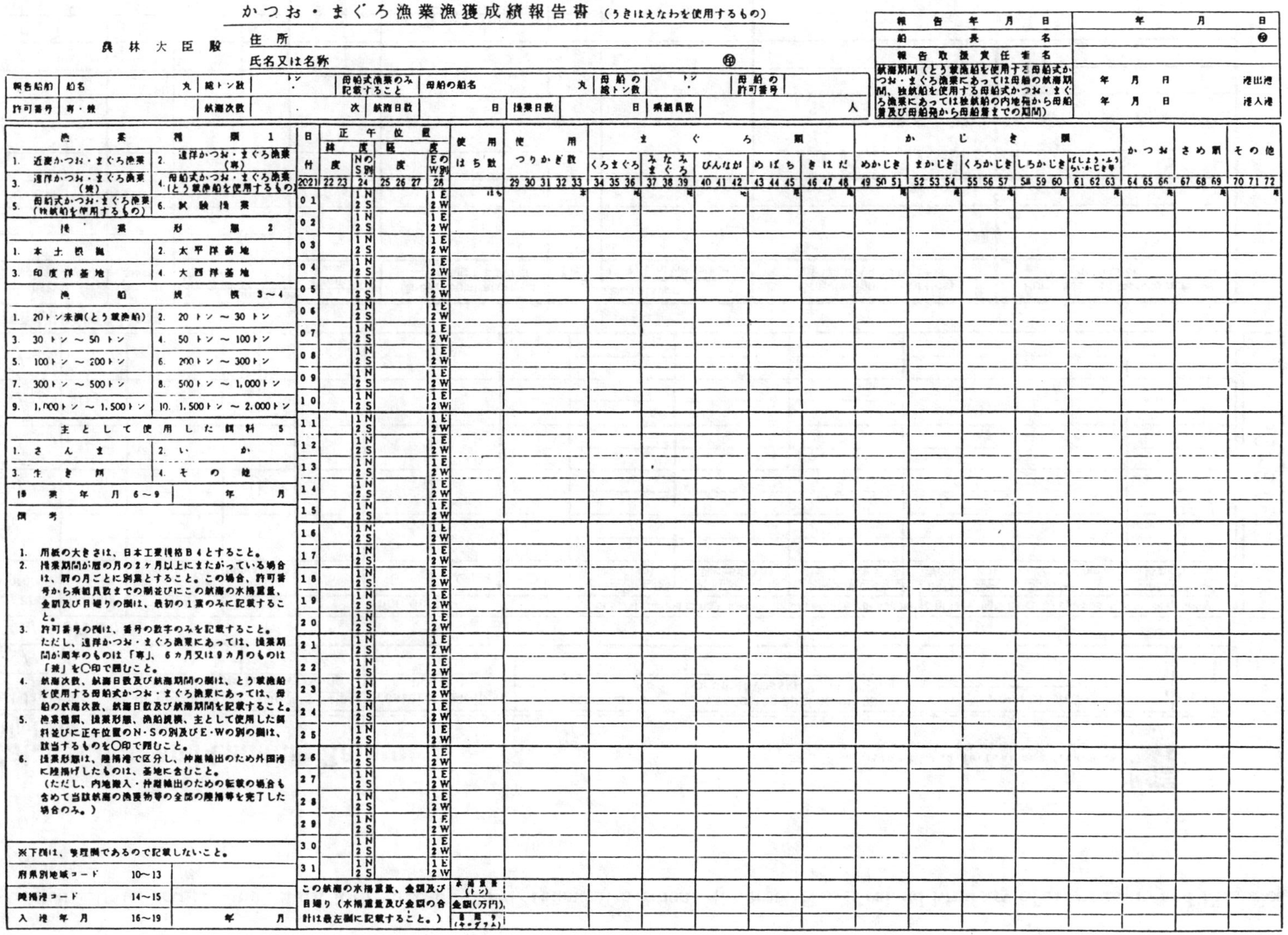

Figure 18.   Catch, effort, and landing return form used by New Zealand fishermen. (Source: New Zealand Ministry of Agriculture and Fisheries.)

Catch, Effort and Landing Return

MINISTRY OF AGRICULTURE AND FISHERIES
TE MANATU AHUWHENUA AHUMOANA
MAF

No: A 0313254

Trip Data

First day of trip	Last day of trip if different from first day of trip	Landing date	Vessel registration number	Vessel name	Vessel registration number of other vessel (if pair fishing)	Point of landing	Page
/ /	/ /	/ /					of

Catch/Effort Data

Day and Month	Method Code	Position Lat/Long or Stat Area		Time hours mins	Effort data A	B	C	D	For each change of day, method or stat area, enter estimated greenweight catch by species in order of quantity Target Species Total (kg)	Species code Weight (kg)	Species code Weight (kg)	Species code Weight (kg)	Species code Weight (kg)	Species code Weight (kg)
/														
/														
/														
/														
/														

Catch Landing Data

Fishstock (Species / Area)	Landed state	Containers Number	Type	Content Weight	Quota registration no. fish caught against	Destination Type	LFR no. or vessel reg no.	Greenweight (kilograms) when advised by LFR	Purchase Tax Invoice number from LFR

Start a new sheet for each landing. It is an offence to fail to complete this return or supply false information or make any material omission.	Permit holders name	Permit holder FIN number	Signature of master or permit holder	Date signed
				/ /

Figure 19.   Prescribed logbook form for use on Japanese longline fishing vessels operating in the 200-nautical-mile New Zealand exclusive economic zone. (Source: New Zealand Ministry of Agriculture and Fisheries.)

Assessments based on VPAs have generally concentrated on trends in the parental biomass and in recruitment at age 1 estimated from VPAs. In particular, attention has focussed on the size of the current parental biomass in relation to its initial level, and on projected stock trajectories based on a fitted stock-recruitment relationship and a range of possible future catches. Annual variations in the Australian fishery for juvenile fish probably prevent reliable use of backwards VPA from ages less than 7 years old to estimate recruitment. Consequently, while parental biomass can be estimated from the VPAs from 1960 to the latest year for which catch data are available, estimates of recruitment at age 1 are not available for the last six years.

Results of VPAs are subject to the uncertainty inherent in input parameters such as natural mortality rate and terminal fishing mortality. Use of a range of feasible values of natural mortality rate generate quite different interpretations on state of the stock. The catch-age structures and terminal ages adopted have been questionable because of uncertainty about growth parameters used for determining age-at-length. Again, the consequences for assessments of state of the stock are significant. In the 1990 assessments, values of terminal fishing mortalities adopted were 'tuned' using fishing mortality rates estimated independently from tag-recapture experiments, by taking account of change over time in the concentration of effort in areas of operation of the longline fishery and by making assumptions about levels of recruitment in the early period of the fishery. In 1992, after more detailed analyses of CPUE and catch length composition data, a broader range of indices were used for tuning, in conjunction with the use of an 'ADAPT' VPA procedure. Even so, it has still not been possible to develop a satisfactory series of effective effort data for the fishery; therefore the tuning remains questionable and the stock assessments uncertain. Despite these shortcomings, VPAs have consistently indicated continuing decline of parental biomass to historically low levels, with the decline persisting to at least 1991 (Figure 20). The nature of trends in recruitment has been less clear, but it appears that recruitment declines had already occurred during the 1970s (Figure 21). The uncertainty in the parent stock-recruit production relationship, and the variability introduced by uncertainties in VPA input parameters, have prevented reliable projection of stock response under various catch regimes. In turn they have led to considerable uncertainty as to which advice to give to managers about appropriate quota levels. In summary, some projections have indicated that the stock will recover with the current quota levels, whereas others have indicated a continuing decline (Figure 22). Even the projections showing recovery indicate that it would be in the order of 20 years before the parent stock returned to levels at which recruitment decline commenced.

Estimated population parameters from the Australian tagging programme do not refer to the global population, but rather to the sub-population that is vulnerable to the Australian fisheries. Tagging of SBT in Australian waters began in 1959. In the 1960s and 1970s, more than 50,000 fish were tagged in the three main fishing grounds (off NSW, SA and WA; Table 3). The purpose was to delineate stock boundaries, show migration paths, confirm growth rates and assist stock assessment. New releases in 1983-84 were directed more at quantitative aspects, such as mortality rates, interactions among fisheries, estimates of the local population and survival to maturity. Special attention was also given to factors such as effects of shrinkage on growth estimates, short-term and long-term effects of tagging on growth, and consequences of tag-shedding.

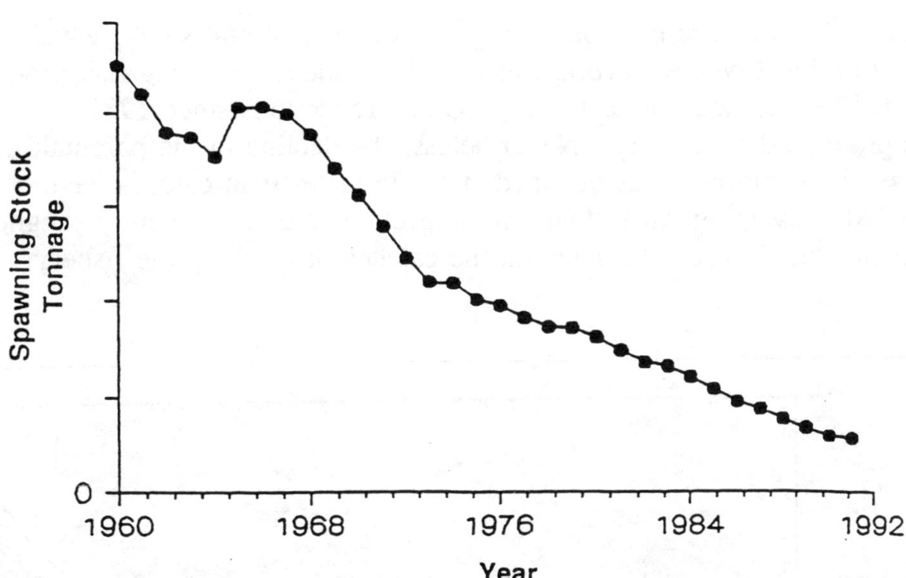

Figure 20.   Indicative virtual population (VPA) estimates of southern bluefin tuna spawning stock tonnage, 1960-1991. Absolute tonnages are not shown because they vary with VPA input parameters and tuning procedures. (Source: Background documentation; Australian, Japan, and New Zealand Scientific Discussions on Southern Bluefin Tuna, Shimizu, Japan, 1992.)

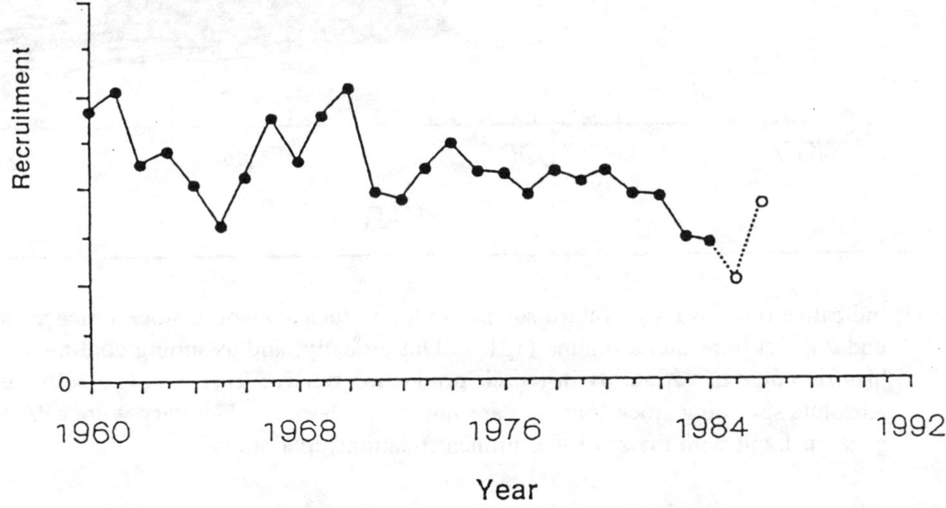

Figure 21.   Indicative virtual population (VPA) estimates of southern bluefin tuna recruitment at age one, 1060-1986. Absolute recruit numbers are not shown because they vary with VPA input parameters and tunin procedures. (Source: Background documentation; Australian, Japan, and New Zealand Scientific Discussions on Southern Bluefin Tuna, Shimizu, Japan, 1992.)

Comparative recovery rates suggest that 2 year old SBT off WA migrate at a higher rate into ocean waters than those off SA, supporting the hypothesis that a significant proportion of fish do not travel from the WA fishing grounds to the more eastern, surface-fishery grounds. Less than 15% of SBT tagged between 1959 and 1980 have been recaptured and reported, the component from the Japanese longline fishery being 0.7% In comparison, over 40% of the fish tagged in 1983-84 were reported as recaptured by Australian fishermen, indicating very high fishing activity during the early 1980s. A substantial fraction of the fish then passing through the Australian fishing grounds as juveniles would not have survived to reproductive maturity. The year-classes of fish tagged in 1983-84 are now becoming part of the parental stock, which has already been reduced to a level where there is concern about adequate young fish production, so the situation is likely to deteriorate further. Quota reductions since 1984 should have reduced this problem dramatically. Nevertheless, the decline in the parental biomass may not be reversed if recruitment has declined at a rate faster than catches have been reduced, or if SBT escaping Australian fishing grounds are subsequently caught at a higher rate than usual in order to maintain the catches of the longline fishery.

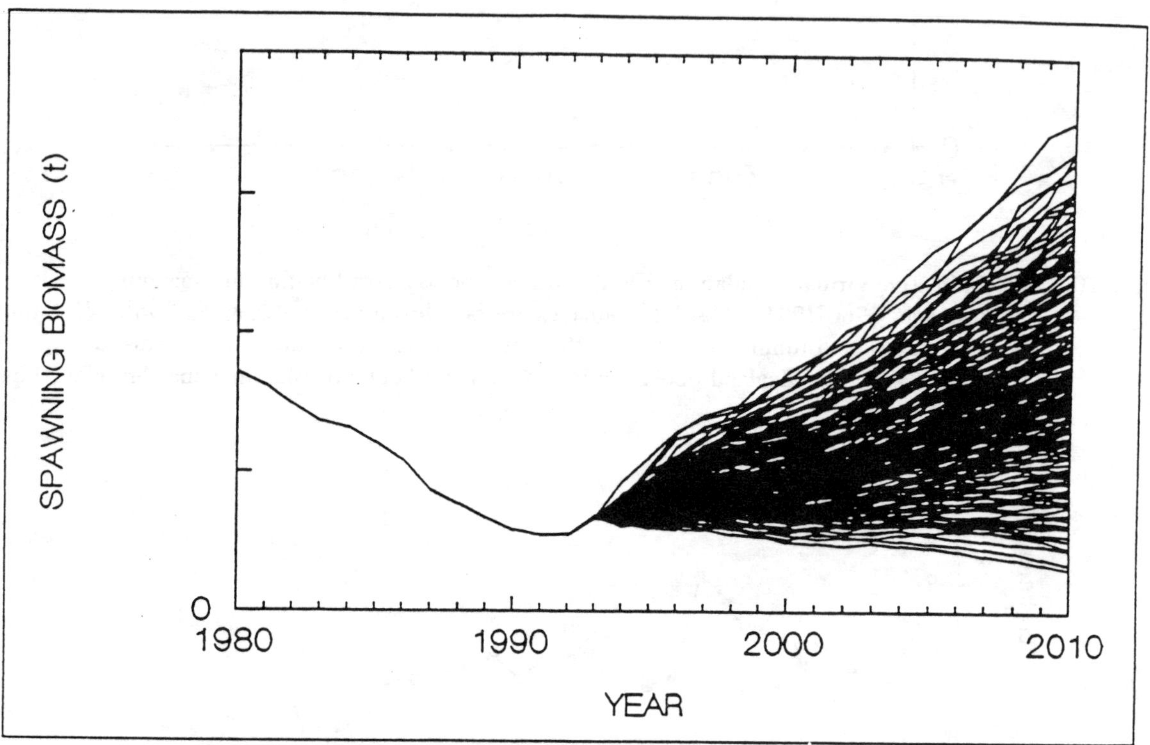

Figure 22.   Indicative trajectories of future southern bluefin tuna spawning stock tonnage to year 2010 under the current quota regime (121,500 mt globally) and assuming continuation of the historic range of variability in recruit production relative to spawning stock tonnage. Absolute spawning stock tonnages are not shown because they vary with VPA estimates generated and with the stock-recruitment relationship assumed.

The number of young fish recruited to Australian fisheries was estimated from the 1983-84 tagging data by the Peterson method. This estimated number could be an underestimate, however, because some recruits may not pass through the WA fishing grounds. The need for a real-time index of recruit abundance has been highlighted. The Japan Marine Fishery Resource Research Center (JAMARC) will complete a 5-year troll and pole-and-line survey in 1992 along the west and south coasts of Australia, which was based on the assumption that all juvenile SBT travel along the WA coast, mainly in November-December. The objectives were to assess if troll and pole-and-line surveys are effective as a means of developing a synoptic recruit abundance index; to determine patterns of occurrence-distribution of recruits; to determine what association exists between physical environmental factors and recruits, and to tag recruits to determine the level of recruitment. It is still unclear as to whether a sufficiently precise index of recruit abundance will be obtained. For the last 3 years of the JAMARC survey, a tagging programme was incorporated, in collaboration with Australia, and extending to the SA fishery region. There is optimism that recruit abundance indices may be estimated with a shorter time lag than estimates derived from VPAs. An aerial survey in progress in SA may provide an additional index if factors influencing fish occurrence at the surface can be clarified. Even so, comprehensive research surveys to determine if there are recruits which migrate directly to the central part of the Indian Ocean from the spawning ground off Java or from the west coast of Australia, are needed.

Table 3.   Numbers of juvenile southern bluefin tuna that were tagged in Australian waters and those subsequently recaptured, according to fishery, year and location of tagging and region of recapture. The Australian fishing grounds are (i) Western Australia (WA), (ii) South Australia (SA) and (iii) Eastern Australia (EA). The longitudes of operation of the Japanese longline fisheries are (iv) West $\leq$ 60○E, (v) 60○E < Mid $\leq$ 120○E and (vi) 120○E $\leq$ East.

Year tagged	Location of tagging	Number tagged	Australian fishery			Japanese fishery			Total
			WA	SA	EA	West	Mid	East	
1959	EA	132	-	2	9	-	-	1	12
1960	EA	43	-	-	-	-	-	-	-
1961	WA	169	-	6	4	-	-	-	10
	SA	18	-	1	-	-	-	-	1
	EA	270	-	1	30	-	-	4	35
1962	WA	4999	104	65	32	1	1	7	210
	EA	323	-	1	17	-	-	-	18
1963	WA	5422	85	108	33	11	8	22	267
	SA	1432	-	19	4	-	1	2	26
	EA	912	-	3	60	-	-	6	69
1964	WA	4606	20	62	55	9	7	9	162
	SA	2859	-	95	13	13	19	45	185
	EA	1056	-	15	239	-	-	10	264
1965	WA	1999	2	28	42	9	1	9	91
	SA	507	-	11	2	-	3	1	17
	EA	381	-	6	30	-	-	6	42
1966	WA	634	1	2	-	1	-	-	4
	SA	1079	-	37	3	7	8	34	89
	EA	2274	-	16	636	-	2	15	669
1967	WA	5018	79	80	50	16	5	13	243
	SA	20	-	-	-	-	-	-	-
	EA	2443	-	40	860	1	2	17	920
1968	SA	502	-	79	28	-	-	2	109
	EA	3988	-	19	2193	5	-	6	2223
1969	SA	2387	8	694	157	9	2	8	878
	EA	1444	-	18	397	-	2	15	432

Continued on the next page.

Table 3 Continued.

Year tagged	Location of tagging	Number tagged	Australian fishery			Japanese fishery			Total
			WA	SA	EA	West	Mid	East	
1970	WA	1328	74	5	3	-	-	-	82
	EA	145	-	4	35	-	-	-	39
1971	WA	14	-	-	-	-	-	-	-
1972	WA	296	1	2	-	-	1	-	4
1973	WA	951	41	13	1	-	-	1	56
1974	WA	796	43	11	6	2	-	1	63
1975	WA	838	24	55	11	2	1	-	93
1976	WA	308	1	-	-	-	-	-	1
1977	WA	219	11	7	1	-	-	-	19
	SA	908	3	199	22	-	-	2	226
1978	WA	1019	39	22	2	-	-	-	63
1980	WA	555	100	43	5	2	-	-	150
1983	WA	6956	1279	1390	4	6	7	5	2691
1984	SA	3223	23	1369	-	13	7	5	1417
1986	SA	6	-	1	-	-	-	-	1
Total	WA	36127	1904	1899	249	59	31	67	4209
	SA	12941	34	2505	229	42	40	99	2949
	EA	13411	-	125	4506	6	6	80	4723
TOTAL		62479	1938	4529	4984	107	77	246	11881

3. REFERENCES CITED

Alvarino, A. 1980. The relation between the distribution of zooplankton predators and anchovy larvae. *Rep.CCOFI*, (21):150-60.

Australian Fisheries Service. 1986. Report of the meeting to discuss feasibility of southern bluefin tuna recruitment surveys off Western Australia. Canberra: Australian Department of Primary Industry, 9 p.

Barndorff-Nielsen, O.E., I.R. James, and G.M. Leigh. 1989a. Asymptotic inference for a semi-parametric estimator of natural mortality. *Res.Rep.Dep.Theoretical Statistics Inst.Math.Univ.Aarhus*, (202):25 p.

Barndorff-Nielsen, O.E., I.R. James, and G.M. Leigh. 1989b. A note on a semi-parametric estimator of mortality. *Biometrika*, (76):803-5.

Behrens Yamada, S., T.J. Mulligan, and D. Fournier. 1987. Role of environment and stock on the elemental composition of sockeye salmon (*Oncorhynchus nerka*) vertebrae. *Can.J.Fish.Aquat.Sci.*, (44):1206-12.

Bell, R.R. 1964. A history of tuna age determinations. *In* Proceedings of the Symposium on Scombroid Fishes, Mandapam Camp, Jan. 12-15, 1962. Part 2. *Mar.Biol.Assoc.India*, pp.693-706.

Beverton, R.J.H., and S.J. Holt. 1957. On the dynamics of exploited fish populations. Fishery Investigations, London, Series 2 19:533 p.

Bourget, E., and D.J. Crisp. 1975. An analysis of growth bands and ridges of barnacle shell plates. *J.Mar.Biol.Assoc.U.K.*, (55):439-61.

Brothers, E.B., C.P. Mathews, and R. Lasker. 1976. Daily growth increments in otoliths from larval and adult fishes. *Fish.Bull.NOAA-NMFS*, 74:1-8.

Brothers, E.B., E.D. Prince, and D.W. Lee. 1983. Age and growth of young-of-the-year bluefin tuna, *Thunnus thynnus*, from otolith microstructure. *NOAA Tech.Rep.NMFS*, 8:49-59.

Bureau of Rural Resources. 1989. Reports of Australia/Japan/New Zealand trilateral scientific discussions on southern bluefin tuna 1982-1988. Working Paper No. 10/89. Canberra: Bureau of Rural Resources, Department of Primary Industries and Energy, 46 p.

Calaprice, J.R. 1980-85. *In* Annual reports of the Inter-American Tropical Tuna Commission. *Annu.Rep.I-ATTC*, (various pagination).

Calaprice, J.R. 1985. X-ray fluorescence study of stock variation in bluefin tuna. *In* Final status report submitted to NMFS, Miami, March 1983, 93 p.

Calaprice, J.R., H.M. McSheffrey, and L.A. Lapi. 1971. Radioisotope X-ray fluorescence spectrometry in aquatic biology: a review. *J.Fish.Res.Board Can.*, (28):1583-94.

Castelnau, F.L. 1872. Contribution to the ichthyology of Australia. *In* Proceedings of the Zoological Acclimatisation Society of Victoria, (1):29-247.

Caton, A.E. 1985. Notes on development of the southern bluefin tuna fishery and its management. *In* Collective Volume of Working Documents presented at the IPTP Expert Consultation on Stock Assessment of Tunas in the Indian Ocean, Colombo, Sri Lanka, 28 November-2 December 1985, pp. 281-285.

Caton, A.E. 1988. The southern bluefin tuna fishery. *In* Proceedings of the Workshop on Scientific Advice for Fisheries Management, Canberra, Australia, 10 July 1987, edited by M.J. Williams. Canberra: Bureau of Rural Resources, Australian Society for Fish Biology, pp.11-22.

Caton, A. 1991. The 1988-89 Australian southern bluefin tuna season. Paper presented at the Eighth Trilateral Scientific Meeting on SBT, Shimizu, Japan, 1989. Canberra: Bureau of Rural Resources, Department of Primary Industries and Energy, W.P.1/91. SBFWS/89/1:27 p.

Caton, A., and J. Majkowski. 1987. Warning issued on global catch limits. *Aust.Fish.*, 46(12):22-4.

Caton, A., K. McLoughlin, and M.J. Williams. 1990. Southern bluefin tuna: scientific background to the debate. *Bull.Bur.Rural Res.Dept.Primary Indust.Energy,* (3):41 p.

Caton, A.E., and K. Williams, 1991. The 1989-90 Australian southern bluefin tuna season. Working Paper No. 2/91. Paper presented at the Ninth Trilateral Scientific Meeting on SBT, Hobart, Australia, 1990. Canberra: Bureau of Rural Resources, Department of Primary Industries and Energy, W.P.2/91. SBFWS/90/5:18 p.

Christensen, M.J. 1964. Burning of otoliths, a technique for age determination of soles and other fish. *J.Cons.CIEM*, 29:73-81.

Collette, B.B. 1978. Adaptations and systematics of the mackerels and tunas. *In* The physiological ecology of tunas, edited by G.D. Sharp and A.E. Dizon. New York, Academic Press. pp.7-39.

Collette, B.B. 1986a. Tunas, mackerels and bonitos. Pp. 831-839, *In* in Smith's sea fishes, edited by M.M. Smith and P.C. Heemstra. Johannesburg, Macmillan South Africa, pp.831-9.

Collette, B.B. 1986b. Scombridae. *In* Fishes of the north-eastern Atlantic and the Mediterranean, edited by P.J.P. Whitehead, M.-L. Bauchot, J.-C. Hureau, J. Nielsen, and E. Tortonese. Paris, UNESCO Press, Vol2:981-97.

Collette, B.B., and L.N. Chao. 1975. Systematics and morphology of the bonitos and their relatives (Scombridae, Sardini). *Fish.Bull.NOAA-NMFS*, 73:516-625.

Collette, B.B., and R.H. Gibbs, Jr. 1963. A preliminary review of the fishes of the family Scombridae. *FAO Fish.Rep.*, 6(3):977-8.

Collette, B. B., and C.E. Nauen. 1983. FAO species catalogue, Vol. 2. Scombrids of the world; an annotated and illustrated catalogue of tunas, mackerels, bonitos and related species known to date. *FAO Fish.Synop.*, 125:137 p.

Collette, B.B., T. Potthoff, W.J. Richards, S. Ueyanagi, J.L. Russo, and Y. Nishikawa. 1984. Scombroidei: development and relationships. *In* Ontogeny and systematics of fishes, edited by Moser, *et al.*, *Spec.Publ.Am.Soc.Ichthyol.Herpetol.*, (1):591-619.

Collette, B.B., and B.R. Smith. 1981. Bluefin tuna, *Thunnus thynnus* orientalis, from the Gulf of Papua. *Jap.J.Ichthyol.*, 28(2):162-8.

Conand, F., and W.J. Richards. 1982. Distribution of tuna larvae between Madagascar and the Equator, Indian Ocean. *Biol.Oceanogr.*, 1:321-36.

Creswell, G.R., and T.J. Golding. 1979. Satellite-tracked buoy data report III. Indian Ocean 1977, Tasman Sea July-December 1977. *Rep.CSIRO Div.Fish.Oceanogr.*, 101.

Creswell, G.R., and T.J. Golding. 1980. Observations of a south-flowing current in the southeastern Indian Ocean. *Deep-Sea Res.*, 27A:449-66.

CSIRO. 1988. Assessment of the southern bluefin tuna stock. Paper presented at the Seventh Trilateral Scientific Meeting on southern bluefin tuna, August 1988, Wellington, New Zealand. SBFWS/88/11.

Cuvier, G. (1816/1817). Le regne animal distribue d'apres son organisation, pour servir de base a l'histoire naturelle des animaux et d'introduction a l'anatomie compare. Paris: Fortin, Masson et Co. 4 volumes. Poissons 2: 104-351. [Roux (1976, Bulletin de Liaison, Museum National d'Histoire, Paris 25: 16) has established that the first edition of Cuvier's 'Regne Animal', which is usually cited as 1817, was actually published in 1816.]

Davis, T.L.O., and L.A. Clementson, 1989. Data report on the vertical and horizontal distribution of tuna larvae in the East Indian Ocean, January-February 1987. *Rep.CSIRO Mar.Lab.*, 206:39 p.

Davis, T.L.O., and L.A. Clementson. (In preparation). Data report on coarse-scale horizontal distribution of tuna larvae in the East Indian Ocean, November-December 1987. *Rep.CSIRO Mar.Lab.*

Davis, T.L.O., G.P. Jenkins, M. Yukinawa, and Y. Nishikawa. 1989. Tuna larvae abundance: comparative estimates from concurrent Japanese and Australian sampling programs. *Fish.Bull.NOAA-NMFS*, 87:976-81.

Davis, T.L.O., G.P. Jenkins, and J.W. Young. 1990. Diel patterns of vertical distribution in larvae of southern bluefin (*Thunnus maccoyii*), and other tuna in the East Indian Ocean. *Prog.Ser.Mar.Ecol.*, 59:63-74.

Davis, T.L.O., G.P. Jenkins, and J.W. Young. (In press). Patterns of horizontal distribution of the larvae of southern bluefin tuna (*Thunnus maccoyii*) and other tuna in the Indian Ocean. *J.Plankton Res.*

Davis, T.L.O., V. Lyne, and G.P. Jenkins. (Submitted). Advection, dispersion and mortality of a patch of southern bluefin tuna larvae, *Thunnus maccoyii*, in the east Indian Ocean. *Prog.Ser.Mar.Ecol.*

de Jager, B. van D., C.S. de V. Negpen, and R.J. van Wyk. 1963. A preliminary report on South African west coast tuna. *Invest.Rep.South Africa Div.Sea Fish.*, 47:1-40.

Dodge, R.E., and J.R. Vaisnys. 1980. Skeletal growth chronologies of recent and fossil corals. *In* Skeletal growth of aquatic organisms, edited by D.C. Rhoads and R.A. Lutz. New York, Plenum Press, pp.493-517.

Dragovich, A. 1969. Review of studies of tuna food in the Atlantic Ocean. *Spec.Sci. Rep.U.S.Fish Wild.Serv. (Fish.)*, 593:21 p.

Eckert, G.J., and J. Majkowski. 1987. Fishermen recapture 40 per cent of tagged southern bluefin tuna. *Aust.Fish.* 46(12):30-3.

Eckert, G.J., J. Kalish, J. Majkowski, and R. Pethebridge. 1987. An indexed bibliography of the southern bluefin tuna [*Thunnus maccoyii* (Castelnau, 1872)]. *Rep.CSIRO Mar.Lab.*, 185:49 p.

Edmonds, J.S., M.J. Moran, and N. Caputi. 1989. Trace element analysis of fish sagittae as an aid in stock identification: pink snapper (*Chrysophrys auratus*) in Western Australian waters. *Can.J.Fish.Aquat.Sci.*, 46:50-4.

Fabens, A.J. 1965. Properties and fitting of the von Bertalanffy growth curve. *Growth* 29:265-89.

Far Seas Fisheries Research Laboratory. 1972. Biological proposition for better management of southern bluefin tuna longline fishery. *S Ser.Far Seas Fish.Res.Lab.*, (6):88 p. (In Japanese with English abstract).

Far Seas Fisheries Research Laboratory. 1986. Report of the fifth meeting of Australian, Japanese and New Zealand scientists on southern bluefin tuna (SBT), June 10-14, 1986. Shimizu, Japan, 11 p.

Fischer, W., and P.J.P. Whitehead (editors). 1974. FAO species identification sheets for fishery purposes. Eastern Indian Ocean (fishing area 57) and Western Central Pacific (fishing area 71). Vol 4, pag. var. Rome: FAO.

Fishery Agency of Japan. 1967a. Annual report of effort and catch statistics by area on Japanese tuna longline fishery, 1964. *Annu.Rep.Effort Catch Stat.Area Jap.Tuna Longline Fish.*, (1964):379 p. (In Japanese).

Fishery Agency of Japan. 1967b. Annual report of effort and catch statistics by area on Japanese tuna longline fishery, 1965. *Annu.Rep.Effort Catch Stat.Area Jap.Tuna Longline Fish.*, (1965):375 p. (In Japanese).

Fishery Agency of Japan. 1968. Annual report of effort and catch statistics by area on Japanese tuna longline fishery, 1966. *Annu.Rep.Effort Catch Stat.Area Jap.Tuna Longline Fish.*, (1966):299 p. (In Japanese).

Fishery Agency of Japan. 1969. Annual report of effort and catch statistics by area on Japanese tuna longline fishery, 1967. *Annu.Rep.Effort Catch Stat.Area Jap.Tuna Longline Fish.*, (1967):293 p. (In Japanese).

Fishery Agency of Japan. 1971. Annual report of effort and catch statistics by area on Japanese tuna longline fishery, 1969. *Annu.Rep.Effort Catch Stat.Area Jap.Tuna Longline Fish.*, (1969):299 p. (In Japanese).

Fishery Agency of Japan. 1972. Annual report of effort and catch statistics by area on Japanese tuna longline fishery, 1970. *Annu.Rep.Effort Catch Stat.Area Jap.Tuna Longline Fish.*, (1970):326 p. (In Japanese).

Fishery Agency of Japan. 1973. Annual report of effort and catch statistics by area on Japanese tuna longline fishery, 1971. *Annu.Rep.Effort Catch Stat.Area Jap.Tuna Longline Fish.*, (1971):319 p. (In Japanese).

Fishery Agency of Japan. 1979. Annual report of effort and catch statistics by area on Japanese tuna longline fishery, 1977. *Annu.Rep.Effort Catch Stat.Area Jap.Tuna Longline Fish.*, (1977):235 p. (In Japanese).

Fishery Agency of Japan. 1980. Annual report of effort and catch statistics by area on Japanese tuna longline fishery, 1978. *Annu.Rep.Effort Catch Stat.Area Jap.Tuna Longline Fish.*, (1978):241 p. (In Japanese).

Fishery Agency of Japan. 1981. Annual report of effort and catch statistics by area on Japanese tuna longline fishery, 1979. *Annu.Rep.Effort Catch Stat.Area Jap.Tuna Longline Fish.*, (1979):243 p. (In Japanese).

Fishery Agency of Japan. 1982. Annual report of effort and catch statistics by area on Japanese tuna longline fishery, 1980. *Annu.Rep.Effort Catch Stat.Area Jap.Tuna Longline Fish.*, (1980):242 p. (In Japanese).

Fishery Agency of Japan. 1986. Summary record of discussions. 1986. Southern bluefin tuna (SBT) trilateral management consultations among Australia, Japan and New Zealand, Tokyo, 17-19 June 1986, 28 p. (Plus Attachments I-V.)

Fishery Agency of Japan, Research Division. 1960. Report on 1959 research cruise of the R/V Shoyo-Maru. Shimizu, Japan, 338 p. (In Japanese)

Fishery Agency of Japan, Research Division. 1962. Report on 1961 research cruise of the R/V Shoyo-Maru. Shimizu, Japan, 188 p. (In Japanese)

Fishery Agency of Japan, Research Division. 1963. Report on 1962 research cruise of the R/V Shoyo-Maru. Shimizu, Japan, 475 p. (In Japanese)

Fishery Agency of Japan, Research Division. 1964. Report on 1963 research cruise of the R/V Shoyo-Maru. Shimizu, Japan, 465 p. (In Japanese)

Fishery Agency of Japan, Research Division. 1968. Report on 1967 research cruise of the R/V Shoyo-Maru. Shimizu, Japan, 199 p. (In Japanese)

Fishery Agency of Japan, Research Division. 1969. Report on 1968 research cruise of the R/V Shoyo-Maru. Shimizu, Japan, 183 p. (In Japanese)

Fishery Agency of Japan, Research Division. 1989. Report on the 1988/89 southern bluefin tuna recruitment monitoring survey. Tokyo: JAMARC.

Fishery Agency of Japan, Research Division. 1989. Report of 1988 research cruise of the R/V Shoyo-Maru. Distribution of juvenile southern bluefin tuna off west coast of Australia, December 1988-January 1989. Shimizu, Japan. Report No. 63, 125 p.

Fishery Agency of Japan, Research Division. 1990. Report of the workshop on SBT recruitment monitoring survey June 6-9 1989, CSIRO Marine Laboratories, Hobart. *In* Report on the 1989/90 southern bluefin tuna recruitment monitoring survey. Tokyo: JAMARC.

Fishery Agency of Japan, Research Division. (In press). Report of 1987 research cruise of the R/V Shoyo-Maru. Shimizu, Japan. (In Japanese)

Fishery Agency of Japan, Research Division. (In press). Report of 1988 research cruise of the R/V Shoyo-Maru. Shimizu, Japan. (In Japanese)

Fournier, D.A., and P.A. Breen. 1983. Estimation of abalone mortality rates with growth analysis. *Trans.Am.Fish.Soc.*, 112:403-11.
Fournier, D.A., J.R. Sibert, J. Majkowski, and J. Hampton. 1990. MULTIFAN a likelihood-based method for estimating growth parameters and age composition from multiple length frequency data sets illustrated using data for southern bluefin tuna (*Thunnus maccoyii*). *Can.J.Fish.Aquat.Sci.*, 47:301-17.

Franklin, P. 1987. Australian southern bluefin tuna fishery. *In* Proceedings and Working Papers of the Symposium on the Exploitation and Management of Marine Fishery Resources in Southeast Asia, Darwin, Australia, 16-19 February 1987. Bangkok: Indo-Pacific Fisheries Commission, FAO Regional Office.

Gauldie, R.W., D.A. Fournier, and D.E. Dunlop. 1986. Atomic emission and proton probe studies of the ion content of otoliths of chinook salmon aimed at recovering the temperature life history of individuals. *Com.Biochem.Physiol.*, 84A:607-15.

Geen, G., and M. Nayar. 1989. Individual transferable quotas and the southern bluefin tuna fishery. Australian Bureau of Agricultural and Resource Economics Occasional Paper No. 105. Canberra: Australian Bureau of Agricultural and Resource Economics, 40 p.

Gibbs, R.H., Jr., and B.B. Collette. 1967. Comparative anatomy and systematics of the tunas, genus *Thunnus*. *Fish.Bull.U.S.Fish Wild.Serv.*, 66(1):65-130.

Gibson, D.M. 1982. A handbook on the preparation of southern bluefin tuna for the Japanese sashimi market. Wellington: New Zealand. Ministry of Agriculture and Fisheries.

Godsil, H.C., and E.K. Holmberg. 1950. A comparison of the blue-fin tunas, genus *Thunnus*, from New England, Australia and California. *Fish Bull.Calif.Dep.Fish Game*, 77:55 p.

Graves, J.E., M.A. Simovich, and K.M. Schaefer. 1988. Electrophoretic identification of early juvenile yellowfin tuna, *Thunnus albacares*. *Fish.Bull.NOAA-NMFS*, 86:835-8.

Gulland, J.A., 1965. Estimation of mortality rates. Annex to Arctic Fisheries Working Group Report. *C.M.Gadoid Fish Comm.Cons.CIEM* (International Council for the Exploration of the Sea), 3:9 p.

Hampton, J. 1984. Southern bluefin tuna population assessment, stabilizing catch analysis, and simulations. Paper presented at the Third Trilateral Scientific Meeting on Southern Bluefin Tuna, Canberra, Australia. Australian Fisheries Service, SBFWS/87/8.

Hampton, J. 1986. Effects of tagging on the condition of southern bluefin tuna, *Thunnus maccoyii* (Castlenau). *Aust.J.Mar.Freshwat.Res.*, 37(6):699-705.

Hampton, J. 1987. Growth of southern bluefin tuna. Paper presented at the Sixth Trilateral Scientific Meeting on southern bluefin tuna, Hobart, Australia. CSIRO, Hobart, SBFWS/87/12.

Hampton, J. 1989. Population dynamics, stock assessment and fishery management of the southern bluefin tuna (*Thunnus maccoyii*). Ph.D. Thesis. University of New South Wales, Kensington, NSW, Australia. 273 p.

Hampton, J., D.A. Fournier, and J.R. Sibert. 1990. MULTIFAN analysis of South Pacific albacore length-frequency data collected by observers, 1988-1990. Paper presented at the Third South Pacific Albacore Research Workshop, South Pacific Commission, New Caledonia. W.P.1:11 p.

Hampton, J., and G.P. Kirkwood. 1990. Tag shedding by southern bluefin tuna (*Thunnus maccoyii*). *Fish.Bull.NOAA-NMFS*, (in press)

Hampton, J., J. Majkowski, and G.I. Murphy. 1983. Population analyses relevant to future management of the southern bluefin tuna fishery. Paper presented at the Second Trilateral Scientific Meeting on Southern Bluefin Tuna, Shimizu, Japan. *Jap.Fish.Agency*, SBFWS/83/1.

Hampton, J., J. Majkowski, and G.I. Murphy. 1984. The 1982 assessment of the southern bluefin tuna (*Thunnus maccoyii*) population and the determination of catch levels which stabilize the parental biomass. *Rep.CSIRO Mar.Lab.*, 165:27 p.

Harden Jones, F.R. 1984. A view from the ocean. *In* Mechanisms of migration in fishes, edited by J.D. McCleave, G.P. Arnold, J.J. Dodson, and W.H. Neill. New York, Plenum Press, pp.1-26.

Hayashi, S. 1958. A review of age determination of the Pacific tunas. Paper presented at the 7th Session of the Indo-Pacific Fisheries Council. *Proc.IPFC*, Section 2-3:53-64.

Hayashi, S., M. Honma, and C. Shingu. 1969. Assessment of the southern bluefin tuna stocks, 1960-66. *Tuna Fish.*, 84:9-23 (In Japanese; English translation by M. A. Hintze available from Australian CSIRO Marine Laboratories.)

Hearn, W.S. 1979. Growth of southern bluefin tuna. Paper presented at the Indo-Pacific Tuna and Billfish Stock Assessment Workshop, June 1979, Shimizu, Japan. *Doc.Far Seas Fish.Res.Lab.*, SAWS/BP/14.

Hearn, W.S. 1982. Fish tagging: data processing, editing and storage. *In* CSIRO data base for southern bluefin tuna (*Thunnus maccoyii* (Castlenau)), edited by J. Majkowski. *Rep.CSIRO Mar.Lab.*, 142:8-9,13-9.

Hearn, W.S. 1986. Mathematical methods for evaluating marine fisheries. Ph.D. Thesis, The University of New South Wales, Kensington, NSW, Australia, 195 p.

Hearn, W.S. 1988. Estimation of the number of SBT recruits from tag release/recovery and commercial catch data. Paper presented at the Seventh Trilateral Scientific Meeting on Southern Bluefin Tuna, August 1988, Wellington, New Zealand, SBFWS/88/12.

Hearn, W.S., G.J. Eckert, and J. Majkowski. (in preparation). Exploitation of southern bluefin tuna off Australia: estimates from the 1983-84 tagging.

Hearn, W.S., G.L. Leigh, and R.J.H. Beverton. (in preparation). An examination of tag shedding assumptions: with application to southern bluefin tuna. *J.Cons.CIEM*.

Hearn, W.S., R.L. Sandland, and J. Hampton. 1987. Robust estimation of the natural mortality rate in a completed tagging experiment with variable fishing intensity. *J.Cons.CIEM*, 43:107-17.

Hisada, K., and H. Kono. 1984. The stock assessment of southern bluefin tuna updated by inclusion of 1981 data. Paper presented at the Third Trilateral Scientific Meeting on Southern Bluefin Tuna, Canberra, Australia Australian Fisheries Service, SBFWS/87/5.

Hunter, J.R., and C.A. Kimbrell. 1980. Egg cannibalism in the northern anchovy, *Engraulis mordax*. *Fish.Bull.NOAA-NMFS*, 78:811-6.

Hunter, J.R., A.W. Argue, W.H. Bayliff, A.E. Dizon, A. Fonteneau, D. Goodman, and G.R. Seckel. 1986. The dynamics of tuna movements: an evaluation of past and future research. *FAO Fish.Tech.Pap.*, (277):78 p.

Hynd, J.S. 1965. Southern bluefin tuna populations in south-west Australia. *Aust.J.Mar.Freshwat.Res.*, 16(1):25-32.

Hynd, J.S. 1968. Sea surface temperature maps as an aid to tuna fishing. *Newsl.Aust.Fish.*, 27:23-9.

Hynd, J.S. 1969. New evidence on southern bluefin tuna stocks and migrations. *Aust.Fish.*, 28(5):26-30.

Hynd, J.S., G.L. Kesteven, and J.P. Robins. 1966. Tuna in southern Australian waters. *Food Tech.Aust.*, 18(4):190-200.

Hynd, J.S., and C. Lucas. 1974. Population dynamics of the southern bluefin tuna. *Proc.Indo-Pac.Fish.Coun.*, 15(3):424-35.

Hynd, J.S., and J.P. Robins. 1967. Tasmanian tuna survey report of first operational period. *Tech.Pap.Div.Fish.Oceanogr. CSIRO Aust.*, 22:1-55.

Inter-American Tropical Tuna Commission. 1988. Annual Report of the Inter-American Tropical Tuna Commission, *Annu.Rep.I-ATTC.*

Ishizuka, Y. 1987. Migration and growth of southern bluefin tuna based on Australian tagging data. Paper presented at the Sixth Trilateral Scientific Meeting on Southern Bluefin Tuna, Hobart, Australia. *CSIRO*, SBFWS/87/10.

Ishizuka, Y., and S. Tsuji. 1990. Assessment of the southern bluefin tuna. Paper presented at the Ninth Trilateral Scientific Meeting on southern bluefin tuna, Hobart, Australia. *CSIRO*, SBFWS/90/2.

Iwai, T., and I. Nakamura. 1964. Olfactory organs of tunas with special reference to their systematic significance. *Bull.Misaki Mar.Biol.Inst., Kiyoto Univ.*, 7:1-8.

Iwai, T., I. Nakamura, and K. Matsubara. 1965. Taxonomic study of tunas. *Spec.Rep.Misaki Mar.Biol.Inst., Kiyoto Univ.*, 2:1-51.

Japan Marine Fishery Resource Research Center. 1989. Report of southern bluefin tuna recruitment monitoring survey. Tokyo, Japan, 182 p. (In Japanese)

Jenkins, G.P., and T.L.O. Davis. 1990. Age, growth rate, and growth trajectory determined from otolith microstructure of southern bluefin tuna, *Thunnus maccoyii*, larvae. *Prog.Series Mar.Ecol.*, 63:93-104.

Jenkins, G.P., J.W. Young, and T.L.O. Davis. (Submitted) Density-dependence of larval growth of a marine fish, the southern bluefin tuna, *Thunnus maccoyii*. *Can.J.Fish.Aquat.Sci.*

Johnson, G.D. 1986. Scombroid phylogeny: an alternative hypothesis. *Bull.Mar.Sci.*, 39(1):1-41.

Jones, S., and M. Kumaran. 1963. Distribution of larval tuna collected by the Carlsberg Foundation's Dana expedition (1928-30) from the Indian Ocean. *FAO Fish.Rep.*, 6(3):1753-73.

Jones, S., and E.G. Silas. 1960. Indian tunas - a preliminary review, with a key for their identification. *Indian J.Fish.*, 7(2):369-93.

Jordan, D.S., and B.W. Evermann. 1926. A review of the giant mackerel-like fishes, tunnies, spearfishes and swordfishes. *Occas.Pap.Calif.Acad.Sci.*, 12:1-113.

Kamimura, T., A. Suda, and S. Hayashi. 1966. Symposium on tuna fishery. Part 2 resources. (Nissuikai) *J.Jap.Fish.Assoc.*, 32(9):756-86.

Kikawa, S. 1964a. Estimated number of eggs spawned out of the ovaries of an indomaguro (*Thunnus maccoyii?*). *Rep.Nankai Reg.Fish.Res.Lab.*, 20:27-35.

Kikawa, S. 1964b. A study of the indomaguro (*Thunnus maccoyii*) in the area south of 20°S. lat. from the examination of ovary weight and maturity. *Rep.Nankai Reg.Fish.Res.Lab.*, 20:37-57.

Kirkwood, G.P. 1981. Generalized models for the estimation of rates of tag shedding by southern bluefin tuna (*Thunnus maccoyii*). *J.Cons.CIEM*, 39:256-60.

Kirkwood, G.P. 1983. Estimation of the von Bertalanffy growth curve parameters using both length increment and age-length data. *Can.J.Fish.Aquat.Sci.*, 40:1405-11.

Kirkwood, G., and N. Klaer. 1990. Assessment of the southern bluefin tuna stock using virtual population analysis. Paper presented at the Ninth Trilateral Scientific Meeting on Southern Bluefin Tuna, Hobart, Australia, SBFWS/90/6:22 p.

Kirkwood, G., J. Majkowski, W. Hearn, and N. Klaer. 1989. Assessment of the southern bluefin tuna stock using virtual population analysis. Paper presented at the Eighth Trilateral Scientific Meeting on SBT, Shimizu Japan, September 1989. *Jap.Fish.Agency*, SBFWS/89/3.

Kirkwood, G.P., and I.F. Somers. 1984. Growth of two species of tiger prawn, *Penaeus esculentus* and *P. semisulcatus*, in the Western Gulf of Carpentaria. *Aust.J.Mar.Freshwat.Res.*, 35:703-12.

Kirkwood, G. P., and M.H. Walker. 1984. A new method for estimating tag shedding rates, with application to data for Australian salmon, *Arripis trutta esper* Whitley. *Aust.J.Mar.Freshwat.Res.*, 35:601-6.

Klaer, N., and G. Kirkwood. 1990. Future projections of southern bluefin tuna biomass incorporating stochastic variation in the stock-recruitment relationship. Paper presented at the Ninth Trilateral Scientific Meeting on Southern Bluefin Tuna, Hobart, Australia, SBFWS/90/7:32 p.

Kono, H. 1985. Re-estimation of southern bluefin tuna catch by Japanese longline fishery, 1980-1983. Paper presented at the Fourth Trilateral Scientific Meeting on Southern Bluefin Tuna, Wellington, New Zealand, 1985, SBFWS/85/11:31 p.

Kono, H. 1987. Assessment of the southern bluefin tuna stock. Paper presented at the Sixth Trilateral Scientific Meeting on Southern Bluefin Tuna, Hobart, Australia. *CSIRO*, SBFWS/87/9.

Kono, H., and Y. Ishizuka. 1989. Assessment of the southern bluefin tuna stock. Paper presented at the Eighth Trilateral Scientific Meeting on Southern Bluefin Tuna, Shimizu, Japan, SBFWS/89/10:17 p.

Kono, H., and Y. Warashina. 1985. Recent status of Japanese longline fishery for southern bluefin tuna, 1979-1983. Paper presented at the Fourth Trilateral Scientific Meeting on Southern Bluefin Tuna, Wellington, New Zealand, SBFWS/85/10:12 p.

Kono, H., and Y. Warashina. 1989. Japanese southern bluefin tuna fishery in recent years. Paper presented at the Eighth Trilateral Scientific Meeting on Southern Bluefin Tuna, Shimizu, Japan, SBFWS/89/7:7 p.

Kristensen, T.K.., 1980. Periodical growth rings in cephalopod statoliths. *Dana*, 1:39-51.

Lapi, L.A., and T.J. Mulligan. 1981. Salmon stock identification using a microanalytic technique to measure elements present in the freshwater growth region of scales. *Can.J.Fish.Aquat.Sci.*, 38:744-51.

Lasker, R. 1981. Marine fish larvae: morphology, ecology, and relation to fisheries. Seattle, Washington Sea Grant Program, 131 p.

Le Cren, E.D. 1951. The length-weight relationship and seasonal cycle in gonad weight and condition in the perch (*Perca fluviatilis*). *J.Anim.Ecol.*, 20:201-19.

Leigh, G.M. 1988. A comparison of estimates of natural mortality from fish tagging experiments. *Biometrika*, 75:347-53.

Leigh, G.M. 1989. Statistical properties of a method of estimating the natural mortality rate in a lightly exploited fishery. *Rep.CSIRO Div.Math.Stat.*, V89/10:44 p.

Linnaeus, C. 1758. *In* Systema Naturae (10th ed.), Vol.1 824 pp. (Reprint, 1956, London.) Nantes and Pisces:230-338.

Lucas, C. 1974. Working paper on southern bluefin tuna population dynamics. *Collect.Vol.Sci.Pap.ICCAT*, 111:110-24.

Lutz, R.A., and D.C. Rhoads. 1980. Growth patterns within the molluscan shell. *In* Skeletal growth of aquatic organisms, edited by D.C. Rhoads and R.A. Lutz. New York, Plenum Press, pp.203-254.

Macdonald, P.D.M., and T.J. Pitcher. 1979. Age groups from size-frequency data: a versatile and efficient method for analysing distribution mixtures. *J.Fish.Res.Board Can.*, 36:987-1001.

Majkowski, J. (editor). 1982. CSIRO data base for southern bluefin tuna (*Thunnus maccoyii* (Castlenau)). *Rep.CSIRO Mar.Lab.*, (142):23 p.

Majkowski, J. 1982. Estimation of the abundance of the southern bluefin tuna [*Thunnus maccoyii* (Castlenau)] subpopulation exploited by the Australian fishery. *Rep.CSIRO Mar.Lab.*, (143):11 p.

Majkowski, J., and J. Hampton. 1983a. The effect of parameter uncertainties in an age-length relationship upon estimating the age composition of catches. *Can.J.Fish.Aquat.Sci.*, 40:272-80.

Majkowski, J., and J. Hampton. 1983b. The estimation of a catch level which stabilizes the parental biomass of an exploited stock. *Fish.Bull.NOAA-NMFS*, 81:723-32.

Majkowski, J., W.S. Hearn, and R.L. Sandland. 1984. An experimental determination of the effect of increasing the minimum age (or size) of fish at capture upon the yield per recruit. *Can.J.Fish.Aquat.Sci.*, 41:736-43.

Majkowski, J., W.S. Hearn, and R.L. Sandland. 1988. A tag-release/recovery method for predicting the effect of changing the catch of one component of a fishery upon the remaining components. *Can.J.Fish.Aquat.Sci.*, 45:675-84.

Majkowski, J., and G. Morris (editors). 1986. Data on southern bluefin tuna (*Thunnus maccoyii* (Castelnau)): Australian, Japanese and New Zealand systems for collecting, processing and accessing catch, fishing effort, aircraft observation and tag release/recapture data. *Rep.CSIRO Mar.Lab.*, (179):95 p.

Majkowski, J., and G.I. Murphy. 1983. CSIRO southern bluefin tuna tagging: scientific achievements and future objectives. *Aust.Fish.* 42(11):26-7.

Mason, J.M. 1976. Food of small, northwestern Atlantic bluefin tuna, *Thunnus thynnus* (L.) as ascertained through stomach content analysis. M.S. Thesis. University of Rhode Island (USA).

Maxwell, J.G.H., and G.R. Cresswell. 1981. Dispersal of tropical marine fauna to the Great Australian Bight by the Leeuwin Current. *Aust.J.Mar.Freshwat.Res.*, 32:493-500.

McCulloch, A.R. 1929. A checklist of the fishes recorded from Australia. *Mem.Aust.Mus.*, 5(2):145-329.

Michael, K.P., K.N. Bailey, P.R. Taylor, and P.B. Sharples. 1989. Report of observer trips on Japanese southern bluefin tuna longliners off East Cape, June-July 1988. *Int.Rep.Fish.Res.Cent.New Zealand Minst.Agri.Fish.*, (111):35 p.

Mimura, K. 1962. Studies on indomaguro, *Thunnus maccoyii*?, (preliminary report). *Occas.Rep.Nankai Reg.Fish.Res.Lab.*, 1:15-22.

Mimura, K., and I. Warashina. 1962. Studies on indomaguro (*Thunnus maccoyii* ?); description of the development of the fishery, geographical difference and seasonal change of distribution and relation which is seen among indomaguro distributions. *Rep.Nankai Reg.Fish.Res.Lab.*, 16:135-54 (In Japanese with English abstract)

Mugiya, Y., and N. Watabe. 1977. Studies on fish scale formation and resorption - II. effect of estradiol on calcium homeostasis and skeletal tissue resorption in the goldfish, *Carassius auratus*, and the killifish, *Fundulus heteroclitus*. *Comp.Biochem.Physiol.*, 57A:197-202.

Mulligan, T.J., F.D. Martin, R.A. Smucker, and D.A. Wright. 1987. A method of stock identification based on the elemental composition of striped bass *Morone saxatilis* (Walbaum) otoliths. *J.Exp.Mar.Biol.Ecol.*, 114:241-8.

Munro, I.S.R. 1958. Handbook of Australian fishes. No. 27. *Aust.Fish.*, 17(9):109-12.

Murphy, G.I. 1976. Some aspects of the dynamics of the Australian fishery. Indo-Pacific Fisheries Council /Indian Ocean Fisheries Commission 4th joint meeting, 1-4. Rome: FAO.

Murphy, G.I. 1977. New understanding of southern bluefin tuna. *Aust.Fish.*, 36(1):2-6.

Murphy, G.I. 1979. Southern bluefin tuna. *Fish.Situation Rep.CSIRO Div.Fish.Oceanogr.*, (1).

Murphy, G.I., and J. Majkowski. 1981. State of the southern bluefin tuna population: fully exploited. *Aust.Fish.*, 40(11):20-9.

Murray, T., and D. Burgess. 1990. Southern bluefin tuna fisheries in New Zealand waters. Paper presented at the Ninth Trilateral Scientific Meeting on SBT, Hobart, Australia, 1990, SBFWS/90/3.

Nakamura, E.L., and W.M. Matsumoto. 1965. Distribution of larval tunas in Marquesan waters. *Fish.Bull.U.S.Fish Wild.Ser.*, 66:1-12.

Nakamura, H. 1969. Tuna distribution and migration. London, Fishing News (Books) Ltd., 76 p.

Nelson, J.S. 1984. Fishes of the world. (2nd ed.) New York, John Wiley and Sons, 528 p.

Nishikawa, Y. 1985. Identification for larvae of three species of genus *Thunnus* by melanophore patterns. *Bull.Far Seas Fish.Res.Lab.*, (22):119-30.

Nishikawa, Y., M. Honma, S. Ueyanagi, S. Kikawa. 1985. Average distribution of larvae of oceanic species of scombrid fishes, 1956-1981. *Contr.Far Seas Fish.Res.Lab.*, 236:1-99.

Nishikawa, Y., and D.W. Rimmer. 1987. Identification of larval tunas, billfishes and other scombroid fishes (suborder Scombroidei): an illustrated guide. *Rep.CSIRO Mar.Lab.*, (186):20 p.

Olive, P.J., 1980. Growth lines in polychaete jaws. *In* Skeletal growth of aquatic organisms, edited by D.C. Rhoads and R.A. Lutz. New York. Plenum Press, pp. 561-92.

Olson, R.J. 1980. Synopsis of biological data on the southern bluefin tuna, *Thunnus maccoyii* (Castlenau 1872). *Spec.Rep.I-ATTC*, (2):151-212.

Palomares, M.L., and D. Pauly. 1989. A multiple regression model for predicting the food consumption of marine fish populations. *Aust.J.Mar.Freshwat.Res.*, 40:259-73.

Pannella, G. 1980. Growth patterns in fish sagittae. *In* Skeletal growth of aquatic organisms, edited by D.C. Rhoads and R.A. Lutz. New York, Plenum Press, pp.519-60.

Pariwono, J. 1986. Wind stress, mean sea level and inter-ocean transport in the Australian region. Ph.D. Thesis. Flinders University, SA, Australia, 196 p.

Pitcher, T.J., and P.D.M. MacDonald. 1973. Two models for seasonal growth in fishes. *J.Appl.Ecol.*, 10:559-606.

Prince, E.D., and L.M. Pulos (editors). 1983. Proceedings of the International Workshop on Age Determination of Oceanic Pelagic Fishes: Tunas, Billfishes, and Sharks. *NOAA Tech.Rep.NMFS*, 8.

Radtke, R.L. 1983. Otolith formation and increment deposition in laboratory-reared skipjack tuna, *Euthynnus pelamis*, larvae. *NOAA Tech.Rep.NMFS*, (8):99-103.

Radtke, R.L., and B. Morales-Nin. 1989. Mediterranean juvenile bluefin tuna: life history patterns. *J.Fish Biol.*, 35:485-96.

Radtke, R.L., and T.E. Targett. 1984. Rhythmic structural and chemical patterns in the otoliths of the Antarctic fish *Notothenia larseni*: their application to age determination. *Polar Biol.*, 3:203-10.

Radtke, R.L., D.F. Williams, and P.C.F. Hurley. 1987. The stable isotopic composition of bluefin tuna (*Thunnus thynnus*) otoliths: evidence for physiological regulation. *Comp.Biochem.Physiol.*, 87A:797-801.

Richards, W.J. 1969. Distribution and relative apparent abundance of larval tunas collected in the Tropical Atlantic during Equalant surveys I and II. Pp. 289-315, *In* Proceedings of the Symposium on the Oceanography and Fisheries Resources of the Tropical Atlantic - Review Papers and Contributions. Paris, UNESCO Press, pp. 289-315.

Richards, W.J., T. Potthoff, and J. Kim. (in press). Problems identifying tuna larvae species (Pisces: Scombridae: *Thunnus*) from the Gulf of Mexico. *Fish.Bull.NOAA-NMFS*.

Richards, W.J., and D.C. Simmons. 1971. Distribution of tuna larvae (Pisces, Scombridae) in the northwestern Gulf of Guinea and off Sierra Leone. *Fish.Bull.NOAA-NMFS*, 69:555-68.

Ricker, W.E. 1975. Computation and interpretation of biological statistics of fish populations. *Bull.Fish.Res.Board Can.*, (191):382 p.

Ride, W.D.L., C.W. Sabrosky, G. Bernardi, and R.V. Melville (editors). 1985. International code of zoological nomenclature (3rd ed.) Los Angeles, University of California Press.

Robins, J.P. 1963. Synopsis of biological data on bluefin tuna, *Thunnus thynnus maccoyii* (Castlenau) 1872. *FAO Fish.Rep.*, 6(2):562 -87.

Robins, J.P. 1975a. Some aspects of tuna and its potential in the oceanic waters off Western Australia. *Rep.Dep.Fish.Wildl.West.Aust.*, (17):59 p.

Robins, J.P. 1975b. Tuna survey in waters off the Western Australian coast during the period August 1973 to August 1974. *Rep.Dep.Fish.Wildl.West.Aust.*, (18):85 p.

Sandland, R.L. 1987. Addendum to: Hearn, W.S., Growth of southern bluefin tuna. Paper presented at the Sixth Trilateral Scientific Meeting on Southern Bluefin Tuna, Hobart, Tasmania. *CSIRO*, SBFWS/87/11.

SBT sets new tag record (1981) *Australian Fisheries* 40(10): 40. (BP#11)

Schneider, R.C., and S.V. Smith. 1982. Skeletal Sr content and density in *Porites* spp. in relation to environmental factors. *Mar.Biol.*, 66:121-31.

Schnute, J., and D.A. Fournier. 1980. A new approach to length frequency analysis: growth structure. *J.Fish.Res.Board Can.*, 37:1337-51.

Serventy, D. L. 1956. The southern bluefin tuna, *Thunnus thynnus maccoyii* (Castlenau), in Australian waters. *Aust.J.Mar.Freshwat.Res.*, 7:1-43.

Sharp, G.D., and S. Pirages. 1978. The distributions of red and white swimming muscles, their biochemistry and the biochemical phylogeny of selected scombrid fishes. *In* The physiology ecology of tunas, edited by G.D. Sharp and A.E. Dizon. New York, Academic Press, pp. 41-78.

Shepherd, J. G. 1982. A versatile new stock-recruitment relationship for fisheries, and the construction of sustainable yield curves. *J.Cons.CIEM*, 40(1):67-75.

Shepherd, S.A., and W.S. Hearn. 1983. Studies on southern Australian abalone (genus *Haliotis*). IV Growth of *H. laevigata* and *H. ruber*. *Aust.J.Mar.Freshwat.Res.*, 34:461-75.

Shingu, C. 1967. Distribution and migration of the southern bluefin tuna. *Rep.Nankai Reg.Fish.Res.Lab.*, (25):19-36 (In Japanese with English abstract)

Shingu, C. 1970. Studies relevant to distribution and migration of the southern bluefin tuna. *Bull.Far Seas Fish.Res.Lab.*, (3):57-114 (In Japanese with English abstract)

Shingu, C. 1978. Ecology and stock of southern bluefin tuna. Fish.Study Jap.Assoc.Fish.Resources Protection, 31:81 p. (In Japanese; English translation in *Rep.CSIRO Div.Fish.Oceanogr.*, 131:79 p., 1981.)

Shingu, C., and S. Hayashi. 1966. The survey of the stock of the southern bluefin. *Chugoka Shikka Branch Jap.Assoc.Fish.Sci.*, May 1966.

Shingu, C., and K. Hisada. 1971. Fluctuations in amount and age composition of catch of southern bluefin tuna in longline fishery, 1957-1969. *Bull.Far Seas Fish.Res.Lab.*, (5):195-218 (In Japanese with English abstract)

Shingu, C., and Y. Warashina. 1965. Studies on the southern bluefin, *Thunnus maccoyii* (Castelnau) - I   Morphometric comparison of southern bluefin. *Rep.Nankai Reg.Fish.Res.Lab.*, (22):85-93 (In Japanese with English abstract)

Sibert, J.R. 1984. A two-fishery tag attrition model for the analysis of mortality, recruitment and fishery interaction. *Tech.Rep.Tuna Billfish Assess.Programme S.Pac.Comm.*, (13):27 p.

Smith, S.V., R.W. Buddemeier, R.C. Redalje, and J.E. Houck. 1979. Strontium calcium thermometry in coral skeletons. *Science,Wash.*, 204:404-7.

South, J.F. 1845. *Thunnus. In* Enclyclopaedia metropolitana, edited by Smedley, Pose and Rose, pp. 620-2.

Suda, A. 1960. Fishing ground investigation. *In* Report of tuna fishing investigation in the Caribbean Sea, northwest Atlantic and the waters of South Australia. *Jap.Fish.Agency*, pp. 5-63 (In Japanese)

Talbot, F.H., and M.J. Penrith. 1962. Tunnies and marlins of South Africa. *Nature, Lond.*, 193(4815):558-9.

Talbot, F.H., and M.J. Penrith. 1963. Synopsis of biological data on species of the genus *Thunnus (senso lato)* (South Africa). *FAO Fish.Rep.*, 6(2):608-46.

Talbot, F.H., and M.J. Penrith. 1968. The tunas of the genus *Thunnus* in South African waters. Part 1. Introduction, systematics, distribution and migrations. *Ann.S.Afr.Mus.*, 52(1):1-41.

Tanaka, S. 1960. Studies on the dynamics and the management of fish populations. *Bull.East Sea Fish.Res.Inst.*, 28:1-200 (Tokai Suisan Kenkyujoho)

Taylor, C.C. 1958. A note on Lee's phenomenon in Georges Bank haddock. Paper presented at the ICNAF/FAO symposium held at Biarritz, France, March 1-10, 1965. Special Publication 1:243-51.

Thorogood, J. 1983. Direct age determination of southern bluefin tuna (*Thunnus maccoyii*). Hobart: CSIRO (Unpubl.)

Thorogood, J., 1986. Aspects of the reproductive biology of the southern bluefin tuna (*Thunnus maccoyii*). *Fish.Res.*, 4:297-315.

Thorogood, J. 1987. Age and growth determination of southern bluefin tuna, *Thunnus maccoyii*, using otolith banding. *J.Fish.Biol.*, 30: 7-14.

Thresher, R.E., J.S. Gunn, and C. Proctor (in press) Aging by means of X-Ray microanalysis. *In* Proceedings of a workshop on southern bluefin tuna research, CSIRO Marine Laboratories, 3-4 November 1988, Canberra: Bureau of Rural Resources.

Tranter, D.J., G.S. Leech, and D. Airey. 1983. Edge enrichment in an ocean eddy. *Aust.J.Mar.Freshwat.Res.*, 34:665-80.

Ueyanagi, S. 1969a. The spawning of southern bluefin tuna (*Thunnus maccoyii*) as indicated by the occurrence of its larvae. *Bull.Far Seas Fish.Res.Lab.*, (1):1-4.

Ueyanagi, S. 1969b. Observations on the distribution of tuna larvae in the Indo-Pacific Ocean with emphasis on the delineation of the spawning areas of the albacore, *Thunnus alalunga*. *Bull.Far Seas Fish.Res.Lab.*, (2):177-254.

Uotani, I., K. Matsuzaki, Y. Makino, K. Noda, O. Inamura, and M. Horikawa. 1981. Food habits of larvae of tunas and their related species in the area northeast of Australia. *Bull.Jap.Soc.Sci.Fish.*, 47:1165-72.

Warashina, Y. 1990. Japanese southern bluefin tuna fishery in recent years. Paper presented at the Ninth Trilateral Scientific Meeting on southern bluefin tuna, Hobart, Australia. *CSIRO*, SBFWS/90/1.

Warashina, I., and K. Hisada. 1970. The ecology of the spawning of southern bluefin, examined by the changes in the meat quality and the body weight. *Bull.Far Seas Fish.Res.Lab.*, (3):147-65.

Warashina, Y., and K. Hisada. 1974. Preliminary evaluation of effort of the voluntary regulation on stock of southern bluefin tuna and the longline fishery. *Bull.Far Seas Fish.Res.Lab.*, (10):193-220 (In Japanese with English abstract)

Webb, B.F. 1972. Report on the investigation of the 'Llore Lopez' 8 January to 2 April 1970, Sect. 6: Tuna catch analysis and seawater temperatures. *Fish.Tech.Rep.New Zealand Minist.Agri.Fish.*, 108:105 p.

Wetherall, J. A. 1982. Analysis of double-tagging experiments. *Fish.Bull.NOAA-NMFS*, 80:687-701.

Williams, K. 1982a. Data collection method. *In* CSIRO data base for southern bluefin tuna (*Thunnus maccoyii* (Castelnau)), edited by J. Majkowski. *Rep.CSIRO Mar.Lab.*, 142:4-5.

Williams, K. 1982b. Tagging method. *In* CSIRO data base for southern bluefin tuna (*Thunnus maccoyii* (Castlenau)), edited by J. Majkowski. *Rep.CSIRO Mar.Lab.*, 142:7-8.

Williams, K. 1983. Big numbers of SBF tuna tagged off Esperance. *Aust.Fish.*, 42(7):26-7.

Yabe, H., S. Ueyanagi, and H. Watanabe. 1966. Studies on the early life history of bluefin tuna *Thunnus thynnus* and on the larva of the southern bluefin tuna *T. maccoyii*. *Rep.Nankai Reg.Fish.Res.Lab.*, (23):95-116.

Yonemori, T., S. Hayashi, and H. Kono, 1985. Assessment and management of southern bluefin tuna stock by Japan, Australia and New Zealand. Japan Association of Fishery Resources Protection. Review on assessment and management of far seas fishery resources No. 1: 277-327 (In Japanese).

Yonemori, T., K. Hisada, and T. Nagai. 1983a. Status of stocks and history of fisheries for southern bluefin tuna. Paper presented at the Second Trilateral Scientific Meeting on Southern Bluefin Tuna, Shimizu, Japan. *Jap.Fish.Agency*, SBFWS/83/12, 46 p.

Yonemori, T., K. Hisada, and T. Nagai. 1983b. The study of the utilization of southern bluefin tuna. Paper presented at the Second Trilateral Scientific Meeting on Southern Bluefin Tuna, Shimizu, Japan. *Jap.Fish.Agency*, SBFWS/83/13, 16 p.

Yonemori, T., and J. Morita. 1978. Report on 1977 research cruise of the R/V Shoyo-Maru. Distribution of tuna and billfishes, and their larvae in the eastern Indian Ocean, October-December, 1977. *Rep.Res.Div.Jap.Fish.Agency*, 52:48 p.

Young, J.W., and T.L.O. Davis. 1990. Feeding ecology of larvae of southern bluefin, albacore and skipjack tunas (Pisces: Scombridae) in the eastern Indian Ocean. *Prog.Ser.Mar.Ecol.*, 61:17-29.

Yukinawa, S. 1970. Age and growth of southern bluefin *Thunnus maccoyii* (Castelnau) determined by their scales. *Bull.Far Seas Fish.Res.Lab.*, (3):229-57.

Yukinawa, M. 1987. Report on 1986 research cruise of the R/V Shoyo-Maru. Distribution of tunas and billfishes larvae and oceanographic observation in the eastern Indian Ocean January-March, 1987. *Rep.Res.Div.Jap.Fish.Agency*, 61:100 p.

Yukinawa, M., and T. Koido. 1985. Report on 1984 research cruise of the R/V Shoyo-Maru third cruise. Distribution of tunas and billfishes and their larvae in the eastern Indian Ocean January-March, 1985. *Rep.Res.Div.Jap.Fish.Agency*, 59:108 p.

Yukinawa, M., and N. Miyabe. 1984. Report on 1983 research cruise of the R/V Shoyo-Maru. Distribution of tunas and billfishes and their larvae in the eastern Indian Ocean October-December, 1983. *Rep.Res.Div.Jap.Fish.Agency*, 58:103 p.

Appendix 1

REVIEW OF ASPECTS OF SOUTHERN BLUEFIN TUNA BIOLOGY, POPULATION AND FISHERIES

CONTENTS

COMMERCIAL AND RECREATIONAL COMPONENTS OF THE SOUTHERN BLUEFIN TUNA (*THUNNUS MACCOYII*) FISHERY

A.E. Caton
Bureau of Resource Sciences
PO Box E11, Queen Victoria Terrace
PARKES A.C.T. 2600, Australia

ABSTRACT

This report identifies, and provides a qualitative description of, the commercial and recreational components of the global southern bluefin tuna fisheries. It describes their distribution in time and space and includes details of catch magnitude and age composition.

1.   INTRODUCTION

The study of interactions among components of the global fishery for southern bluefin tuna (SBT; *Thunnus maccoyii*) in the first instance requires the identification of those components, their distribution in time and space, and their relative significance. This summary provides a qualitative description, augmented by generalized indications of catch magnitude and size composition. Comprehensive time series of SBT catch data for commercial vessels from Australia, Japan and New Zealand are available but the data for catches by other countries, and for recreational fisheries, are much less certain.

2.   COMMERCIAL FISHERIES

The major commercial fisheries for SBT have been the Australian surface fishery (predominantly purse-seine and pole-and-line) for juveniles, the Japanese longline fishery for older juveniles and adults, an expanding Taiwanese longline fishery and, on a smaller scale, New Zealand's surface and longline fisheries (Caton, 1991). Less substantial commercial activities are, or have been, carried out by longliners from Indonesia and Korea, and there is a bycatch of SBT in the Taiwanese gillnet fishery.

2.1   Australia

2.1.1   Western Australian pole-and-line fishery

A pole-and-line fishery with SBT as the target species, locating fish by trolling and using dead pilchards (*Sardinops neopilchardus*) as chum, was located year-round along the western south coast of Western Australia (Figure 1). It commenced at the end of the 1960s, the catch steadily increasing to 6,000 mt in 1982/83 (representing some 1.2 million fish), then rapidly declining as a result of quota regulations. The annual catch is now less than 20 mt (4,000 fish). The catch was used almost exclusively for canning, then there was an emphasis on exporting fresh fish to the Japanese sashimi market, but the main use remaining is for local processing as smoked tuna. The fleet, which exceeded 100 vessels at the peak of activity, currently consists of three or four vessels fishing for SBT on a part-time basis. Average length of fish in the catch has usually been about 65 cm (5 kg; 2 1/2 years old), the main age classes taken being 1-3 year olds.

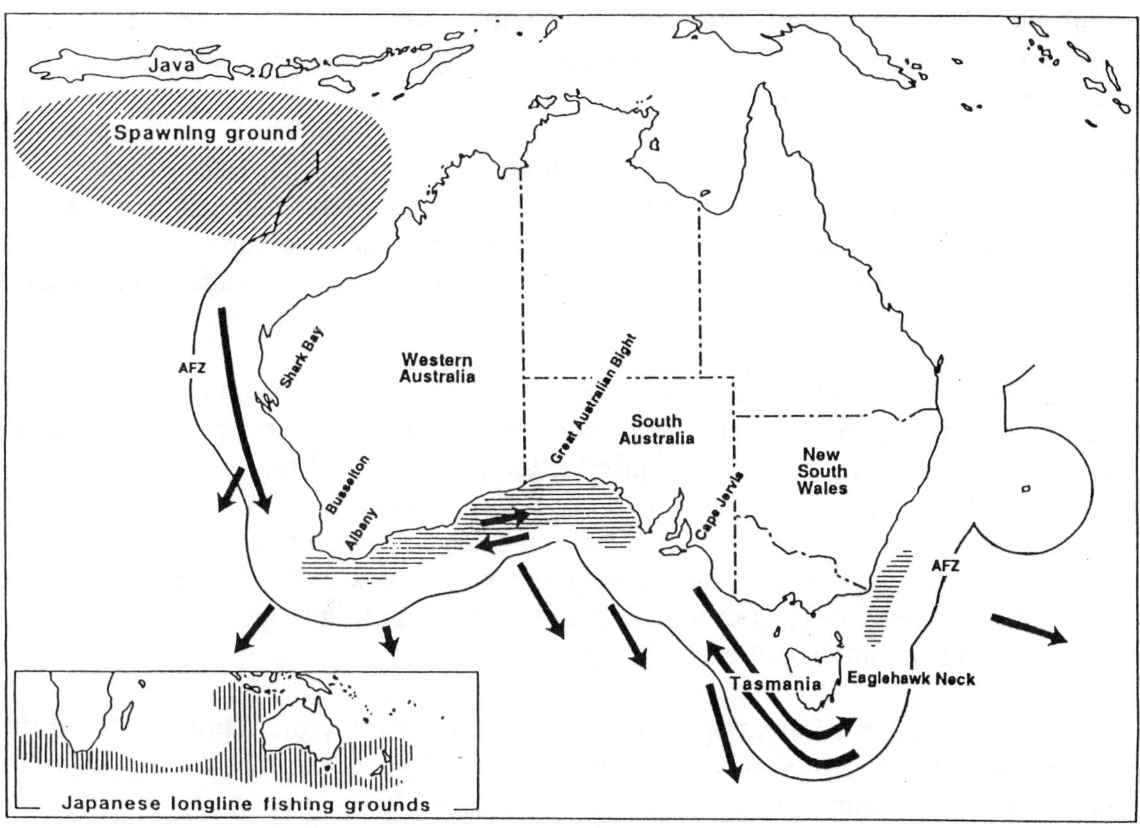

Figure 1. Southern bluefin tuna spawning ground and migration pattern within and out of Australian waters. The 200-mile Australian fishing zone is indicated by the solid line and the horizontal hatching indicates the composite distribution of the Australian surface fishery. The general distribution of Japanese longline fishing is inset. (Modified from Majkowski *et al.*, 1988).

A very small amount of skipjack tuna (*Katsuwonus pelamis*), albacore (*Thunnus alalunga*) and bigeye (*T. obesus*) were taken incidentally. The only other pelagic fishery in the region, and one in which some of the multi-purpose SBT vessels operated, is a small-vessel, purse-seine fishery for pilchards, but no incidental catches of SBT are taken by it. At the peak of the fishery, the Western Australian fleet operations were almost continuous with, but generally inshore of, those of South Australia. The demarcation between the fisheries was approximately 125°E. The remaining Western Australian activities are now in the Albany area (Figure 1).

Some occasional SBT-directed trolling activities occurred off the far southern west coast of Western Australia in the late 1970s. There were also reports around the same time that 1-2 year old SBT were taken by trolling for bait by rock lobster fishermen between Shark Bay and the Abrolhos Islands (26°S-28°S; Figure 1). However, the taking of SBT north of 34°S off the west coast has been prohibited since the early 1980s to protect small fish.

2.1.2 South Australian purse-seine/pole-and-line fishery

Experimental pole-and-line fishing during the early 1950s led to steady development of a commercial fishery off the west coast of South Australia (Figure 1) from the mid 1950s, the season extending from mid-December to May, and landings being canned, or exported for canning. Catches averaged 5,000-6,000 mt annually during the later 1970s when experimental purse seining was introduced, at first in competition with the pole-and-line operations. It soon became apparent that catches were enhanced when the vessels worked together, a pole-and-line vessel holding fish at the surface with live bait while a purse seiner set around the fish and pole vessel. This cooperative operation continued until the end of the 1980s, whereupon the involvement of purse-seine vessels diminished in conjunction with the decline in the proportion of the catch destined for canning.

Catches in the South Australian fishery increased to about 14,000 mt in 1982-83 (representing some 1.6 million fish). Subsequently, concern about the extent of reduction in SBT parental biomass prompted the introduction of a quota. Progressive reduction of quotas for the remainder of the 1980s, and diversion of some of the surface activity to longline operations brought the catch to 2,500 mt (some 125,000 fish) in 1990-91. The fleet has declined from a peak of more than 70 vessels to less than 20. The fishery at times took opportunistic catches of albacore and skipjack but there are no other commercial domestic tuna fisheries in the region.

There have been different age groups targeted over the years. Initially (until the late 1960s) the catch consisted predominantly of 3 and 4 year olds, but during the 1970s 2- and 3- year olds dominated. In the 1980s, effort was directed away from very young fish and numbers of 2 year olds in the catch progressively diminished. Fishermen sought to divert product from canning to the Japanese market for fresh or frozen sashimi. This was accompanied by a temporary improvement in representation of 5 to 7 year olds, but by the late 1980s, 3 and 4 year olds were again dominant (Figure 2). Finally, with diversion of surface fishery quota to longlining, those surface operations remaining have avoided 3 year olds and now concentrate on 4 and 5 year olds.

A renewed activity in South Australia in 1990-91, but one unlikely to expand significantly, was the operation of a small commercial troll fishery in the southeast of the state in April-May, air freighting fresh SBT to Japan. Five rock lobster vessels took approximately 10 mt during June and July. The fish were mainly 3-4 year olds.

The most recent developments have been the conversion of some of the surface fleet to longlining, and the establishment of experimental, cage-rearing operations with SBT in South Australia. Longline catches were minor (about 15 mt) in 1991-92, but about 130 mt were held in rearing cages. Fish introduced to cages are captured by surface gear and are in the 3-4 year old age range. It seems probable that expansion of both these alternatives will occur. However, most Australian quota is now committed to Australia-Japan joint-venture longlining operations by Japanese vessels.

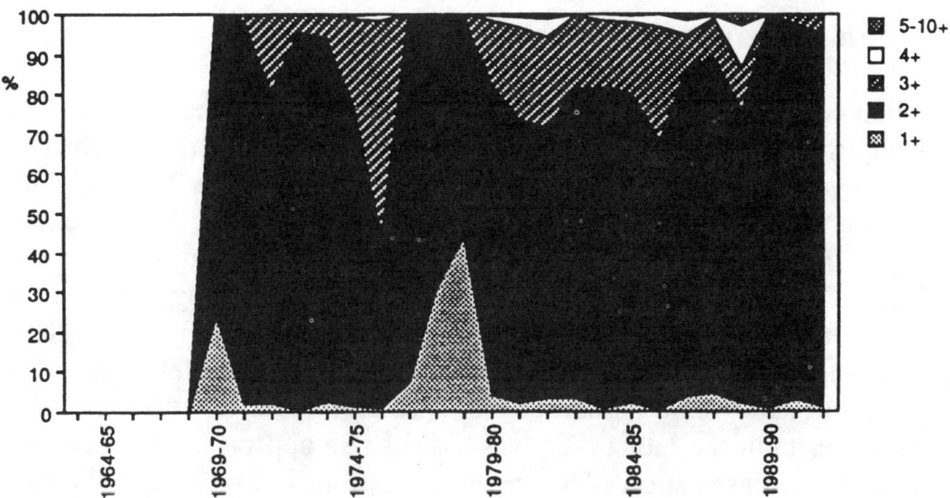

Figure 2a. Percentage catch in numbers of 1+, 2+, 3+, 4+ and 5+ to 10+ year old fish in the Western Australia SBT catch, quota years 1959-60 to 1991-92.

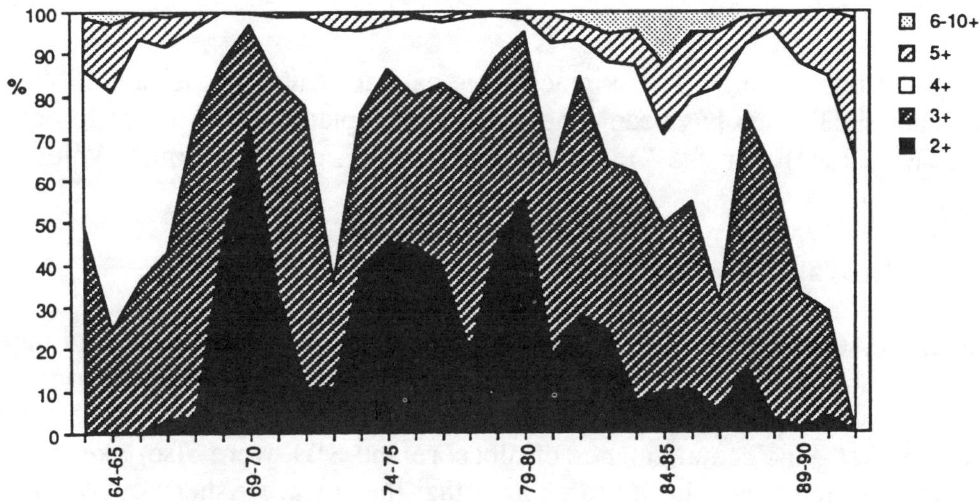

Figure 2b. Percentage catch in numbers of 2+, 3+, 4+ and 5+ and 6+ to 10+ year old fish in the South Australia SBT catch, quota years 1963-64 to 1991-92. (Quota year is 1 October to 30 September, with the exception of 1991-92, which extends from 1 October 1991 to 31 October 1992.)

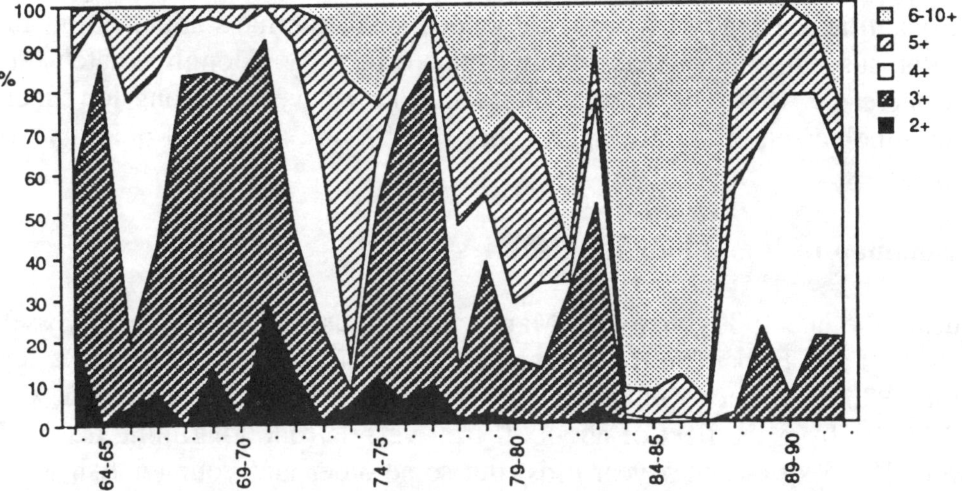

Figure 2c. Percentage catch in numbers of 2+, 3+, 4+, 5+ and 6+ to 10+ year old fish in the New South Wales SBT catch, quota years 1963-64 to 1991-92.

2.1.3 New South Wales purse-seine/pole-and-line fishery

Australian commercial pole-and-line tuna fishing commenced off New South Wales (Figure 1) during the early 1950s, after many years of incidental troll catches of SBT and other pelagics. Production reached approximately 5,000 mt (some 275,000 fish) by the early 1970s with a fleet at times in excess of 100 vessels. The catch was used for canning. Despite introduction of collaborative purse-seine/pole-and-line operations during the late 1970s, and the much more extensive coverage of the fishing grounds by fish-spotting aircraft during that period, the New South Wales surface fishery catch of SBT declined, failing altogether in 1985 and for the remainder of the 1980s. Southern bluefin tuna 2 to 7 years old had been common in the catch from year to year. The dominant age group in number until the late 1970s was that of fish approaching 3 years of age, after which time the representation of younger fish declined (Figure 2). By the 1980s they were absent from the New South Wales catch, and it was only in 1991 that scattered, small, surface schools of SBT were again reported off New South Wales. A small bycatch of SBT was also taken by fishers trolling for albacore, yellowfin (*T. albacares*) and skipjack.

Poling and purse seining for skipjack have expanded during the late 1980s but to date no incidental SBT catch has been reported. The skipjack season tends to occur later (December to May/June) than the 'traditional' SBT season off New South Wales (September to January).

2.1.4 New South Wales longline fishery

A small vessel (10-25 m), domestic longline fishery with yellowfin and bigeye tuna as the target species developed off the central and south coast of New South Wales in the mid 1980s with the successful establishment of air-freight outlets to the Japanese fresh sashimi market. Incidental catches of albacore and SBT were also taken, the latter between March and September in the same area that the surface fishery operated. Up to 175 vessels are licensed in the fishery but only 20 or so vessels reported landing SBT in 1991-92. Whereas albacore are unsuitable for export as sashimi and their market value is low, some of the SBT taken have commanded very high prices (to ¥10,000 per kg). The size of the SBT varied from around 15 kg processed weight to over 120 kg (4-15 years old). The quantities currently involved have been small (perhaps about 60 mt in 1991-92) but their value suggests that increases may occur, especially if longline catch rates of SBT continue to increase. It is possible that interaction may develop among longliners, trollers and pole-and-line vessels if surface aggregations of SBT re-appear in quantity off southern New South Wales.

2.1.5 Tasmanian troll and longline fisheries

During the late 1980s, between March and August, small (7-15 m) vessels from Eaglehawk Neck and Hobart on the Tasmanian east coast (Figure 1) trolled a few miles offshore for SBT. The catch, which was air-freighted fresh-chilled to Japan, was about 50 mt in 1991-92 from the fleet of about 20 troll vessels (one longlining also). Most of the fish were 15-25 kg (4 and 5-year olds) but some older and younger fish were also taken. The level of activity is likely to increase, with further diversification to longlining.

2.1.6 Japanese-style longlining in the Australian fishing zone

The waters of what, after 1979, became the 200-nautical-mile Australian Fishing Zone (AFZ; Figure 1) have been fished since the 1950s by the Japanese fleet of large (30-60 m) distant-water longliners. Those vessels operating south of 30°S seasonally targeted SBT, with incidental catches of albacore, broadbill swordfish (*Xiphias gladius*), bigeye tuna, and sharks. In more northern regions of the AFZ, SBT were incidental to catches of yellowfin, bigeye, marlins (*Makaira* spp.), albacore, and broadbill.

The size of fish taken by Japanese longliners in the AFZ has varied seasonally and by area. Younger SBT (from 3 years old) were taken from the region of the Great Australian Bight (Figure 1). Off New South Wales, SBT from 4 years to 15-20 years old were taken. Fish of the same general range of ages have been taken recently off eastern Tasmania. Off south western Tasmania, the younger ages have been much less strongly represented, with most of the catch consisting of adult SBT (8 years and older).

When the AFZ was established, Japanese longliners were given access under a licensing arrangement developed bilaterally between Australia and Japan. In the mid-1980s, two Australian-owned, ex-Japanese longliners operated. Since 1990, some of the 'bilateral' longliners have participated in an Australia/Japan joint-venture arrangement. There were 42 joint-venture licenses issued in 1991-92, and an allocation of 2,069 mt of Australian quota. From 1990 to early 1992, there were three Australian-chartered, Japanese-style longliners. Each of these groups of longliners was subject to separate licensing conditions and permissible fishing areas.

Access to SBT in the AFZ by 'bilateral' longliners has been reduced virtually to the waters surrounding Tasmania, where there is a May to August season off the eastern and southern coasts, and a November to January season off the south west. For the 1991-92 season, a limit of 400 mt (debited against the Japanese global quota) applied, and only 12 'bilateral' vessels operated there. Off the eastern Australian mainland, 'bilateral' vessel numbers were limited to 50 in 1991-92 for a maximum of 3,150 days fishing. The western region of the AFZ has a limit of 40 'bilateral' vessels but no limit on total fishing days. These eastern and western activities target other species, and recently the annual SBT catches there have been less than 50 mt and 10 mt respectively.

The Japanese 'bilateral' and Australia/Japan joint-venture longline activities off eastern Tasmania are immediately adjacent to, and concurrent with, the Eaglehawk Neck to Hobart domestic troll/longline fishery. Off southwest Tasmania, there is no domestic SBT fishing activity adjacent to the November-to-January bilateral and joint-venture longlining activity.

A major, Japanese longline fishery for SBT occurred annually between April and August off southern New South Wales but it was phased out by access restriction after 1982. It was centred just seaward of the same area, and during the same months as the New South Wales domestic longline vessels. In 1990, 15 Australia-Japan joint-venture vessels and three Australian charter vessels fished the wider New South Wales grounds again, in both cases using Australian quota. Joint-venture vessels were excluded in 1991 and subsequently, but the charter vessels continued fishing there until early 1992, when their operations ceased altogether.

For the 1991-92 season, some of the joint-venture vessels had access to waters of the Great Australian Bight off southern Australia. This area had been closed to Japanese longliners since the mid 1980s, and prior to that (from 1971) had been subject to a voluntary October-to-March closure established by the Japanese fleet to reduce catch of 3- and 4-year-old SBT. The region is immediately adjacent to the South Australian surface-fishery area, which operates during the same months, and size composition of the longline catch there is similar to that in the surface fishery. The experimental longlining mentioned above conducted by modified, Australian, surface-fishery vessels has also commenced in the area, as the Australian fleet seeks to divert activities away from the surface fishery and towards bigeye tuna and larger SBT.

2.2   Japan

Japan is involved in the SBT fishery by way of a distant-water longline fishery which developed in the 1940s to satisfy a demand for tuna exports to canning markets. The fishery moved into the Southern Hemisphere during the 1950s, with SBT activities commencing in the region of the spawning grounds (the Indian Ocean south of Java and off north western Australia) taking post-spawning adult SBT (8-20 year olds). Catch peaked at 78,000 mt (representing some 1,225,000 fish) in 1961, when approximately 150 vessels operated. The catch, destined for canning, was generally held in refrigerated brine. However, SBT was also used in Japan for 'kirimi' (fish filleted, sliced and cooked, usually in oil), and this led to expanded use of freezer storage.

At the end of the 1960s activity moved south of the spawning grounds (west of North West Cape; 22°S) and finally spread broadly in the region of the West Wind Drift of the Southern Ocean (35°S-45°S; Figures 3 and 4), extending from the western South Atlantic, eastward to the South Pacific adjacent to the New Zealand east coast (Shingu, 1978). The final product of the catch changed to sashimi because the development of very low temperature freezer storage (<-50°C) enabled vessels to land high-quality SBT even after lengthy cruises.

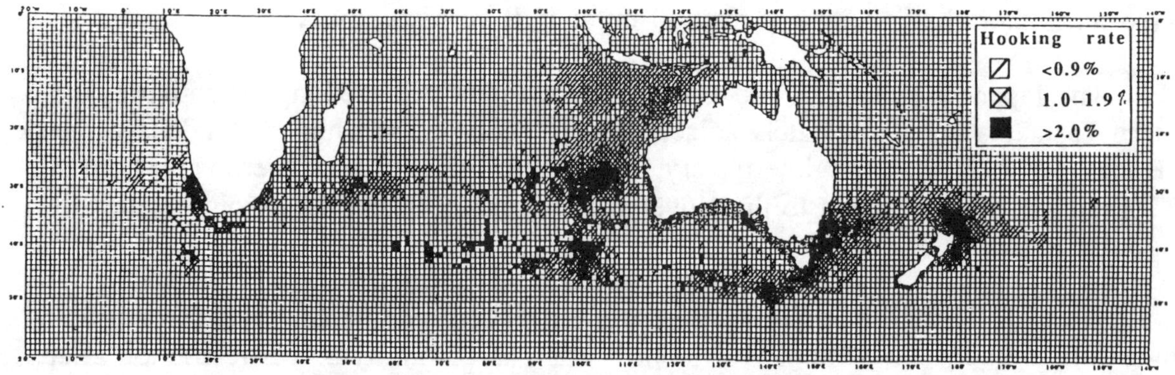

Figure 3.   Fishing grounds of Japanese longliners in the mid-1960s, as shown by average year's hooking rate for each 1 degree square, after operations had spread south from the northeastern Indian Ocean. Coastal distribution of Australia from Robins (1963) and that of New Zealand from McKenzie (1962). Subsequent operations spread more broadly between 40oS and 50oS. (Source: Shingu, 1970.)

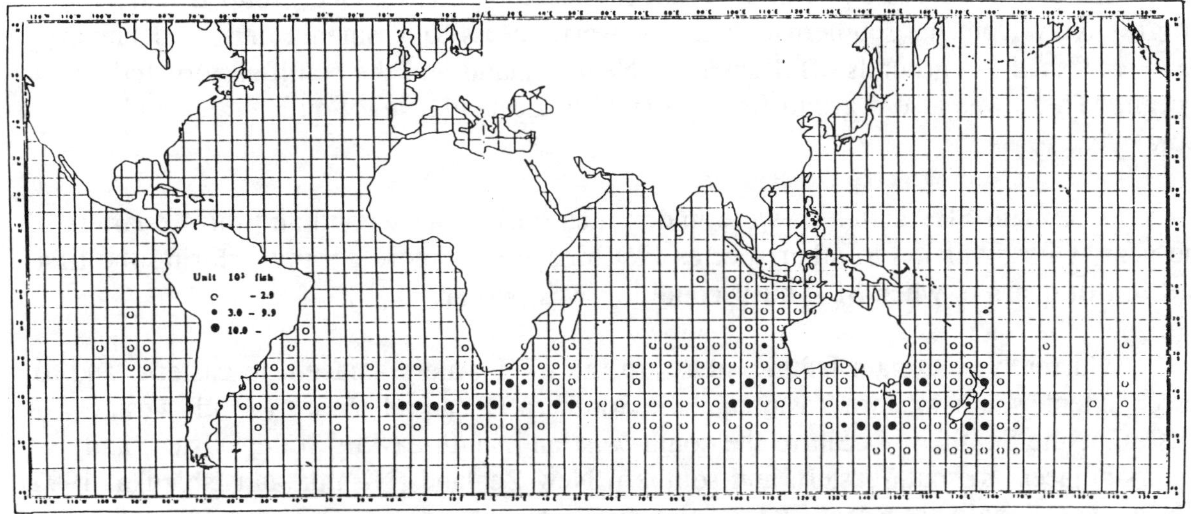

Figure 4.   Distribution of southern bluefin tuna catch in numbers by Japanese longline tuna fishery, 1970. (Source: Fisheries Agency of Japan, 1972.)

The southern operations extended the range of ages of fish caught to approximately 4 years and older in general. More specifically, younger-aged fish (*i.e.* including sub-adults) were taken north of 35°S. The main areas were south and east of Australia, off north-eastern New Zealand and off South Africa. Older (*i.e.* generally adult) fish were more prevalent south of 35°S. The main areas were off southern New Zealand, off Tasmania, off south-western Australia, east of South Africa, and on and south of the spawning grounds.

During the 1970s, SBT catch was in the order of 25,000 mt to 40,000 mt (some 450,000 - 800,000 fish). The fleet increased to a maximum of 250 vessels in 1980, after which economic rationalization of operations, and more recently the advent of quotas and operational restraints, brought about a reduction to around 200 active vessels annually. By the end of the 1980s catch had declined to under 10,000 mt (175,000 fish), and areas of operation had contracted substantially (Nishida and Ishizuka, 1992).

Some components of the fleet moved seasonally to bigeye grounds in the Indian and Atlantic Oceans. Others had moved to yellowfin and marlin grounds in the Tasman and Coral Seas, but this was curtailed by changed 200-mile-zone access conditions in the region. Nevertheless, the capacity to divert seasonal operations to other species, and the increasing price for high quality SBT sashimi, enabled continued seasonal operations on SBT despite the very low catch rates associated with the substantially-diminished parental biomass level.

Prior to contraction of the fishing seasons and areas, there was a range of seasonal patterns of Japanese longline operations. The main seasons by area were February to July off New Zealand; May to August off New South Wales; October to January off south-western Tasmania; year round in the Great Australian Bight (until establishment of the voluntary closure between October and March); October to November and then January to March off Albany, Western Australia; August to December in the south-eastern Indian Ocean; and March to October off South Africa. The SBT fleet consisted of several groups of vessels with different fishing campaigns. One group fished progressively from the southern Atlantic in April across to Tasmania until mid-January. Another fished the grounds off Tasmania, New Zealand and the south-eastern Indian Ocean. There was also a group fishing off New Zealand, New South Wales and in the Coral Sea.

The main SBT activities were undertaken during the second and third quarters of the year, with diversion to other species like bigeye tuna more common during the first and fourth quarters when SBT meat condition was poorer.

Currently, the main fishing areas are the AFZ waters adjacent to eastern, and to a lesser extent, south-western Tasmania, the Southern Ocean adjacent to south-western Australia, the Southern Ocean to the south-west and south-east of South Africa, and waters adjacent to south-eastern and southern New Zealand (Nishida and Ishizuka, 1992). No Japanese longliners target SBT on the spawning grounds or the region immediately to its south, but incidental catches are taken during bigeye- and yellowfin/striped marlin-directed operations. Size composition varies among the main fishing areas (Figure 5). The 1991-92 fishing seasons were March to August off New Zealand, May to September off Tasmania, April to July off South Africa, and August to October off southwestern Australia and the southern Indian Ocean. As part of a cooperative, SBT size-composition monitoring programme conducted by Australia, Japan, and New Zealand a small number (17 in 1991-92) of special-quota vessels continued operations after the main seasons were closed.

During the 1970s and 1980s the establishment of 200-mile economic zones proliferated, with the result that access to previously-important fishing areas was increasingly restricted. Consequently, Japan sought other countries in order to establish joint-venture longline arrangements. However, as regards SBT activities, only Australia and New Zealand were involved in these ventures.

2.3   New Zealand

2.3.1 South Island handline/troll fishery

The observation that large (15-20+ year old) adult SBT were attracted to offal from factory trawlers off the west coast of the South Island led to development in 1980 of a handline/troll fishery between 41°S and 43°S (Figure 6), supplying SBT to the Japanese sashimi market. Catch increased to 305 mt in 1982 then declined steadily, reflecting the decline in SBT parental biomass. Approximately 25-30 vessels are engaged in the fishery, with a season which has contracted to July to August (Murray and Burgess, 1992).

2.3.2 South Island longline fishery

Declining catches in the handline/troll fishery led some fishermen to diversify into small vessel (15-30+ m) longlining in 1988. Now about six to eight multi-purpose vessels (one a 50 m former Japanese longliner) operate, some with other species as target, and extending in range and in season beyond that of the handline/troll fleet (Murray and Burgess, 1992). Their catch in 1990 was 286 mt, and size composition was similar to that of the catch from licensed Japanese longliners and joint-venture charter longline vessels in the New Zealand Exclusive Economic Zone (NZEEZ) (Figure 6).

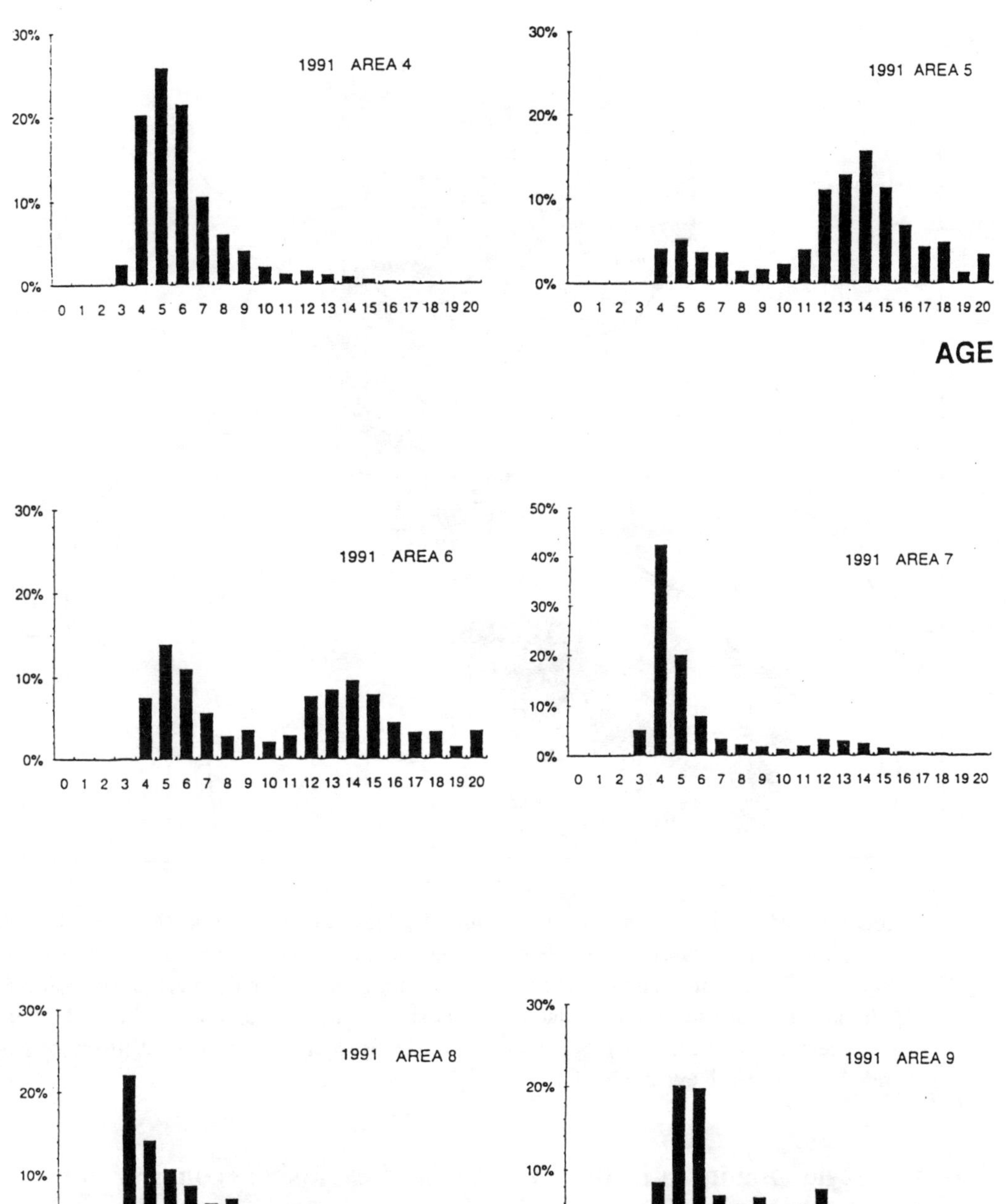

Figure 5.   Preliminary 1991 percent-frequency distribution of catch-at-age in Japanese southern bluefin tuna longline statistical areas (Area 4: New South Wales; Area 5: northern New Zealand; Area 6: southern New Zealand; Area 7: southern Australia; Area 8: southeastern Indian Ocean; Area 9: south of Africa). Age given in years. (Source: Nishida and Ishizuka, 1992.)

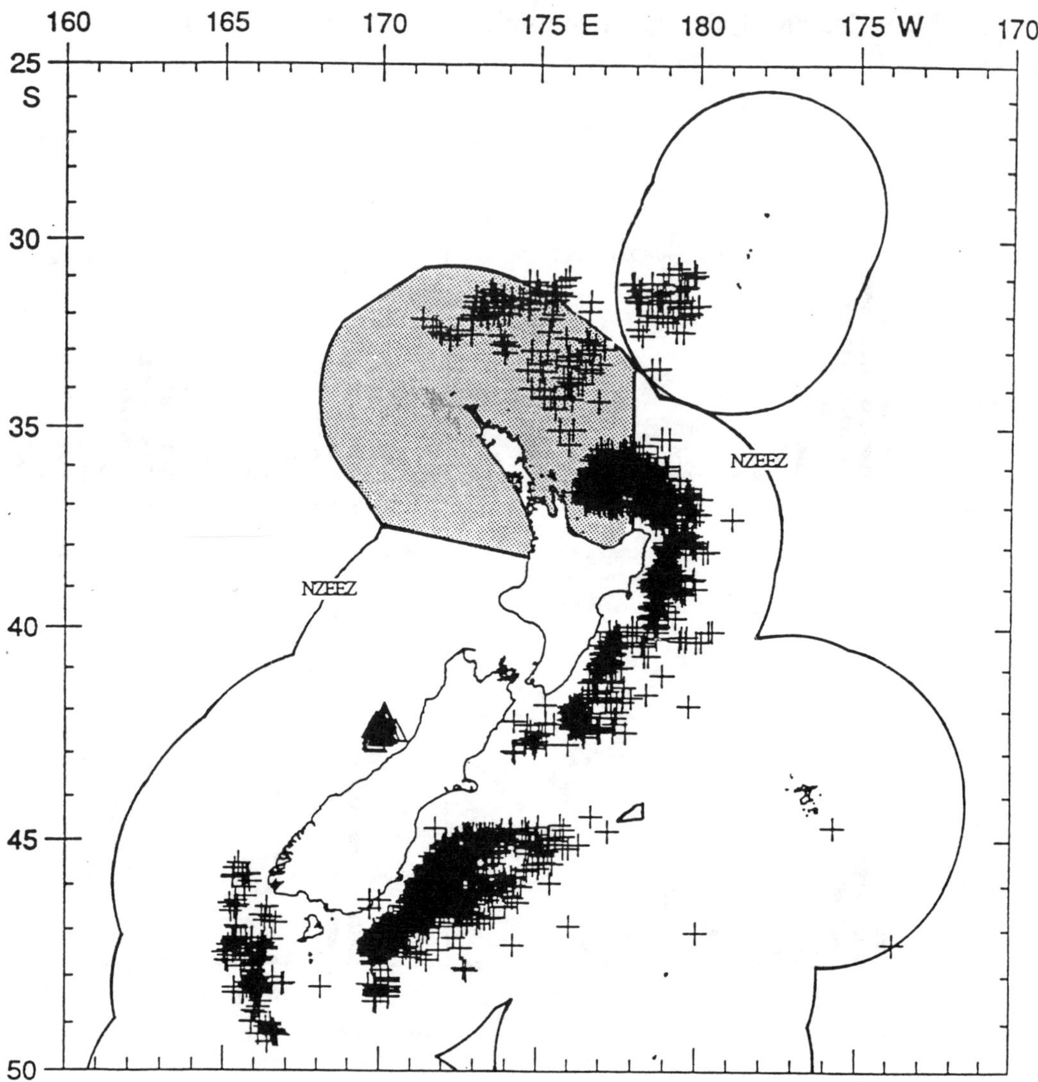

Figure 6.   Locations of catch of southern bluefin tuna by New Zealand domestic (triangles) and foreign (crosses) vessels. Positions represent locations of daily position reports by domestic vessels and start-of-set locations for longliners, for the period 1980-88. The general outline of the New Zealand EEZ is shown, as well as the Auckland Fisheries Management Area (shaded). (Source: Fisheries Research Division, Ministry of Fisheries and Agriculture, New Zealand.)

2.3.3 Japanese-style longlining in the New Zealand exclusive economic zone

The waters of what became the NZEEZ in 1978 have been fished since 1956 by the Japanese distant-water longline fleet and, perhaps since the late 1950s and mid 1960s, by longliners from Korea and Taiwan, until the establishment of the NZEEZ. Currently Japanese (and until 1989 Korean) longliners have been licensed to operate in the zone (Murray and Burgess, 1992). Two fishery components are involved. The 'southern', which operates between February and August, involves Japanese longliners which target SBT and to a much lesser extent bigeye and swordfish. It is located east and west of the South Island (February to June) and east of the North Island (May to August) (Figure 6). SBT catches are incidental in the other (northern) component, which is located north of the North Island between June and August, and which previously targeted albacore (Korean vessels) but which now target only bigeye and swordfish (Japanese vessels). It has been restricted in season since 1987 by an October-to-May closure in the Auckland Fisheries Management Area (Figure 6) this enclosure is designed to reduce competition with a domestic game fishery for striped marlin. All Japanese-style longlining is excluded from the 12-mile territorial sea at all times.

Over 7,000 mt of SBT were taken annually in the NZEEZ by Japanese longliners when the NZEEZ was first established, but now the catch is less than 1,000 mt, the fleet having decreased from around 90 vessels to less than 40, only 25-35 of which target SBT. The size composition of the catch (Figure 7) shows progressive absence of younger individuals during the 1980s but an increase in 1990 and 1991, in conjunction with an increase in activities west of the South Island and east of the North Island. The SBT catch reported from northern waters by Korean and Japanese longliners targeting other species, is less than five mt; no details of size composition, or indication of the completeness of these bycatch reports are available.

Since 1989, four to five New Zealand/Japan joint-venture charter vessels have operated in the NZEEZ. The four which operated in 1991 fished alongside the small, domestic longline vessels for part of their season, and among the Japanese licensed longliners for the remainder. Their catch in 1990 was 233 mt, and the size composition of the catches from the domestic fleet was virtually identical to that from the Japanese licensed longliners (Murray and Burgess, 1992).

Historic SBT catch rates in the 'northern' fishery area were high (Figure 8), peaking from September to November (Shingu, 1970). Part of this season is now curtailed by the October-to-May closure, but if the availability of smaller (<40 kg) SBT increases in the NZEEZ, then an increased by-catch of smaller SBT from the region might be anticipated.

2.3.4 Albacore surface fishery

A troll fishery for albacore has operated off the west coast of the South Island from January to March since the 1960s, but has no by-catch of SBT. During the mid to late 1980s a major drift gillnet fishery developed in the Tasman Sea between the AFZ and NZEEZ, and also in the region of the Sub-Tropical Convergence Zone in the western South Pacific. Reports from monitored transhipments and observers indicate that, despite substantial incidental catches of a range of pelagic species, the catch of SBT was extremely low and the fish were large. This is consistent with the reduced abundance of surface aggregations of young SBT off New South Wales and their poor representation in NZEEZ longline catches at that time. If the abundance of young SBT in the Tasman region increases, there may be potential for an increased by-catch of SBT from the troll fishery. There would also have been potential for increased drift gillnet by-catch, but since 1 July 1991 the South Pacific has been subject to a closure for drift gillnetting in accordance with UN Resolution 44/225.

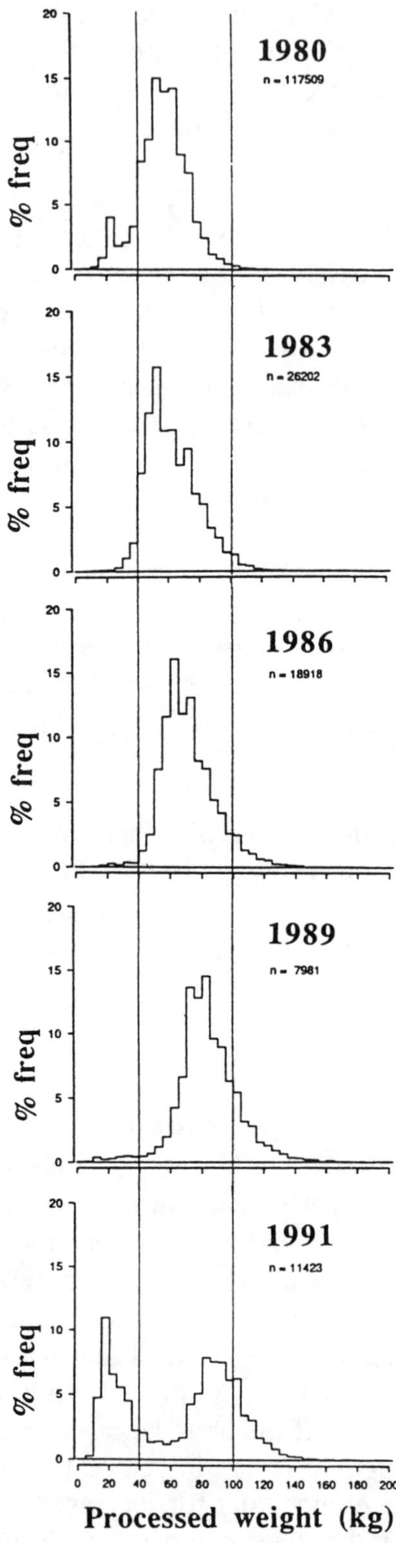

Figure 7.   Weight distribution (5 kg size classes) of southern bluefin tuna caught by foreign licensed longline vessels in the New Zealand EEZ, 1980-91. (Source: Murray and Burgess, 1992.)

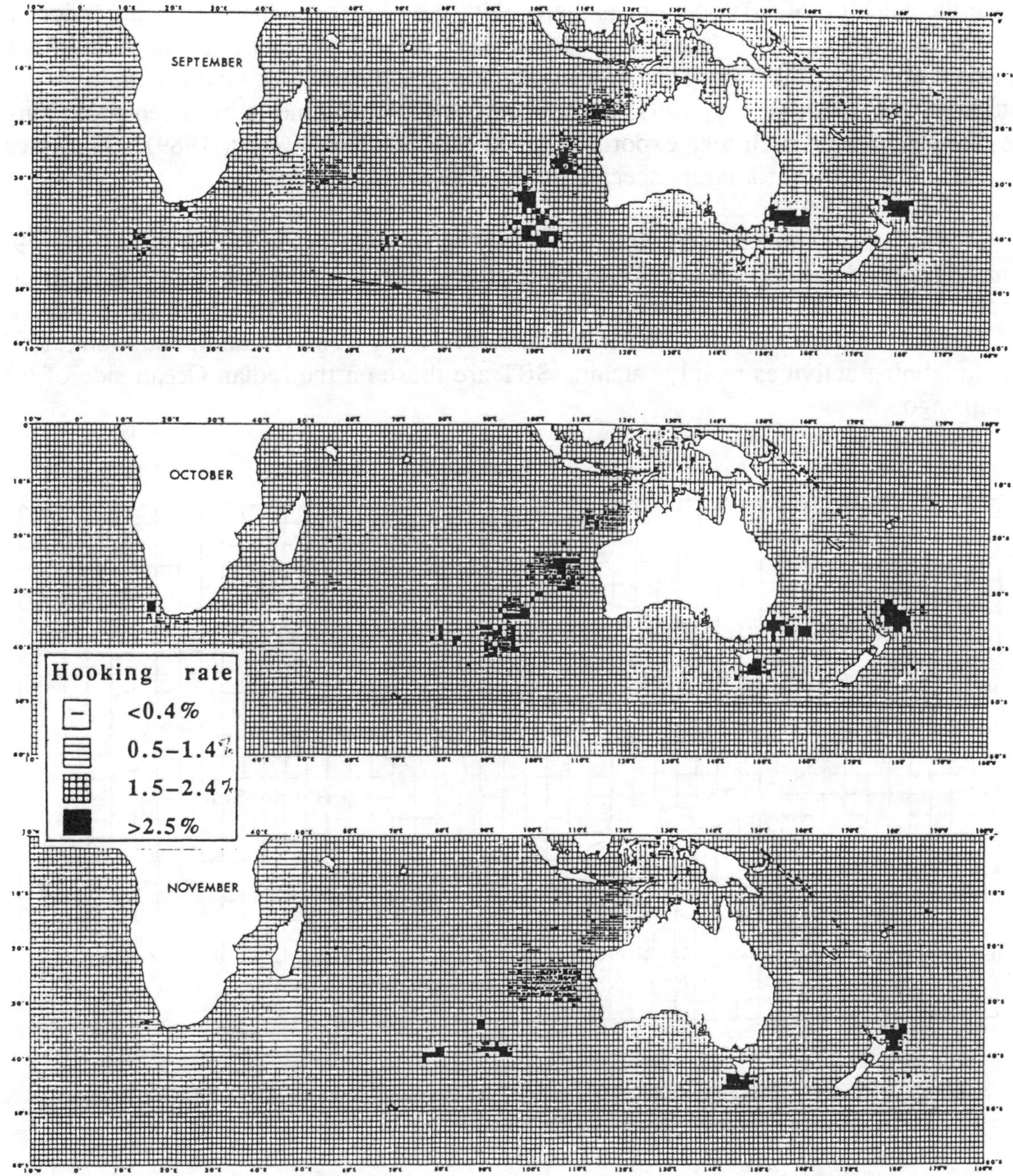

Figure 8.   September to November changes in extent of longline fishing grounds and in average hooking rate for 1965-67. (Source: Shingu, 1970.)

2.4   Indonesia

Indonesia lies immediately north of the SBT spawning grounds. The spawning season extends from September to March with a peak in January and February. Japanese distant-water longline vessels commenced operations on SBT there, but the poorer meat quality of post-spawning fish prompted dispersal of Japanese SBT target activities further south. While adult SBT can be anticipated seasonally on the spawning grounds, it is a region where the other pelagics targeted by longlines are also prevalent. Longlining, and hence the catching of SBT, continues there, with vessels from Japan, Taiwan, Indonesia and Korea involved. The Japanese vessels target bigeye tuna, and this and yellowfin are the main targets of the Taiwanese and Korean vessels (Institute of Oceanography, National Taiwan University, 1991; Park, *et al.*, 1991). Indonesia has been promoting the development of fresh tuna exports to Japan (Sodikin and Siregar, 1989). The price for fresh SBT makes it a target species for Indonesian longliners.

Indonesian domestic longlining commenced in 1972 in the Banda Sea (Figure 9). It has developed rapidly since 1985 with the strong demand for fresh tuna especially in Japan. The domestic fleet increased from 18 in 1979 to 151 in 1990, but there were also 341 foreign longliners operating in Indonesian waters by 1989 (Naamin and Gafa, 1991). The longlining activities mainly catching SBT are those on the Indian Ocean side of the archipelago.

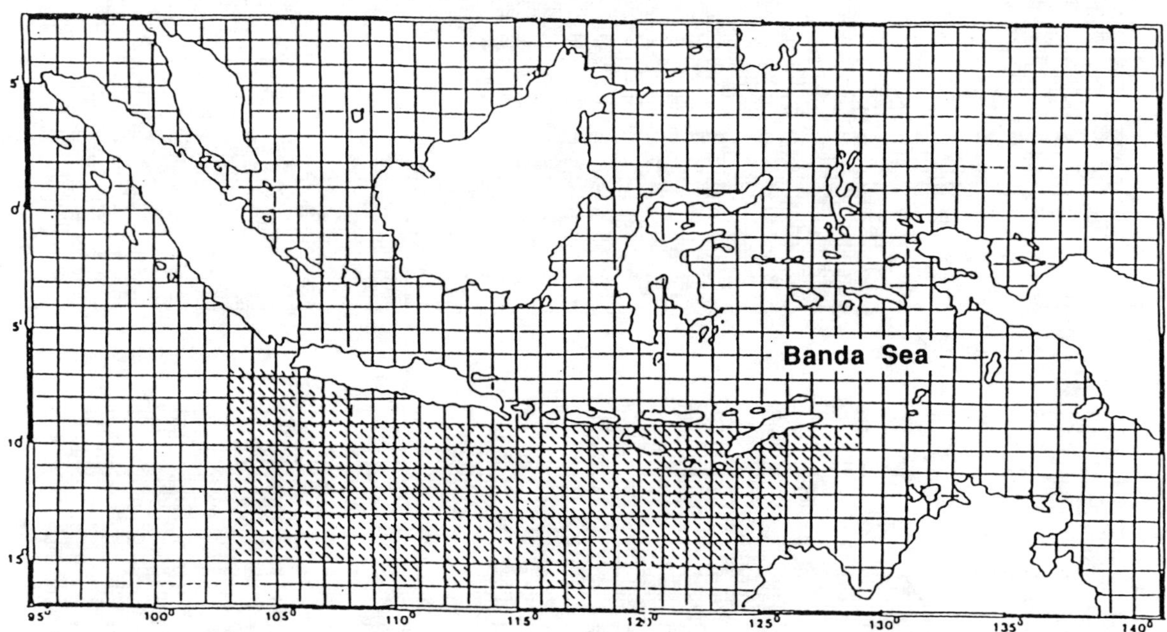

Figure 9.   Yearly fishing grounds of southern bluefin tuna in southern Java and Nusatenggara waters, 1978-88. (Source: Bahar and Naamin, 1989.)

The main SBT fishing ground for Indonesian vessels is located between 103°E and 128°E , and 7°S and 17°S (Figure 9), with activities moving closer to the 12-mile territorial waters in August-October. Catches averaged four mt annually between 1978 and 1988, with fish ranging in size from 61-164 kg (Bahar and Naamin, 1989).

Indonesian domestic vessels undertake trips of short duration (up to 14 days) and air freight their catch fresh to Japan. The domestic catch from the Indian Ocean region is augmented by the catches of Indonesia/Taiwan joint-venture longliners. These range from smaller longliners returning fresh fish after trips of short duration, to freezer vessels carrying out broader-ranging operations in the Indian Ocean. The extent of the SBT catch of these vessels, which may have a range of target species, is not known. It is a valuable component and may have been in the order of 100 mt in 1990 (Naamin, pers. commun.).

Japanese imports of fresh and frozen bluefin from Indonesia amounted to 150 mt and 15 mt (whole weight) respectively in 1991 (Japan Tariff Association, various). There is no indication of the proportion of northern and southern bluefin on these imports but it can be assumed reasonably that the imports from Indonesia are all SBT. Half of the Indonesian SBT catch destined for export fresh to Japan is classified as unsuitable and consumed locally (Davis, 1992). On this basis, the additional 75 mt would suggest a total Indonesian domestic and joint venture SBT catch of the order of 240 mt for 1991.

Waters within the Indonesian Archipelago support a component of the southwest Pacific tuna fishery. Yellowfin and skipjack tuna are the main target species. No SBT by-catch is reported. The waters are further north and more tropical than those where SBT are distributed.

2.5   Taiwan

Taiwanese involvement in the SBT fishery has probably occurred predominantly through distant-water longline operations in the South Atlantic and Indian Oceans, and to a lesser extent through bycatch in drift gillnet operations. This latter catch will have ceased if the United Nations resolution requiring global cessation of driftnetting by the end of 1992 has been effective.

Longliners from Taiwan commenced operations in the Indian Ocean in 1963, at first with albacore as the target species, then also yellowfin and bigeye (Hsu and Liu, 1990). In the mid- to late-1980s, 'regular' longline operations for albacore were concentrated between 10°S and 40°S latitudes. Deep longlining for bigeye was located mainly between 20°N and 15°S, but increasing activity occurred from 1986 to 1988 between 25°S and 40°S, especially in the south western Indian Ocean (Figure 10). In 1988 monthly mean weights of bluefin (assumed to be SBT) in Indian Ocean longline catches ranged between 31.58 kg (July) and 68.77 kg (September). A fleet of 187 vessels operated.

There were also drift gillnet vessels operating in the northern and southern Indian Ocean from 1983, targeting albacore tuna (Figure 11). Vessel numbers reached 150 by 1987-88. Hsu and Liu (op. cit.) reported the average weight of albacore taken by drift gillnet as nine kg, and of bigeye as 26 kg, but no details were given for SBT.

Japanese longliners achieved good SBT catch rates between 30°S and 40°S when their SBT operations spread to the southern Indian Ocean (Figure 3). The catches probably consisted of a larger proportion of young SBT in these latitudes than between 40°S and 45°S where operations are now concentrated. Broad availability of younger SBT in the southern Indian Ocean during the late 1970s and early 1980s was diminished because of the build-up in Australian catch of juveniles. The subsequent major reductions in Australian juvenile catch might be expected to have resulted in substantial escapement of younger fish to the area, given earlier examples of rapid westward movement of some of the SBT tagged on the Australian fishing grounds. Representation of small SBT increased in Japanese catches off south-west Australia and South Africa at the end of the

1980s. If this results from the broader distribution of young fish between 30°S and 45°S westwards in the southern Indian Ocean, then increased Taiwanese SBT catches might be expected.

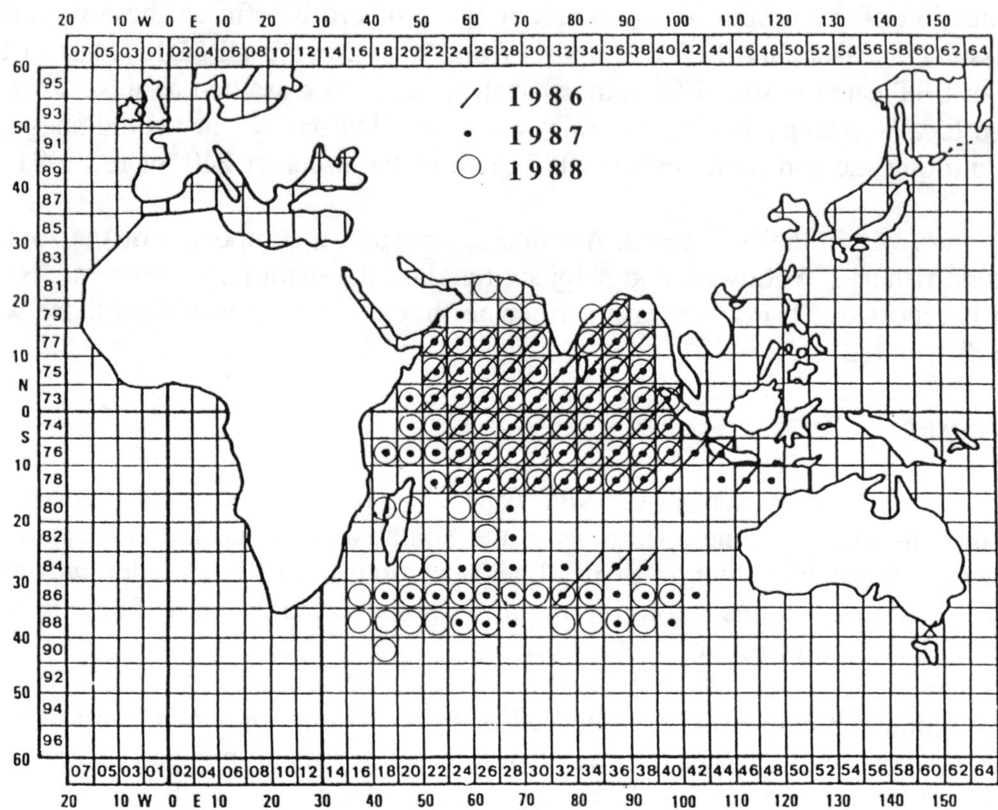

Figure 10.   Distribution of annual Taiwanese "deep" longlining effort in the Indian Ocean, 1986-88. (Source: Hsu and Liu, 1990.)

Catch statistics published by the Indo-Pacific Tuna Programme indicate that the total annual catches of SBT in the Indian Ocean ranged between 4 mt and 266 mt from 1970 to 1990 for vessels from Taiwan (Indo-Pacific Tuna Management and Development Programme, 1992). This seems inconsistent with Taiwan Statistical Yearbook figures for 'bluefin' (the statistics do not discriminate between northern and southern bluefin) which, if landing ports and ocean areas are considered, would suggest annual catches of SBT considerably in excess of 500 mt for the late 1980s (Taiwan Fisheries Bureau, various). Japanese imports of fresh and frozen bluefin from Taiwan amounted to 232 mt and 1,157 mt respectively in 1991 (Japan Tariff Association, various; processed weights). There is no indication of the breakdown between northern and southern bluefin but fresh imports from Taiwan entering Japan from April to July are presumably northern bluefin, whereas frozen bluefin arriving year-round would be from the distant-water longline fleet. The most evident feature of the pattern of bluefin imports by Japan is the major increase from Taiwan after 1988. An examination of price for frozen SBT from Taiwan shows a similar large increase at that time. This change coincides with the imposition of quotas on the Japanese SBT fishery. The first quota involving a reduction of Japanese catch was agreed in late 1988 to come into effect in 1989. The increased Taiwanese imports, at sashimi prices, commenced in that latter year. There are anecdotal reports of Taiwanese vessels moving onto traditional Japanese SBT fishing grounds after closed seasons forced Japanese-flag vessels to move. This is consistent with the sudden increase in supplies of sashimi-grade bluefin from Taiwan to Japan, most of which might reasonably be assumed to be SBT.

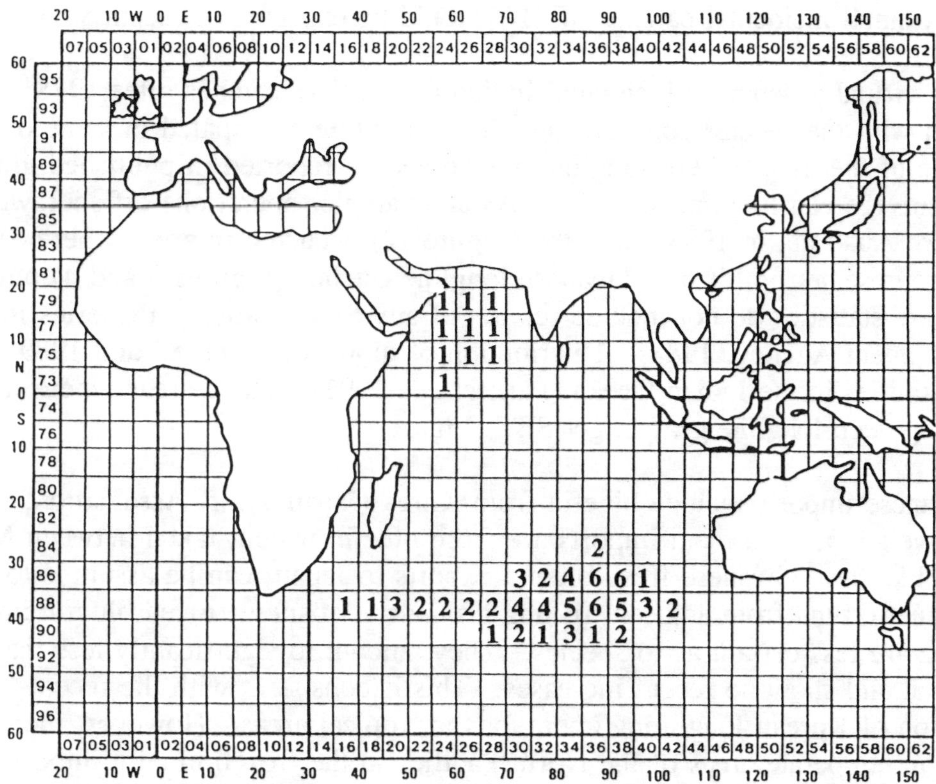

Figure 11.   Accumulated number of months of Taiwanese large-scale pelagic drift gillnet effort in 5○ squares in the Indian Ocean, 1987-88. (Source: Hsu and Liu, 1990.)

Actual SBT catch by Taiwan is difficult to determine because it appears that some Taiwanese bluefin catches by-pass the statistical collection. For example, the 1980 Taiwanese production statistics report a global bluefin catch of 179 mt, whereas correspondence in 1983 from the South African Sea Fisheries Research Institute to the Australian Trade Commissioner in South Africa advised that in 1980, 14,800 mt of tuna were transhipped in South Africa by vessels from Taiwan with SBT representing 7% of 11,982 mt of that tuna (*i.e.* 1,001 mt). The correspondence advised that the bulk of the catch transhipped would have been made in the mid-Atlantic and even off the coast of Brazil, and that "the SBT were caught by boats fishing more to the south than most of the fleet". Again, the 1991 Taiwanese production statistics report a total bluefin catch of 960 mt (Weng, pers. commun.), whereas 1991 bluefin imports to Japan from Taiwan and Singapore (the latter presumably also originating from Taiwanese vessels) were 1,389 mt and 67 mt processed weight respectively. As these would not include the Indian Ocean drift gillnet bluefin catch (which would not be suitable for export to Japan as sashimi) the overall 1991 bluefin production would have exceeded 1,456 mt.

2.6 Korea

The Korean distant-water longline fleet has taken SBT as an incidental component of catches. Published statistics for the Indian Ocean report that Korean SBT catches ranged between 7 mt and 500 mt per year from 1971 to 1978, and then ceased (Indo-Pacific Tuna Management and Development Programme, 1992). As, progressively, Korean longliners have concentrated activities in tropical areas for bigeye tuna during the 1980s, only small incidental catches of SBT would be expected in recent years.

The virtual absence of incidental Indian Ocean SBT catches after 1978 seems inconsistent with the species composition which might be anticipated in some of the areas and at some of the times where longline operations are reported in published global five degree-square Korean longline statistics. As an example, operations off southwestern and northwestern Australia in 1983 and 1984 (Figure 12) occurred in areas where SBT have been well represented in historic Japanese longline catches (Figures 3 and 4) but the 5 degree-square statistics do not include reports of any SBT (National Fisheries Research and Development Agency, 1988). Distribution of more recent (1988 and 1989) Korean longline activities in the Indian Ocean (Park *et al.*, 1991) include areas where Indonesian vessels using deep longline gear target SBT (Figure 9).

Japanese imports include bluefin from Korea (Japan Tariff Association, various; processed weights). The main imports are fresh bluefin which, taken between May and November like the Taiwanese fresh bluefin exports to Japan, can be assumed to be northern bluefin tuna from adjacent fishing grounds. Japanese frozen bluefin imports from Korea are less certain as to species. They amount to significantly less than those from Taiwan and show no recent increases. This is consistent with the recent concentration of Korean longlining operations in tropical areas. However, there are still operations in temperate areas of the North Pacific, so the frozen bluefin imports to Japan, given their high price, probably represent northern bluefin from those activities.

2.7 South Africa

South African catches of SBT were uncommon until the 1950s (de Jager *et al.*, 1963), when handlining, and then in the early 1960s experimental Japanese-style longlining, by research vessels indicated promising catches of SBT, albacore, yellowfin, and bigeye tunas off the west coast. Southern bluefin tuna was described incorrectly as *Thunnus thynnus orientalis*, which does not occur in the South Atlantic or southwestern Indian Oceans. Commercial longlining developed rapidly within 240 km of the coast in the area between 32°S and 35°30'S (Negpen, 1970), but then rapidly declined after 1963 because returns were uneconomic. This was partly because catch rates had declined. In the interim, the SBT catch, taken predominantly during the southern winter, reached around 400 mt in 1963, with fish ranging in size from 90-170 cm. Japanese longliners fished on the same grounds (and over a wider area) during the same period.

Recent catches of SBT by domestic South African fisheries appear to be insignificant (Molteno, 1986; Vere Shannon *et al.*, 1989). Correspondence in 1982 to CSIRO from the South African Department of Agriculture and Fisheries indicated that about 20 mt of SBT had been taken by pole-and-line in the vicinity of the Vema and Tripp seamounts (30°S,8°E and 29°S,14°E respectively). Molteno reports: "The southern bluefin tuna is caught in the southern Atlantic entirely by longline and surface schools have not been observed in recent years". Vere Shannon *et al.* (1989) indicate "Southern bluefin tuna (nowadays 25 kg commercially but up to 80 kg in the early 1960s) are more common off southern Africa than northern bluefin and were extremely abundant off the Cape of Good Hope during the early 1960s, mainly during the winter but also in early spring." A South African catch of one mt of 'bluefin' in 1989 (in a total South African tuna catch of 2,600 mt) and zero bluefin catches in 1988 and 1987 are reported in South African statistics (Anon., 1990).

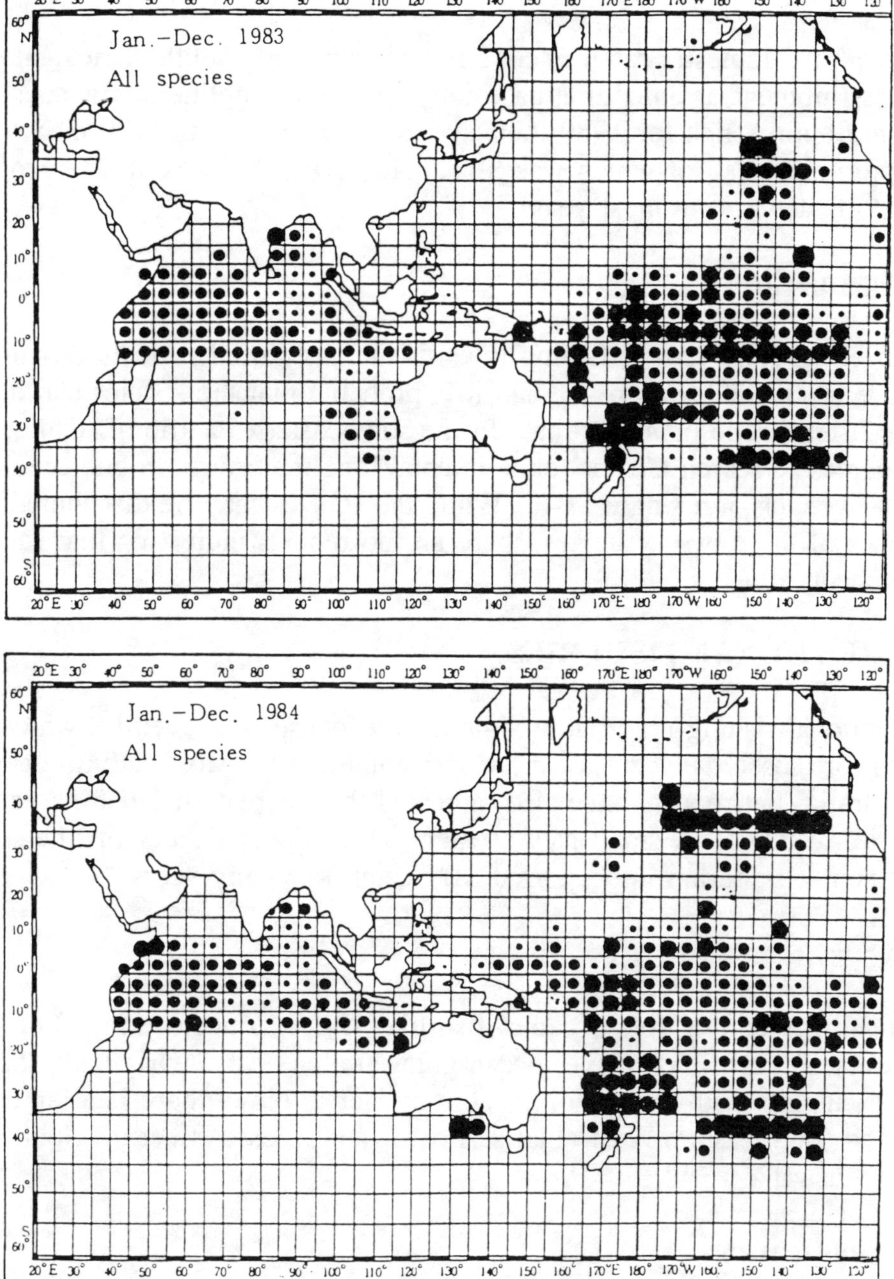

Figure 12.   Distribution of Korean longline effort in the Indian and western Pacific Oceans in 1983 and 1984 as reflected by catches of all species. (Source: National Fisheries Research and Development Agency, 1988.)

There are indications of previous high SBT catch rates by the Japanese longline fishery from areas now encompassed by the South African 200-mile fishing zone, especially closely adjacent to south western South Africa (Figure 3). The size composition reported included fish from 90 cm to 150 cm (Shingu 1970), with a large proportion from 90-120 cm (3-7 year olds). If abundance of younger age classes of SBT increases in regions away from southern Australia subsequent to the reduction in surface fishing activities there, it is possible that incidental catches of SBT in regions like southwestern South Africa might increase. However, Japanese longlining activity tends to be concentrated in more southern areas and beyond the South African 200-mile fishing zone.

Although no domestic catch of SBT is now reported, South Africa, like Singapore, acts as a trans-shipment or unloading base for Taiwanese longliners. Japanese import statistics from South Africa for 1989 indicate a substantial quantity (99 mt whole weight) of bluefin, presumably all of which was SBT. However, this was an isolated instance, and SBT imports are rare in most years.

2.8   South America

Japanese longline vessels fishing towards the eastern edge of the South American continent have taken SBT (Figure 4), and it is probable that the species can occur within the 200-mile fishing zone of Argentina. In the southeastern Pacific off Chile, a 'Japan Marine Resources Research Center' experimental longline cruise around the beginning of 1980s took a by-catch of a single SBT. While the species may be distributed within the 200-mile zones of these two countries, there are no reports indicating that catches of any significance occur.

3.   RECREATIONAL FISHERIES

Recreational fishing for SBT predominantly focuses on juveniles which are easily accessible on, or closely adjacent to, the Continental Shelf. Most activity occurs around southern Australia, but reports have also indicated that recreational fishing has taken place in South Africa at times. A small New Zealand recreational fishery off the southwest coast of the South Island has not reported SBT catch for many years.

3.1   Australia

Recreational angling for 'oceanic' fish like SBT was not common until the increased availability of inexpensive, ocean-going trailer boats. Originally the sport was the domain of anglers with ocean-going cabin cruisers. Now, there is a large fleet along the eastern and south-eastern Australian seaboard, and a lesser fleet off South Australia and Western Australia.

3.1.1 New South Wales

From as early as 1936 (Roughley, 1951) the southern coast of New South Wales has been an important recreational fishing area for SBT. Troll fishing along the Continental Shelf routinely provided catches of 7-15 kg SBT from July/August to January, in the same general area as commercial operations but inshore of them. Catch is likely to have been in the order of 50-100 mt in the 1970s, but gamefishing clubs report a virtual cessation of catches after 1980. The absence persisted until July 1991, when small, scattered catches were taken again by recreational anglers in some of the traditional areas. Recreational anglers belonging to fishing clubs have been concerned about the state of SBT stocks to the extent that the species no longer scores competition points on landing. Instead, club anglers are encouraged to tag and release the species as part of the New South Wales recreational, gamefish-tagging programme. The greater proportion of recreational anglers do not belong to clubs, but their operations are subject to legislation (not specifically directed at SBT) which imposes a daily bag limit of 2 fish per angler. Trailer vessels have been increasing in popularity along the southeastern seaboard of Australia, and the number capable of operating in the SBT area in quiet weather is probably in the order of 1,000 now. There is thus a potential for high recreational fishing effort if SBT concentration builds up off New South Wales.

3.1.2 Western Australia

The main centre for recreational tuna fishing is the area adjacent to Busselton, south of Perth. Smaller-scale activities operate off the western south coast (from Albany) and in the Perth region. Up to 100 small vessels might participate seasonally (summer-holiday months), especially on weekends, with four SBT, 1-2 years old, representing a reasonable catch. On this basis, the annual recreational catch is estimated to be less than 50 mt for the State.

3.1.3 South Australia

The main centre of recreational tuna fishing in South Australia is near Port Lincoln where commercial pole-and-line fishing commenced. However, the number of vessels is small because the waters are exposed, and weather conditions restrict access to the generally small vessels involved. Recreational tuna fishing activities were also carried out on a small scale near Cape Jervis, south of Adelaide, up to the early 1980s, but progressive contraction of fish aggregations westwards led to their cessation. A similar decrease in availability occurred in the southeast of the state. In that area, the number of vessels was again small, except each Easter when a fishing tournament for SBT took place, involving around 20 or 30 trailer vessels.

In some years during the late 1970s and early 1980s, the commercial purse-seine and pole-and-line fisheries extended to this southeast region of South Australia later in the season. However, these commercial operations also contracted westwards during the late 1980s. Following the introduction of restrictive quotas in the Australian surface fishery, the westwards contraction was reversed to the extent that, in 1990 and 1991 (but not in 1992), recreational fishing reports suggested a reappearance of fish in the southeast of the state.

The annual recreational catch for the State is unlikely to have exceeded 20 mt annually, and the fish have usually been 5-20 kg (2-4 year olds).

3.1.4 Tasmania

The SBT has been the target of seasonal (January to August) recreational tuna fishing activities, mainly adjacent to Eaglehawk Neck in eastern Tasmania. One trophy for the heaviest tuna caught each year has been presented since 1958. A gamefishing charter-boat operator at Eaglehawk Neck has maintained records of his daily SBT catches since 1965. Concurrent with the rapid increase in catches of 2-3 year old SBT by the Australian commercial surface fishery at the end of the 1970s, and the collapse of the New South Wales commercial fishery, his daily catch rate (3.5-14.5 fish per day) declined until 1984 when no SBT were caught. Subsequently, when quotas reduced the commercial catches, his daily catch rates increased, but remain lower (0.5-4.8 fish per day) than the average catch rate prior to the decline. The number of vessels probably does not exceed 50, with an annual catch probably less than 25 mt.

3.2   South Africa

Smith and Heemstra (1986) comment that in South African waters SBT is "known only from long-line catches off the Cape region during winter", whereas for northern bluefin tuna (*Thunnus thynnus*) "schools congregate during summer in the Agulhas Bank area and frequently enter False Bay". Negpen (1970) makes reference to "big Atlantic bluefin tuna *Thunnus thynnus thynnus* caught in False Bay by sport fishermen during the summer months". However, Molteno (1986) reports "In the late 1950s bluefin tuna schools entered False Bay and anglers using spinners caught a number of these fish by casting off the rocks at Cape Point. Most of these tuna were young southern bluefin." Anglers may have problems discriminating between the two bluefins, but it would seem likely that winter catches would involve SBT. In a personal communication, Malcolm Bertani of the Bureau of Resource Sciences reports that as a recreational angler he trolled for and caught SBT in winter within sight of land from Cape Town during the 1960s. However, their abundance declined to the extent that by the end of the 1970s they were no longer taken there.

If 'young' southern bluefin are involved in recreational catches off South Africa, their availability during the 1980s would likely have been poor if recruitment to the area was subject to the same influences that led to the experiences off south-eastern Australia. If the reappearance in recreational troll catches off Australia is indicative of broader escapement of juveniles, then South African recreational catches might also be anticipated to increase.

3.3   New Zealand

Recreational fisheries for striped marlin and yellowfin tuna operate in northern New Zealand waters in summer months. These fisheries catch other tunas (*e.g.*, albacore, skipjack, and bigeye) but do not catch SBT. A small, recreational fishery for SBT operated off the south west coast of the South Island in the early 1980s but caught few fish. No recreational catches have been reported in that area for many years (Murray, pers. commun.).

4. REFERENCES CITED

Anonymous. 1990. Catches by fishery 1989. *South African Shipping News and Fishing Industry Review*, August 1990, p. 41.

Bahar, S. and N. Naamin. 1989. Fishing ground and distribution of southern bluefin tuna (*Thunnus maccoyii*) in southern Java and Nusatenggara waters. *In* Report of the Third Southeast Asian Tuna Conference, Bali, Indonesia, 22-24 August, 1989., Indo-Pacific Tuna Management and Development Programe, Colombo, Sri Lanka, pp. 225-33.

Caton, A.E. (editor). 1991. Review of aspects of southern bluefin tuna biology, population and fisheries. *In* World Meeting on Stock Assessment of Bluefin Tunas: Strengths and Weaknesses, edited by R.B. Deriso and W.H. Bayliff. *Spec.Rep.I-ATTC*, 7:181-357.

Davis, T. 1992. Reject and export southern bluefin tuna caught by the Indonesian/Taiwanese longline fishery. Paper presented at the Eleventh Australia, Japan and New Zealand Trilateral Scientific Meeting on Southern Bluefin Tuna, Shimizu, Japan, 1992; Information Paper, 4:2 p.

de Jager, B.van D., C.S.de V. Negpen, and R.J. van Wyk. 1970. A preliminary report on South African west coast tuna. *Invest.Rep.Div.Sea Fish.*, 47:39 p.

Fisheries Agency of Japan. 1972. Annual report of effort and catch statistics by area on Japanese tuna longline fishery 1970. Shimizu, Japan: Far Seas Fisheries Research Laboratory, Fisheries Agency of Japan, 1972.

Hsu C-C., and H-C. Liu. 1990. Taiwanese longline and gillnet fisheries in the Indian Ocean. Paper presented at the Expert Consultation on Stock Assessment of Tunas in the Indian Ocean, Bangkok, Thailand, July 1990. Background document, FAO/IPTP/TWS/90/54:30 p.

Indo-Pacific Tuna Management and Development Programme. 1992. Indian Ocean and Southeast Asian tuna fisheries data summary for 1990. *Data Summ.Indo-Pac.Tuna Dev.Mgt.Programme*, Colombo, Sri Lanka, 12:94 p.

Institute of Oceanography, National Taiwan University. 1991. National report of Taiwan. *In* Collective Volume of Working Documents presented at the Workshop on Stock Assessment of Yellowfin Tuna in the Indian Ocean, 7-12 October, 1991. *Coll.Vol.Indo-Pac.Tuna Dev.Mgt.Programme*, Colombo, Sri Lanka, 6:149-53.

McKenzie, M.K. 1962. A review of present knowledge related to a possible tuna fishery in New Zealand. *Tech.Rep.New Zealand Mar.Dep.Fish.*, 4:48 p.

Molteno, C.J. 1986. The southern African tunas and billfishes: A handbook of taxonomy, economic importance, catching and handling procedures and practices. South African Fishing Industry Research Institute, University of Cape Town, Cape Town.

Murray, T., and D. Burgess. 1992. Southern bluefin tuna fisheries indicators in the New Zealand Exclusive Economic Zone, 1980-1992. Paper presented at the Eleventh Australia, Japan and New Zealand Trilateral Scientific Meeting on Southern Bluefin Tuna, Shimizu, Japan, 1992. Ministry of Agriculture and Fisheries, Wellington, New Zealand, Working Paper:14 p.

Naamin, N. and B. Gafa. 1991. Present status of yellowfin tuna fishery in Indonesia. *In* Collective Volume of Working Documents presented at the Workshop on Stock Assessment of Yellowfin Tuna in the Indian Ocean, 7-12 October, 1991. *Coll.Vol.Indo-Pac.Tuna Dev.Mgt.Programme*, Colombo, Sri Lanka, 6:154-62.

National Fisheries Research and Development Agency. 1988. Annual report of catch and effort statistics and fishing grounds for the Korean tuna longline fishery 1983-1985. Pusan, Republic of Korea: National Fisheries Research and Development Agency, 1988.

Negpen, C.S.de V. 1970. Exploratory fishing for tuna off the South African west coast. *Invest.Rep.Div.Sea Fish.*, 87:26 p.

Nishida, T., and Y. Ishizuka. 1992. Japanese southern bluefin tuna (*Thunnus maccoyii*) fishery in recent years (1985-90). Paper presented at the Eleventh Australia, Japan and New Zealand Trilateral Scientific Meeting on Southern Bluefin Tuna, Shimizu, Japan, 1992; Working Paper 10:22 p.

Park, Y.C., W.S. Yang and T.I. Kim. 1991. Status report of the Korean tuna longline fishery for yellowfin tuna in the Indian Ocean. *In* Collective Volume of Working Documents presented at the Workshop on Stock Assessment of Yellowfin Tuna in the Indian Ocean, 7-12 October, 1991. *Coll.Vol.Indo-Pac.Tuna Dev.Mgt.Programme*, Colombo, Sri Lanka, 6:138-42.

Robins, J.P. 1963. Synopsis of biological data on bluefin tuna *Thunnus thynnus maccoyii* (Castelnau) 1872. *FAO Fish.Rep.*, 2(6):562-87.

Roughley, T.C. 1951. Fish and fisheries of Australia. Angus and Robertson, Melbourne, 343 p.

Shingu, C. 1970. Studies relevant to distribution and migration of southern bluefin tuna. *Bull.Far Seas Fish.Res.Lab.*, 3:57-113.

Shingu, C. 1978. Ecology and stock of southern bluefin tuna. Japan Association of Fishery Resources Protection. *Fisheries Study*, 31:81 p. (In Japanese; English translation in CSIRO Division of Fisheries and Oceanography, *Report* 131:79 p., 1981.)

Smith, M.M., and P.C. Heemstra. (editors). 1986. Smiths' Sea Fishes. Macmillan South Africa (Publishers), Johannesburg, pp. 837-8.

Sodikin, D., and S. Siregar. 1989. Some notes on industrial scale tuna fisheries in Indonesia. *In* Report of the Third Southeast Asian Tuna Conference, Bali, Indonesia, 22-24 August, 1989. *Indo-Pac.Tuna Dev.Mgt.Programme*, Colombo, Sri Lanka. pp. 234-238.

Taiwan Fisheries Bureau. (various years). Fisheries Yearbook Taiwan. Taiwan: Taiwan Fisheries Bureau, Department of Agriculture and Forestry.

Vere Shannon, L., R.P. Van der Elst, and R.J.M. Crawford. 1989. Tunas, bonitos, Spanish mackerels and billfish. *In* Oceans of life off southern Africa, edited by A.I.L. Payne and R.J.M. Crawford, pp. 188-97, Cape Town: Vlaeberg Publishers.

A REVIEW OF THE BIOLOGY AND FISHERIES FOR LONGTAIL TUNA (*THUNNUS TONGGOL*) IN THE INDO-PACIFIC REGION

Mitsuo Yesaki
Indo-Pacific Tuna Development and Management Programme
Colombo, Sri Lanka

1. INTRODUCTION

This review is abridged from the species synopsis compiled by Yesaki (1987). Information published subsequent to 1987 has been added.

2. CLASSIFICATION

Kingdom Animalia
  Phylum Chordata
    Subphylum Vertebrata
      Superclass Gnathostomata
        Class Osteichthyes
          Division Teleostei
            Cohort Acanthopterygii
              Order Perciformes
                Suborder Scomberoidei
                  Family Scombridae
                    Subfamily Scombrinae
                      Tribe Thunnini
                        Genus *Thunnus*
                          Species *tonggol*

Thunnus tonggol is included in the subgenus *Neothunnus* with two other tropical species of this genus, *T. albacares* and *T. atlanticus* (Collette, 1978).

3. EARLY LIFE HISTORY

There is no information on the fertilized eggs on longtail tuna. Rao (1964) collected ripe residual ova from a spent ovary of an 81-cm fish captured in September in India. These ova were translucent, with average diameter of 1.09 mm and an oil globule varying in diameter from 0.31 to 0.33 mm in fresh condition.

Chayakul and Chamchang (1988) described the morphological development of longtail tuna larvae captured in the Gulf of Thailand. Larvae of this species have been collected in the Gulf of Thailand in January, February, April, May, and June (Chamchang and Chayakul, 1988) and off the west coast of Thailand in February, March, and April (Boonragsa, 1987). Larvae have also been reported from Malacca Straits (Wilson, 1981b) and South China Sea (Chen and Wei, 1981). There are relatively few reported collections of longtail tuna larvae; the paucity of larval and postlarval specimens of this species may be due either to the fact that its distribution is restricted to the neritic regime or to misidentification as a more common, better-known species.

Preadult longtail tuna, as small as 20 cm, are captured by luring purse seines off the west coast of Thailand. Preadults were also captured during exploratory fishing cruises by pole and line and troll line off this coast. The smallest fish captured by these gears were 24 and 28 cm, respectively. Preadults with mode at 30 cm were captured during exploratory fishing cruises in March 1980 and February 1981 (Yesaki, 1982). Many schools of preadults were sighted during these months, but were fished only for samples because of the small size of the fish (Lee, 1982).

Twenty-two-cm preadults are recruited to the purse-seine fishery in the Gulf of Thailand (IPTP, 1985) and slightly larger fish (26 cm) enter the troll fishery off the east coast of Peninsular Malaysia (IPTP, 1986). Preadults are captured in drift gillnets at Cochin in February (Silas, *et al.*, 1986a) and at Mangalore in October (Muthiah, 1986).

4.   AGE AND GROWTH

Several growth studies based on length data have been completed for longtail tuna. These include studies where modal lengths have been assigned ages (Serventy, 1956; Chiampreecha, 1978; Klinmuang, 1978; Yesaki, 1982), modal progressions were followed by the "pen and paper" method (Wilson, 1981b; Supongpan an Saikliang, 1987; Yesaki, 1989) and modal progressions were determined by the ELEFAN (Electronic Length-Frequency Analysis) software package (Silas *et al.*, 1986b; Prabhakar and Dudley, 1989). Only one study based on hard parts has been completed to date for longtail tuna. Wilson (1981b) described the growth of this species by counting increments on otoliths of fish captured in Papua New Guinea. He did not validate increments as daily events in this study. Growth parameters and/or lengths-at-age derived by the various authors are summarized in Table 1 and Figure 1.

The growth curves derived by Wilson (1981b) from otolith increment counts and length frequencies and by Yesaki (1989) from length frequencies grouped by 10-day intervals are similar up to age 3. These growth curves differ from the remaining curves in having larger lengths-at-age. Derivation of similar results by two different methods gives credence to the fast-growth curve as best describing the growth of longtail tuna. The results of tagging experiments lend further support for the fast-growth curve as the most appropriate. Release and recapture lengths of three longtail tuna recovered after relatively long intervals at liberty were superimposed on the growth curves derived by Wilson (1981b) and Yesaki (1989) (Figure 2). Increases in length of tagged fish during the interval at liberty correspond closely with the growth predicted by these curves.

5.   MATURATION, SPAWNING, AND SEX RATIO

5.1   Maturation

Yesaki (1982) determined sexual maturity of longtail tuna by visual examination of ovaries and assigned maturity stages according to a 5-point scale. The characters used to define maturity stages were size of gonad, colour, degree of softness, and turgidity. Of 796 females examined, the smallest mature female (stage-IV) was 43 cm. Cheunpan

Table 1. Summary of age and growth studies on longtail tuna by length-frequency distributions and hard parts.

Author	Area	Method	Growth parameters			Length-at-age[1]				
			K	L_{oo}(cm)	to	1	2	3	4	5

(a) Length-frequency distributions

Author	Area	Method	K	L_{oo}(cm)	to	1	2	3	4	5
Serventy, 1956	Australia	Modal lengths	–	–	–	38	51	62	–	–
Chiampreecha, 1978	Gulf of Thailand	Modal lengths	–	–	–	27	35	45	–	–
	east coast Malaysia	Modal lengths	–	–	–	30	35	–	–	–
Klinmuang, 1978	Gulf of Thailand	Modal lengths	–	–	–	31	49	–	–	–
Wilson, 1981b	Gulf of Papua	Modal progres.	0.41	122.9	–.032	42	69	87	99	107
Yesaki, 1982	west coast Thailand	Modal lengths	–	–	–	30	47	–	–	–
Silas et al, 1985	India	ELEFAN	0.49	93.0	–.240	42	62	74	81	86
Supongpan and Saikliang, 1987	Gulf of Thailand	Modal progres.	1.44	58.2	–.027	45	55	57	58	–
Prabhakar and Dudley, 1989	Oman	ELEFAN	0.228	133.6	–	30	51	68	–	–
Yesaki, 1989	Gulf of Thailand	Modal progres.	0.55	108.0	–	46	72	87	96	101

(b) Hard parts

Author	Area	Method	K	L_{oo}(cm)	to	1	2	3	4	5
Wilson, 1981b	Gulf of Papua	otoliths	0.395	131.8	–.035	44	73	92	105	114

1/ - relative age

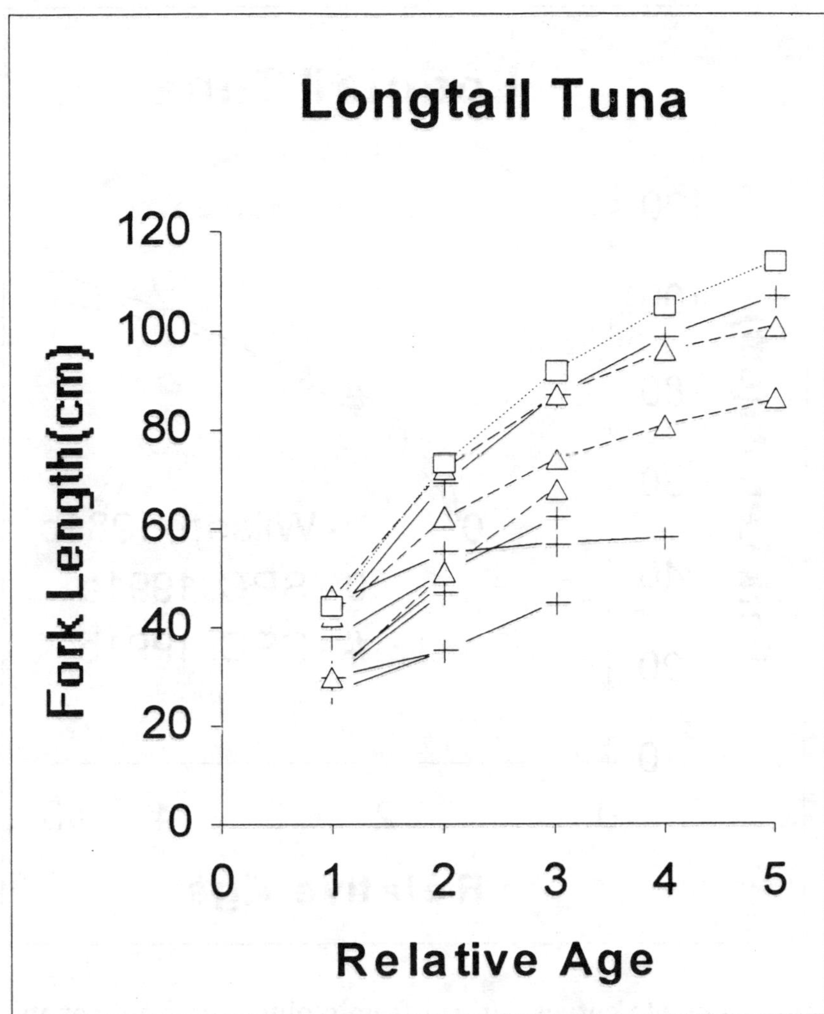

Figure 1.   Estimates of growth and relative age (yr) of longtail tuna
from length-frequency analyses (crosses: manual method;
triangles: computer programmes; squares: hard parts) (from
Yesaki, 1989).

(1984) examined 939 females for ovary weight and used gonad index to determine sexual maturity with size. Fifty percent of the females were sexually mature at 396 mm.

Observations of fish captured in Australia prompted Serventy (1956) to conclude that fish in the size group with mode at 38 cm were immature and in the group with mode at 51 cm were maturing, as these had enlarged, developing gonads. Wilson (1981b) found maturing ovaries first appeared in fish of about 60 cm in Papua New Guinea. These studies suggest that longtail tuna from Southeast Asia mature at a smaller size than do fish from Australia-Papua New Guinea.

5.2   Spawning

Klinmuang (1978) estimated numbers of eggs in ovary pairs of four fish. Fecundity of longtail tuna ranging in size from 43.8 to 49.1 cm varied from 1.2 to 1.9

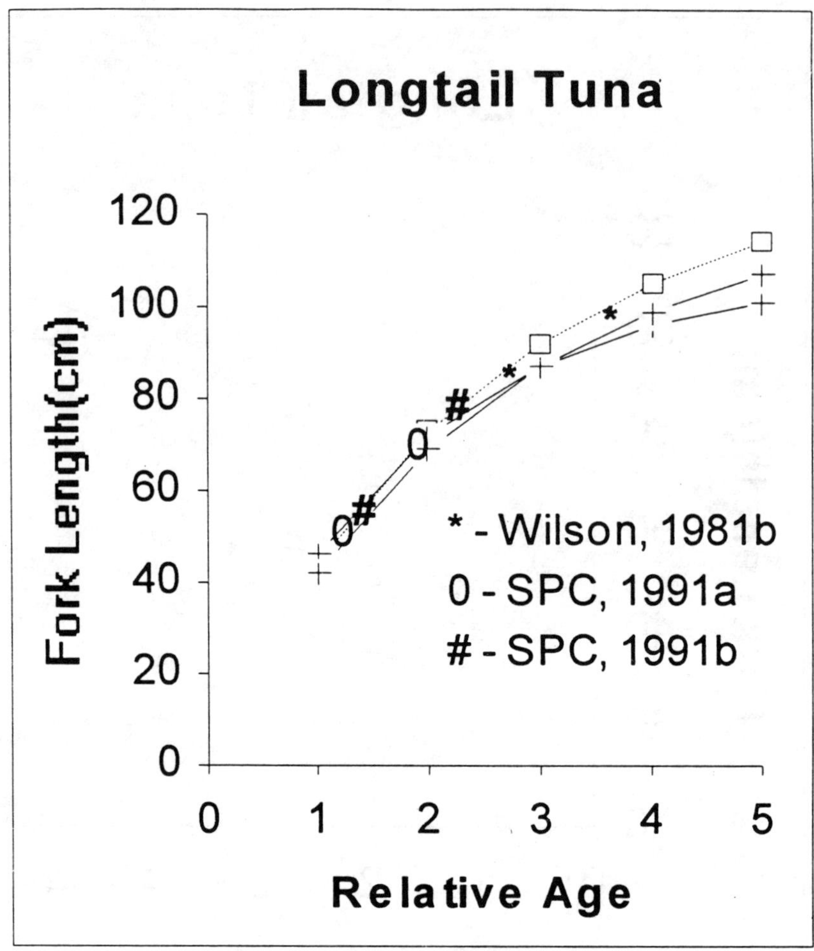

Figure 2.　Growth curves derived from otolith increment counts (squares) and length-frequency distributions (crosses) with growth increments of long-term tag recoveries of longtail tuna.

million eggs and averaged 1.4 million eggs. Wilson (1981b) determined fecundity by counting oocytes in the most advanced mode of six maturing females (ranging in size from 75.5 to 98.0 cm) captured in Papua New Guinea. Oocyte counts ranged from 0.8 to 1.9 million per female.

Longtail tuna, like other species of tuna, probably spawn more than once a year.

Mature (stage-IV) and spent (stage-V) female longtail tuna were captured in the outer-neritic and restricted inner-neritic regime north of Phuket Island during the northeast monsoon (Yesaki, 1982). Mature females were also captured in the outer-neritic regime during the southwest monsoon. It was postulated that longtail tuna spawn principally in the outer-neritic regime along the entire west coast of Thailand. Stage of maturity indices collected during a 2-year period off the west coast of Thailand were grouped by month to show development of maturity through the year and to define spawning seasons. The proportion of maturing females (stage-III) was high in January-March, decreased to lowest levels in May-August and thereafter increased in highest level in December. Mature females were found in January-April and again in August-September. Spent females were observed in January and March. There appear to be two distinct spawning seasons for longtail tuna; a major spawning during the northeast monsoon from January through April, with probable peak in March, and a minor spawning during the southwest monsoon in August-September off the west coast of Thailand.

Cheunpan (1984) also suspected two spawning seasons for longtail tuna in the Gulf of Thailand from peaks in gonad indices in March-May and July-December.

Wilson (1981a) surmised from gonadal development of fish captured in Papua New Guinea and the absence of adult fish during October-April that longtail tuna spawned during the austral summer. Serventy (1956) observed fish with definitive ova and others with milt flowing in spring (September, October) and examined fish with spent gonads in autumn (April) off New South Wales. He assumed longtail tuna spawned during the southern hemisphere summer in Australia.

5.3   Sex Ratio

There are few reports in the literature on sex ratios of longtail tuna. Klinmuang (1978) reports a 1:1 ratio for 141 fish from the South China Sea . Yesaki (1982) found no significant differences in sex ratio for fish captured by pole and line (1:1 and troll line (0.97:1) off the west coast of Thailand. Numbers of males and females by 10-cm intervals and for total numbers were not significantly different for longtail tuna captured in Papua New Guinea (Wilson, 1981b).

6.   STOCK STRUCTURE, DISTRIBUTION, AND MIGRATION

6.1   Subpopulations and Stocks

Several authors have noted differences in characters of longtail tuna from various areas throughout its range. Serventy (1956) found body depth and distance from tip of snout to origin of first dorsal fin of fish from western Australia to differ significantly from those measurements for fish of northern and eastern Australia. Wilson (1981b) concluded from a study of 13 morphometric measurements that fish from western Australia had larger head, deeper body, and longer pectoral fin than fish from Papua New Guinea. Gibbs and Collette (1967) found total gillraker counts for fish from Southeast Asia and Australia to be similar to one another and less than counts of fish from India and Red Sea. Silas (1967) found significant differences in the number of gillrakers, especially on the lower limb, for fish from western and northern Australia versus fish from India. Significant differences were also found in the number of gillrakers for fish from Oman, as compared to those from the Gulf of Mannar in the southeastern part of India (Abdulhaleem, 1989). Lewis (1981) found fixed and large differences in enzyme frequencies for longtail tuna from Malaysia and Australia. These differences in morphometric, meristic, and electrophoretic characters may reflect distinct stocks. Longtail tuna is essentially a neritic species with most phases of its life cycle confined to the continental shelf. Therefore, it is highly probable that there are numerous stocks (self-sustaining units) throughout the distributional range of the species.

6.2   Distribution

The general distribution of longtail tuna extends from the coast of Somalia, Gulf
of Aden, Red Sea of Japan, Papua New Guinea, Australia, and New Zealand (Figure 3).
Longtail tuna is basically confined to the neritic regime (Anon., 1975).

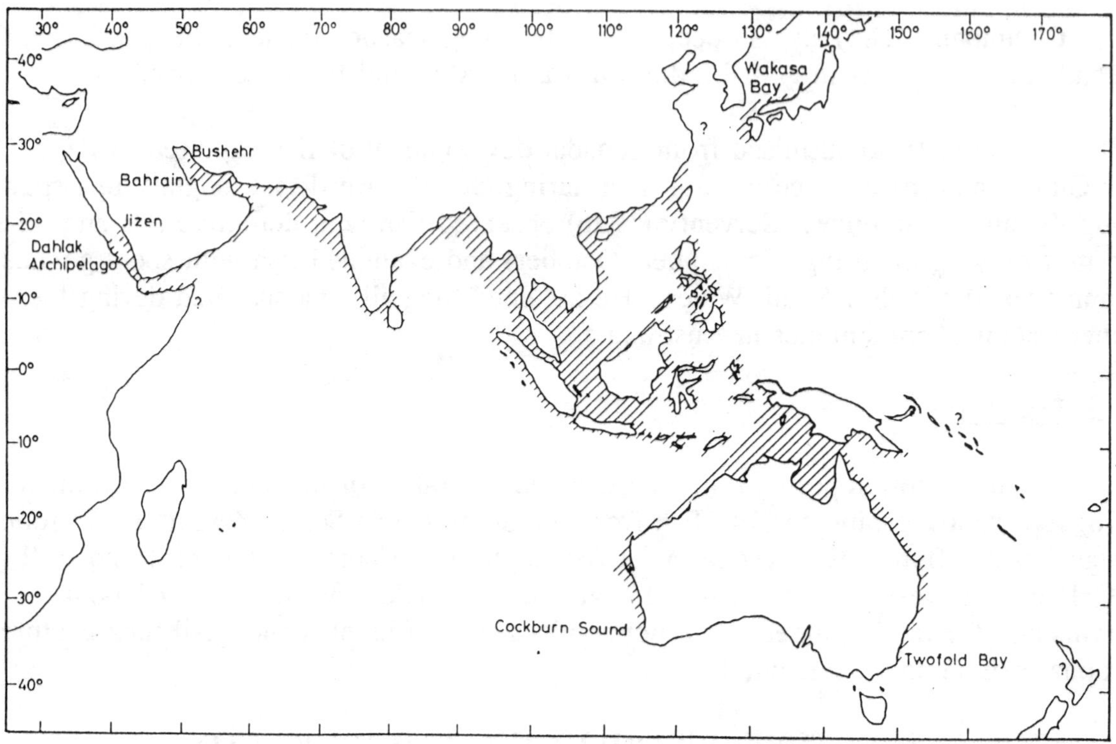

Figure 3.   Distribution of longtail tuna.

In the Indian Ocean, the southernmost distribution of longtail tuna off the east
coast of Somalia is at about Ras Hafun. It occurs in the Gulf of Aden along the north
coast of Somalia (Losse, 1970) and coast of PDR Yemen and in the Gulf of Oman off
Oman and United Arab Emirates. In the Red Sea, this species has been reported from
around the Dahlak Archipelago of Ethiopia (Ben Yami, 1968) and from Jizen, Saudi
Arabia (Peacock and Alam, 1977), but probably extends further north along both coasts.
The distribution of longtail in the Persian Gulf extends along the Iranian coast to Bushehr
and along the Arabian side to Bahrain. It has a continuous distribution along the
continent of Asia from Iran to mainland China. The northeastern extent of its range off
China is unknown, but longtail tuna have been reported from Wakasa Bay on the Japan
Sea coast of Japan (Nakamura, 1969) and off Taiwan (Chi and Yang, 1971).

Longtail tuna do not occur in the Lakshadweep Islands of India (Jones and
Kumaran, 1980) and have not been found in the Maldives. The species is, however,
found in Sri Lanka (Munro, 1955), Nicobar, and Andaman Islands. The latter island
groups are nearer to the Asian continent and large island (Sumatra) than the former
groups, though the difference in distances is slight. Distance does not appear to be the
limiting factor, but rather the extent of the continental shelf. Lakshadweep and Maldives

are comprised of very small islands, with restricted continental shelves so the waters over them would be essentially oceanic in character and unsuitable for longtail tuna.

This species occurs throughout the northwest and south coasts of Irian Jaya and south coast of Papua New Guinea (Munro, 1967). Its distribution probably does not extend further eastward. Longtail tuna have not been reported in the Solomon Islands chain and its presence off the north coast of Papua New Guinea and in New Britain (Serventy, 1956) has been questioned (A. Lewis, person. commun.). Longtail tuna have also been reported from New Zealand (Wilson, 1981a), but this has to be confirmed as there may be misidentification with bigeye tuna (*Thunnus obesus*) (Gibson[1]).

Troll-line catches off the west coast of Thailand averaged 0.31 kg/line-hour in the neritic regime and only 11 of a total of 604 troll-hooked longtail tuna were captured in the oceanic regime. However, these fish were all captured in grid areas which included small segments of the continental edge (Yesaki, 1982). This species was not found under a FAD (fish aggregating device) deployed in 432 m west of Ranch Yai Island. On the other hand, kawakawa (*Euthynnus affinis*) were captured around the FAD after 2 months soaking and kawakawa and skipjack tuna (*Katsuwonus pelamis*) after 3 months soaking (Lee, 1982). Longtail tuna have been reported occurring on the edge of shelf areas of eastern Indonesia and in the Philippines (A. Lewis, pers. commun.).

There are few reported catches on longlines, which support the contention that longtail tuna is a neritic species. One longtail tuna was captured in the Andaman Sea during cruises conducted by the Exploratory Fishing Division of the Department of Fisheries, Thailand (Poreeyanond and Kambud, 1985). Exploratory fishing trials with modified longlines on the north coast of Somalia captured yellowfin tuna (*Thunnus albacares*) and a few kawakawa, but no longtail tuna (Losse, 1975).

6.3 Migrations and Local Movements

In Australia, longtail tuna are found in the tropical province along the northern coast throughout the year. The species migrates into the subtropical province along the east coast during the austral summer and autumn. No definite seasonal movement is apparent on the coast of Western Australia (Serventy, 1956). There are differences in the geographic distribution of size groups; fish in the 30- to 70-cm interval are found off northern Australia, whereas fish with a mean length of 94 cm are encountered at the southern extremity of its range off the east and west coasts. Longtail tuna captured during summer in the northern extremity of its range off Japan, however, are smaller. Fish ranging in size from 28 to 30 cm have been reported from Tsushima Island (Fukusho and Fujita, 1972) and 49 to 51 cm from Wakasa Bay (Nakamura, 1969) in September.

Pole-and-line fishing explorations off the west coast of Thailand delineated juvenile longtail tuna in the inner-neritic regime southwest of Phuket (Yesaki, 1982). These fish move offshore with increasing size and have essentially all departed from the inner-neritic by 40 cm. Longtail tuna in the 40- to 49-cm size range are distributed principally in the

---

[1] Gibson, D. (pers. comm.) Marine Resources Assessment Group, Imperial College of Science and Technology, 48, Princess Gardens, London, SW7 1LU, U.K.

outer-neritic and is the dominant species in this regime. Larger fish (more than 50 cm) were rarely captured during the investigations and it was speculated that these fish emigrate out from the coast.

7.   NATURAL MORTALITY

Pauly's (1980) empirical equation based on growth parameters and mean water temperatures have been used by various authors to estimate natural mortality coefficients of tropical species. Prabhakar and Dudley (1989) obtained M of 0.429 for longtail tuna (sizes ranging from 22 to 116 cm, fork length) in Oman with input values for K of 0.228, L of 133.6 cm and mean water temperatures of 25.5°C.

8.   OCEANOGRAPHIC FEATURES ASSOCIATED WITH THE SPECIES

Longtail tuna were captured with pole and line and /or troll line throughout the year during exploratory fishing cruises off the west coast of Thailand. Sea-surface temperatures during these cruises ranged from 28.0° to 31.3°C (Yesaki and Jantarapagdee, 1981). Pole-and-line catches of longtail tuna were highest from January to April, with peak in March (Lee, 1982). During this period, mean sea-surface temperatures were rising from 28.6°C in December-January to a high of 30.2°C in June. Water over the mid and outer continental shelf was strongly stratified with a thin veneer of warm water and rapidly decreasing temperatures with depth. The 25°C isotherm was stationary at 50 m from December to March. During the southwest monsoon, when longtail tuna catches were poor, temperatures were essentially uniform from surface to bottom over the mid and outer shelf. Temperatures during September were 29°C at the surface and 25°C at 110 m.

9.   INTERACTIONS WITH OTHER SPECIES

Longtail tuna are generally found in the neritic regime with kawakawa and frigate tuna (*Auxis thazard*). Ben Yami (1968) gives the relative importance of pelagic fish around the Dahlak Archipelago, in decreasing order of importance, as kawakawa, longtail tuna, Indian mackerel (*Rastrelliger kanagurta*), queenfish (*Scomberoides lysan*), and frigate tuna. The subordinate rank of longtail tuna in the area to kawakawa may be attributed to the western extremity of its geographic range.

The chartered purse-seine vessels *Southward Ho* and *Royal Venture* conducted several exploratory fishing cruises for small tuna and other pelagic species in the eastern Andaman Sea and South China Sea from November 1975 to April 1977 (Simpson and Chikuni, 1978). They reported all three species of small tuna (longtail, kawakawa, frigate) were common throughout the coastal waters, with longtail tuna more abundant off Peninsular Malaysia and Thailand than off Sabah and Sarawak. Small tuna schools were common from the coast out to some 40 miles offshore and were occasionally seen more than 60 miles from shore, but were not encountered beyond the continental shelf.

The relative abundance of large pelagic fish off the west coast of Thailand was estimated by Yesaki (1982) from troll catches and/or visual identification of fish schools and the composition of pole-and-line catches. Longtail tuna was the dominant species, accounting for 60-70%, and kawakawa an abundant species, representing 17-19% of all fish schools off this coast. Frigate tuna were not as common, though the relative abundance of this species was probably underestimated because of the selectivity of the pole-and-line for larger fish. Species composition of schools fished with pole and line suggests a non-preferential association of longtail tuna with the other three tuna species (kawakawa, frigate, and skipjack).

Longtail tuna were found with only 5 out of a total of 510 porpoise pods observed and with none of the 6 whales sighted off the west coast of Thailand. However, this species was found under a school of whale sharks (*Rhincodon typus*) and successfully fished with pole-and-line (Yesaki, 1982).

Seventy-three percent of all fish schools sighted during pole-and-line activities were accompanied with birds. Percentages of longtail, kawakawa, skipjack, and frigate tuna schools accompanied by bird flocks were similar and ranged from 68% to 75% (Yesaki, 1982). Sooty terns were dominant and especially common in the oceanic and outer-neritic regimes. White terns and gulls were more common in the inner-neritic, whereas frigate birds and shearwaters were common throughout the neritic regime (Lee, 1982).

10. GENERAL DESCRIPTION OF THE FISHERIES

There are two major fishing grounds for longtail tuna, one off the South China Sea coast of Thailand and Malaysia and the other off the countries bordering the North Arabian Sea.

10.1 South China Sea

Descriptions of the purse-seine and gillnet fisheries of Thailand and the purse-seine, gillnet, and troll fisheries of Malaysia are given in Yesaki (1993).

10.2 North Arabian Sea

Countries exploiting longtail tuna in the North Arabian Sea include Oman, United Arab Emirates, Iran, and Pakistan. Gillnet is the most important gear used to capture longtail tuna in all these countries. Troll lines are also used in the United Arab Emirates and Oman. Two types of vessels are engaged in fisheries for longtail tuna. One type is the traditional design dhow constructed of wood ranging in length from 10 to 25 m. These vessels are powered by inboard diesel engines of 33-240 HP. The other type is the planing hull skiff constructed of glass-reinforced plastic (GRP) ranging in length from 5 to 9 m. These skiffs are powered by outboard engines of 25 to 120 HP. The large wooden dhow is the predominant type used in Pakistan and Iran, whereas the small GRP skiff is more common in the United Arab Emirates and Oman.

Gillnets fished by the small GRP skiffs are 1,000 to 1,700 m long and those used by the large wooden dhows are 2,500 to 9,000 m long. These gillnets are approximately 12 m deep with mesh size between 10 and 16 cm, with the most common being 13 cm. One line, usually with a single hook, is fished off the stern of the small GRP skiffs engaged in the troll fishery.

Pakistani gillnetters fish throughout the North Arabian Sea to the coasts of Oman, P.D.R. Yemen, and Somalia. Gillnetters of Oman and the United Arab Emirates confine fishing operations to continental shelf waters of their respective coasts.

11. TRENDS IN CATCH, FISHING EFFORT, AND CATCH PER UNIT OF EFFORT

11.1 Annual Catches

Nominal catches of longtail tuna in the Indo-Pacific region increased from 19,000 mt in 1979 to 116,000 mt in 1988 (Table 2). Thailand is the most important producer of longtail tuna, accounting for 54% of the total catch in 1988, followed by Iran with 15% and Oman with 13%.

11.2 Fishing Effort and Catch Per Unit of Effort

A relatively long time series of catch and effort statistics is available only for the purse-seine and gillnet fisheries of Thailand and the troll fishery of Malaysia. However, catch rates for the Thai purse-seine and gillnet fisheries are not reliable indices of abundance because these fisheries target a variety of species and/or species groups. Also, information of luring, one-boat and tuna purse seines are included in the purse-seine fishery. These purse seines are fished differently and target different species groups. Luring purse seines are fished in conjunction with FADs and night lights to capture small pelagic species, whereas one-boat purse seines are used to capture free-swimming schools of small pelagic species. Small tunas are generally captured only incidentally by these purse seines. Tuna purse seines are used to capture free-swimming schools of large pelagic fish, including small tunas, torpedo scad (*Megalaspis cordyla*), yellowtail scad (*Caranx mate*), etc. (Supongpan and Saikliang, 1987).

The Malaysian troll fishery targets small tuna so catch rates are reliable indices of abundance for these species. A tuna sampling programme was initiated by the Terengganu Marine Fisheries Resources Research Center in 1982. This programme was continued at various levels of intensity through 1986. This Center initiated another tuna sampling programme with assistance from the Indo-Pacific Tuna Development and Management Programme (IPTP) in January 1987. Effort, catch. and species composition are available for most months from August 1982 through December 1989, except for all months of 1986.

Catch and effort for the Thai purse seine and gillnet and Malaysia troll fisheries show different trends (Figure 4). The Thai purse-seine fishery shows increasing catch and effort from 1979 to 1988, whereas the Thai gillnet fishery shows increasing effort after 1985 with decreasing catch after 1986. Catch and effort of the Malaysian troll fishery was higher during the 1982-1985 interval than during the 1987-1989 interval.

Catch rates of longtail tuna by the Thai purse-seine fishery were at low levels till 1982, than increased markedly to highest level in 1983 after this fishery started targeting for small tuna and developed a net specifically for large pelagic species (Figure 5). Purse seine catch rates have fluctuated between 200 and 300 kg/day since 1984. Gillnet catch rates of longtail tuna were low till 1980, then increased to highest level in 1982 after this

Table 2. Landings in metric tons of longtail tuna in the Indo-Pacific region by country and year.

Country/Year	1979	1980	1981	1982	1983	1984	1985	1986	1987	1988	1989
Australia	19	5	0	0	9	131	138	135	16	40	19
China (Taiwan)1/	208	179	259	366	323	0	0	0	0	7290	NA
India	0	0	83	43	18	220	5540	185	444	55	55
Iran	846	969	2229	2924	5894	6389	11848	11710	12069	16907	19746
Malaysia	5875*	7081*	9605*	8677*	16158	12141	8760	7510	10034	7795	5651*
Oman	0	0	0	0	0	0	0	0	17259	15669	6527
Pakistan	3908*	1804*	2770*	3467*	1171*	1264*	5268	3275	3402	3010	4856
Thailand	12505	8671	11564	23416	57159	44378	48000	48303	38789	61062*	65926*
U. A. E.	0	0	350	4000	3000	3000	2830	3973	3654	3930	3938
Yeman, A. R.	1800	1530	1440	1260	328	678	840	307	420	414	414
Yeman, P. D. R	0	0	0	0	0	3	87	90	77	138	149
Total	25161	20239	28300	44153	84060	68204	83311	75488	86164	116310	107281

Source - IPTP, 1991
1/   - SEAFDEC, 1981; 1982; 1983; 1984; 1985; 1986; 1987; 1988; 1989; 1990
*   - estimated - MALAYSIA - 1979/1982 - 49% of TUN (% from 1984 - LOT+KAW+TUN)
                    - 1989 - 33% of TUN (% from 1988 - LOT+KAW+FRZ+TUN)
              - PAKISTAN - 1979/1984 - 32% of KAW+SKJ (% from 1986 - YFT+LOT+KAW+SKJ+FRZ+TUN)
              - THAILAND - 1988/1989 - Gulf of Thailand landigs from logbook survey

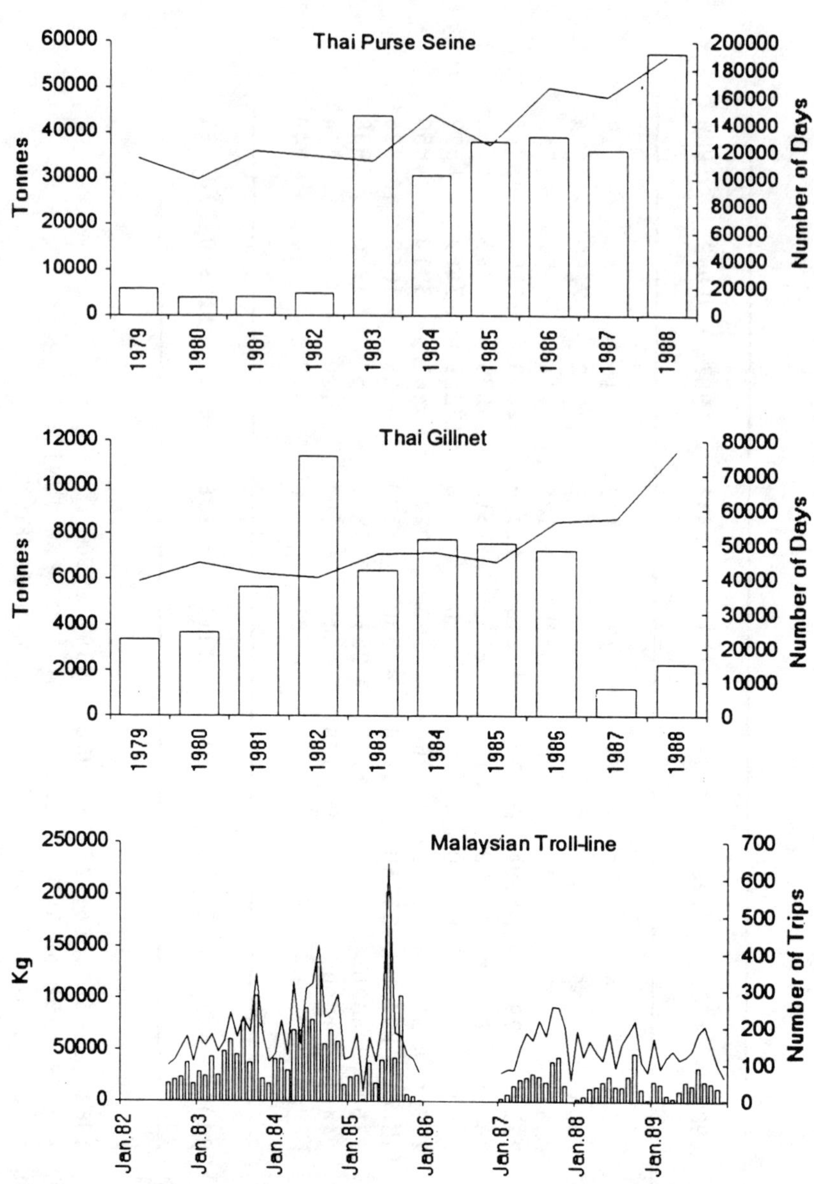

Figure 4.   Catch and effort of longtail tuna for purse-seine and gillnet fisheries of Thailand and troll-line fishery of Malaysia.

fishery started targetting for small tuna. Gillnet catch rates decreased to moderate levels in the 1983-1986 interval, then decreased to lowest levels in 1987-1988. These low catch rates for the most recent years may reflect a shift back to targetting for king mackerel. Average troll-line catch rates of longtail tuna decreased from 240 kg/trip during the 1982-1985 interval to 115 kg/trip during the 1987-1989 interval.

12.   POPULATION DYNAMICS

Stock assessment studies have not been completed for longtail tuna because of the lack of a long time series of reliable catch, effort, species, and size composition data.

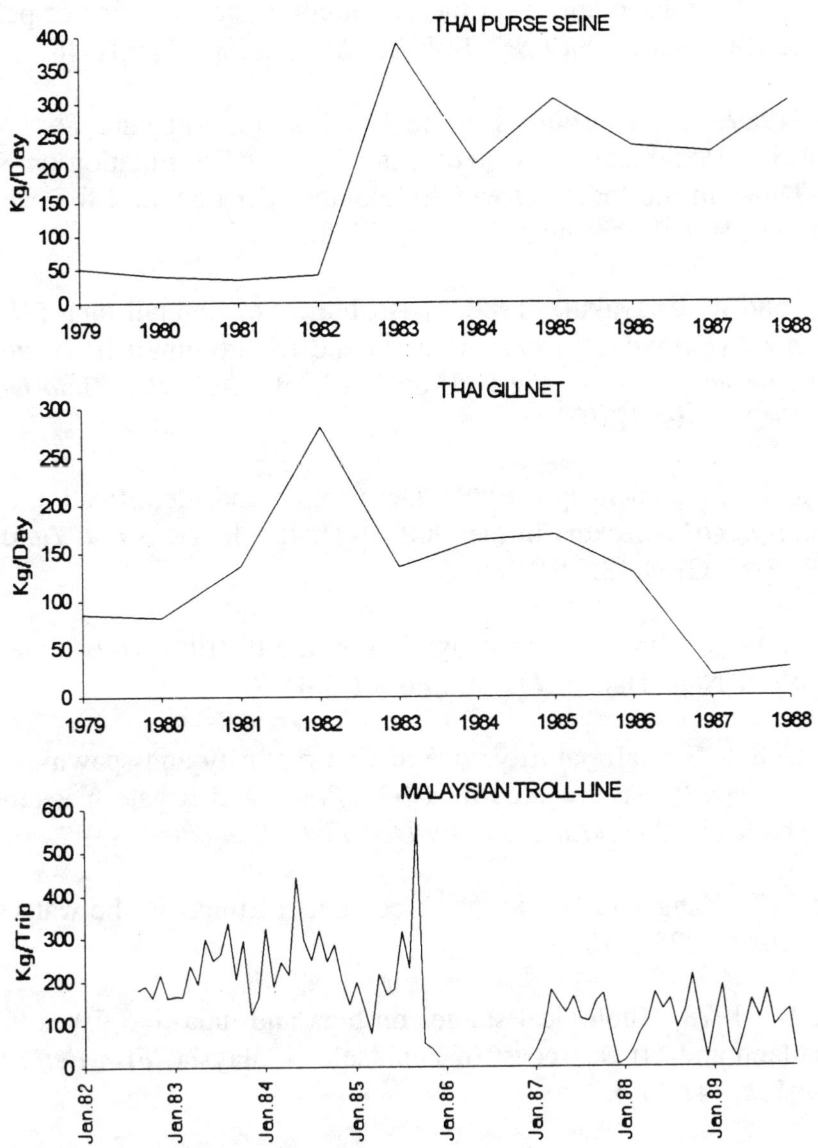

Figure 5.   Catch per unit effort of longtail tuna for the purse-seine and gillnet fisheries of Thailand and troll-line fishery of Malaysia.

13.   REFERENCES CITED

Abdulhaleem, S. Z. 1989. Gill raker counts: a possible means of stock separation for longtail tuna (*Thunnus tonggol*) in the Indian Ocean. *Indo-Pac.Tuna Dev.Mgt. Programme,* IPTP/89/GEN/16:64-8.

Anonymous. 1975. A summary of available data on the long-tailed tuna, *Thunnus tonggol* (Bleeker), in Papua New Guinea. Fisheries Research Laboratory, Dept. of Agriculture, Stock and Fisheries, Papua New Guinea, Interim draft report, 9 p.

Ben Yami, M. 1968. Observations on the distribution and behavior of pelagic schooling fish in the southern Red Sea. *Bull.Sea Fish.Res.Sta.*,Haifa, (51):31-46.

Boonragsa, V. 1987. Tuna resources in the Thai waters, Andaman Sea. *In* Collective Volume of Working Documents presented at the "Expert Consultation on Stock Assessment of Tunas in the Indian Ocean," Colombo, Sri Lanka, 4-8 December 1986. *Indo-Pac.Tuna Dev.Mgt.Programme,* Vol. 2:267-80.

Chamchang, C. and R. Chayakul. 1988. Distribution of longtail tuna (*Thunnus tonggol,* Bleeker), kawakawa (*Euthynnus affinis,* Cantor) and frigate tuna (*Auxis thazard,* Lacepede) in the western coast of the Gulf of Thailand. *Indo-Pac.Tuna Dev.Mgt. Programme,* IPTP/87/GEN/13:80-8.

Chayakul, R., and C. Chamchang. 1988. Description and identification of longtail tuna larvae, *Thunnus tonggol* (Bleeker) in the Gulf of Thailand. *Indo-Pac.Tuna Dev.Mgt. Programme,* IPTP/87/GEN/13:71-9.

Chen, Z., and S. Wei. 1981. An investigation on the distribution of tuna fish larvae in the central region of Nan. Hai. *J.Fish.China,* 5(1):41-7.

Cheunpan, A. 1984. Sexual maturity, size at first maturity and spawning season of longtail tuna (*T. tonggol*), eastern little tuna (*E. affinis*) and frigate mackerel (*A. thazard*) in the Gulf of Thailand. *Rep.Mar.Fish.Div.Dep.Fish.,Bangkok,* (43):22 p.

Chi, K-S., and R-T. Yang. 1971. Stomach contents of tunas in the waters southwest off Taiwan. *China Fish.,* 225:3-18.

Chiampreecha, B. 1978. Biological studies on tuna and tuna-like fishes in the west of the Gulf of Thailand and off east coast of Peninsular Malaysia. *Fish.Rep.Mar.Fish.Div. Dep.Fish., Bangkok,* (4):25 p.

Collette, B. B. 1978. Adaptations and systematics of the mackerels and tuna. *In* The Physiological Ecology of tunas, edited by G. D. Sharp and A. E. Dizon. New York, Academic Press, Inc., pp. 7-39.

Fukusho, K., and S. Fujita. 1972. Notes on the young tunas referred to the Genus *Thunnus* caught in the waters of Tsushima Island. *Jap.J.Ichthyol.,* 19(1):34-5.

Gibbs, R. H., Jr., and B. B. Collette. 1967. Comparative anatomy and systematics of the tunas, genus *Thunnus*. *Fish.Bull.U.S.Fish Wildl.Serv.,* 66(1):65-130.

IPTP. 1985. Report on the Joint Tuna Research Group Meeting of the Philippines and Indonesia. *Indo-Pac.Tuna Dev.Mgt.Programme,* IPTP/85/GEN/8:85 p.

IPTP. 1986. Report of the Meeting of Tuna Research Group in the Southeast Asian Region. *Indo-Pac.Tuna Dev.Mgt.Programme,* IPTP/86/GEN/10:75 p.

Jones, S., and M. Kumaran. 1980. Fishes of the Laccadive Archipelago. Cochin, Mathrubhumi Press, 757 p.

Klinmuang, H. 1978. Preliminary studies on the biology of tunas in the west of the Gulf of Thailand and off the east coast of Peninsular Malaysia. *Pelagic Fish.Rep.Mar.Fish. Div.Dep.Fish., Bangkok,* (5):27 p.

Lee, R. E. K. D. 1982. Thailand. Fishing for tuna. A report prepared for the pole-and-line fishing in Southern Thailand Project, FAO. FI: DP/THA/77/008: Field doc. 1: 65 p.

Lewis, A.D. 1981. Population studies of northern Australian pelagic species utilizing the electrophoretic approach. *In* Northern Pelagic Fish Seminar, edited by C. J. Grant and D. G. Walter. *Aust.Gov.Publ.Serv.*, pp. 35-44.

Losse, G.F. 1970. The tuna fishery of the Gulf of Aden. *FAO* FI: SF/SOM/13 Tech. Rep. 1:86 p.

Losse, G.F. 1975. Exploratory fishing in the Gulf of Aden and observations on the biology of large pelagic fishes. *Rep.East Afr.Mar.Fish.Res.Organ.*, L/7:61 p.

Munro, I.S.R. 1955. The Marine and Fresh Water Fishes of Ceylon. Sydney, Halstead Press, 351p + 56 plates.

Munro, I.S.R. 1967. The Fishes of New Guinea. *Dep.Agri.Stock Fish.,Port Moresby*, 650p + 84 plates.

Muthiah, C. 1986. Fishery and bionomics of tunas at Mangalore. *In* Tuna fisheries of the exclusive economic zone of India: biology and stock assessment, edited by E. G. Silas. *Bull.Cent.Mar.Fish.Res.Inst.,Cochin*, (36):51-70.

Nakamura, I. 1969. Big catches of longtail tuna in Wakasa Bay, Japan. *Jap.J.Ichthyol.*, 15(4):160-1.

Pauly, D. 1980. On the interrelationships between natural mortality, growth parameters and mean environmental temperature in 175 fish stocks. *J.Cons.CIEM*, 39(3):175-92.

Peacock, N., and K. Alam. 1977. Progress report: Assessment of the fisheries and fish landing patterns of the Red Sea coast, Saudi Arabia. Fish. Dev. Proj. Field Rep. No. 12:24 p.

Poreeyanond, D., and W. Kambud. 1985. Tuna resources survey by longline in the Andaman Sea. *In* Collective Volume of Working Documents presented at the "Expert Consultation on Stock Assessment of Tunas in the Indian Ocean". Colombo, Sri Lanka, 28 Nov. - 2 Dec. 1985. *Indo-Pac.Tuna Dev.Mgt.Programme*, Vol. 1:242-5.

Prabhakar, A., and R.G. Dudley. 1989. Age, growth and mortality rates of longtail tuna *Thunnus tonggol* (Bleeker) in Omani waters based on length data. *Indo-Pac.Tuna Dev.Mgt.Programme*, IPTP/89/GEN/16:90-6.

Rao, K.V.N. 1964. An account of the ripe ovaries of some Indian tunas. *In* Proceedings of the Symposium on Scombroid Fishes, Part 2. *Symp.Ser.Mar. Biol.Assoc.India*, 1:733-43.

Serventy, D.L. 1956. Additional observations on the biology on the northern bluefin tuna, *Kishinoella tonggol* (Bleeker) in Australia. *Aust.J.Mar.Freshwat.Res.*, 7(1):44-63.

Silas, E.G. 1967. Tuna fishery of the Tinnevelly coast, Gulf of Mannar. *In* Proceedings of the Symposium on Scombroid Fishes, part 3. *Symp.Ser.Mar. Biol.Assoc.India*, 1:1083-118.

Silas, E.G., P.P. Pillai, A.A. Jayaprakash, and M.A. Pillai. 1986a. Fishery and bionomics of tunas at Cochin. *In* Tuna fisheries of the exclusive economic zone of India: biology and stock assessment, edited by E. G. Silas. *Bull.Cent.Mar.Fish.Res.Inst., Cochin*, (36):28-44.

Silas, E.G., P.P. Pillai, M. Srinath, A.A. Jayaprakash, Muthiah, V. Balan, C.T.M. Yohannan, P. Siraimeetan, M. Mohan, P. Livingston, K.K. Kunhikoya, M.A. Pillai, and P.S.S. Sarma. 1986b. Population dynamics of tunas: stock assessment. *In* Tuna fisheries of the exclusive economic zone of India: biology and stock assessment, edited by E. G. Silas. *Bull.Cent.Mar.Fish.Res.Inst.,Cochin*, (36):20-7.

Simpson, A.C., and S. Chikuni. 1978. Test purse seine fishing for small tunas and other small pelagic fish off the coasts of Thailand, Peninsular Malaysia, Sabah and Sarawak (1975-1977). Part II. *In* Test Fishing for Tuna and Small Pelagic Species. *South China Sea Fish.Dev.Coord.Programme*, SCS/DEV/78/18:28 p.

SEAFDEC. 1981. Fishery Statistical Bulletin for South China Sea Area 1979. *Southeast Asian Fish.Dev.Center*, 227 p.

SEAFDEC. 1982. Fishery Statistical Bulletin for South China Sea Area 1980. *Southeast Asian Fish.Dev.Center*, 268 p.

SEAFDEC. 1983. Fishery Statistical Bulletin for South China Sea Area 1981. *Southeast Asian Fish.Dev.Center*, 275 p.

SEAFDEC. 1984. Fishery Statistical Bulletin for South China Sea Area 1982. *Southeast Asian Fish.Dev.Center*, 231 p.

SEAFDEC. 1985. Fishery Statistical Bulletin for South China Sea Area 1983. *Southeast Asian Fish.Dev.Center*, 178 p.

SEAFDEC. 1986. Fishery Statistical Bulletin for South China Sea Area 1984. *Southeast Asian Fish.Dev.Center*, 231 p.

SEAFDEC. 1987. Fishery Statistical Bulletin for South China Sea Area 1985. *Southeast Asian Fish.Dev.Center*, 178 p.

SEAFDEC. 1988. Fishery Statistical Bulletin for South China Sea Area 1986. *Southeast Asian Fish.Dev.Center*, 175 p.

SEAFDEC. 1989. Fishery Statistical Bulletin for South China Sea Area 1987. *Southeast Asian Fish.Dev.Center*, 162 p.

SEAFDEC. 1990. Fishery Statistical Bulletin for South China Sea Area 1988. *Southeast Asian Fish.Dev.Center*, 183 p.

SPC. 1991a. Regional tuna tagging survey. Monthly tagging summary - April 1991. *Tuna Billfish Assess.Programme, S.Pac.Comm.*, 8 p.

SPC. 1991b. Regional tuna tagging survey. Monthly tagging summary - July 1991. *Tuna Billfish Assess.Programme, S.Pac.Comm.*, 11 p.

Supongpan, S., and P. Saikliang. 1987. Fisheries status of tuna purse seiners (using sonar) in the Gulf of Thailand. *Rep.Mar.Fish.Div.Dep.Fish.,Bangkok*, (3):78 p.

Wilson, M.A. 1981a. Some aspects of the biology and production of longtail tuna in Oceania. *In* Northern Pelagic Fish Seminar, edited by C.J. Grant and D.G. Walter. *Aust.Gov.Publ.Serv.*, 24-44.

Wilson, M.A. 1981b. The biology, ecology and exploitation of longtail tuna, *Thunnus tonggol* (Bleeker) in Oceania. M.Sc. Thesis. School of Biological Sciences, Macquarie University, Sydney, 195 p.

Yesaki, M. 1982. Thailand. Biological and Environmental Observations. A report prepared for the Pole-and-Line Tuna Fishing in Southern Thailand Project. *FAO*. FI: DP/THA/77/008: Field doc. 3, 46 p.

Yesaki, M. 1987. Synopsis of biological data on longtail tuna, *Thunnus tonggol*. *Indo-Pac.Tuna Dev.Mgt.Programme*, IPTP/87/WP/16:56 p.

Yesaki, M. 1989. Estimates of age and growth of kawakawa (*Euthynnus affinis*), longtail tuna (*Thunnus tonggol*) and frigate tuna (*Auxis thazard*) from the Gulf of Thailand based on length data. *Indo-Pac.Tuna Dev.Mgt.Programme*, IPTP/89/GEN/17:94-108.

Yesaki, M. 1993. Interactions between fisheries for small tunas off the South China Sea coast of Thailand and Malaysia. *In* Proceedings of the FAO Expert Consultation on Interactions of Pacific Tuna Fisheries, edited by R.S. Shomura, J. Majkowski, and S. Langi, 3-11 December 1991, Noumea New Caledonia. [See this document.]

Yesaki, M., and P. Jantarapagdee. 1981. Wind stress and sea temperature changes off the west coast of Thailand. *Res.Bull.Mar.Biol.Cent.,Phuket*, (28):27-41.

A REVIEW OF THE BIOLOGY AND FISHERIES FOR KAWAKAWA (*EUTHYNNUS AFFINIS*) IN THE INDO-PACIFIC REGION

Mitsuo Yesaki
Indo-Pacific Tuna Development and Management Programme
Colombo, Sri Lanka

1. INTRODUCTION

A synopsis of the genus *Euthynnus* was completed by Yoshida (1979) and a synopsis of kawakawa (*E. affinis*) was compiled by Yesaki (1989a). This review was abridged from the synopsis by the latter author. Some new information available since publication of this synopsis have been incorporated in this review.

2. CLASSIFICATION

Phylum Chordata
   Subphylum Vertebrata
      Superclass Gnathostomata
         Class Osteichthyes
            Subclass Actinopterygii
               Order Perciformes
                  Suborder Scomberoidei
                     Family Scombridae
                        Subfamily Scombrinae
                           Tribe Thunnini
                             Genus *Euthynnus*
                                Species *affinis*

3. EARLY LIFE HISTORY

Buñag (1956) and Rao (1964) described the development of ovarian ova of kawakawa, and Kaya *et al.* (1981) described the development of ova during the first artificially-induced spawning experiments with this species. The first cleavage divisions were observed at about 1 h after fertilization and the first hatching at about 31 h in temperatures that ranged between 21° and 26°C. The yolk sac was absorbed over 2 days.

The larvae of kawakawa have been described in detail by Matsumoto (1958) and the juveniles by Wade (1950a) and Jones (1960).

Kawakawa larvae appear to be very patchy and are widely distributed and generally taken close to land masses (Matsumoto, 1958; Nishikawa *et al.*, 1985). Nishikawa *et al.* (1985) shows the highest concentrations of *Euthynnus* from the central and western Pacific off northwestern Australia, Java, and Papua New Guinea and off the Ryukyu and Solomon Islands.

Two kawakawa larvae were captured in plankton tows at a diel variability station about 15 miles southwest of the island of Hua Pou in the Marquesas Islands (Nakamura and Matsumoto, 1967).

A small-mesh mid-water trawl was fished 83 times at various locations around Hawaii to capture juvenile tunas (Higgins, 1970). Kawakawa were captured only off Oahu during these explorations. Of the 25 juveniles captured, 23 were captured in the inshore areas (7-9 km from land), whereas, only two were captured in the offshore area (56 km from land). The catches per tow averaged 0.1, 0.1, and 2.5 larvae during July, August, and September, respectively, in the inshore area.

Chamchang and Chayakul (1988) studied the distribution and abundance of small tuna larvae in the western Gulf of Thailand. Kawakawa larvae were the least abundant of the three kinds of tuna larvae captured in this area and were captured only in April and June.

Kawakawa larvae accounted for 24 percent of all tuna larvae (146 specimens) captured during a survey of the Ragay Gulf, Burias Pass, and Ticao Pass region of the Philippines (Abuso, 1988). These kawakawa larvae were captured in temperatures and salinities ranging from 24.7° to 28.9°C and 33.53°/·· to 34.66°/··, respectively.

Juvenile kawakawa are commonly captured in fisheries of many countries. Yabe *et al.* (1953) found juvenile kawakawa ranging in size from 18.3 to 33.2 cm during August to October among fish caught in Aburatsu, Japan. Jones (1960) described juveniles of kawakawa (2.5 to 22.2 cm) captured by beach seines off the southwest coast and by boat seines off the south coast of India. Siraimeetan (1985) collected 106 specimens ranging in size from 6.2 to 19.8 cm from the sardine-gillnet fishery off Tuticorin, India, during June-September. Wade (1950a) obtained many of his juvenile and preadult kawakawa from various fish markets in the Philippines. Juvenile kawakawa ranging in size from 16 to 32 cm make up the bulk of the catch by the troll-line fishery in Sri Lanka (Sivasubramaniam (1970). A high percentage of the kawakawa captured by Thai purse seiners in the Gulf of Thailand is comprised of fish in the 12- to 32-cm interval (IPTP data files). The frequent capture of kawakawa juveniles is attributed to the neritic habit of the species.

4.   AGE AND GROWTH

Numerous studies have been completed to determine the age and growth of kawakawa. These include 13 studies based on length-frequency distributions and one study each on vertebrae, dorsal spines, and otoliths (Table 1).

Available evidence suggests rapid growth during the juvenile stage of kawakawa. Yabe *et al.* (1953) sampled juvenile kawakawa landed at Aburatsu, Japan, during August-October 1950 (Figure 1). The mean size of the juveniles increased from 20.0 to 27.2 cm in 53 days.

Kawakawa ranging from 12 to 57 cm are captured in the purse-seine fisheries of the Gulf of Thailand. Length-frequency distributions of juveniles (12-30 cm) captured during August-December 1987 were examined by Yesaki (1989b). Frequency

Table 1. Summary of age and growth studies on kawakawa by length-frequency distributions and hard parts (from Yesaki, 1989b).

Author	Area	Method	Growth parameters			Length at relative age				
			K	L_{oo}(cm)	to	1	2	3	4	5
(a) Length-frequency distributions										
Ommanney (from Yoshida, 1979)	Seychelles	Modal lengths	–	–	–	25	45	65	65	–
Williamson, 1970	Hong Kong	Modal lengths	–	–	–	44	62	–	–	–
Chiampreecha, 1978	Gulf of Thailand	Modal lengths	–	–	–	21	30	41	–	–
	east coast Malaysia	Modal lengths	–	–	–	19	29	35	50	–
Klinmuang, 1978	Gulf of Thailand	Modal progres.	–	–	–	27	41	53	–	–
Yesaki, 1982	west coast Thailand	Modal progres.	0.46	76.0	–	29	45	57	65	–
Silas et al., 1985b	India	ELEFAN	0.37	81.0	-.344	31	47	57	64	70
Joseph et al., 1987	Sri lanka	ELEFAN	0.63	59.6	–	28	43	51	55	57
	Sri Lanka	ELEFAN	0.61	63.0	–	29	44	53	58	60
	Sri Lanka	Bhattacharya	0.69	59.5	–	30	45	52	56	58
Supongpan and Saikliang, 1987	Gulf of Thailand	Modal progres.	2.23	55.1	-.015	49	55	55	55	–
Yesaki, 1989b, 10-day intervals	Gulf of Thailand	Modal progres.	0.96	76.0	–	47	65	72	–	–
Yesaki, 1989b, monthly intervals	Gulf of Thailand	Modal progres.	0.56	76.0	–	33	51	62	68	71
(b) Hard parts										
Landau, 1965	Red Sea	vertebrae	–	–	–	35	46	51	54	55
Shabotinets, 1968	Gulf of Aden	dorsal spines	–	–	–	–	–	58[1]/	65[1]/	73[1]/
Uchiyama, 1980	Hawaii	otoliths	0.42	117.8	-.030	41	68	85	96	104

[1]/ - median

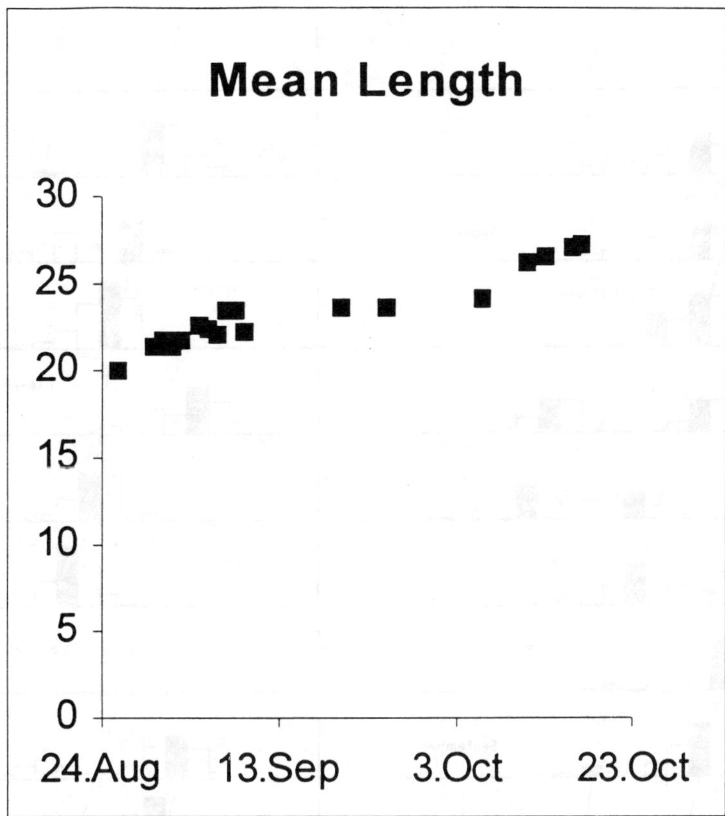

Figure 1.   Mean lengths of juvenile kawakawa captured off Aburatsu, Japan during August-October (data from Yabe, 1953).

distributions were grouped by 5-day intervals to show entry and progression of cohorts. Six cohorts were identified in the 15- to 25-cm interval. Modal progressions of these cohorts ranged from 4.80 to 7.20 cm per month, and averaged 5.76 cm per month (Figure 2).

The initial length-based studies, including those of Ommanney (cited in Yoshida, 1979), Williamson (1970), and Chiampreecha (1978), subjectively assigned ages to prominent modal lengths. In later studies, Klinmuang (1978), Yesaki (1982; 1989b), and Supongpan and Saikliang (1987) estimated growth by "pen and paper" tracing of modal progressions. More recently, Silas *et al.* (1985b) and Joseph *et al.* (1987) used computer programmes to derive growth parameters and relative ages. The results for seven of the 11 length-based studies are very similar (Figure 3). Lengths at age derived from these seven studies range from 25 to 33, 43 to 51, and 52 to 65 cm for ages 1, 2, and 3, respectively.

Landau (1965) and Shabotinets (1968) determined age of kawakawa by counting annuli on vertebrae and dorsal spines, respectively. Their results are similar to the majority of the length-based estimates with lengths of 35, 46, and 51-58 cm for ages 1, 2, and 3, respectively (Figure 3).

Uchiyama (1980) examined the microstructure of otoliths for age and growth of kawakawa in the 20- to 70-cm range. Increment counts were validated as daily events in

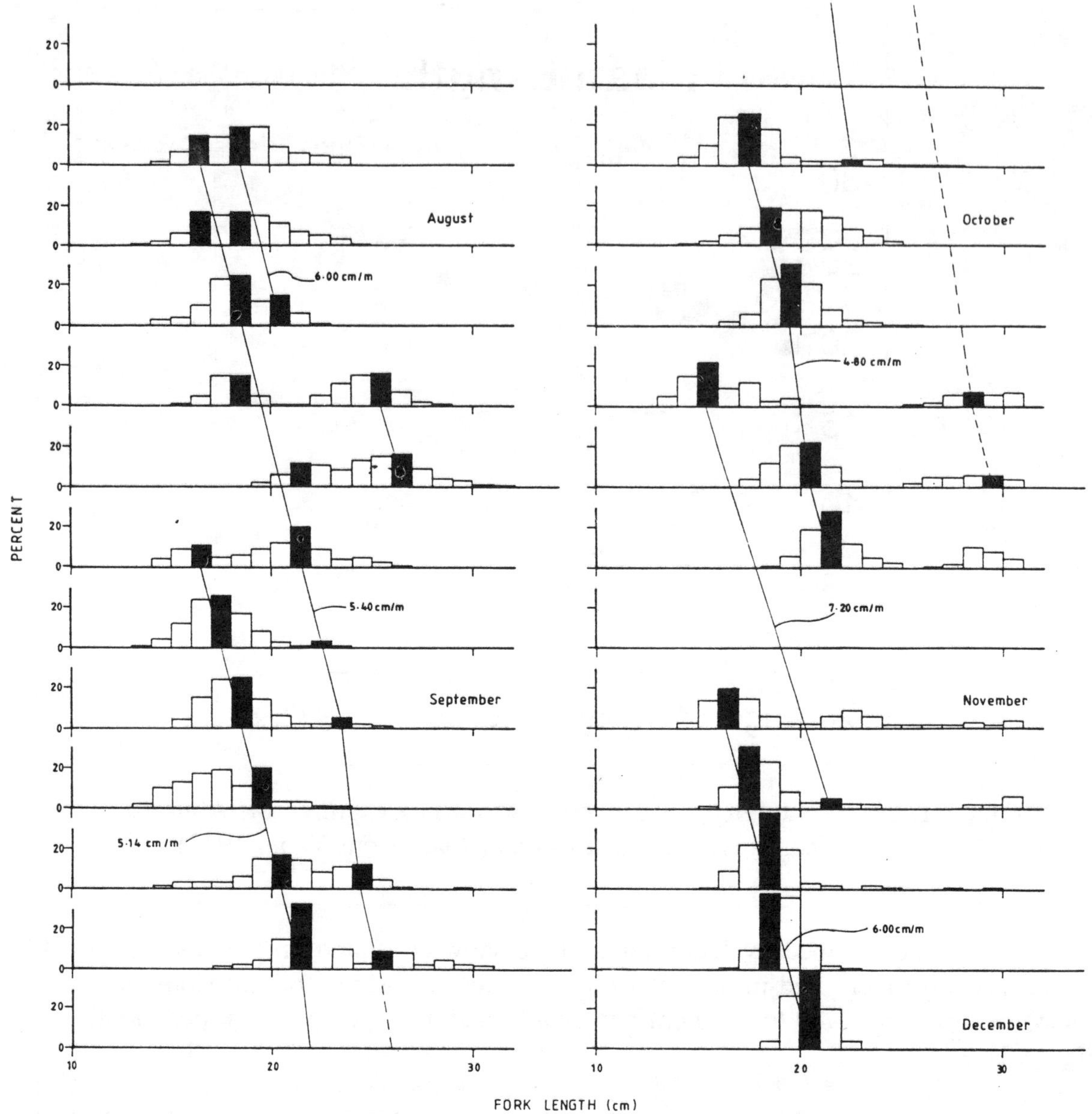

Figure 2.   Modal progressions of juvenile kawakawa captured by the purse-seine fishery of the Gulf of Thailand (from Yesaki, 1989).

this study by either injecting or feeding fish with oxytetracycline; these fish were maintained at the Kewalo Research Facility of the National Marine Fisheries Service, Honolulu Laboratory. Uchiyama (1980) obtained lengths of 41 cm for age 1, and 68 cm for age 2. Yesaki (1989b) obtained lengths of 47 and 65 cm for these respective ages by following modal progressions by the "pen and paper" method in frequency distributions grouped by 10-day intervals (Figure 3).

Age and growth studies of kawakawa completed to date give conflicting results. Length-frequency distributions of juveniles show rapid growth during early life. Also, otolith-increment counts of adults give length-at-age 1 of about 41 cm and length distributions grouped by 10-day intervals give 47 cm at age 1. Conversely, length distributions of adults grouped by monthly intervals give length-at-age 1 of between 25 and 33 cm.

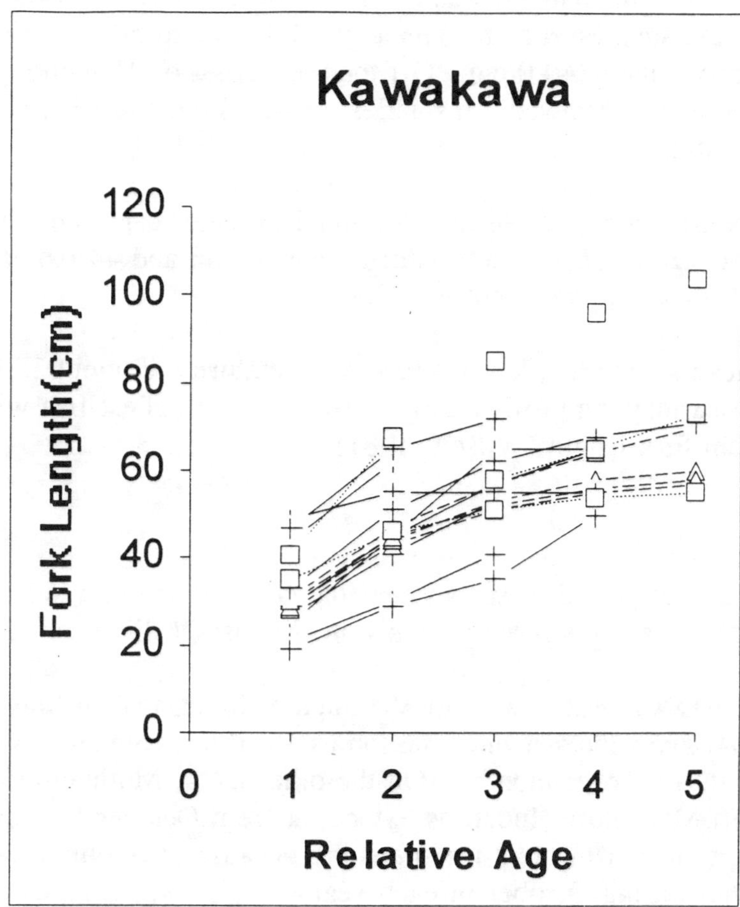

Figure 3. Estimates of growth and relative age (yr) for kawakawa from length-frequency analyses (crosses: manual method; triangles: computer programmes) and hard parts (squares) (from Yesaki, 1989).

5. MATURATION, SPAWNING, AND SEX RATIO

5.1 Maturation

Tunas captured principally by trolling during exploratory fishing activities in the Philippines were examined for maturity (Wade, 1950b). The smallest ripe, spawning, and spent female kawakawa found during this survey were 40, 45, and 40 cm, respectively.

Buñag (1956) measured ova diameters of 30 fish to determine maturity of kawakawa captured in the Philippines, and correlated the development of ova to the various maturity stages. The smallest mature and spent females examined were 49.0 cm and 47.7 cm, respectively. A larger sample of 144 kawakawa, ranging in length from 33.1 to 65.2 cm, was examined by Ronquillo (1963) in the Philippines. He observed gonad indices greater than 3.0 (mature) in females ranging from 38.5 to 65.0 cm.

Klinmuang (1978) suggested that kawakawa from the Gulf of Thailand and east coast of Peninsular Malaysia attained sexual maturity at between 37 and 42 cm. A total of 1,010 specimens from the Gulf of Thailand was examined over a 4-year period by Cheunpan (1984). She found the smallest mature female at 33.4 cm and 50 percent of the females to be mature at approximately 40.0 cm. Off the west coast of Thailand, a sample of 222 maturing (stage-III) females showed the smallest to be 38 cm and 93 percent to be 43 cm and larger (Yesaki, 1982).

Muthiah (1985) examined 198 females and 183 males captured off Mangalore, India, and found the size of 50-percent maturity was 43 cm and 44 cm, respectively, though males less than 43 cm were not examined.

These studies indicate that kawakawa attains maturity at about 38 cm. Kawakawa in Papua New Guinea may mature at a larger size, as the smallest fish with maturing oocytes was 48.9 cm fork length (Wilson, 1981).

5.2   Spawning

Kawakawa are suspected to spawn near the coast on the east side of Hong Kong. The spawning in this area takes place during June-August (Williamson, 1970).

Six ripe kawakawa were found off Vizhingam, India, two in May and four in August. Rao (1964) gives the spawning season as April to September, but does not discount spawning during other months. On the other hand, Muthiah (1985) gives the spawning season off Mangalore, India, as extending from October to May. He speculated that kawakawa enters inshore waters for spawning because of the high incidence of ripe fish in gillnet catches during October of each year.

Kawakawa in all stages of maturity were captured during most months in the Philippines. Therefore, Wade (1950b) concluded that spawning occurred throughout the year. A total of 110 spawning and/or spent females were found in January-July and October-November. The highest gonad index for kawakawa in the Philippines occurred in March (Ronquillo, 1963).

The highest percentage of maturing (stage-III) females was observed during September-April off the west coast of Thailand (Yesaki, 1982). Mature 9 (stage-IV) kawakawa were found in March and September and a spent (stage-V) female in February. Two spawning seasons were indicated by maturity stages; the principal season from February to April and the second season in September. Gonad indices for kawakawa in the Gulf of Thailand were consistently higher during April-July and December-January (Cheunpan, 1984).

These studies indicate that there are two spawning seasons for kawakawa in the equatorial regions. In the northern hemisphere, the principal spawning season generally occurs during the first half of the year and the secondary spawning season during the latter half. At higher latitudes of the tropical zone, there appears to be only one spawning season near the middle of the year.

Few studies have been published on the fecundity of kawakawa. Rao (1964) counted the number of opaque maturing ova and ripe translucent ova in ripe gonads of five kawakawa captured off Vizhingam, India. Counts of all ova and of ripe ova were used to estimate numbers released during the spawning season during each spawning, respectively. The numbers estimated to be released during each spawning ranged from 210,000 to 680,000 ova and during the spawning season from 790,000 to 2,500,000 ova. These fish ranged in size from 48.0 to 65.0 cm. A total of 38 fish captured off Vizhingam, India, were examined for fecundity by Muthiah (1985). Fecundity estimates for mature and ripe ovaries for fish in the 39.4 to 67.0 cm interval ranged from 202,000 to 1,570,000 ova.

Klinmuang (1978) estimated the fecundity of five kawakawa ranging in size from 39.5 to 51.0 cm from the South China Sea region. Fecundity for these fish averaged 1,730,000 ova and ranged from 585,000 to 2,593,000 ova.

These studies probably underestimate fecundity of kawakawa as the numbers of oocytes produced at the beginning of the spawning season is not fixed. Kawakawa is a multiple spawner releasing ova at frequent intervals during a spawning season.

Buñag (1956) and Rao (1964) concluded from the polymodal distribution of ova diameters that kawakawa spawned more than once during a spawning season. However, they were not able to determine frequency of spawning. An indication of spawning frequency can be obtained from a study on the reproductive biology of *Euthynnus lineatus* from the eastern Pacific. Schaefer (1987b) estimated the frequency of spawning from the proportion of sampled females with ovaries containing hydrated oocytes. *E. lineatus* was found to spawn at average intervals of 2.1, 5.7, and 3.6 days off the coasts of Mexico, Central America, and South America, respectively.

Kawakawa was the first tuna species to be successfully spawned in captivity (Kaya *et al.*, 1981). Females and males kept in captivity for 2-4 weeks and about 4 months, respectively, at the Kewalo Research Facility, National Marine Fisheries Service, Honolulu Laboratory, were used in the experiments. Fish were induced to spawn with injections of salmon pituitary (SP), human chorionic gonadotropin (HCG) and pregnant-mare serum gonadotropin (PMS). Fish spawned after the second treatment with the higher dosages of the hormones.

5.3   Sex Ratio

Sex ratios for kawakawa from various areas are given in Table 2.

Williamson (1970) reports kawakawa in the size range from 24 to 73 cm occurring in the Hong Kong area. He found males and females of all lengths within this range, but the males accounted for 67 percent of the fish in the 50- to 73-cm interval. The dominance of males at sizes larger than 50 cm also occurs in the Philippines, Seychelles, and India (Table 3).

Table 2. Sex ratios of kawakawa.

| | | | Sex | | | |
Area	No. fish	Size range cm	?	Males	Females	Author
Philippines	460	28-68	17	239	204	Wade, 1950b
South China Sea	402	32-45	-	234	168	Klinmuang, 1978
West coast Thailand	659	-	80	250	329	Yesaki, 1982
Seychelles	5,728	29-87	-	3,095	2,633	Steinberg et al., 1982
West coast India	398	34-71	-	196	202	Muthiah, 1985

Table 3. Numbers of male and female kawakawa by length intervals.

Locality Length interval cm	Philippines[1]		Seychelles[2]		India[3]	
	Male	Female	Male	Female	Male	Female
20-29	-	-	3	1	-	-
30-39	9	13	135	126	28	34
40-49	85	85	1,888	1,832	38	50
50-59	113	93	800	490	74	100
60-69	32	13	186	161	54	18
70-79	-	-	76	22	2	-
80-89	-	-	7	1	-	-
	239	204	3,095	2,633	196	202

[1] - from Wade, 1950b
[2] - from Steinberg et al., 1982
[3] - from Muthiah, 1985

6. DISTRIBUTION AND MIGRATION

6.1 Distribution

Kawakawa are widely distributed in the tropical and subtropical waters of the Indo-Pacific region (Figure 4). In the Indian Ocean, this species extends from Cape St. Francis, South Africa (Smith and Heemstra, 1986), along the coasts of east Africa, Arabian Peninsula, the Indian sub-continent, and Malaysian Peninsula. It is also found in the Red Sea, Persian Gulf, and off islands in the Indian Ocean, including Madagascar, Comoros Islands, Mauritius, Reunion, Seychelles, Lakshadweep, Andaman Islands, Nicobar Islands, Sri Lanka, and Maldives (Williams, 1963).

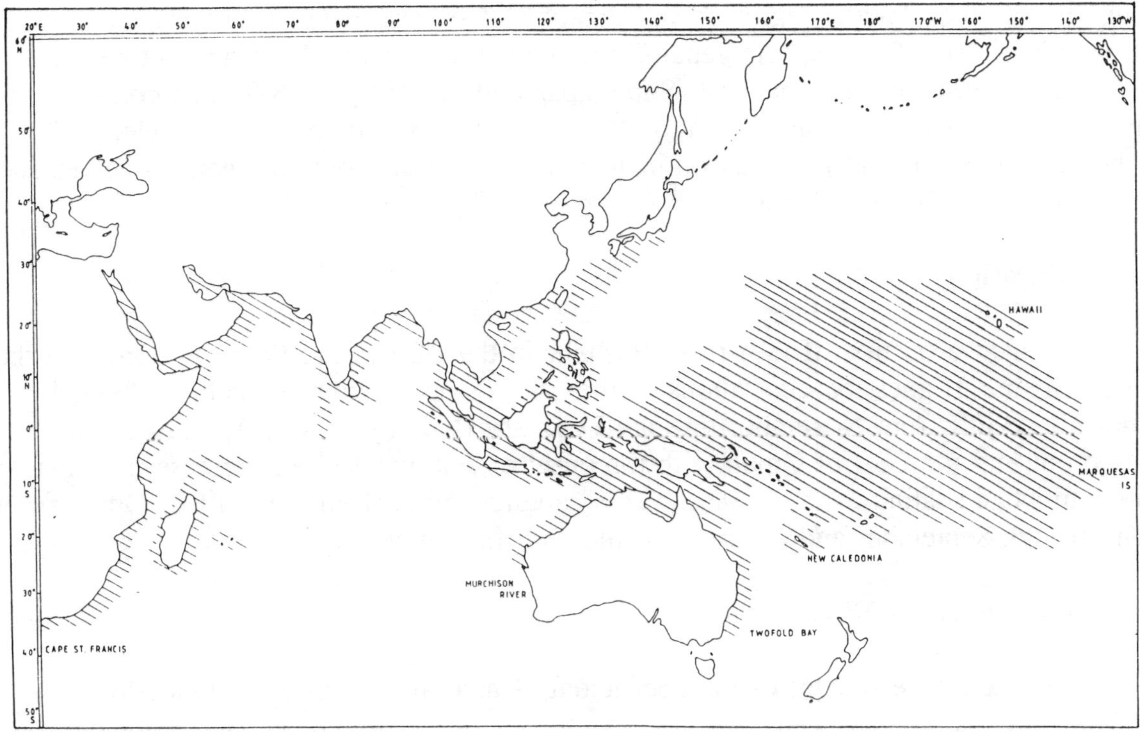

Figure 4. Distribution of kawakawa in the Indo-Pacific region.

Kawakawa are distributed throughout the Indonesian and Philippine archipelagos and off Papua New Guinea.

In the western Pacific Ocean, this species is distributed along the Asian continent from the Malaysian Peninsula eastward to southern mainland China. It is also found off southern Japan, where it is common along the Pacific Ocean coast to about 34°N latitude, but rare along the Japan Sea coast (Kikawa *et al.*, 1963). This species is found throughout the islands of the Pacific Ocean including Caroline Islands, Marshall Islands, Kiribati, Hawaii, Midway Islands, Line Islands, Marquesas Islands, New Caledonia, Vanuatu, and Tahiti (Yoshida, 1979). The southern, northern, and eastern limits of kawakawa in the Pacific Ocean are probably New Caledonia, Hawaii, and Marquesas Islands, respectively.

Kawakawa occurs off the west, north, and east coasts of Australia. The southern most limit of this species on the west coast is Murchison River in western Australia and on the east coast is Twofold Bay in New South Wales (Whitley, 1964).

Kawakawa is replaced by *Euthynnus lineatus*, in the eastern Pacific, but stray specimens have been captured in this region. A 361-mm kawakawa (fork length) was reported from Los Angeles harbour, California, in 1952 (cited in Schaefer, 1987a). A second specimen of 920 mm and 13.15 kg was captured with sport fishing gear off Clarion Island in the Revillagigedo group in December 1986. This specimen exceeds by 1.36 kg the previous record of the heaviest kawakawa (Schaefer, 1987a).

Kawakawa and *Thunnus tonggol* occur together off continents and large islands throughout the Indo-Pacific region from Somalia in the west to Japan and Papua New Guinea in the east. *T. tonggol* is generally dominant in areas with broad continental shelves such as the Gulf of Thailand (Supongpan and Saikliang, 1987), east coast of Peninsular Malaysia (Chee and Yasin, 1988) and north coast of Australia (Lyle, 1988). On the other hand, kawakawa are dominant in areas with narrow continental shelves such as Sri Lanka and the Philippines.

6.2   Migration

Kawakawa occurs off Iran from Bushehr in the Gulf to the Pakistan border in the Gulf of Oman. Highest concentrations of this species occur from Bushehr to Lengeh Island in the Gulf from November through February (Nikouyan, 1988). Kawakawa migrate out into the Gulf of Oman with the onset of summer and an increase in seawater temperatures and salinities in the Gulf. It is captured off Jask in the Gulf of Oman from April through September and returns into the Gulf in winter.

7.   NATURAL MORTALITY

There are no estimates of the coefficient of natural mortality for kawakawa.

8.   OCEANOGRAPHIC FEATURES ASSOCIATED WITH THE SPECIES

Kawakawa are essentially confined to the continental shelf. Of 309 kawakawa captured by trolling off the west coast of Thailand, 47 were taken in the inner-neritic (0 to approximately 50 m), 260 in the outer-neritic (approximately 50-200 m) and only two in the oceanic zone (depths greater than 200 m). Distribution of pole-and-line and troll catches indicated that small kawakawa in the 20- to 40-cm interval were distributed principally in the inner-neritic. Small fish begin leaving the inner-neritic with increasing size, and essentially all have moved into the outer-neritic after 40 cm (Yesaki, 1982).

Kawakawa are rarely captured beyond the edge of the continental shelf. Anderson (1988) reports that the bulk of the kawakawa catch in the Maldives is taken by traditional sailing craft trolling in and around the atolls. On the west coast of Thailand, kawakawa were found under a fish-aggregating device (FAD) deployed in 432-m depths (Lee, 1982). Frusher (1986) notes kawakawa were the most abundant species captured around FADs anchored in 160 m and 390 m off the northwestern coast of Papua New Guinea.

Kishinouye (cited in Yoshida, 1979) reported that kawakawa were found in waters with temperatures ranging from 18° to 28°C off the coast of Japan.

Williamson (1970) reports that sea-surface temperatures off Hong Kong range from 14° to 29°C throughout the year. The waters off the west side of Hong Kong are brackish because of the West River estuary, with surface salinities 33°/.. in February and 6°/.. in July. The waters on the east side are more oceanic, with surface salinities of 34°/.. and 26°/.. during these respective months. Concentrations of kawakawa are reported close inshore on the east side of Hong Kong during June-August and scattered fish occur in deep waters south of Hong Kong throughout the year. This species was not seen in inshore waters during the coldest months nor in brackish waters on the west side.

Mean sea-surface temperatures off the west coast of Thailand increased from a low of 28.6°C in December 1979-January 1980 to a high of 32.2°C in June, and thereafter decreased again to 28.6°C in November 1980 (Yesaki, 1982). Kawakawa were captured throughout the year off this coast, but the highest catches by pole-and-line fishing were made during February, March, and April (Lee, 1982) when sea-surface temperatures were rising.

9. INTERACTIONS WITH OTHER SPECIES

Kawakawa are captured with *Thunnus tonggol* and *Auxis thazard* throughout much of the Indo-Pacific region, including the north coast of Australia (Lyle and Read, 1985), Gulf of Thailand (Spongpan and Saikliang, 1987), west coast of Thailand (Yesaki, 1982), east coast of Peninsular Malaysia (Chee and Yasin, 1988), Bangladesh (Rashid, 1988), west coast of India (Silas *et al.*, 1985a), Gulf of Oman coast of United Arab Emirates (Ali and Thomas, 1979), Oman (Prabhakar and Dudley, 1989), and southern Red Sea (Ben-Yami, 1968). This species is captured with *Sarda orientalis* off the north coast of Australia (Lyle and Read, 1985), west coast of India (Silas, *et al.*, 1985a), and Oman (Prabhakar and Dudley, 1989). Kawakawa are associated with *Megalaspis cordyla* in the Gulf of Thailand (Supongpan and Saikliang, 1987), southern Red Sea (Ben-Yami, 1968) and east Africa (Williams, 1963). Kawakawa also co-occurs with *T. albacares* and *Katsuwonus pelamis* in the coastal waters of Sri Lanka (Sivasubramaniam, 1970) and with *K. pelamis* and *Elagatis bipinnulata* off Papua New Guinea (Wilson, 1981). All these species are probably competitors of kawakawa.

Argue *et al.* (1983) examined stomach contents of pelagic species captured by pole and line for occurrence of juvenile tunas. They found juveniles kawakawa in stomachs of *K. pelamis*, *T. albacares*, and kawakawa (Table 4). The occurrence of juvenile kawakawa in *K. pelamis* stomachs was significantly higher after 1500 hours than earlier in the day. However, the occurrence of juvenile kawakawa in *K. pelamis* stomachs did not differ with respect to distance from land.

Table 4.　　Number and incidence of kawakawa juveniles in the stomachs of predators sampled from tropical waters (modified from Argue *et al.*, 1983).

Predator	Predators examined for full stomach content	tuna juveniles	No of juveniles	Predators with juveniles	Juveniles per 100 predators	% predators with juveniles
Katsuwonus pelamis	3,896	8,175	31	19	0.38	0.23
Thunnus albacares	1,018	1,711	30	2	1.75	0.12
Euthynnus affinis	145	233	2	1	0.90	0.45

Other scientists have reported predation of juvenile kawakawa by tunas. Ronquillo (1953) and Nakamura (1965) noted the presence of kawakawa in the stomachs of *K. pelamis*. Kumaran (1967) found a 10.0-cm kawakawa with a 3.8-cm kawakawa in its stomach. Kawakawa appears to be cannibalistic during much of its life cycle.

Tetrapterus audax and *Carcharinus* sp. are predators of kawakawa off the east coast of Africa (Williams, 1963). *Carcharinus sorrah* preys on kawakawa off the north coast of Australia (Stevens and Wiley, 1986).

10. GENERAL DESCRIPTION OF THE FISHERIES

There are no specific fisheries for kawakawa, but rather fisheries that aim for small tunas, which generally include kawakawa, *T. tonggol*, *A. thazard*, and *S. orientalis*. Descriptions of purse-seine, gillnet, and troll fisheries for small tuna in the South China Sea are given in the "Interactions between fisheries for small tunas off the South China Sea coast of Thailand and Malaysia" included in this volume (Yesaki, 1993).

11. TRENDS IN CATCH, FISHING EFFORT, AND CATCH PER UNIT OF EFFORT

11.1 Annual Catches

Nominal catches of kawakawa have increased from 55,000 mt in 1979 to 128,000 mt in 1988 (Table 5). This doubling of the reported catches in the 11-year period resulted in part from improved statistics collecting and reporting systems in several countries, but for the most part from higher catches in most countries. The most pronounced increases in kawakawa catches were made in Thailand, Malaysia, and Philippines.

11.2 Fishing Effort and Catch per Unit of Effort

The only fishery for which there is a relatively long time-series of effort and catch statistics for kawakawa is the troll fishery off the South China Sea coast of Malaysia. The average number of trips per month by troll-line vessels based in Terengganu declined from 194 trips during the 1982-1985 interval to 146 trips during the 1987-1989 interval (Figure 5). There was a corresponding decrease in kawakawa catches from 22.6 mt to 12.4 mt and in catch rates from 117 to 85 kg/trip during these respective intervals.

12. POPULATION DYNAMICS

No information is available on this subject.

13. INTERACTIONS BETWEEN FISHERIES

Kawakawa differs from most of the other tunas in that almost the entire life cycle is completed in the neritic regime. Larvae, juveniles, preadults, and adults are found over the continental shelf, where fisheries are generally highly developed. Distribution in areas of intense fishing effort for small-pelagic species results in incidental catches of juvenile kawakawa. Juvenile kawakawa are captured in small-mesh gillnets in Sri Lanka,

Table 5. Landings in metric tons of kawakawa in the Indo-Pacific region by country and year.

Country/Year	1979	1980	1981	1982	1983	1984	1985	1986	1987	1988	1989
Australia	0	0	0	0	0	0	0	0	0	0	1
China (Taiwan)1/	5233	7026	4938	7826	8769	10299	13051	7689	12897	0	NA
Comoros	1000	1000	1110	1160	1210	1260	1300	1300	1300	1360	1360
India	14962*	19746*	11648	11619	8849	11224	16529	18116	13880	19045	19045
Iran	198	242	429	716	2620	4135	1707	1870	647	2165	766
Malaysia	1199*	1445*	1960*	1771*	2093*	2503	4380	3755	8164	10126	7363*
Maldives	721	1063	1274	1887	2087	1715	2177	1071	1186	1257	1323
Oman	0	0	0	0	0	0	1041	1173	2544	3420	1708
Pakistan	1466*	676*	1039*	1300*	439*	474*	1309	1225	1067	2937	1407
Philippines	23094	24730	30891	46524	48880	41899	41060	42445	46933	56266	57899
Seychelles	297	474	410	448	314	544	326	323	274	153	232
Sri Lanka	3639*	3813*	5016*	5507*	4129*	3040*	2771*	1360	1541	1580	2170
Thailand	2649*	3057*	6532*	15794*	19813*	19801*	23717*	27739*	39879	26020	26020
United Arab Emirates	0	0	140	2000	1045	1045	980	1396	2108	2260	2347
U.S.A.2/	385	24	23	23	18	0	12	7	4	NA	NA
Yemen, A.R.	500	0	0	0	272	366	378	438	301	272	272
Yemen, P.D.R.	0	0	0	2000	2000	2494	1272	1270	751	1043	980
Total	55343	63296	65410	98575	102538	100799	112010	111177	133476	127904	122893

Source - IPTP, 1991
1/   - SEAFDEC, 1981; 1982; 1983; 1984; 1985; 1986; 1987; 1988; 1989; 1990
2/   - FAO, 1981; 1984; 1989
*   estimated - INDIA   - 1979/1980 - 65% of TUN (% from 1981 - LOT+KAW+SKJ+FRZ+TUN)
              - MALAYSIA - 1979/1983 - 10% of TUN (% from 1984 - LOT+KAW+TUN)
              -           - 1989 - 43% of TUN (% from 1988 - LOT+KAW+FRZ+TUN)
              - PAKISTAN - 1979/1984 - 12% of KAW+SKJ (% from 1986 - YFT+LOT+KAW+FRZ+TUN)
              - SRI LANKA - 1979/1983, 1985 - 44% of TUN (% from 1984 - KAW+FRZ)
              - THAILAND - 1979/1986 - 61% of TUN (% from 1987 - KAW+FRZ)

by purse-seine fisheries off the South China Sea coast of Thailand and Malaysia, and by ringnet (Jabat and Dalzell, 1988), and bagnet (Arce, 1987) fisheries of the Philippines. The incidental captures of juvenile and preadult kawakawa may be the most significant factor affecting the productivity of kawakawa stocks.

Figure 5.   Catch, effort and catch per unit of effort of kawakawa for troll-line vessels based in Terengganu, Malaysia.

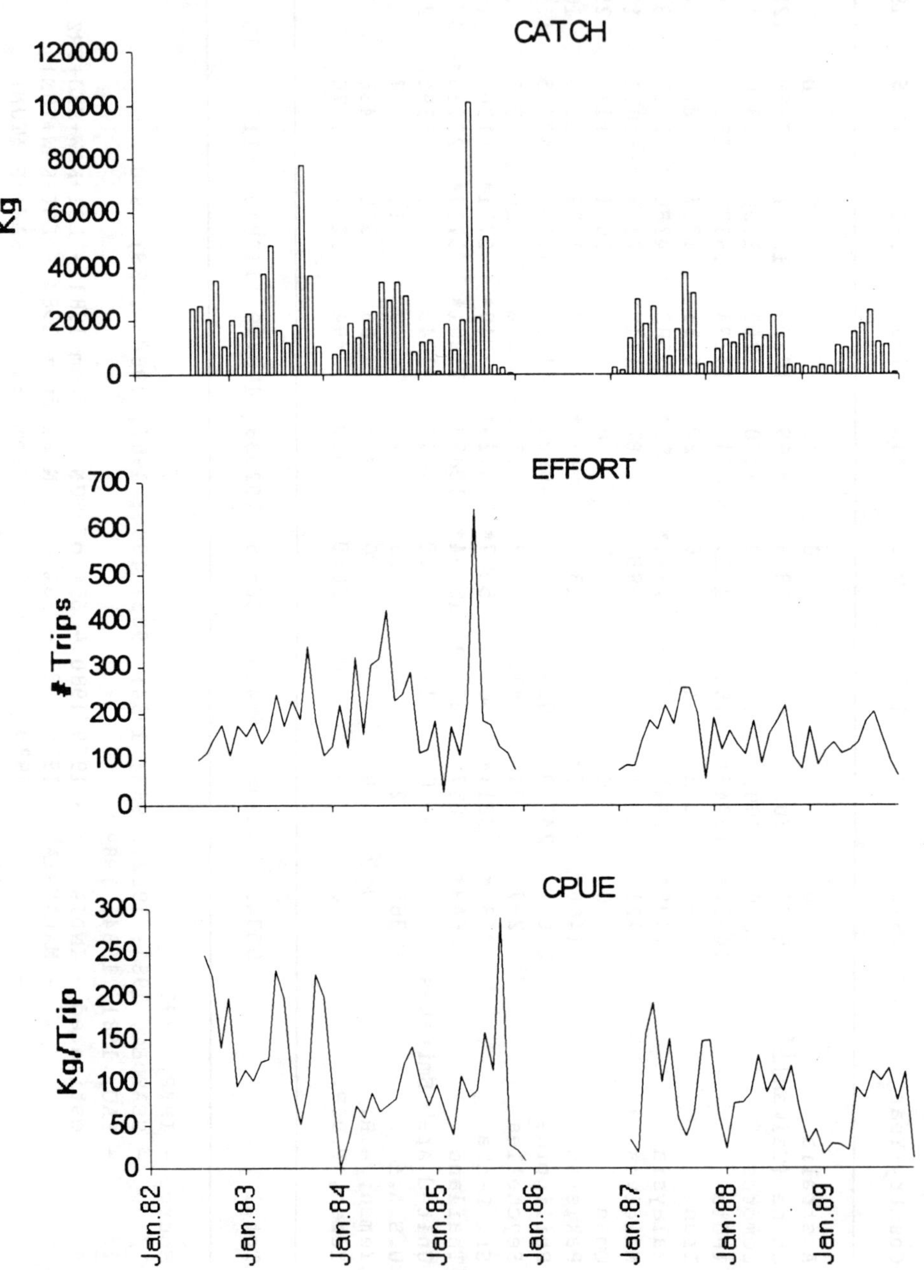

14. REFERENCES CITED

Abuso, Z.V. 1988. Occurrence and distribution of larval tunas in Ragay Gulf, Burias Pass, Ticao Pass and waters north of Samar Sea. *Indo-Pac.Tuna Dev.Mgt.Programme*, IPTP/88/GEN/15:164-79.

Ali, R.M., and P.J. Thomas. 1979. Fish landing survey in Khorfakkan, November 1976 to October 1977. *Tech.Rep.Dep.Fish.Minist.Agri.Fish., United Arab Emirates*, (3):58 p.

Anderson, R.C. 1988. Small tunas, seerfishes and billfishes in the Maldives. *Indo-Pac.Tuna Dev.Mgt.Programme*, IPTP/87/GEN/13: 38-45.

Arce, F.M. 1987. The *Auxis* spp. fisheries of Batangas, Philippines. *Indo-Pac.Tuna Dev.Mgt.Programme*, IPTP/87/GEN/12:137-44.

Argue, A.W., F. Conand, and D. Whyman. 1983. Spatial and temporal distributions of juvenile tunas from stomachs of tunas caught by pole-and-line gear in the central and western Pacific Ocean. *Tech.Rep.Tuna Billfish Assess.Programme, S.Pac.Comm.*,9:47 p.

Ben-Yami, M. 1968. Observations on the distribution and behavior of pelagic schooling fish in the southern Red Sea. *Bull.Sea Fish.Res.Sta., Haifa*, (51):31-46.

Buñag, D.M. 1956. Spawning habits of some Philippine tuna based on diameter measurements of the ovarian ova. *J.Philipp.Fish.*, 4:145-77.

Chamchang, C., and R. Chayakul. 1988. Distribution of longtail (*Thunnus tonggol*, Bleeker), kawakawa (*Euthynnus affinis*, Cantor) and frigate tuna (*Auxis thazard*, Lacepede) in the western coast of the Gulf of Thailand. *Indo-Pac.Tuna Dev.Mgt. Programme*, IPTP/87/GEN/13:80-8.

Chee, P.E., and A.H. Yasin. 1988. Status report of Malaysia. *Indo-Pac.Tuna Dev. Mgt.Programme*, IPTP/88/GEN/15:83-7.

Cheunpan, A. 1984. Sexual maturity, size at first maturity and spawning season of longtail tuna (*T. tonggol*), eastern little tuna (*E. affinis*) and frigate mackerel (*T. thazard*) in the Gulf of Thailand. *Rep.Mar.Fish.Div.Dep.Fish., Bangkok*, 43:22 p.

Chiampreecha, B. 1978. Biological studies on tuna and tuna-like fishes in the west of the Gulf of Thailand and off east coast of Peninsular Malaysia. *Pelagic Fish.Rep.Mar. Fish.Div.Dep.Fish., Bangkok*, (4):25 p.

FAO. 1981. Catches and landings, 1980. *FAO Yearb.Fish.Statist.*, (50):386 p.

FAO. 1984. Catches and landings, 1982. *FAO Yearb.Fish.Statist.*, (54):393 p.

FAO. 1989. Catches and landings, 1987. *FAO Yearb.Fish.Statist.*, (64):490 p.

Frusher, S.D. 1986. Utilization of small-scale fish aggregation devices by Papua New Guinea's artisanal fishermen. *In* The First Asian Fisheries Forum, edited by J.L. Maclean, L.B. Dizon and L.V. Hosillos. *Asian Fish.Soc., Manila,* pp. 371-4.

Higgins, B.E. 1970. Juvenile tunas collected by midwater trawling in Hawaiian waters, July-September 1967. *Trans.Am.Fish.Soc.,* 99:60-9.

IPTP. 1991. Indian Ocean and Southeast Asian Tuna Fisheries Data Summary for 1989. *Data Summ.Indo-Pac.Tuna Dev.Mgt.Programme,* (11):96 p.

Jabat, M. and P. Dalzell. 1988. Preliminary stock assessment of the Danao ringnet fishery for bullet tunas and small pelagic fishes in the Camotes Sea, Central Visayas, Philippines. *Tech.Pap.Bur.Fish.Aquat.Resour. Phillip.,* Vol. XI, No. 1:34 p.

Jones, S. 1960. Notes on eggs, larvae and juveniles of fishes from Indian waters. V. *Euthynnus affinis* (Cantor). *Indian J.Fish.,* 7:101-6.

Joseph, L., R. Maldeniya, and M. Van der Knaap. 1987. Fishery and age and growth of kawakawa (*E. affinis*) and frigate tuna (*A. thazard*). *In* Collective Volume of Working Documents presented at the "Expert Consultation on Stock Assessment of Tunas in the Indian Ocean", Colombo, Sri Lanka, 4-8 December, 1986. *Indo-Pac.Tuna Dev.Mgt. Programme,* Vol. 2:113-23.

Kaya, C.M., A.E. Dizon, and S.D. Hendrix. 1981. Induced spawning of a tuna, *Euthynnus affinis*. *Fish.Bull.NOAA-NMFS,* 79:185-7.

Kikawa, S. and Staff. 1963. Synopsis on the biology of little tuna *Euthynnus yaito* Kishinouye 1923. *FAO Fish.Rep.* 6:218-40.

Klinmuang, H. 1978. Preliminary studies on the biology of tunas in the west of the Gulf of Thailand and off the east coast of Peninsular Malaysia. *Pelagic Fish.Rep.Mar. Fish.Div.Dep.Fish.,* Bangkok, 5:27 p.

Kumaran, M. 1967. Studies on the food of *Euthynnus affinis* (Cantor), *Auxis thazard* (Lacepede), *Auxis thynnoides* (Bleeker) and *Sarda orientalis* (Temminck and Schlegel). Proc. Symp. Scombroid Fishes, Part 2. *Symp.Ser.Mar.Biol.Assoc.India,* 1:599-606.

Landau, R. 1965. Determination of age and growth rate in *Euthynnus alleteratus* and *E. affinis* using vertebrae. *Rapp.P.-V.Reun.Comm.Int.Explor.Sci.Mer.Medit.,* 18:241-3.

Lee, R.E.K.D. 1982. Thailand. Fishing for tuna. A report prepared for the pole-and-line fishing in Southern Thailand Project. *FAO.* FI: DP/THA/77/008:65 p.

Lyle, J.M. 1988. Fishery situation report for small tunas: Australia. *Indo-Pac.Tuna Dev.Mgt.Programme,* IPTP/87/GEN/13:46-9.

Lyle, J.M. and A.D. Read. 1985. Tuna in northern Australian waters: a preliminary appraisal. *Fish.Rep.Dep.Ports Fish.,Northern Territory,* 14:41 p.

Matsumoto, W.M. 1958. Description and distribution of larvae of four species of tuna in central Pacific waters. *Fish.Bull.U.S.Fish Wildl.Serv.*, 58:31-72.

Muthiah, C. 1985. Maturation and spawning of *Euthynnus affinis*, *Auxis thazard* and *A. rochei* in the Mangalore inshore area during 1979 to 1982. *In* Tuna fisheries of the exclusive economic zone of India: biology and stock assessment, edited by E. G. Silas. *Bull.Cent.Mar.Fish.Res.Inst.*, *Cochin*, 36:71-85.

Nakamura, E.L. 1965. Food and feeding habits of skipjack (*Katsuwonus pelamis*) from the Marquesas and Tuamoto Islands. *Trans.Am.Fish.Soc.*, 94: 236-42.

Nakamura, E.L., and W.M. Matsumoto. 1967. Distribution of larval tunas in Marquesan waters. *Fish.Bull.U.S.Fish Wildl.Serv.*, 66:1-12.

Nikouyan, A. 1988. Tuna catch and its distribution along the Iranian coast. *In* Collective Volume of Working Documents presented at the "Expert Consultation on Stock Assessment of Tunas in the Indian Ocean", Mauritius, 22-27 June 1988. *Indo-Pac.Tuna Dev.Mgt.Programme*, Vol. 3:265-8.

Nishikawa, Y., M. Honma, S. Ueyanagi, and S. Kikawa. 1985. Average distribution of larvae of oceanic species of scombroid fishes, 1956-1981. *S.Ser.Far Seas Fish.Res.Lab.*, (12):99 p.

Prabhakar, A., and R.G. Dudley. 1989. Age, growth and mortality rates of longtail tuna *Thunnus tonggol* (Bleeker) in Omani waters based on length data. *Indo-Pac.Tuna Dev.Mgt.Programme*, IPTP/89/GEN/16:90-6.

Rao, K.V.N. 1964. An account of the ripe ovaries of some Indian tunas. Proc. Symp. Scombroid Fishes, Part 2. *Symp.Ser.Mar.Biol.Assoc.India*, 1:733-43.

Rashid, M.H. 1988. Tuna and tuna-like fishes in Bangladesh. *Indo-Pac.Tuna Dev.Mgt. Programme*, IPTP/87/GEN/13:122-3.

Ronquillo, I.A. 1953. Food habits of tunas and dolphins based upon the examination of their stomach contents. *J.Philipp.Fish.*, 2:71-83.

Ronquillo, I.A. 1963. A contribution to the biology of Philippine tunas. *FAO Fish.Rep.*, 6:1683-752.

Schaefer, K.M. 1987a. Second record of the kawakawa *Euthynnus affinis* from the eastern Pacific Ocean. *Fish.Bull.NOAA-NMFS*, 85:647-8.

Schaefer, K.M. 1987b. Reproductive biology of black skipjack, *Euthynnus lineatus*, an eastern Pacific tuna. *Bull.I-ATTC*, 19:169-260.

Shabotinets, E.I. 1968. Age determination of Indian Ocean tunas. Tr. VINRO 64, *Tr. AzcherNIRO* 28: 374-376. [Engl. trans. by W. L. Klawe, 1968. Inter-Am. Trop. Tuna Comm., La Jolla, Calif.]

Silas, E.G., P.P. Pillai, A.A. Jayaprakash, and M.A. Pillai. 1985a. Fishery and bionomics of tunas at Cochin. *In* Tuna fisheries of the exclusive economic zone of India: biology and stock assessment, edited by E. G. Silas. *Bull.Cent.Mar.Fish.Res.Inst., Cochin*, 36:28-50.

Silas, E.G., P.P. Pillai, M. Srinath, A.A. Jayaprakash, C. Muthiah, V. Balan, T.M. Yohannan, P. Siraimeetan, M. Mohan, P. Livingston, K.K. Kunhikoya, M.A. Pillai, and P.S. Sarma. 1985b. Population dynamics of tunas: stock assessment. *In* Tuna fisheries of the exclusive economic zone of India: biology and stock assessment, edited by E. G. Silas. *Bull.Cent.Mar.Fish.Res.Inst., Cochin*, 36:20-7.

Siraimeetan, P. 1985. On the occurrence, size distribution, morphometry and feeding habits of the juveniles of *Euthynnus affinis* (Cantor), *Auxis thazard* (Lacepede) and *Sarda orientalis* (Temminck and Schlegel) along the Tuticorin coast in the Gulf of Mannar, south-east coast of India. *In* tuna fisheries of the exclusive economic zone of India: biology and stock assessment, edited by E. G. Silas. *Bull.Cent.Mar.Fish.Res.Inst., Cochin*, 36:104-14.

Sivasubramaniam, K. 1970. Biology of the exploited stock of mackerel tuna *E. affinis* (Cantor) off the south-west region of Ceylon. *Bull.Fish.Res.Stn., Ceylon*, 21(1):7-16.

Smith, M.M., and P.C. Heemstra (eds.). 1986. Smith's sea fishes. Springer-Verlag Berlin, Heidelberg, New York, London, Paris, Tokyo, 1047 p.

SEAFDEC. 1981. Fishery Statistical Bulletin for South China Sea Area 1979. *Southeast Asian Fish.Dev.Center*, 227 p.

SEAFDEC. 1982. Fishery Statistical Bulletin for South China Sea Area 1980. *Southeast Asian Fish.Dev.Center*, 268 p.

SEAFDEC. 1983. Fishery Statistical Bulletin for South China Sea Area 1981. *Southeast Asian Fish.Dev.Center*, 275 p.

SEAFDEC. 1984. Fishery Statistical Bulletin for South China Sea Area 1982. *Southeast Asian Fish.Dev.Center*, 231 p.

SEAFDEC. 1985. Fishery Statistical Bulletin for South China Sea Area 1983. *Southeast Asian Fish.Dev.Center*, 178 p.

SEAFDEC. 1986. Fishery Statistical Bulletin for South China Sea Area 1984. *Southeast Asian Fish.Dev.Center*, 231 p.

SEAFDEC. 1987. Fishery Statistical Bulletin for South China Sea Area 1985. *Southeast Asian Fish.Dev.Center*, 178 p.

SEAFDEC. 1988. Fishery Statistical Bulletin for South China Sea Area 1986. *Southeast Asian Fish.Dev.Center*, 175 p.

SEAFDEC. 1989. Fishery Statistical Bulletin for South China Sea Area 1987. *Southeast Asian Fish.Dev.Center*, 162 p.

SEAFDEC. 1990. Fishery Statistical Bulletin for South China Sea Area 1988. *Southeast Asian Fish.Dev.Center*, 183 p.

Steinberg, R., W. Weber, U. Lowenberg, and T. Kunzel. 1982. Final report of the Joint Fisheries Project of the Government of the Republic of Seychelles and the Government of the Republic of Germany. Federal Research Center for Fisheries, Hamburg and Bremerhavener Fischversorgung, Heinrich Abelmann OHG, Bremerhaven, 101 p.

Stevens, J.D. and P.D. Wiley. 1986. Biology of two commercially important carcharhinid sharks from northern Australia. *Aust.J.Mar.Freshwat.Res.*, 37:671-88.

Supongpan, S. and P. Saikliang. 1987. Fisheries status of tuna purse seiners (using sonar) in the Gulf of Thailand in 1983. *Rep.Mar.Fish.Div.Dep.Fish., Bangkok*, 3:78 p.

Uchiyama, J.H. 1980. Survey of the pelagic fishes of the northwestern Hawaiian Islands. *In* Status of resource investigations in the northwestern Hawaiian Islands, edited by R. W. Grigg and R. T. Pfund. Univ. Hawaii Sea Grant College Prog., Honolulu, Hawaii, pp. 251-63.

Wade, C.B. 1950a. Juvenile forms of *Neothunnus macropterus*, *Katsuwonus pelamis* and *Euthynnus yaito* from Philippine seas. *Fish.Bull.U.S.Fish Wildl.Serv.*, 51:395-404.

Wade, C.B. 1950b. Observations on the spawning of Philippine tuna. *Fish.Bull.U.S.Fish Wildl.Serv.*, 51:409-23.

Whitley, G.P. 1964. Scombroid fishes of Australia and New Zealand. Proc. Symp. Scombroid Fishes, Part 1. *Symp.Ser.Mar.Biol.Assoc.India*, 1:221-53.

Williams, F. 1963. Synopsis of biological data on little tuna *Euthynnus affinis* (Cantor) 1850 (Indian Ocean). *FAO Fish.Rep.*, 6:167-79.

Williamson, G.R. 1970. Little tuna *Euthynnus affinis* in the Hong Kong area. *Bull.Jap. Soc.Sci.Fish.*, (36):9-18.

Wilson, M.A. 1981. Aspects of the biology and production of mackerel tuna in Oceania. *In* Northern pelagic fish seminar, edited by C. J. Grant and D. G. Walter. *Aust.Gov.Publ.Serv.*, pp. 45-50.

Yabe, H., N. Anraku, and T. Mori. 1953. Scombroid youngs found in the coastal seas of Aburatsu, Kyushu, in summer. *Contr.Nankai Reg.Fish.Res.Lab.*, 11:10 p.

Yesaki, M. 1982. Thailand. Biological and environmental observations. A report prepared for the Pole-and-Line Tuna Fishing in Southern Thailand Project. *FAO.* FI:DP/THA/77/008:46 p.

Yesaki, M. 1989a. Synopsis of biological data on kawakawa, *Euthynnus affinis*. *Indo-Pac.Tuna Dev.Mgt.Programme*, IPTP/89/WP/19:55 p.

Yesaki, M. 1989b. Estimates of age and growth of kawakawa (*Euthynnus affinis*), longtail tuna (*Thunnus tonggol*) and frigate tuna (*Auxis thazard*) from the Gulf of Thailand based on length data. *Indo-Pac.Tuna Dev.Mgt.Programme*, IPTP/89/GEN/17:94-108.

Yesaki, M. 1993. Interactions between fisheries for small tunas off the South China Sea coasts of Thailand and Malaysia. *In* Proceedings of the FAO Expert Consultation on Interactions of Pacific Tuna Fisheries, edited by R.S. Shomura, J. Majkowski, and S. Langi, 3-11 December 1991, Noumea, New Caledonia. [See this document.]

Yoshida, H.O. 1979. Synopsis of biological data on tuna of the Genus *Euthynnus*. *NOAA Tech.Rep.NMFS Circ.*, (429):57 p.

A REVIEW OF THE *AUXIS* FISHERIES OF THE PHILIPPINES AND SOME ASPECTS OF THE BIOLOGY OF FRIGATE (*A. THAZARD*) AND BULLET (*A. ROCHEI*) TUNAS IN THE INDO-PACIFIC REGION

Mitsuo Yesaki
Indo-Pacific Tuna Development and Management Programme
Colombo, Sri Lanka

and

Flerida Arce
Bureau of Fisheries and Aquatic Resources
Quezon City, Philippines

1. INTRODUCTION

The "Synopsis of biological data on frigate tuna, *Auxis thazard*, and bullet tuna, *A. rochei*," was published in 1981 (Uchida, 1981). This synopsis summarizes all available information concerning these species throughout the tropical and subtropical oceans and seas. There has been little new information published on the biology of these species since this date.

The world catch of *Auxis* species has increased 62% since publication of the above synopsis and in 1988 was 176,000 mt, of which 107,000 mt (61%) was caught in the Philippines (FAO, 1988). The only other country reporting catches exceeding 10,000 mt was Japan with 25,000 mt (14%). The total world catch is low considering it is generally acknowledged that *Auxis* is the most abundant tuna, in numerical terms, in the world's oceans (FAO, 1976). There are several reasons for the present low catch of *Auxis* including the following:

(a) Low demand because there is no international market for these species.

(b) Under-reporting of catches, especially by the tuna purse-seine fleets operating in the Pacific, Atlantic, and Indian Oceans.

(c) Confinement of fisheries in many developing countries to near-shore areas of low *Auxis* abundance.

(d) Primitive level of development of purse-seine fisheries in most developing countries.

The present review of *Auxis* consists of two parts. The first examines in detail the fisheries of the Philippines. The Philippines is the only country in the world with highly developed fisheries for *Auxis*, so a review of its fisheries will cover many essential aspects of *Auxis* fisheries. The second part summarizes current information for the Indo-Pacific region on the biology of *Auxis* species, especially on reproduction and population characteristics which are essential for the assessment of stocks.

2. THE AUXIS FISHERIES OF THE PHILIPPINES

2.1 Source of Information

Information presented in this review of the *Auxis* fisheries in the Philippines was obtained primarily from three sources. These were:

(a) Fisheries Statistics of the Philippines, Volumes 28-37 published by the Bureau of Fisheries and Aquatic Resources (BFAR), Ministry of National Resources, Metro Manila, Philippines.

(b) The BFAR, under the supervision of the South China Sea Fisheries Programme and with funding from Norway through the Programme of Assistance to Developing Countries in Planning the Utilization of the Living Resources in the Expanded Economic Zones (EEZs), initiated a 1-year tuna-sampling programme in late 1979. The objectives of this programme were to obtain data on species and length compositions, and seasonality of tuna and associated species landed at four sites on Mindanao Island (White and Yesaki, 1982). These included General Santos (Moro Gulf), Santa Cruz (Davao Gulf), Labuan (East Sulu Sea) and Opol (Bohol Sea). The BFAR continued the tuna sampling programme at these and other sites after 1980 with assistance from the Japan Trust Fund Project through the Indo-Pacific Tuna Development and Management Programme (IPTP). Sampling effort was directed primarily towards monitoring the catches of ringnets and handlines and secondarily for purse seine, bagnet, gillnet, and troll lines. However, only information for ringnets at General Santos, Santa Cruz, and Opol and for purse seine at General Santos was summarized in the present analysis, as *Auxis* catches by the other gears were either insignificant or erratic. Also, other landing sites besides the above mentioned four were monitored for varying periods of time. Information from some of these sites were included with the major sampling sites if the sites were close to each other and the vessels were similar in size. In the present analysis, Malita was included with Santa Cruz and Initao with Opol.

(c) The BFAR initiated a small-pelagic species sampling programme at various locations throughout the Philippines in 1983. These included Wawa and Lemery in Batangas Coast, Danao in Camotes Sea, Daliao, and Talomo in Davao Gulf. The bagnet fishery at Wawa and the ringnet fishery at Lemery were monitored for 19 and 21 months respectively, whereas, the ringnet fisheries at the other sites were monitored until the end of 1989. *Auxis* constituted an important component of the landings at all these sites.

2.2 Tuna Landings

The Philippine landings of tuna increased from 184,000 mt in 1978 to 302,000 mt in 1989 (Figure 1). *Auxis* is the most important tuna, in terms of landed weight, accounting for 39% of the 1989 total. Skipjack (*Katsuwonus pelamis*), yellowfin

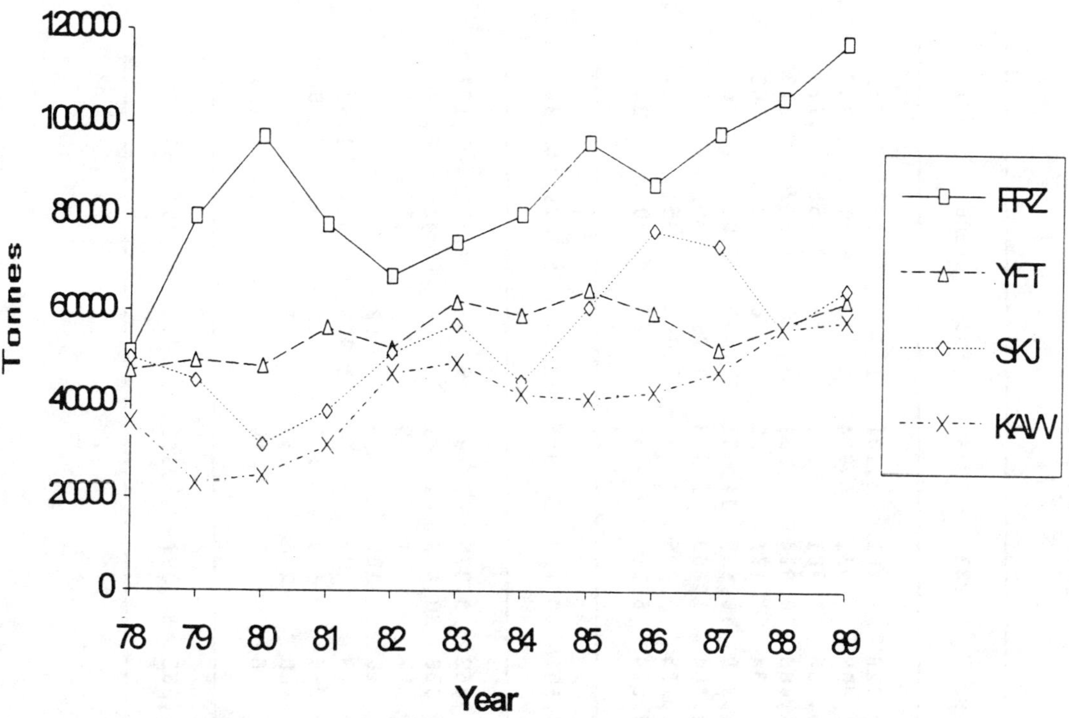

Fig. 1. Tuna landings by species in the Philippines from 1978 to 1989 (FRZ includes frigate and bullet tunas; YFT includes yellowfin and bigeye tunas).

(*Thunnus albacares*), and kawakawa (*Euthynnus affinis*) each accounted for approximately 20% of the remainder. *Auxis* and kawakawa landings increased by 131% and 59%, respectively from 1978 to 1989. On the other hand, yellowfin and skipjack landings increased by only 32% and 30%, respectively, during this 11-year interval. Landings of yellowfin peaked in 1985 and those of skipjack in 1986. Landings statistics for these two species in recent years may be overestimated by inclusion of fish captured in extra-territorial waters. Tuna purse seiners from at least three Philippine companies started fishing in the western Pacific Ocean in 1985. Yellowfin and skipjack captured by these vessels may be included in the Philippine landing statistics.

2.3 Landings by Gear Type

Philippine landings statistics are classified into commercial and municipal fisheries on the basis of fishing craft size. Crafts less than 3 gross tons (GT) are considered as municipal and larger crafts as commercial. In 1978, *Auxis* landings by the municipal sector exceeded that by the commercial sector and in the following year landings by these two sectors were about equal (Table 1). In all subsequent years except 1983, the landings by the commercial sector exceeded those of the municipal sector.

Twelve municipal fishing gears are listed in Philippine statistics as capturing *Auxis* (Table 1). The most important is the hook and line (including troll line, and pole and line), which accounted for 32-57% of the municipal landings during the 1978-87 interval. Other important gears are gillnet (8-29%) and purse seine/ringnet (5-40%).

Table 1. Landings of _Auxis_ by gear type in the Philippines from 1978 to 1987 (metric tonnes).

Fishing gear		1978	1979	1980	1981	1982	1983	1984	1985	1986	1987
Municipal	Bagnet (BAG)	404	540	679	231	128	1,462	2,281	1,368	244	442
	Purse seine/ringnet (PS/RN)	8,174	15,070	11,275	3,748	1,858	2,031	4,964	6,520	4,511	5,903
	Other seines (SEN)1/	433	406	7,599	676	1,106	381	611	386	2,255	1,213
	Gillnet (GILL)	4,208	4,525	6,426	7,708	4,483	8,924	9,521	8,128	11,132	7,319
	Trawl (TRAW)	10	–	32	2	534	171	25	6	33	62
	Hook-and-line (HOOK)2/	10,965	19,070	14,150	16,855	12,369	16,508	14,051	21,430	23,020	23,161
	Longline (LL)	393	334	567	433	2,264	1,603	1,358	1,268	1,382	1,697
	Fish corral (TRAP)	473	101	2,625	1,034	1,135	2,805	79	3,066	351	341
	Others (UNCL)3/	18	169	211	420	3,624	6,237	55	68	101	224
	Total	25,078	40,215	43,564	31,107	27,501	40,122	32,945	42,240	43,029	40,362
Commercial	Bagnet (BAG)	1,728	7,537	19,431	15,082	4,452	3,075	5,781	9,681	7,271	7,167
	Purse seine/ringnet (PS/RN)	18,865	31,990	32,904	31,638	35,038	30,403	41,081	41,689	36,435	48,320
	Other seines (SEN)4/	73	64	170	23	27	1	9	5	144	28
	Gillnet (GILL)	89	–	–	75	18	125	–	52	–	–
	Trawl (TRAW)	–	1	446	11	23	109	1	1,978	301	1,892
	Hook-and-line (HOOK)	102	81	356	204	236	123	117	19	31	80
	Longline (LL)	–	14	–	99	60	254	357	36	5	59
	Others (UNCL)	2	7	3	9	8	7	14	18	10	124
	Total	20,859	39,694	53,310	47,141	39,862	34,097	47,360	53,478	44,196	57,670

1/ - includes beach and round-haul seines
2/ - includes handline and troll-line and pole-and-line
3/ - includes others and driven-in net
4/ - includes round-haul seine

Eight commercial fishing gears are identified as capturing *Auxis* (Table 1). Purse seine/ringnet is the most important gear accounting for 62-90% of the commercial landings from 1978 to 1987. The other important gear is the bagnet with catches making up 8-36% of the commercial landings of *Auxis*.

2.4 Landings by Fishing Areas

Twenty-four fishing areas are identified for the Philippine landing statistics (Figure 2). The bulk of the *Auxis* landings is made in archipelagic waters, with negligible catches from the fishing areas along the exposed Pacific Ocean and South China Sea coasts (Table 2).

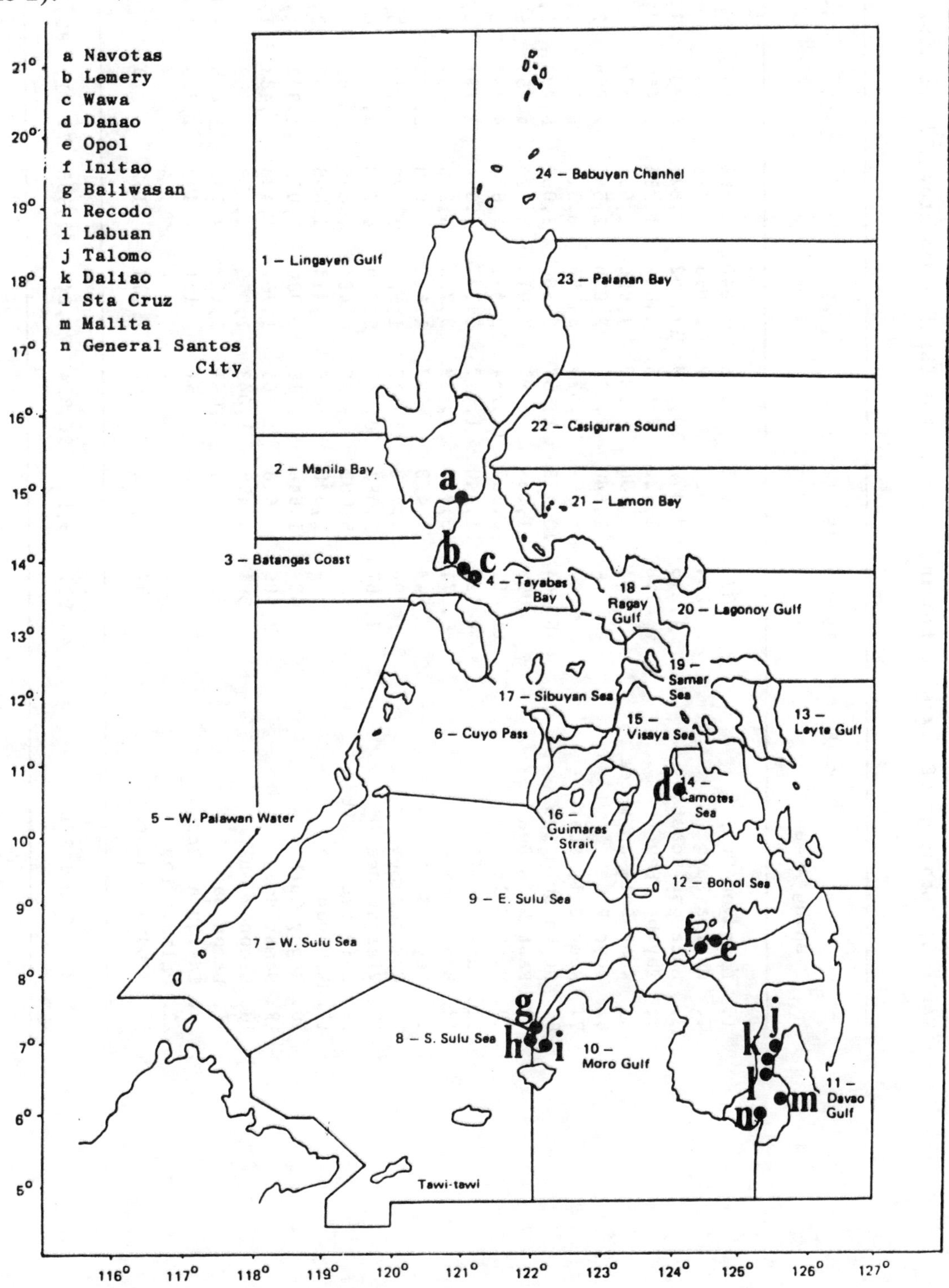

Table 2. Landings of _Auxis_ by fishing areas of the Philippines from 1978 to 1987 (metric tonnes).

Fishing Area	1978	1979	1980	1981	1982	1983	1984	1985	1986	1987
1. Lingayen Gulf	321	675	256	640	600	650	537	942	618	657
2. Manila Bay	254	356	942	2,055	191	235	703	1,175	725	2,361
3. Batangas Coast	1,215	7,472	10,229	11,962	3,826	4,288	4,989	6,933	8,408	10,793
4. Tayabas Bay	758	1,120	2,825	2,882	1,435	1,459	3,094	4,492	3,935	2,346
5. West Palawan water	81	990	120	1,056	1,633	377	827	407	205	288
6. Cuyo Pass	-	1,640	2,329	2,025	2,598	2,855	2,614	1,094	2,410	2,365
7. West Sulu Sea	552	-	10,629	8,766	8,315	5,516	4,014	5,550	7,213	6,518
8. South Sulu Sea	10,444	10,173	1,337	2,660	3,052	6,494	2,389	5,397	2,986	4,035
9. East Sulu Sea	1,093	1,259	997	7,014	2,126	4,959	6,523	5,300	4,609	8,137
10. Moro Gulf	11,451	16,307	12,254	12,491	14,690	14,437	22,960	26,664	21,857	27,713
11. Davao Gulf	2,123	1,440	1,577	2,777	1,574	1,781	2,212	1,187	1,810	1,729
12. Bohol Sea	6,571	15,957	14,539	5,780	6,856	12,557	11,643	14,182	10,586	10,665
13. Leyte Gulf	844	468	337	1,642	840	2,415	528	412	626	753
14. Camotes Sea	1,196	4,652	3,135	1,683	1,520	3,714	3,454	2,081	3,165	4,039
15. Visayan Sea	3,187	2,598	10,448	5,209	4,533	3,601	4,251	6,687	3,892	2,993
16. Guimaras Strait	8	8,002	926	484	1,626	680	552	1,074	1,363	2,562
17. Sibuyan Sea	-	1,348	829	2,414	4,058	1,167	2,057	1,704	2,264	2,316
18. Ragay Gulf	683	1,646	17,788	1,900	1,032	1,902	1,011	420	442	1,948
19. Samar Sea	3,586	618	1,065	1,762	1,423	1,483	1,434	4,273	4,816	1,201
20. Lagonoy Gulf	504	455	1,043	1,055	2,178	559	416	1,338	1,471	1,258
21. Lamon Bay	929	2,614	3,061	1,417	2,746	1,488	3,005	3,091	2,321	2,731
22. Casiguran Sound	-	78	51	101	409	1,323	474	581	425	425
23. Palanan Bay	-	-	-	-	-	-	-	-	-	-
24. Babuyan Channel	137	41	157	473	102	279	618	734	1,078	199
T O T A L	45,937	79,909	96,874	78,248	67,363	74,219	80,305	95,718	87,225	98,032

Fishing areas in archipelagic waters were classified as either shallow-water or deep-water, depending upon whether the extent of the continental shelf was more or less than half, respectively, of the entire area. Munro (1986) gives the areas of neritic and oceanic zones of those fishing areas bounded by land masses. Yesaki (1983) estimated the extent of the Moro Gulf and Davao Gulf fishing areas.

The catch trends of *Auxis* and kawakawa from 1978 to 1987 for shallow-water and deep-water fishing areas are shown in Figures 3 and 4, respectively. Kawakawa landings are generally higher in the shallow-water than in the deep-water fishing areas. *Auxis* landings in shallow-water fishing areas show marked annual fluctuations, especially in Guimaras Strait, Ragay Gulf, Samar Sea and, to a lesser extent, in Tayabas Bay and Visayan Sea. On the other hand, *Auxis* landings in the deep-water fishing areas are more stable with a slight increase during the decade under review. The most productive shallow-water and deep-water fishing areas for *Auxis* are Visayan Sea and Moro Gulf, respectively.

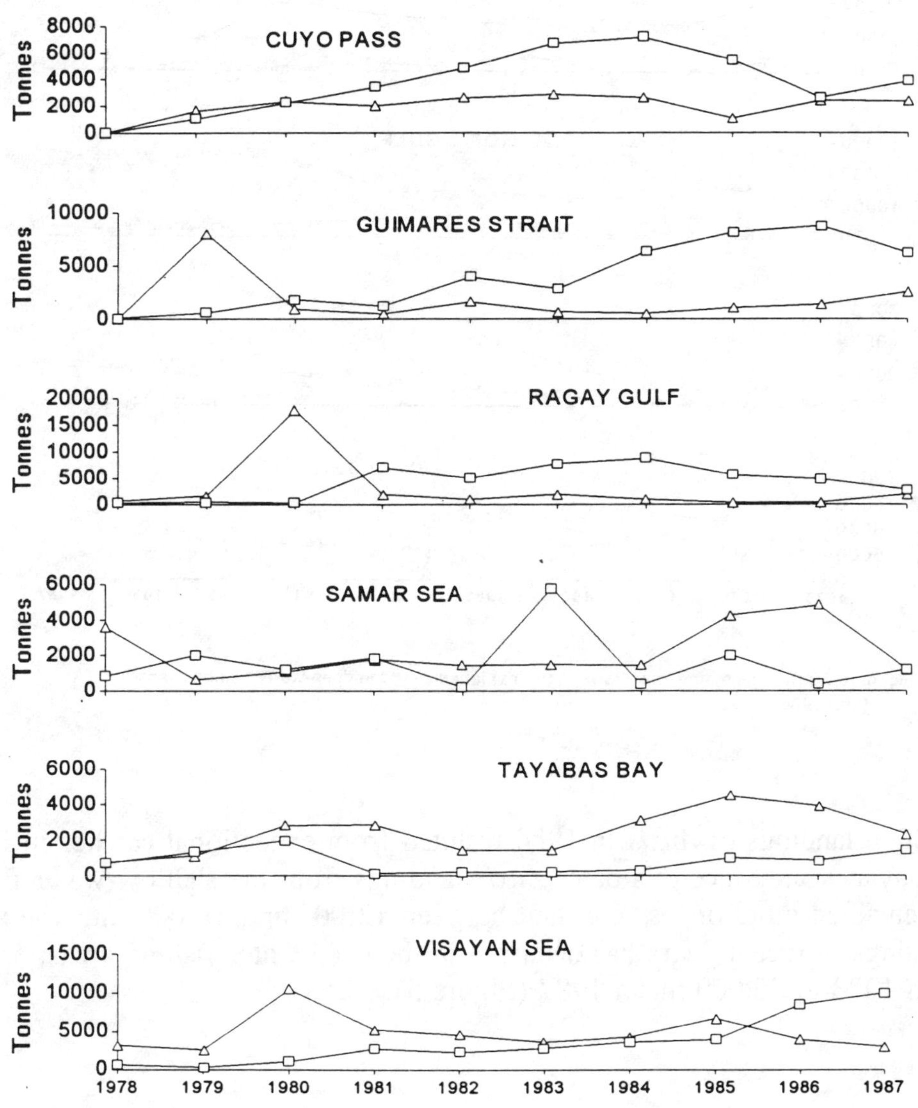

Fig.3. Auxis (triangles) and kawakawa (squares) landings in the shallow-water fishing areas from 1978 to 1987.

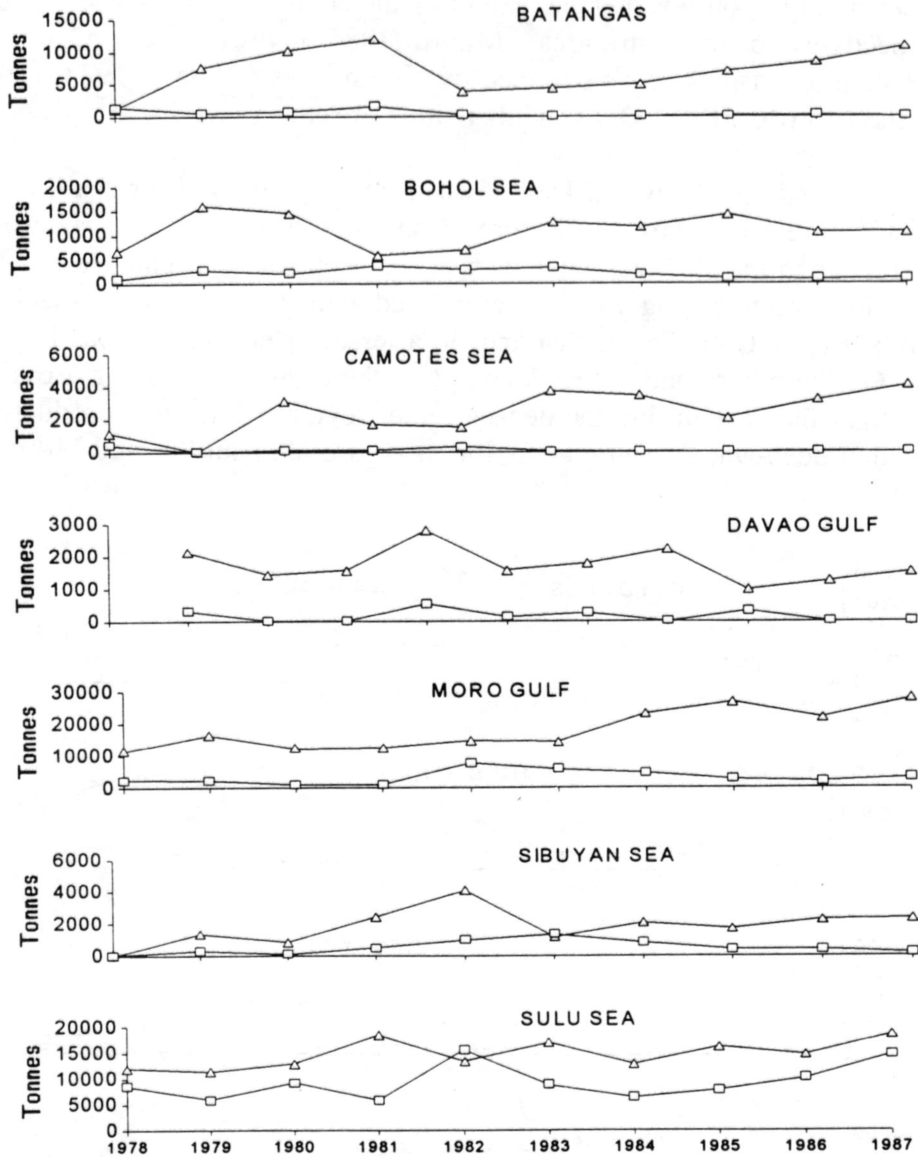

Fig. 4. Auxis (triangles) and kawakawa (squares) landings in the deep-water fishing areas from 1978 to 1987.

The high landings of *Auxis* in 1980 resulted from exceptional catches in Ragay Gulf and Visayan Sea. Since this date, *Auxis* landings from the shallow-water fishing areas have remained more or less constant between 12,000 and 18,000 mt. the increase in *Auxis* landings in recent years has occurred in the deep-water fishing areas; from 35,000 mt in 1978 to 76,000 mt in 1987 (Figure 5).

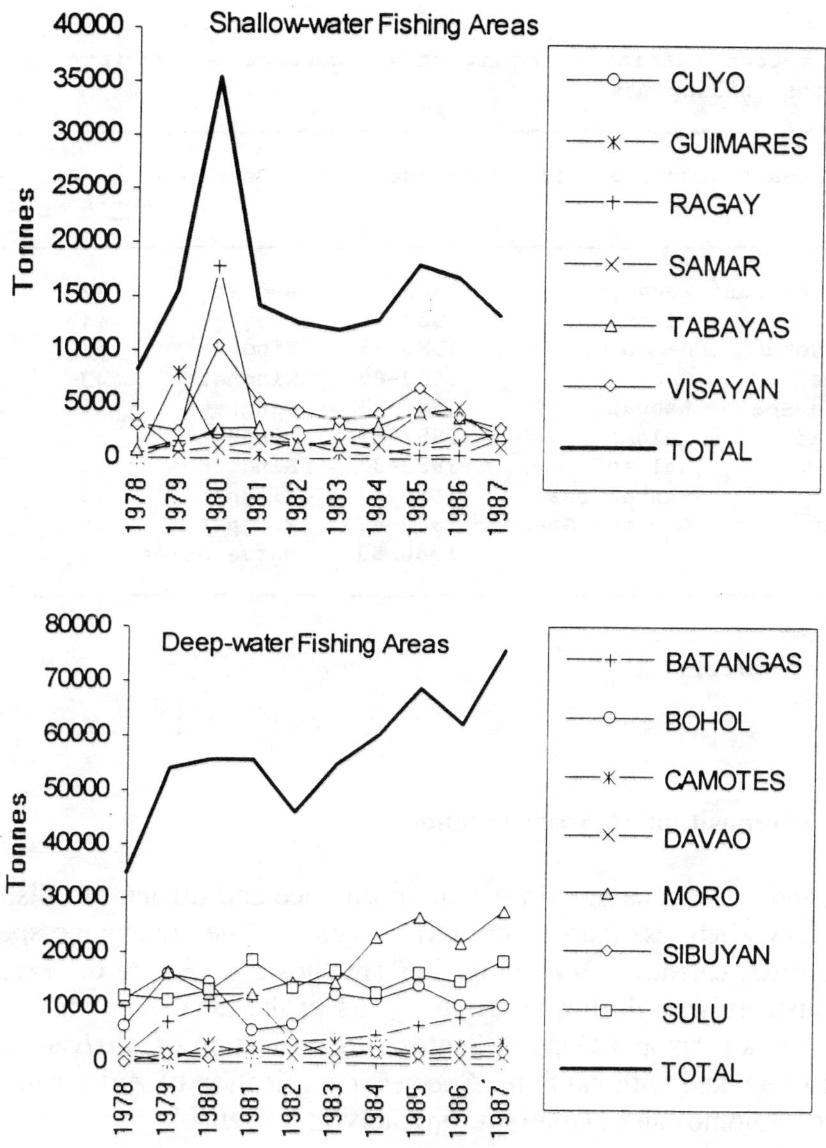

Fig. 5. Auxis landings in the shallow-water and deep-water fishing areas from 1978 to 1987.

2.5   Landings by Sites

2.5.1 Species composition of landings

Table 3 shows the percentage species composition of catches by various gears at nine landing sites in the Philippines. The percentage of *Auxis* was high (43-72%) for enclosed archipelagic fishing areas (Batangas Coast, Camotes Sea, Bohol Sea) and low (15-22%) for the exposed archipelagic fishing areas (East Sulu Sea, Davao Gulf, Moro Gulf). Tunas comprised 60-65% of the catch in the East Sulu Sea and Moro Gulf, a considerably higher percentage than in the other fishing areas.

Table 3. **Percentage species composition of catches at various landing sites in the Philippines.**

Fishing area	Landing site	Time interval	Gear	Species groups		
				Auxis	Tuna	Others
3. Batangas Coast[1]	Wawa	1983	Bagnet	72	1	27
	Lemery	1983-84	Ringnet	44	0	56
14. Camotes Sea[2]	Danao	1983-89	Ringnet	57	1	42
12. Bohol Sea	Opol	1980-89	Ringnet	43	19	38
9. East Sulu Sea	Labuan	1983-87	Ringnet	16	65	19
11. Davao Gulf	Talomo	1983-87	Ringnet	16	10	74
	Daliao	1983-87	Ringnet	22	22	56
	Santa Cruz	1980-87	Ringnet	15	43	42
10. Moro Gulf	General Santos	1980-89	Ringnet	6	19	75
		1980-89	purse seine	6	27	67

1/ - Arce, 1987
2/ - Jabat and Dalzell, 1988

2.5.2 Species composition of *Auxis* catches

Frigate and bullet tuna are similar in appearance and difficult to distinguish from external characters when less than 15 cm (Arce, 1987). The percentage species composition of *Auxis* catches at 9 landing sites are shown in Figure 6. Frigate and bullet tuna were not distinguished during the first 2 years of the tuna sampling programme at Opol, Labuan, Santa Cruz and General Santos. Identification of *Auxis* species at these sites should still be taken with caution. Species identification of *Auxis* catches at Wawa, Lemery, Danao, Talomo, and Daliao are probably more reliable.

Bullet tuna accounts for almost the entire *Auxis* catch by bagnet at Wawa (Batangas Coast) and ringnet at Danao (Camotes Sea). The percentage of frigate tuna is higher at the other landing sites. The ratio of bullet to frigate tuna can reverse from year to year; for example at Lemery, Talomo, and Daliao. In general, bullet tuna accounts for a high percentage of the *Auxis* catch in the enclosed archipelagic fishing areas (Batangas Coast, Camotes Sea, and Bohol Sea) and about half of the *Auxis* catch in the exposed archipelagic fishing ground (Moro Gulf).

2.5.3 *Auxis* catch rates by landing sites

Catch rates by ringnet, purse seine and bagnet at seven landing sites in the Philippines were examined to detect changes in relative abundance. These three gears account for over 50% of the Philippine catch of *Auxis*. Ringnetters, purse seiners, and bagnetters focus their effort on a complex of pelagic species, including tuna (skipjack, yellowfin, and kawakawa), *Auxis* species and small-pelagic species [big-eye scads (*Selar* spp.), round scads (*Decapterus* spp.), Indian mackerels (*Rastrelliger* spp.), sardines

(*Sardinella* spp.), etc.]. These species are either captured under payaos (Fish Aggregating Devices) or in free-swimming schools. As these gears catch a complex of species, it is necessary when reviewing catch rates for a specific species, to do so in relation to the total catch.

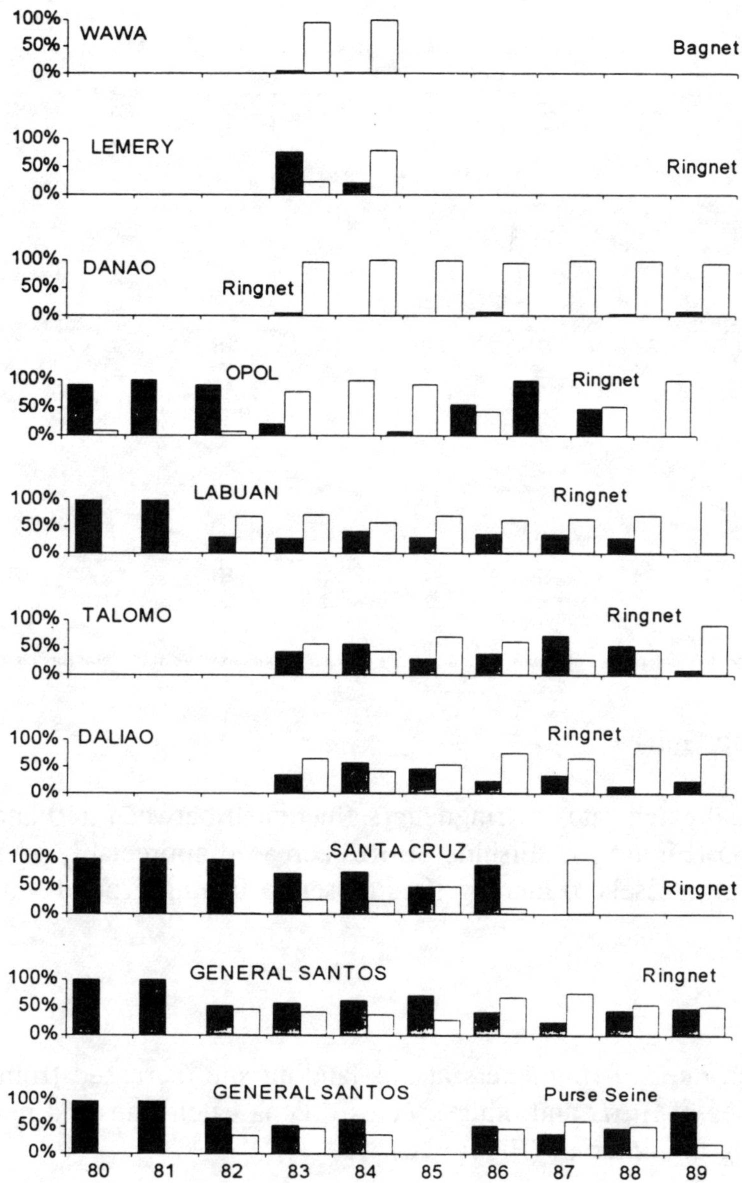

Fig. 6. Percentage species composition of Auxis catches by various fishing gears at nine landing sites in the Philippines (filled bar-frigate tuna; clear bar-bullet tuna).

General Santos (Moro Gulf)

Purse-seine catch rates of *Auxis* and other species were low in 1980-81, but have since increased (Figure 7). Annual catch rates of tuna have increased during the 10-year interval (Table 4).

Ringnet catch rates of all three species groups were low till mid-1982, but have since increased markedly (Figure 8). Annual catch rates of *Auxis* and tuna show a generally increasing trend since mid-1982 (Figure 8). Larger vessels and greater efficiency probably accounts for the pronounced increases in ringnet catch rates.

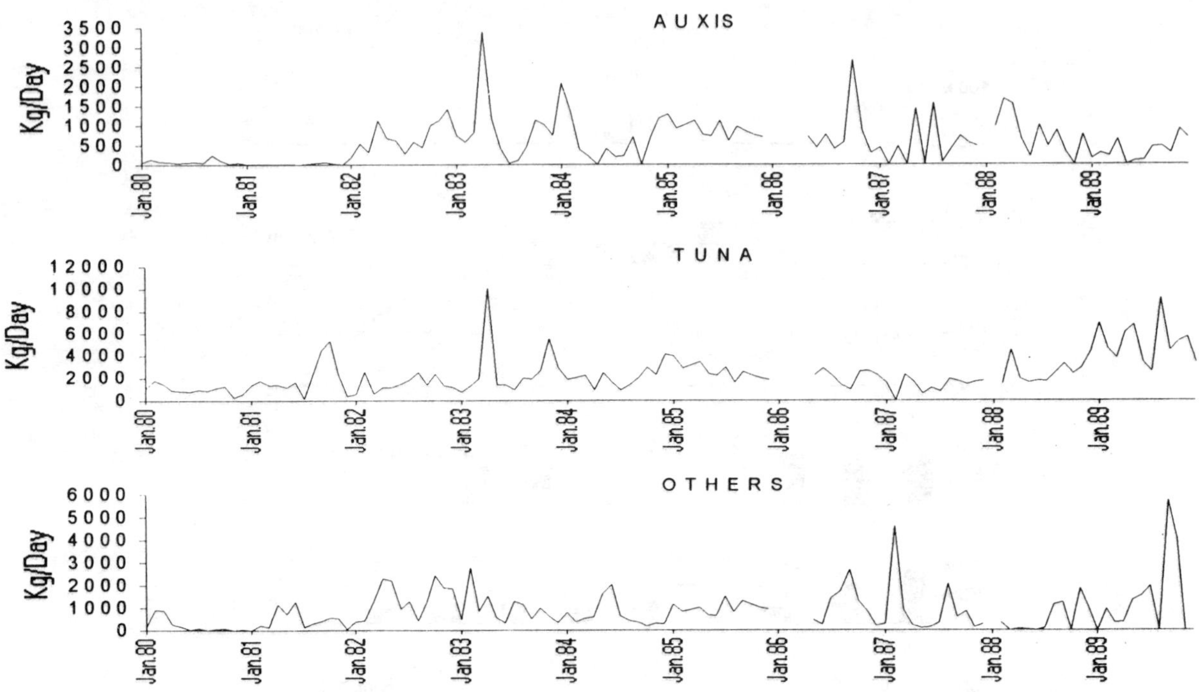

Fig. 7. Monthly catch rates of Auxis, tuna and other species by purse seines at General Santos, Moro Gulf from 1980 to 1989.

Santa Cruz (Davao Gulf)

Total annual catch rates of ringnetters fluctuated between 240 and 400 kg/day from 1980 to 1985 (Figure 9). Fishing effort decreased appreciably after 1984 with relocation of fishing vessels to landing sites closer to fishing grounds outside the Davao Gulf (Table 4).

Daliao (Davao Gulf)

Total catch rates of ringnetters at this landing site increased from 1983 to 1989 with higher catches of *Auxis* and other species. Tuna catch rates did not change appreciably during this interval (Figure 10).

Talomo (Davao Gulf)

Catch rates of *Auxis* and other species varied markedly from year to year with lowest rates in 1987 for both groups. Tuna catch rates were more stable at low levels during the 6-year interval (Figure 10).

Total catch rates at Talomo were the lowest in 1987 and were also low at Daliao during this year. Low abundance of *Auxis*, tuna and other species in the Davao Gulf in 1987 may have contributed to the abandonment of Santa Cruz as a landing site by ringnetters.

Table 4. **Average annual catch rates (kg/day) by species groups of purse seiners and ringnetters at General Santos, Santa Cruz and Opol.**

Landing site	Gear	Year	No. Trips	FRZ	TUN	OTH	TOTAL
General Santos	PS	1980	168	96	1,001	183	1,280
		1981	145	18	1,732	520	2,270
		1982	308	765	1,646	1,555	3,966
		1983	274	938	2,904	1,029	4,871
		1984	197	697	2,151	626	3,474
		1985	292	?	?	4,476	4,476
		1986	230	837	2,161	1,116	4,114
		1987	309	502	1,399	860	2,781
		1988	359	760	2,595	543	3,898
		1989	413	319	5,103	1,582	7,004
	RN	1980	689	80	311	181	572
		1981	730	64	179	271	514
		1982	328	673	1,226	1,918	3,817
		1983	400	694	2,798	834	4,326
		1984	406	535	2,809	1,453	4,797
		1985	440	757	4,038	2,165	6,960
		1986	333	1,171	3,492	1,488	6,151
		1987	393	2,004	3,759	1,437	7,200
		1988	408	1,306	6,217	856	8,379
		1989	482	1,275	7,306	2,200	10,781
Santa Cruz	RN	1980	704	64	112	144	319
		1981	360	58	73	106	238
		1982	225	49	168	166	383
		1983	240	86	183	128	397
		1984	233	32	211	135	378
		1985	109	20	117	114	251
		1986	72	5	46	80	132
		1987	1	0	16	31	47
Labuan	RN	1980	117	138	136	347	621
		1981	178	46	81	311	438
		1982	410	86	126	389	601
		1983	404	356	348	192	896
		1984	350	438	1,742	198	2,379
		1985	302	423	2,509	274	3,206
		1986	609	147	817	121	1,085
		1987	470	296	874	211	1,381
		1988	294	235	1,758	340	2,333
		1989	392	9	220	193	422
Opol	RN	1980	284	105	267	142	514
		1981	620	115	195	78	388
		1982	1,107	183	57	96	336
		1983	1,046	523	252	15	791
		1984	362	417	90	350	857
		1985	289	458	57	366	881
		1986	252	154	45	213	412
		1987	205	63	18	45	126
		1988	184	263	58	294	614
		1989	269	165	68	583	815

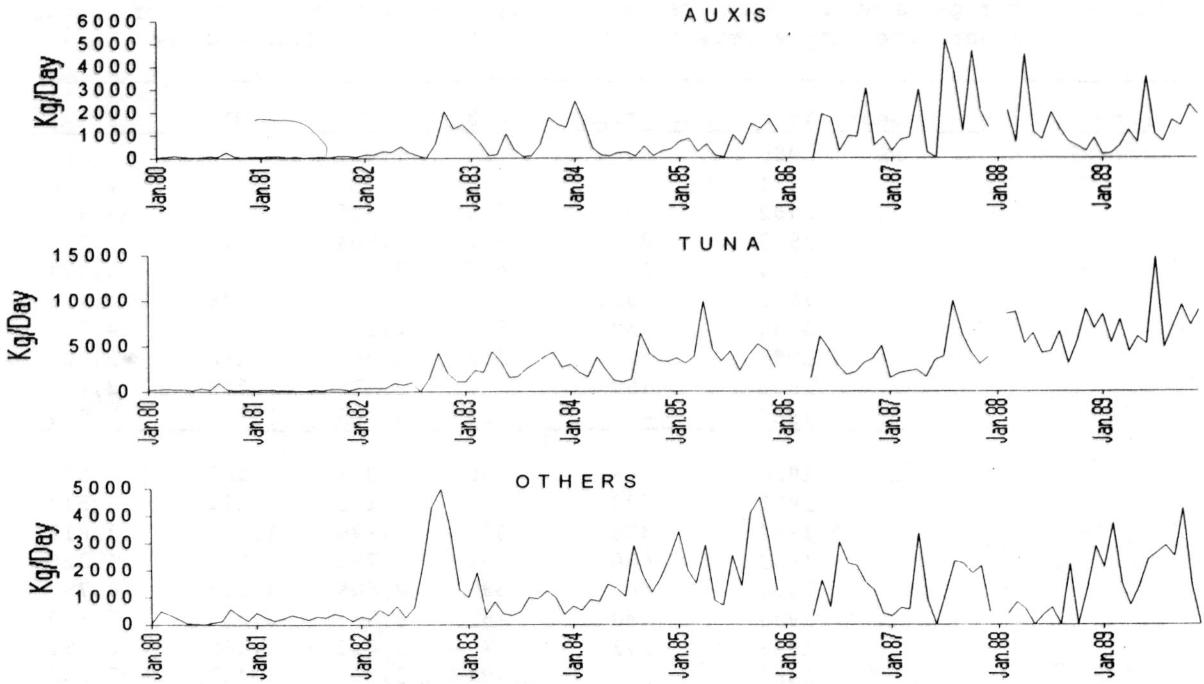

Fig. 8. Monthly catch rates of Auxis, tuna and other species by ringnets at General Santos, Moro Gulf from 1980 to 1989.

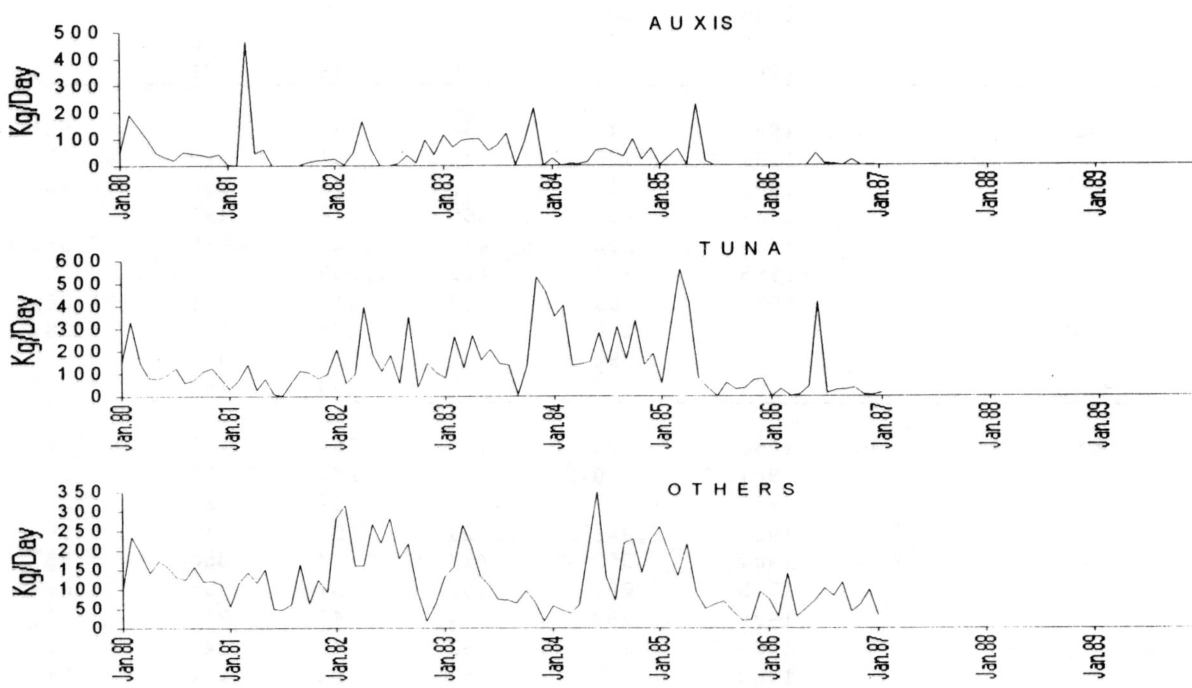

Fig. 9. Monthly catch rates of Auxis, tuna and other species by ringnets at Santa Cruz, Davao Gulf from 1980 to 1989.

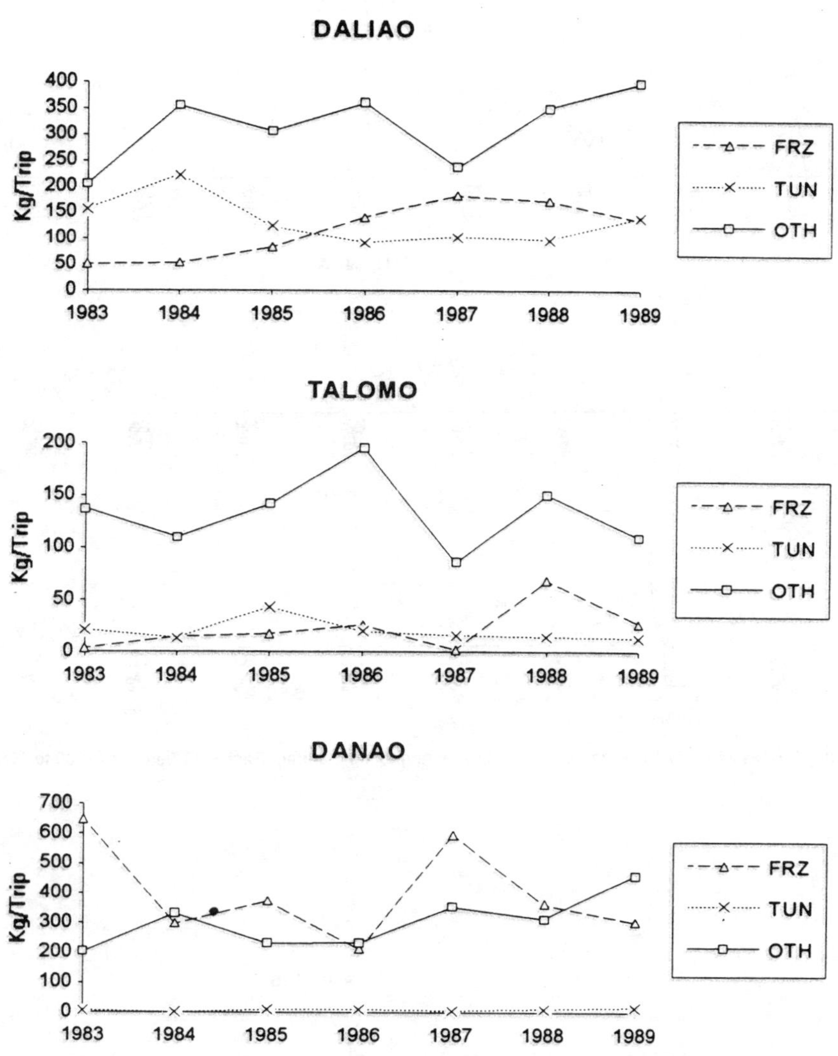

Fig. 10. Annual catch rates of Auxis (FRZ), tuna and other species by ringnets at three landing sites from 1983 to 1989 (1989 catch rates for Daliao and Talomo includes information only from June to December).

Labuan (East Sulu Sea)

Catch rates of *Auxis* increased markedly in late 1982 and remained high to the end of 1988. *Auxis* catches were not reported after February 1989. Tuna catch rates also increased markedly in 1984 and remained high to the end of 1988 (Figure 11). The exceptional total catch rates for 1984 to 1988 resulted primarily from high tuna catches and to a lesser extend to the *Auxis* catches (Table 4). Annual catch rates of other species were more constant and fluctuated between 120 and 390 kg/day (Table 4).

Opol (Bohol Sea)

Auxis catch rates of ringnetters were highest during the 1983-85 interval. Catch rates of other species increased, while those of tuna declined from 1980 to 1989 (Figure 12 and Table 4).

424

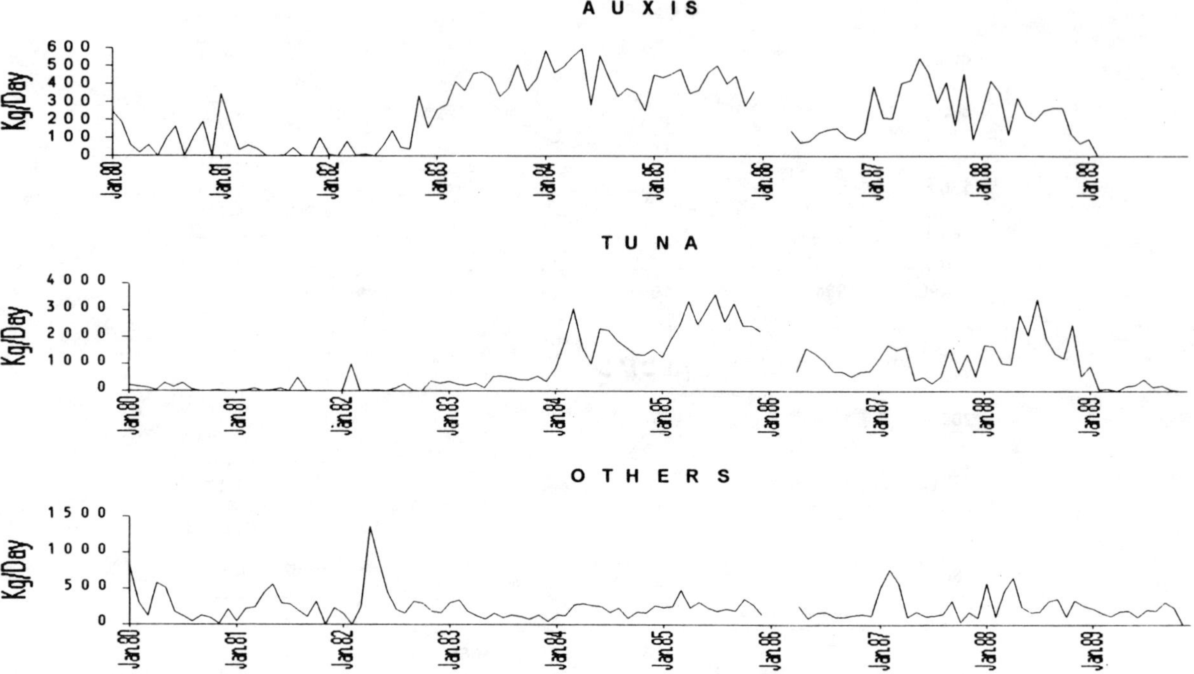

Fig. 11. Monthly catch rates of Auxis, tuna and other species by ringnets at Labuan, East Sulu Sea from 1980 to 1989.

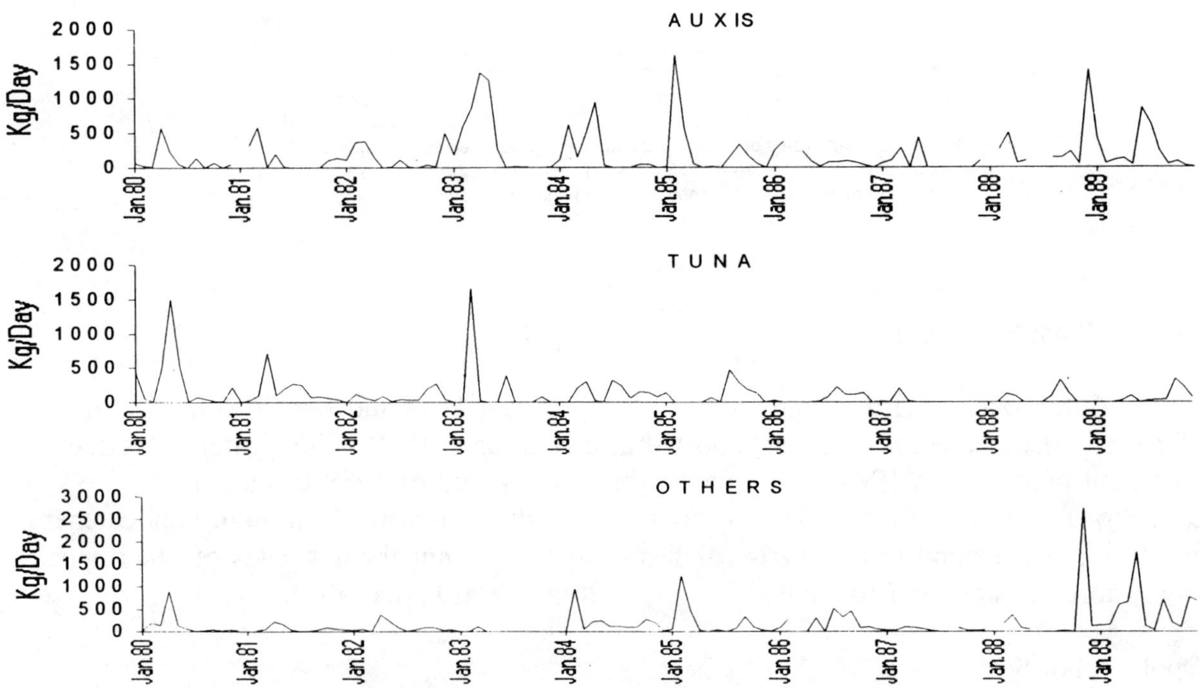

Fig. 12. Monthly catch rates of Auxis, tuna and other species by ringnets at Opol, Bohol Sea from 1980 to 1989.

Danao (Camotes Sea)

Ringnet catch rates of *Auxis* were most variable, with prominent peaks in 1983 and 1987. Catch rates of other species and tuna were relatively stable at moderate and low levels, respectively (Figure 10).

Auxis catch rates show wide annual fluctuations at all landing sites. High annual catch rates of *Auxis* occurred during 1983 at all sites, except Daliao and Talomo in the Davao Gulf (Table 4 and Figure 10), suggesting an exceptional year-class during this year throughout much of the Philippines. There is no evidence of declining *Auxis* catch rates at any of the landing sites.

2.6   Size Composition of Landings

Length-frequency distributions grouped by 1-year intervals of frigate and bullet tuna captured in ringnets at General Santos, Labuan, and Opol are given in Figures 13-15, respectively. Figure 13 shows little variation in size composition of frigate and bullet tuna at General Santos from 1980 to 1988. Frigate ranged from 16 to 38 cm and bullet from 16 to 30 cm. Size compositions of these two species were most varied at Labuan, especially in the early eighties, when frigate ranged from 20 to 54 cm and bullet from 14 to 50 cm (Figure 14). Bullet length-frequency distributions probably also include some frigate tuna.

Annual length-frequency distributions of bullet tuna capture by ringnets at Danao (Camotes Sea) show the dominant mode to be in the 20-23 cm interval for all years (Figure 16). A minor mode in the 13-16 cm interval was also observed from 1985 to 1987.

3.   REVIEW OF THE BIOLOGY OF FRIGATE AND BULLET TUNAS IN THE INDO-PACIFIC REGION

This review summarizes information from the Indo-Pacific region which for the most part has become available since publication of the synopsis by Uchida (1981) on some aspects of the biology of frigate and bullet tunas. The topics covered in this brief review include reproduction and population characteristics of these species.

3.1   Reproduction

3.1.1   Frigate tuna

Various authors have used the maturity-stage method based on the general appearance of the gonads (size, colour, ova development, firmness) to determine the maturity of frigate tuna. Chiampreecha (1978) determined maturity of 79 frigate tuna from the east coast of Peninsular Malaysia with a 5-stage scale and found more than 50% of the fish around 37 cm to be mature (stage-IV).

Klinmuang (1978) used a similar maturity scale for 100 fish from the Gulf of Thailand and east coast of Peninsular Malaysia and concluded that this species attains maturity between 37 and 41 cm.

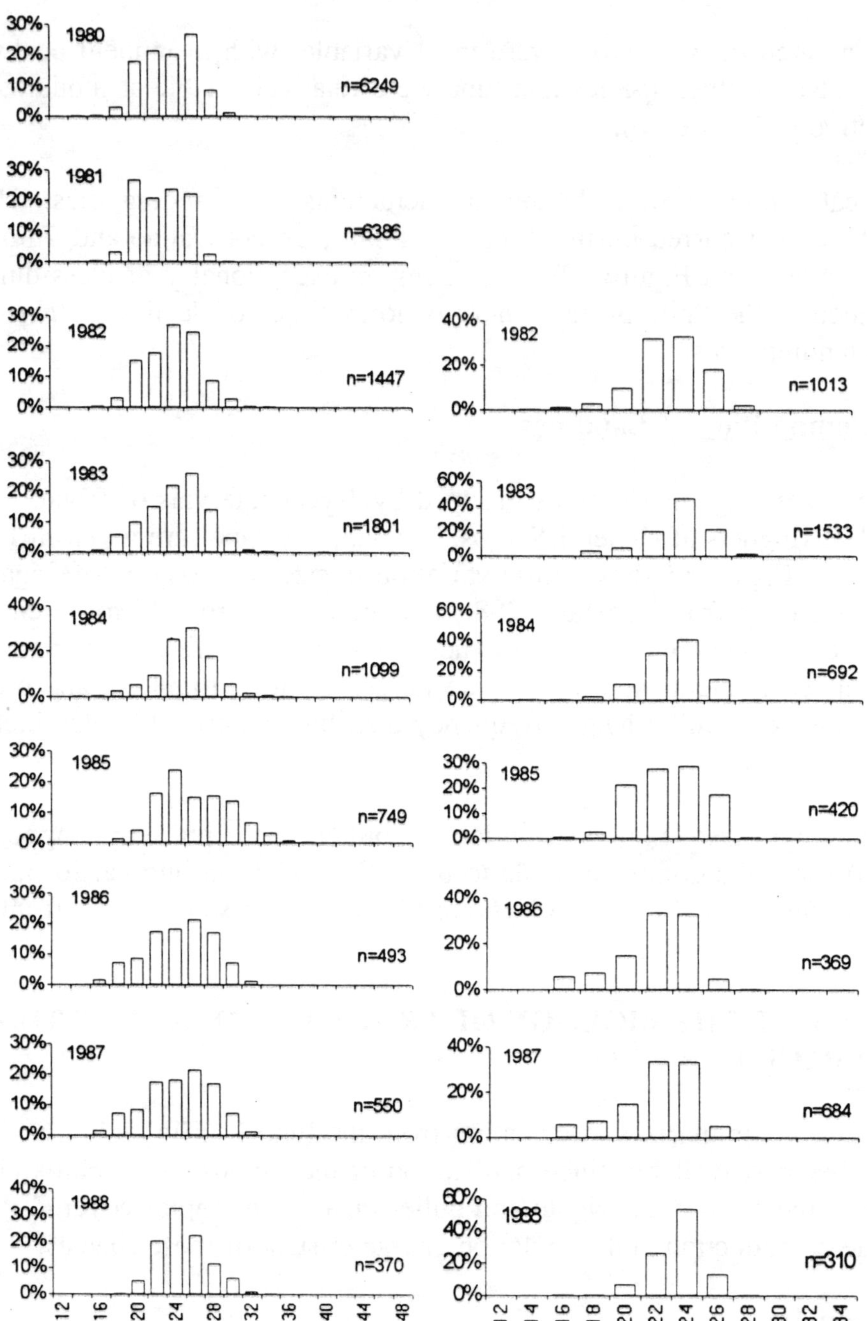

Fig. 13. Annual length- frequency distributions of frigate tuna (left) and bullet tuna (right) captured by ringnets at General Santos, Moro Gulf.

A 5-stage scale was also used by Yesaki (1982) to determine maturity of 120 frigate tuna off the west coast of Thailand. The smallest maturing female (stage-III) was 33 cm and 79% of the fish in this maturity stage were over 38 cm.

Gonad indices of 767 fish from the Gulf of Thailand were determined by Cheunpan (1984) over a 4-year period. She found the minimum size of maturity to be 310 mm and size at 50% maturity to be 341 mm (Table 5).

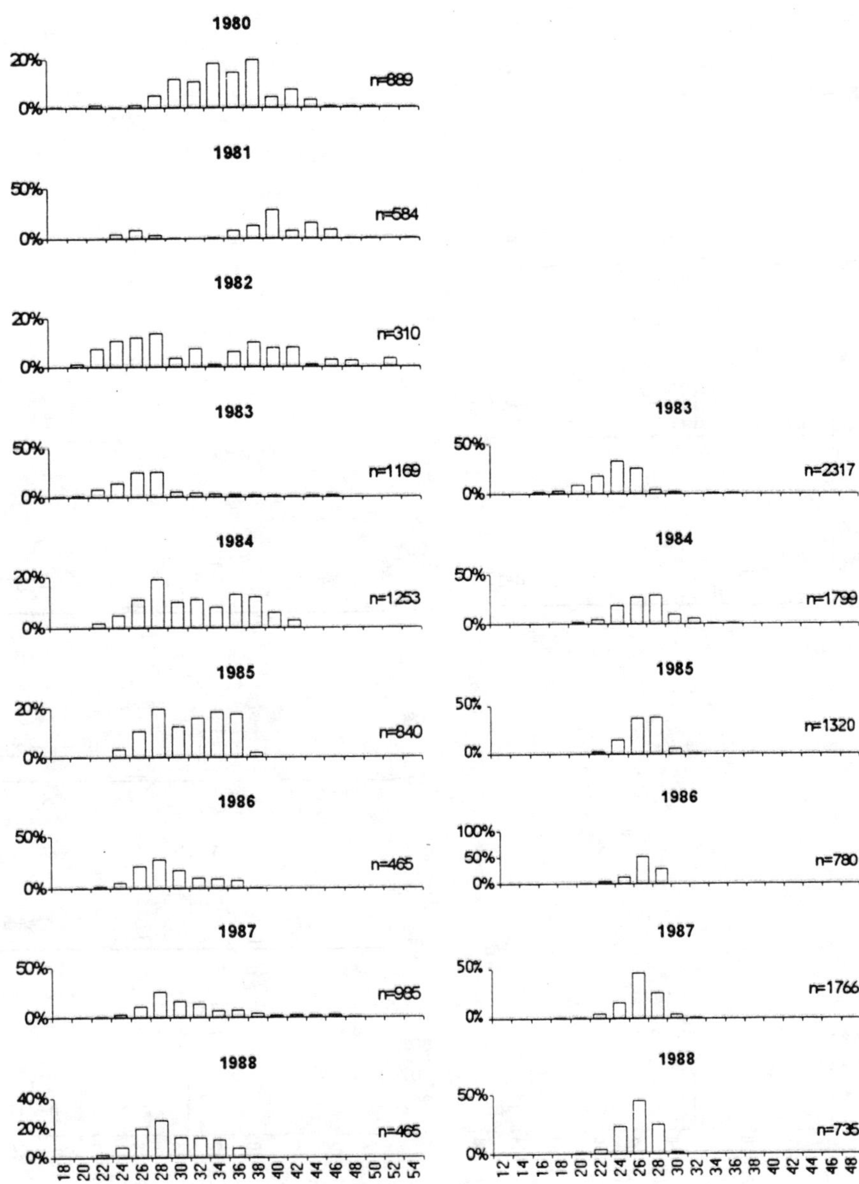

Fig. 14. Annual length- frequency distributions of frigate tuna (left) and bullet tuna (right) captured by ringnets at Labuan, East Sulu Sea.

Table 5. **Number of immature and mature female frigate tuna by length class (from Cheunpan, 1984).**

Length (mm.)	Immature	Mature	% Mature
301	18	–	–
301 – 315	33	15	31.3
316 – 330	78	34	30.4
331 – 345	50	43	46.3
346 – 360	61	69	53.1
361 – 375	33	108	76.6
376 – 390	23	96	80.7
391 – 405	15	54	78.3
406 – 420	2	14	87.5
– 421	–	21	100.0

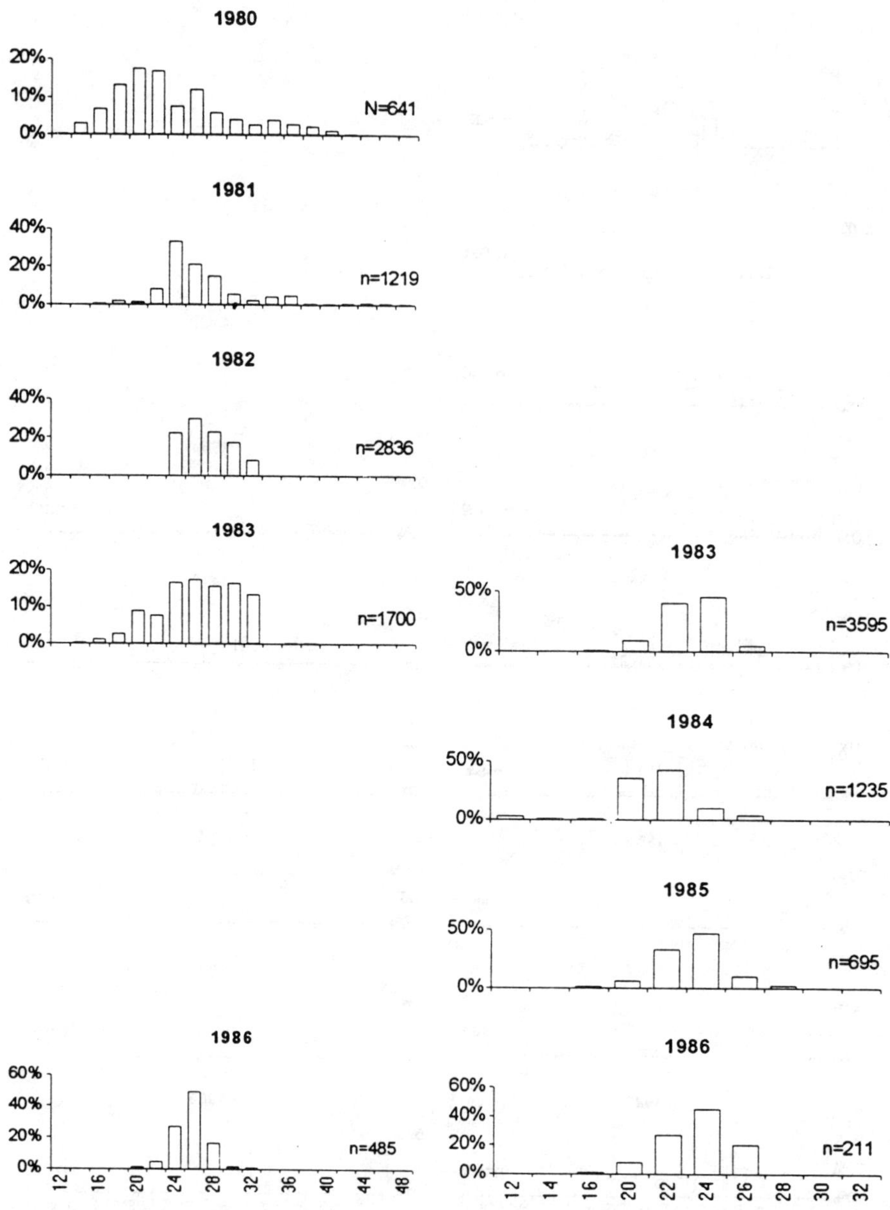

Fig. 15. Annual length- frequency distributions of frigate tuna (left) and bullet tuna (right) captured by ringnets at Opol, Bohol Sea.

Muthiah (1985) used a 7-point scale for determining maturity of fish from off Mangalore, India. He examined 336 males and 410 females and found the size at 50% maturity to be 30.0 and 30.5 cm for these respective sexes. However, fish less than 28 cm were not examined in this study.

The sizes at first and 50% maturity for frigate tuna are probably about 31.0 cm and 34.0 cm, respectively.

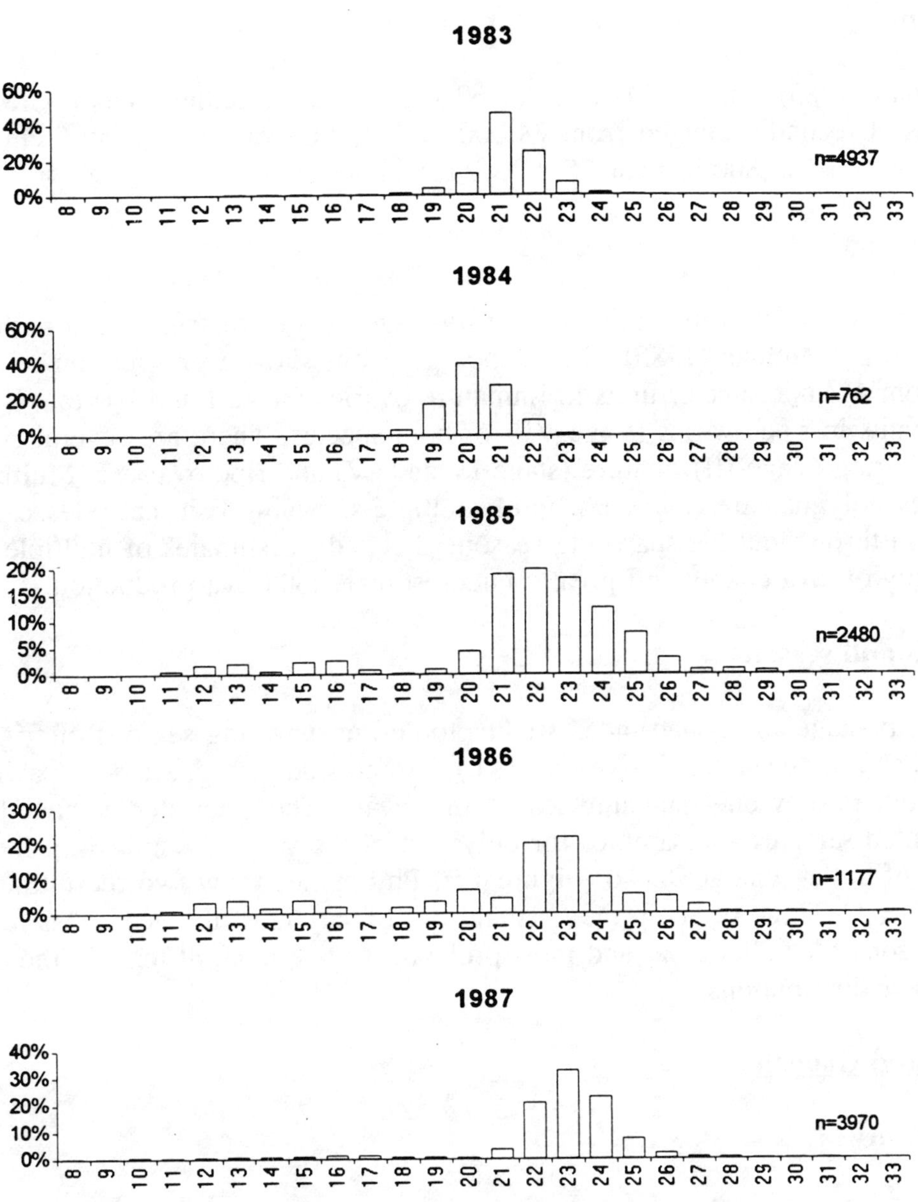

Fig. 16. Annual length- frequency distributions of bullet tuna captured by ringnet at Danao, Camotes Sea (from Jabat and Dalzell, 1988)

3.1.2 Bullet tuna

Muthiah (1985) determined maturity stages of 292 males and 354 females from off Mangalore, India, and found the sizes at 50% maturity for these respective sexes were 24.0 cm and 23,8 cm. Fish in the 20-32 cm interval were examined in this study.

Arce (1987) examined fish from the Batangas coast of the Philippines and found the size at first maturity was 17.0 cm and size at 50% maturity was 18.8 cm.

3.2 <u>Spawning</u>

3.2.1 Fecundity

Muthiah (1985) examined ovaries of 69 frigate and 85 bullet tunas for fecundity. His estimates of fecundity ranged from 78,000 to 717,900 ova for 31.5-39.0 cm frigate tuna and 52,600 to 162,800 ova for 25.2-33.7 cm bullet tuna.

3.2.2 Frequency

Diameters of ova in the various maturity-stage ovaries of frigate and bullet tunas were measured by Muthiah (1985). Frequency polygons show a progression of ova diameters from 4-8 micrometer units in immature ovaries (stage-I and -II) to 44-64 micrometer units in ripe ovaries (stage-VI) for both species. There are several size groups of ova in maturing (stage-III), mature (stage-IV and -V) and ripe ovaries. Multimodal ova frequency polygons are characteristic of multiple-spawning fish that release successive batches of ova throughout the spawning season. Fecundity estimates of multiple-spawning fish based only on ova counts will greatly underestimate total ova production.

3.2.3 Areas and seasons

Maturity-stage and gonad-index studies to define spawning seasons of frigate and bullet tunas have not been conclusive (Table 6). Some studies suggest two spawning seasons and others only one spawning season in a year. The latter studies have been based on limited samples and samples for only part of the year. Length-frequency distributions of bullet tuna at Danao (Figure 17), Philippines show two recruitment periods in a year. These occur in the first and third quarters of the year. There are two spawning seasons for bullet tuna, and most probably frigate tuna, at least in the equatorial regions of their distributions.

3.3 <u>Age and Growth</u>

3.3.1 Longevity

Frigate tuna

The largest frigate tuna measured in the Maldives from 1985 to 1989 was 53 cm (IPTP data files). Sivasubramaniam (1973) reports the presence of this size fish in a length-frequency distribution for this species captured off Sri Lanka. Similarly, Silas *et al.* (1985a) show occurrence of frigate tuna up to 52 cm in length-frequency distributions of fish captured off Cochin, India. The maximum size of frigate tuna in the tropical regions of its distribution is probably about 53 cm. A first approximation of the longevity of this species is 2 years.

Bullet tuna

Anderson (1988) reports a rare occurrence of 30-35 cm bullet tuna among catches of similar-sized frigate tuna in the Maldives. A 33.7 cm bullet tuna was examined by Muthiah (1985) during his study of fecundity. Silas *et al.* (1985a) and Sivasubramaniam

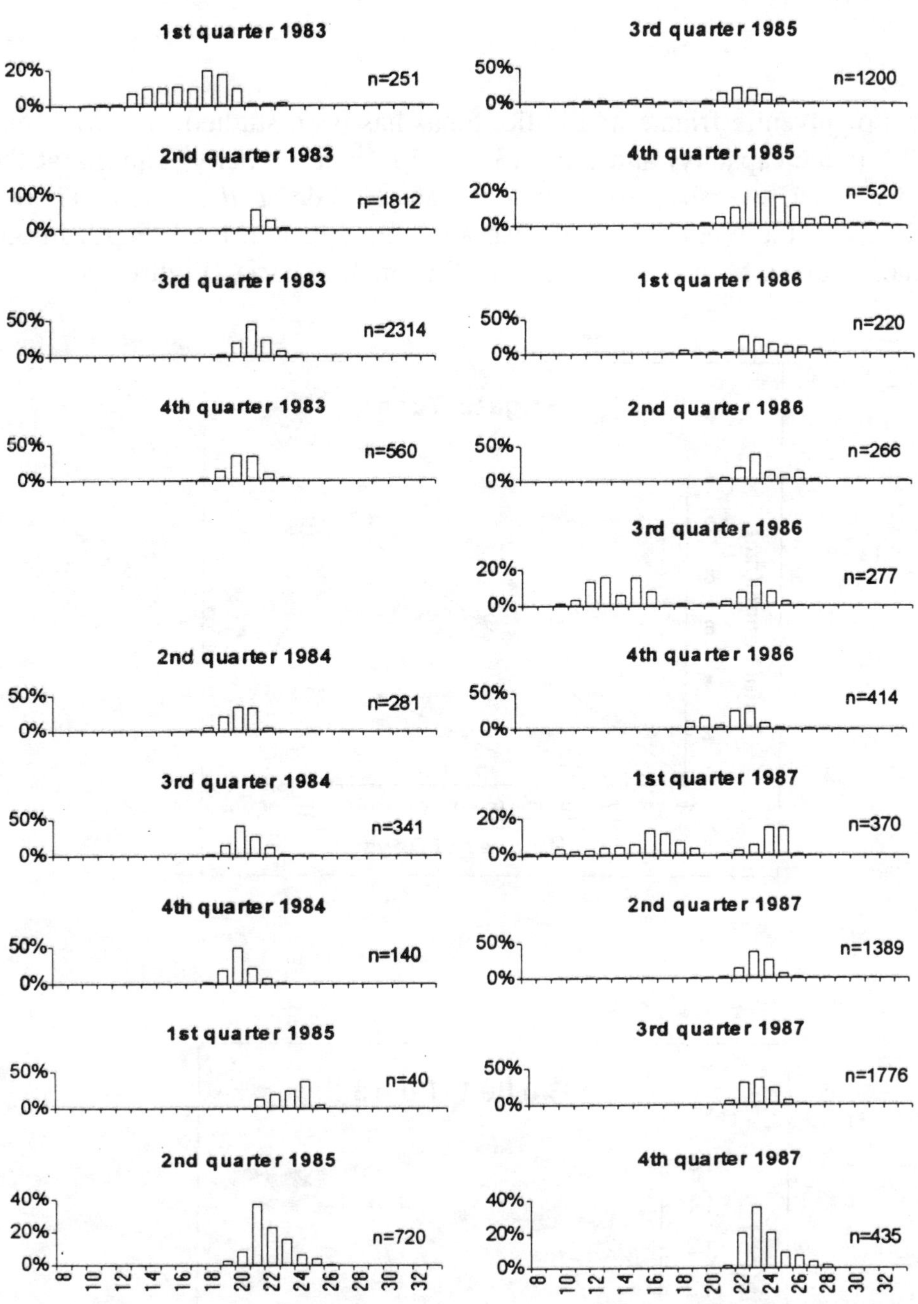

Fig. 17. Length- frequency distributions grouped by three month intervals of bullet tuna captured by ringnet at Danao, Camotes Sea (from Jabat and Dalzell, 1988).

(1973) show in length-frequency distributions the incidence of this species in the 34-36 cm and 31-35 cm intervals, respectively. The maximum size of bullet tuna in the tropical regions of its distribution is probably about 35 cm. A first approximation of the longevity of bullet tuna is 1 year.

3.3.2 Growth rate

Frigate tuna

Growth of juvenile frigate and bullet tunas has been studied in rearing experiments with small (0.5 tonne capacity) and large (3 and 70 tonne capacity) aquaria at the Marine Laboratory of Kinki University, Shirahama, Japan (Harada *et al.*, 1973). Growth rates differed markedly in these aquaria, with fish attaining 120 mm total length in 33 days in the large aquaria versus 55 mm in 35 days in the small aquaria (Figure 18).

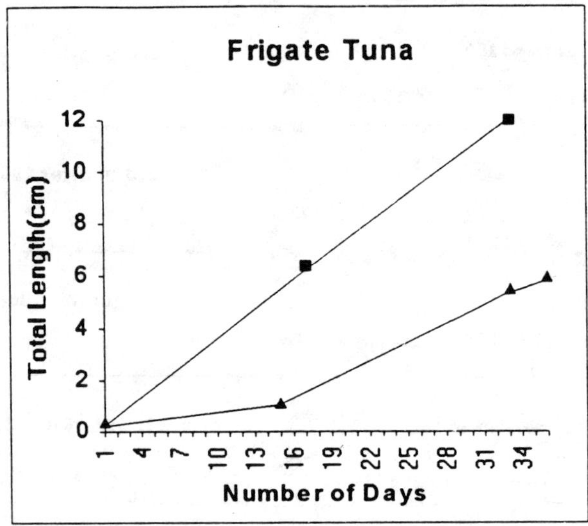

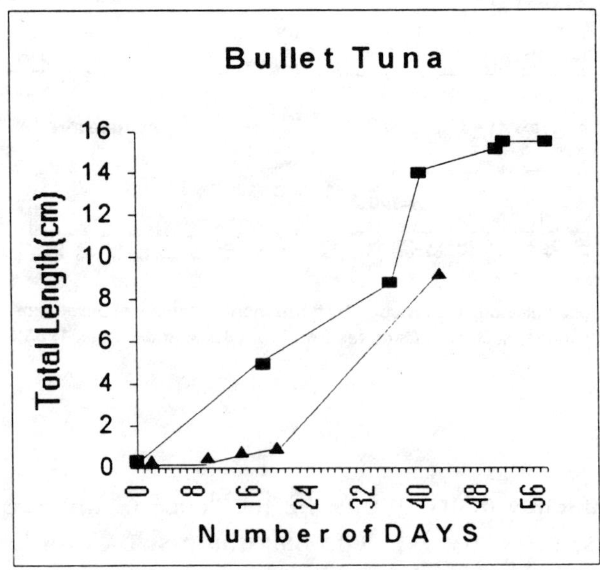

Fig. 18. Growth of juvenile frigate and bullet tunas in small (triangles) and large (squares) aquaria (from Harada et al.,1973).

Growth of adult frigate tuna has been studied by various authors (Table 7). All these studies have been based either on the "pen and paper" method or ELEFAN programme analyses of length-frequency distributions grouped by 1-month intervals, except the analysis by Yesaki (1989). He grouped length-frequencies by 10-day intervals and obtained lengths-at-ages 1 and 2 of 39 and 49 cm, respectively. All the other studies give lower predicted lengths, mostly between 24-27 cm for age 1 and 30-42 cm for age 2 (Figure 19).

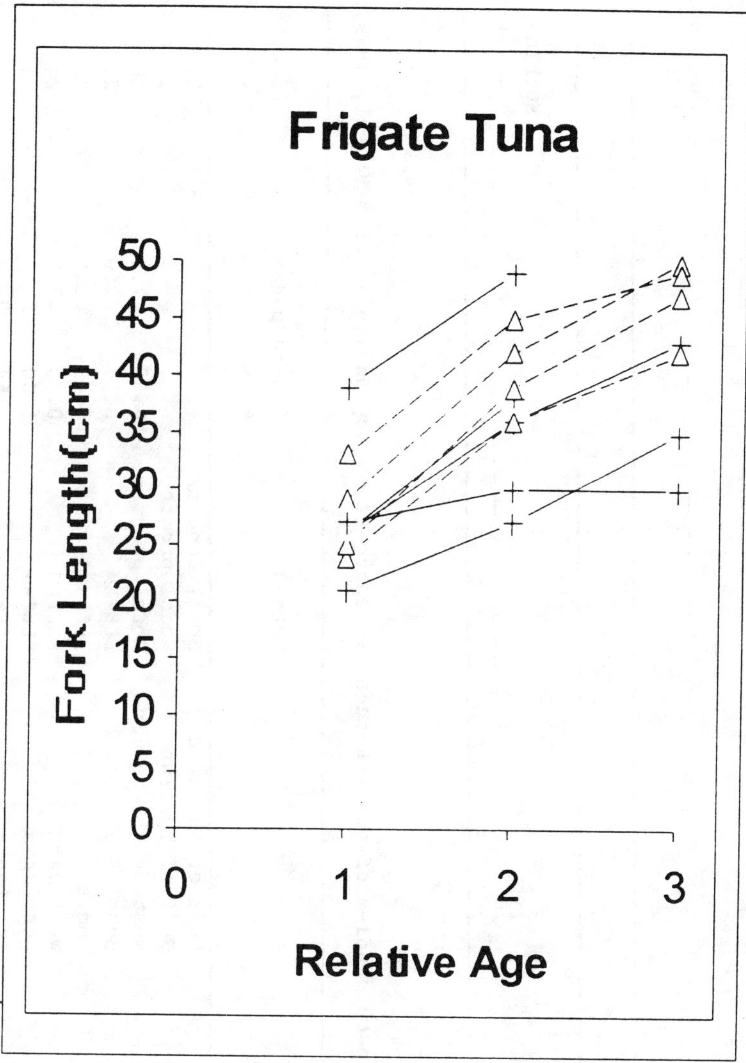

Fig. 19. Estimates of growth and relative age(years) of frigate tuna from length- frequency analyses (crosses-manual method; triangles-computer programmes).

Rearing experiments of frigate tuna show rapid growth after hatching, supporting the growth curve giving length-at-age 1 as 39 cm. Maturity studies of frigate tuna indicate size at 50% maturity to be about 34 cm, which also lends credence to the growth curve that gives 39 cm at age 1.

Table 6. Spawning seasons of frigate and bullet tuna at various locations in the Indo-Pacific region.

(a) Frigate tuna

Locality	Month	Source
	Jan Feb Mar Apr May Jun Jul Aug Sep Oct Nov Dec	
E. coast Peninsular Malaysia		Chiampreecha, 1978
Gulf of Thailand and E. coast Peninsular Malaysia		Klinmuang, 1978
W. coast Thailand		Yesaki, 1982
Gulf of Thailand		Cheunpan, 1984
Mangalore, India		Muthiah, 1985

b) Bullet tuna

Locality	Month	Source
Mangalore, India		Muthiah, 1985
Batangas Bay, Philippines		Arce, 1987

Table 7. Summary of age and growth studies of frigate tuna by length-frequency analyses (from Yesaki, 1989).

Author	Area	Method	Growth parameters			Length-at-age[1]		
			K	Lx	to	1	2	3
Chiampreecha, 1978	Gulf of Thailand	Modal lengths	–	–	–	21	27	35
Klinmuang, 1978	Gulf of Thailand	Modal progres.	–	–	–	26	38	–
Yesaki, 1982	west coast Thailand	Modal progres.	0.80	47.2	–	26	37	43
Ingles and Pauly, 1985	Bohol Sea	ELEFAN	0.73	47.0	–	24	36	42
Silas et al., 1985b	India	ELEFAN	0.49	63.0	-.270	29	42	50
Dwipongo et al., 1986	West Java	ELEFAN	0.70	47.5	–	24	36	42
	West Java	ELEFAN	1.00	51.5	–	33	45	49
Joseph et al., 1986	Sri Lanka	ELEFAN	0.54	58.0	–	25	39	47
Supongpan and Saikliang, 1987	Gulf of Thailand	Modal progres.	0.20[2]	43.9	-.290	27	30	30
Yesaki, 1989	Gulf of Thailand	Modal progres.	1.40	52.0	–	39	49	–

1/ - relative age
2/ - monthly basis.

Bullet tuna

Harada *et al.* (1973) found that growth rates of bullet tuna also varied markedly in aquaria of different sizes. Fish reared in small aquaria (0.5 tonne capacity) grew to 92 mm (total length) in 43 days, whereas fish reared in large aquaria (3 and 70 tonne capacity) attained 140 mm in 40 days and 155 mm in 58 days (Figure 18).

There have been no studies on the growth of bullet tuna from equatorial regions of its distribution.

3.4   Population

3.4.1   Sex ratios

Sex ratios of frigate and bullet tunas determined by various authors are given in Table 8.

3.4.2   Length-weight relationships

Length-weight relationships for frigate and bullet tunas are given in Table 9.

Table 8. Sex ratios of frigate and bullet tunas from various locations in the Indo-Pacific region.

(a) Frigate tuna

Location	Male	Female	Total	Source
Gulf of Thailand and E. coast Peninsular Malaysia	58	66	124	Klinmuang, 1978
W. coast Thailand	106	120	226	Yesaki, 1982
Mangalore, India	336	410	746	Muthiah, 1985

(b) Bullet tuna

Location	Male	Female	Total	Source
Mangalore, India	292	354	646	Muthiah, 1985
Batangas coast, Philippines	277	217	494	Arce, data files

Table 9. Length-weight relationships ($w = aL^b$) of frigate and bullet tunas.

(a) Frigate tuna

Locality	No. fish	Size range of fish Weight(kg)	Length(cm)	Constants a	b	Unit Wt.	Len.	Predicted wt. at 40 cm (kg)	Source
E. coast Peninsular Malaysia	271	–	–	4.900×10^{-3}	3.3651	g	cm	1.21	Chiampreecha, 1978
Gulf of Thailand	364	–	21–49	2.000×10^{-5}	2.9900	kg	cm	1.23	Klinmuang, 1981
W. coast of Thailand	–	–	–	1.100×10^{-5}	3.1190	kg	cm	1.09	Yesaki, 1982
Cochin, India	–	–	–	1.501×10^{-5}	3.0433	kg	cm	1.13	Silas et al., 1985a
Mangalore, India	–	–	–	1.121×10^{-6}	3.4649	kg	cm	0.39	Muthiah, 1985
Calicut, India	–	–	–	2.196×10^{-6}	3.5923	kg	cm	1.190	Balan and Yohannan, 1985
Lakshadweep Islands, India	120	–	20–52	2.462×10^{-5}	2.8969	kg	cm	1.080	Silas and Pillai, 1985

(b) Bullet tuna

Locality	No. fish	Size range of fish Weight(kg)	Length(cm)	Constants a	b	Unit Wt.	Len.	Predicted wt. at 30 cm (kg)	Source
Cochin, India	–	–	–	1.487×10^{-5}	2.9265	kg	cm	0.311	Silas et al., 1985a
Mangalore, India	–	–	–	5.187×10^{-6}	3.1711	kg	cm	0.246	Muthiah, 1985
Lakshadweep Islands, India	126	–	19–29	8.200×10^{-7}	3.8758	kg	cm	0.425	Silas and Pillai, 1985
Wawa, Philippines	–	–	–	4.529×10^{-3}	3.3600	g	cm	0.416	Arce, 1987
Lemery, Philippines	–	–	–	1.663×10^{-3}	3.6760	g	cm	0.447	Arce, 1987

4.   REFERENCES CITED

Anderson, R.C. 1988. Small tunas, seerfishes and billfishes in the Maldives. *Indo-Pac.Tuna Dev.Mgt.Programme*, IPTP/87/GEN/13:38-45.

Arce, F.M. 1987. The *Auxis* spp. fisheries of Batangas, Philippines. *Indo-Pac.Tuna Dev.Mgt.Programme*, IPTP/87/GEN/12:137-44.

Balan, V., and T.M. Yohannan. 1985. Fishery and bionomics of tunas at Calicut. *In* Tuna fisheries of the exclusive economic zone of India: biology and stock assessment, edited by E.G. Silas. *Bull.Cent.Mar.Fish.Res.Inst., Cochin*, 36:115-21.

Cheunpan, A. 1984. Sexual maturity, size at first maturity and spawning season of longtail tuna (*T. tonggol*), eastern little tuna (*E. affinis*) and frigate mackerel (*A. thazard*) in the Gulf of Thailand. *Fish.Rep.Mar.Fish.Div.Dep.Fish., Bangkok*, 43:33 p.

Chiampreecha, B. 1978. Biological studies on tuna and tuna-like fishes in the west of the Gulf of Thailand and off the east coast of Peninsular Malaysia. *Fish.Rep.Mar. Fish.Div.Dep.Fish., Bangkok*, 4:25 p.

Dwipongo, A., T. Hariati, S. Banon, M.L. Palomares, and D. Pauly. 1986. Growth, mortality and recruitment of commercially important fishes and penaeid and penaeid shrimps in Indonesian waters. *ICLARM Tech.Rep.*, 17:91 p.

FAO. 1976. The biology and status of stocks of small tunas. Report of an *Ad Hoc* Committee of Specialists, Honolulu, Hawaii, U. S. A., 15-18 December, 1975. *FAO Fish.Tech.Pap.*, 154:21 p.

FAO. 1988. Catches and landings. Food and Agriculture Organization. *FAO Yearb. Fish.Statist.*, (62): 479 p.

Harada, T., O. Murata, H. Furatani, and S. Miyashita. 1973. Report on experiments on the development of tuna culturing techniques (April 1970 - March 1973). *S Ser.Far Seas Fish.Res.Lab.*, (8):48-63.

Ingles, J. and D. Pauly. 1985. An atlas of the growth, mortality and recruitment of Philippines fishes. *ICLARM Tech.Rep.*, (13):127 p.

Jabat, M. and P. Dalzell. 1988. Preliminary stock assessment of the Danao ringnet fishery for bullet tunas and small pelagic fishes in the Camotes Sea, Central Visayas, Philippines. *Tech.Pap.Ser.Bur.Fish.Aquat.Resour., Philipp.*, Vol. XI, No.1:34 p.

Joseph, L., R. Maldeniya, and M. Van der Knaap. 1987. Fishery and age and growth of kawakawa (*E. affinis*) and frigate tuna (*A.thazard*). *In* Collective Volume of Working Documents presented at the "Expert Consultation on Stock Assessment of Tunas in the Indian Ocean", Colombo, Sri Lanka, 4-8 December, 1986. *Indo-Pac.Tuna Dev.Mgt.Programme*, Vol. 2:113-23.

Klinmuang, H. 1978. Preliminary studies on the biology of tunas in the west of the Gulf of Thailand and off the east coast of Peninsular Malaysia. *Fish.Rep.Mar.Fish.Div. Dep.Fish., Bangkok*, (5):27 p.

Klinmuang, H. 1981. Studies on the size frequency distribution and length-weight relationship of the tunas in the Gulf of Thailand. *Fish.Rep.Mar.Fish.Div.Dep.Fish., Bangkok*, (24):39 p.

Muthiah, C. 1985. Maturation and spawning of *Euthynnus affinis, Auxis thazard* and *A. rochei* in the Mangalore inshore area during 1979 to 1982. *In* Tuna fisheries of the exclusive economic zone of India: biology and stock assessment, edited by E. G. Silas. *Bull.Cent.Mar.Fish.Res.Inst., Cochin*, (36):71-85.

Munro, J. L. 1986. Marine fishery resources of the Philippines: catch and potentials. *In* Resources, management and socio-economics of Philippines marine fisheries, edited by D. Pauly, J. Saeger, and G. Silvestre. *Tech.Rep.Dep.Mar.Fish.Coll.Fish.Univ. Philippines*, (10):19-45.

Silas, E.G., and P.P. Pillai. 1985. Exploratory fishing by oceanic drift gillnetting and purse seining in the Lakshadweep. *In* Tuna fisheries of the exclusive economic zone of India: biology and stock assessment, edited by E.G. Silas. *Bull.Cent.Mar.Fish.Res.Inst., Cochin*, pp. 165-75.

Silas, E.G., P.P. Pillai, A.A. Jayaprakash, and M.A. Pillai. 1985a. Fishery and bionomics of tunas in Cochin. *In* Tuna fisheries of the exclusive economic zone of India: biology and stock assessment, edited by E. G. Silas. *Bull.Cent.Mar.Fish.Res.Inst., Cochin*, (36):28-43.

Silas, E.G., P.P. Pillai, M. Srinath, A.A. Jayaprakash, C. Muthiah, V. Balan, T.M. Yohannan, P. Siraimeetan, M. Mohan, P. Livingston, K.K. Kunhikoya, M.A. Pillai, and P.S.S. Sarma. 1985b. Population dynamics of tunas: stock assessment. *In* tuna fisheries of the exclusive economic zone of India: biology and stock assessments, edited by E.G. Silas. *Bull.Cent.Mar.Fish.Res.Inst., Cochin*, (36):20-7.

Sivasubramaniam, K. 1973. Co-occurrence and the relative abundance of narrow and broad corseletted frigate mackerels *Auxis thazard* (Lacepede) and *Auxis rochei* (Risso), around Ceylon. In Proceedings of the Symposium on Living Resources of the Seas around India. *Cent.Mar.Fish.Res.Inst., Cochin*, pp. 537-47.

Supongpan, S., and P. Saikliang. 1987. Fisheries status of tuna purse seiners (using sonar) in the Gulf of Thailand. *Rep.Mar.Fish.Div.Dep.Fish., Bangkok*, (3):78 p.

Uchida, R.N. 1981. Synopsis of biological data on frigate tuna, *Auxis thazard* and bullet tuna, *A. rochei. NOAA Tech.Rep.NMFS Circ.*, (436):63 p.

White, T.F., and M. Yesaki. 1982. The status of tuna fisheries in Indonesia and the Philippines. *Indo-Pac.Tuna Dev.Mgt.Programme*, IPTP/82/WP/3:64 p.

Yesaki, M. 1982. Thailand. Biological and environmental observations. A report prepared for the pole-and-line tuna fishing in southern Thailand Project. *FAO. DP/THA/77/008*:46 p.

Yesaki, M. 1983. The pelagic fisheries of the Philippines. *Indo-Pac.Tuna Dev.Mgt.Programme*, IPTP/83/WP/6:15 p.

Yesaki, M. 1989. Estimates of age and growth of kawakawa (*Euthynnus affinis*), longtail tuna (*Thunnus tonggol*) and frigate tuna (*Auxis thazard*) from the Gulf of Thailand based on length data. *Indo-Pac.Tuna Dev.Mgt.Programme*, IPTP/89/GEN/17:94-108.